WITHDRAWN
NDSU

THE ROLE OF POTASSIUM IN AGRICULTURE

THE ROLE OF POTASSIUM
IN AGRICULTURE

Proceedings of a symposium sponsored and financed by the Tennessee Valley Authority and the American Potash Institute and cosponsored by the American Society of Agronomy, Crop Science Society of America, and Soil Science Society of America, and held at the National Fertilizer Development Center, TVA, Muscle Shoals, Alabama, June 18-19, 1968.

Edited by

V. J. KILMER

Chief, Soils & Fertilizer Research Branch, Division of Agricultural Development, Tennessee Valley Authority, Muscle Shoals, Alabama

S. E. YOUNTS

Vice President, American Potash Institute, Inc., Atlanta, Georgia

N. C. BRADY

Director of Research and of the Cornell University Agricultural Experiment Station, Cornell University, Ithaca, New York

Managing Editor: RICHARD C. DINAUER

Assistant Editor: DAVID M. KRAL

Published by

American Society of Agronomy

Crop Science Society of America

Soil Science Society of America

Madison, Wisconsin USA

1968

The cover design is by M. Thigpen of the Office of Agricultural and Chemical Development, Tennessee Valley Authority, Muscle Shoals, Alabama. Mr. Thigpen studied at the Ringling School of Art, Sarasota, Florida

Copyright © 1968, by the American Society of Agronomy, Inc., the Crop Science Society of America, Inc., and the Soil Science Society of America, Inc., 677 South Segoe Road, Madison, Wisconsin, USA 53711

ALL RIGHTS RESERVED
No part of this book may be reproduced in any form, by photostat, microfilm, or any other means, without written permission from the publisher.

Library of Congress Catalog Card Number: 68-28346

Printed in the United States of America

FOREWORD

In succession with nitrogen and phosphorus, potassium is now given its full recognition as a fertilizer element. While the proposition of Baron Justus von Liebig that the nitrogen requirements of plants could be completely satisfied from ammonia in the air was contested through the field and pot experiments of Sir J. B. Lawes and J. H. Gilbert, the soil was largely accepted as a sufficient source of potassium for plants. Phosphorus availability from phosphate rock was assured by Lawes' acidulation process. Water-soluble mineral deposits as the sources of fertilizer potassium were developed when the process of cation exchange in soils, not understood by Liebig, came to be recognized as the soil characteristic that prevents Liebig's feared loss of potassium by leaching.

As agricultural need for fertilizer potassium became known, exploration and industrial processing provided adequate sources of water-soluble potassium from salt deposits accumulated by fractional crystallization in ancient sea evaporites stored in the geologic column. The potassium cycle, consisting of weathering of feldspar and mica minerals in soil, leaching of potassium to the sea, and returning it to the land as fertilizer, was thus completed.

It is appropriate that the industrial production and crop uses of potassium should be reviewed at the present time, because vastly increased quantities of fertilizer potassium will be required to supply food for increasing billions in the human population. Industry has served commendably well thus far by providing adequate production of suitable fertilizer material to meet current demands. Our mission is to determine the needs for this plant nutrient and to make them known in order that the increasing demands for potassium fertilizer may be anticipated.

Due recognition is given to Dr. J. Fielding Reed, President of the American Potash Institute, and Dr. Lewis B. Nelson, Manager, Office of Agricultural and Chemical Development, TVA, for their initiative in organizing and financing the symposium on potassium and for the invitation extended to the associated societies to become cosponsors of the symposium and now publishers of the proceedings which constitute this book. This symposium represented the most comprehensive treatment of the subject relating to potassium in agriculture ever undertaken in this hemisphere.

May 1968

D. C. SMITH, *President*
American Society of Agronomy
F. L. PATTERSON, *President*
Crop Science Society of America
M. L. JACKSON, *President*
Soil Science Society of America

PREFACE

The Role of Potassium in Agriculture includes contributions by a number of outstanding national and international scientists who have firsthand knowledge of the use of potassium and the indispensable role it plays in modern agriculture. The "from mine to man" arrangement of the book illustrates the role of potassium as a vital element from the time it is extracted from the mine or the brine solution to the time it is consumed in foodstuffs or feed. The authors not only explain how potassium is processed into fertilizer, but they evaluate and compare the different forms of potassium used in today's agricultural industry. Experienced researchers describe the chemical and physical behavior of potassium in the soil and in the plant itself. Other researchers explain the nutritive effects of potassium not only on the more common forage, fruit, vegetable, and field crops of the United States but also on the tropical crops grown in the Southern Hemisphere.

Though this book is by no means an exhaustive treatise on the role of potassium in agriculture, the most important facets of potassium technology and use are thoroughly explored. The book is directed to extension workers, farm managers, college and high school teachers, applied research groups, members of the fertilizer industry, teams concerned with the use of potassium in developing countries, and others with specific interests in potassium and its use in fertilizer.

The papers published here were presented at a symposium held at Muscle Shoals, Alabama, June 18 and 19, 1968. The symposium was sponsored by the American Potash Institute, the Tennessee Valley Authority, the American Society of Agronomy, the Crop Science Society of America, and the Soil Science Society of America. The editorial committee is grateful to the authors and to these organizations for their aid and cooperation. The advice and tireless efforts of Matthias Stelly, Richard C. Dinauer, and David M. Kral on the staff of the headquarters office of the three scientific societies are worthy of special acknowledgement.

May 1968

    The Editorial Committee
    V. J. KILMER, *Tennessee Valley Authority, Muscle Shoals, Alabama*
    S. E. YOUNTS, *American Potash Institute, Atlanta, Georgia*
    N. C. BRADY, *Cornell University, Ithaca, New York*

CONTRIBUTORS

Samuel S. Adams — Geologist, Geological Department, The Anaconda Company, Salt Lake City, Utah (formerly Geologist, International Minerals & Chemical Corporation, Skokie, Illinois)

Stanley A. Barber — Professor of Agronomy, Department of Agronomy, Purdue University, Lafayette, Indiana

Roy E. Blaser — Professor of Agronomy, Department of Agronomy, Virginia Polytechnic Institute, Blacksburg, Virginia

George A. Cummings — Associate Professor, Department of Soil Science, North Carolina State University, Raleigh, North Carolina

Orvis P. Engelstad — Agronomist, Soils & Fertilizer Research Branch, Tennessee Valley Authority, Muscle Shoals, Alabama

Harold J. Evans — Professor of Plant Physiology, Department of Botany & Plant Pathology, Oregon State University, Corvallis, Oregon

Roy L. Goss — Associate Agronomist and Extension Specialist in Agronomy, Western Washington Research & Extension Center, Washington State University, Puyallup, Washington

Billy W. Hipp — Assistant Professor of Soil Chemistry, Lower Rio Grande Valley Research & Extension Center, Texas A & M University, Weslaco, Texas

William A. Jackson — Professor of Soil Science, Department of Soil Science, North Carolina State University, Raleigh, North Carolina

Edwin C. Kapusta — Technical Sales Director, Potash Company of America, New York, New York

E. Lamar Kimbrough — Graduate Research Assistant, Department of Agronomy, Virginia Polytechnic Institute, Blacksburg, Virginia

Robert C. J. Koo — Associate Horticulturist, Institute of Food & Agricultural Sciences, University of Florida, Citrus Experiment Station, Lake Alfred, Florida

W. C. Liebhardt — Agronomist, Standard Fruit Company, LaCeiba, Honduras

Robert E. Lucas — Extension Specialist in Soils, Department of Soil Science, Michigan State University, East Lansing, Michigan

K. Mengel — Professor, Landwirtschaftliche Forschungsanstalt Büntehof, Hannover, Germany.

Robert D. Munson	Midwest Director, American Potash Institute, Inc., St. Paul, Minnesota
Werner L. Nelson	Senior Vice President, American Potash Institute, Inc., West Lafayette, Indiana
John Pesek	Professor of Agronomy, Department of Agronomy, Iowa State University, Ames, Iowa
Charles I. Rich	Professor of Agronomy, Department of Agronomy, Virginia Polytechnic Institute, Blacksburg, Virginia
Merle R. Teel	Formerly Managing Director, American Farm Research Association, West Lafayette, Indiana (now Chairman, Plant Science Department, University of Delaware, Newark, Delaware)
Grant W. Thomas	Professor of Soil Chemistry, Department of Soil & Crop Sciences, Texas A & M University, College Station, Texas
H. R. von Uexkuell	Director, Kali Kenkyu Kai (Potash Research Association), Tokyo, Japan
Richard J. Volk	Professor of Mineral Nutrition, North Carolina State University, Raleigh, North Carolina
Gerald E. Wilcox	Associate Pofessor of Horticulture, Department of Horticulture, Purdue University, Lafayette, Indiana
Walter S. Wilde	Professor of Physiology, Department of Physiology, University of Michigan, Ann Arbor, Michigan
Richard H. Wilson	Research Associate, Department of Botany, University of Illinois, Urbana, Illinois (formerly at the Department of Botany & Plant Pathology, Oregon State University, Corvallis, Oregon)
Ronald D. Young	Chief, Process Engineering Branch, Division of Chemical Development, Tennessee Valley Authority, Muscle Shoals, Alabama

CONTENTS

FOREWORD v
PREFACE vii
CONTRIBUTORS ix

1 Potassium Reserves in the World

SAMUEL S. ADAMS

I. Introduction 1
II. Sources of Potassium 3
III. World Potassium Reserves 4
IV. Projected Potassium Reserves 17
V. Recovery of Potassium 17
VI. Potassium Supply and Demand 19

2 Potassium Fertilizer Technology

EDWIN C. KAPUSTA

I. Introduction 23
II. The World Potash Industry 24
III. Potassium Raw Materials 24
IV. Potassium Raw Material Retrieval 25
V. Production of Potassium Fertilizers 33
VI. Shipping Potassium Fertilizers 50

3 Preparation of Finished Fertilizers Containing Potassium

RONALD D. YOUNG

I. Introduction 53
II. Types of Materials Used 54
III. Preparation of Fertilizers Containing Potassium . . 55
IV. Sensitivity of Nitrate Fertilizers Containing Potassium Chloride 58
V. Two-Component Fertilizers Containing Potassium . . 59
VI. Analytical Methods 60

4 Agronomic Evaluation of Potassium Polyphosphate and Potassium Calcium Pyrophosphates as Sources of Potassium

ORVIS P. ENGELSTAD

    I. Introduction 63
   II. Chemical Characteristics 63
  III. Agronomic Results 67
  IV. Discussion and Conclusions 75

5 Mineralogy of Soil Potassium

CHARLES I. RICH

    I. Introduction 79
   II. Structures of Soil Minerals Involved in Potassium Reactions . 80
  III. Release of Soil Potassium By Weathering 85
  IV. Release of Nonexchangeable Potassium by Drying . . 92
   V. Potassium Fixation 94
  VI. Potassium Selectivity of Mica-Vermiculites as Affected by Mineral Properties and Hydrogen Ions 96

6 Role of Potassium in Photosynthesis and Respiration

WILLIAM A. JACKSON AND RICHARD J. VOLK

    I. Introduction 109
   II. Measurement of Gas Exchange Components . . . 110
  III. Photosynthesis and Photorespiration 114
  IV. Respiration 130
   V. Summary 140

7 Effect of Potassium on Carbohydrate Metabolism and Translocation

W. C. LIEBHARDT

    I. Introduction 147
   II. Carbohydrate Metabolism 147
  III. The Carbohydrate Fraction 148
  IV. Enzyme Systems 149
   V. Translocation 150
  VI. Parenchyma Breakdown 156

8 The Effect Of Potassium on the Organic Acid and Nonprotein Nitrogen Content of Plant Tissue

MERLE R. TEEL

I. Introduction	165
II. Variation in Organic Acid Composition	171
III. Variation in NPN	179
IV. Implications	183

9 The Effect of Potassium and Other Univalent Cations on the Conformation of Enzymes

RICHARD H. WILSON AND HAROLD J. EVANS

I. Introduction	189
II. Mechanisms of Action of Univalent Cations in Enzyme Activation	191
III. General Conclusions	199

10 Role of Potassium in Human and Animal Nutrition

WALTER S. WILDE

I. Introduction	203
II. Concentration of Potassium in Cells and Body Fluids	204
III. Mechanism of Accumulation of Potassium Inside Cells	205
IV. Membrane ATPase	206
V. Potassium Fluxes During Activity	209
VI. Potassium Deficiency	212
VII. Other Alkali Metals	217
VIII. Other Reviews	218
IX. Concluding Remarks	218

11 The Effects of Potassium on Disease Resistance

ROY L. GOSS

| I. Introduction | 221 |
| II. The Effect of Soluble Forms of Potassium on Plant Diseases | 222 |

12 Effect of Potassium on Quality Factors — Fruits and Vegetables

GEORGE A. CUMMINGS AND GERALD E. WILCOX

I. Introduction	243
II. Fruit	244
III. Vegetables	254

13 Soil Factors Affecting Potassium Availability

GRANT W. THOMAS AND BILLY W. HIPP

I.	Introduction	269
II.	Chemical Factors Affecting Potassium Availability	270
III.	Physical Factors Affecting Potassium Availability	280

14 Mechanism of Potassium Absorption by Plants

STANLEY A. BARBER

I.	Introduction	293
II.	Movement of Potassium Into the Plant Root	294
III.	Dependence of Plant Growth Rate on Potassium Concentration of the Substrate	297
IV.	Influence of Calcium on Potassium Absorption	298
V.	Exchange Capacity of Roots and Potassium Uptake	299
VI.	Quantity-Intensity Measurements and Potassium Absorption	300
VII.	Effect of Clay Suspensions on K Uptake	301
VIII.	Exchange Diffusion and Potassium Uptake	302
IX.	Potassium Uptake From Soils	303

15 Exchangeable Cations of Plant Roots and Potassium Absorption By the Plant

K. MENGEL

I.	Introduction	311
II.	Materials and Methods	312
III.	Results	313
IV.	Discussion	317
V.	Conclusions	318

16 Interaction of Potassium and other Ions

ROBERT D. MUNSON

I.	Introduction	321
II.	Characteristics and Expression of Potassium and other Cations	322
III.	Factors Which Influence the Interrelations Among Ions	325
IV.	Critical Level of Potassium Shifted by Interacting Ions	342
V.	Cation Relations and Organic Acids	342
VI.	Cation-Anion Balance	347
VII.	Summary and Conclusions	348

17 Plant Factors Affecting Potassium Availability and Uptake

WERNER L. NELSON

I.	Introduction	355
II.	Uptake of Potassium	355
III.	Role of Root Systems	357
IV.	Variety	368
V.	Yield Level	371
VI.	Species	376
VII.	In The Future	380

18 Potassium Nutrition of Tropical Crops

H. R. VON UEXKUELL

I.	Introduction	385
II.	General Problems of Fertilizer Use in the Tropics	386
III.	Tropical Crops	388
IV.	Conclusions	414

19 Potassium Nutrition of Forage Crops with Perennials

ROY E. BLASER AND E. LAMAR KIMBROUGH

I.	Introduction	423
II.	Potassium Composition and Yield	426
III.	Potassium Applications for Perennial Forages	441

20 Potassium Nutrition of Soybeans and Corn

JOHN PESEK

I.	Introduction	447
II.	Some General Concepts	449
III.	Potassium Nutrition of Corn	452
IV.	Potassium Nutrition of Soybeans	459

21 Potassium Nutrition of Tree Crops

ROBERT C. J. KOO

I.	Introduction	469
II.	Potassium and Tree Crops	470
III.	Leaf Analysis and Potassium Nutrition	480

22   Potassium Nutrition of Vegetable Crops

Robert E. Lucas

I.	Introduction	489
II.	Potassium Composition and Uptake	489
III.	Plant Response to Potassium Fertilizer	490
IV.	Effect of Excess Potassium on Vegetables	495

Glossary—Common and Scientific Names of Crops Referred to in this Book   499

Subject Index   503

Potassium Reserves in the World

SAMUEL S. ADAMS

The Anaconda Company
Salt Lake City, Utah

I. INTRODUCTION

This survey attempts to estimate the magnitude of world potassium reserves. Such an assessment seems timely as the world potassium demand is currently increasing at a rate greater than 7%/year, while the production capacity is due to increase 60% by 1975. Reference will be made not only to the current economically recoverable reserves, but also to marginal deposits which are destined to be exploited as technological developments and fertilizer demands permit.

Discussions of the potassium industry, its ores and products, employ two conventions which must be explained — the word "potash" and the units in which it is usually measured, K_2O equivalents. "Potash" referred originally to potassium carbonate produced by the leaching of wood ashes. It is a confusing term because it is now applied to a variety of crude potassium-bearing ores as well as refined products. It has been customary in the industry to refer to the grade of potassium ores and products in terms of percent K_2O. This is an unfortunate unit of measurement, as K_2O does not occur naturally. Percent K_2O might best be abandoned in favor of percent K. The K_2O unit is nonetheless, the industry standard and will be used throughout this paper. The potassium content and K_2O equivalents of the common potassium minerals, ores, and refined products are given in Table 1 (1.0% K is equivalent to 1.2046% K_2O).

The world potassium reserves are presented in terms of estimated recoverable metric tons of K_2O. Potassium oxide tons were chosen rather than product tons because the various grades and types of products contain different potassium concentrations and hence, nutrient value. By expressing the total recoverable product in terms of K_2O, the reserves can be conveniently compared with projected world fertilizer demands which are expressed in the same terms.

Table 1—Common minerals, ores, and products of potassium deposits

Name	Formula	% K$_2$O	% K	Remarks
MINERALS				
Chlorides				
Sylvite	KCl	63.17	52.44	principal ore mineral
Carnallite	KCl·MgCl$_2$·6H$_2$O	16.95	14.07	ore mineral and contaminant
Kainite	4KCl·4MgSO$_4$·11H$_2$O	19.26	15.99	important ore mineral
Halite	NaCl	0	0	principal ore contaminant
Sulfates				
Polyhalite	K$_2$SO$_4$·MgSO$_4$·2CaSO$_4$·2H$_2$O	15.62	12.97	ore contaminant
Langbeinite	K$_2$SO$_4$·2MgSO$_4$	22.69	18.84	important ore mineral
Leonite	K$_2$SO$_4$·MgSO$_4$·4H$_2$O	25.68	21.32	ore contaminant
Schoenite	K$_2$SO$_4$·MgSO$_4$·6H$_2$O	23.39	19.42	rare
Glaserite	3K$_2$SO$_4$·Na$_2$SO$_4$	42.51	35.29	rare
Syngenite	K$_2$SO$_4$·CaSO$_4$·H$_2$O	28.68	23.81	rare
Alunite	K$_2$Al$_6$(OH)$_{12}$(SO$_4$)$_4$	11.4	9.46	
Kieserite	MgSO$_4$·H$_2$O	0	0	common ore contaminant
Anhydrite	CaSO$_4$	0	0	common ore contaminant
Nitrates				
Niter	KNO$_3$	46.5	38.60	
ORES				
Sylvinite*	KCl + NaCl	10-35	8-29	Canada, U.S.A., U.S.S.R.
Hartsalz	KCl + NaCl + CaSO$_4$ (MgSO$_4$·H$_2$O)	10-20	8-17	Germany
Carnallitite†	KCl·MgCl$_2$·6H$_2$O + NaCl	10-16	8-14	Germany, Congo, Brazil
Langbeinitite‡	K$_2$SO$_4$·2MgSO$_4$ + NaCl	7-12	5-10	U.S.A., U.S.S.R.
Mischsalz	Hartsalz + Carnallite	8-20	6-17	Germany
Kainite	4KCl·4MgSO$_4$·11H$_2$O + NaCl	13-18	10-14	Italy, Ethiopia
Niter (Caliche)	KNO$_3$ + NaNO$_3$ + NaSO$_4$ + NaCl	0.6-1.9	0.5-1.5	Chile
PRODUCTS				
Potassium Chloride	KCl	63.17	52.44	principal potassium product
Potassium Sulfate	K$_2$SO$_4$	54.05	44.87	artificial product
Potassium-magnesium sulfate (langbeinite)	K$_2$SO$_4$·2MgSO$_4$	22.69	18.84	natural product (New Mexico)
Manure salts	KCl + NaCl	40-60	33-50	U.S.S.R. and East Germany
Niter	KNO$_3$	46.5	38.6	
Potassium nitrate	KNO$_3$ + NaNO$_3$	10-14	8-11	

* May contain one or more sulfate minerals or carnallite. † May contain sylvite.
‡ May contain sylvite and leonite, kainite, or other sulfate minerals.

As with any world commodity survey, the estimated reserves are at best a first approximation. The necessary basic data, such as ore grades, mine extractions, process recoveries, and reserves are rarely publicized by mining companies. This summary has been derived almost entirely from published data interpreted in the context of a familiarity with the geology, mining, and refining of potassium ores. Should the estimates fall within 20% of the realized recoverable reserves, the writer will indeed be fortunate.

Reserves, as the term is used in the mining industry, are generally ore reserves and may be described as proven, probable, possible, known, indicated, inferred, or similar restrictive terms. The word "ore" implies that the reserves can be mined and refined at a profit. Other adjectives are used to qualify the probability that the reserves can indeed be mined and refined at a profit and consider such factors as grade, continuity, depth, and mineralogy of the deposit as they affect the costs of mining and beneficiation.

In this discussion, "ore" is specifically not used with respect to world potassium reserves because the factors which determine the profitability of a deposit are subject to change. Rather, all reserves are included which can likely be exploited profitably at some time in the future in a world of exploding population, escalating fertilizer demands, and limited arable land. Potassium deposits which are at depths of less than 3,500 feet, at least 4 feet in thickness, and containing 10% or more K_2O as soluble potassium minerals are the most attractive for mining by conventional underground or by solution mining methods. Quite obviously, the deeper, thinner, lower grade deposits will not be developed as soon.

The reliability of and potential recovery from the known world potassium deposits is rather difficult to assess with available information. In general, the reputed in-place reserves must be discounted from 25 to 60% in estimating the recoverable K_2O tons. These discounts allow for mining and refining losses which are a function of the depth, grade, and mineralogy of the ore. The discounts, therefore, must be applied to each deposit separately. The probablility of the ore reserves equalling their estimated magnitude has been based on the extent of exploration drilling and predictability of the ore type in question. Additional discounts have been applied where considered appropriate.

II. SOURCES OF POTASSIUM

A. Common Ores

Potassium is produced principally from bedded marine evaporite deposits by underground and solution mining methods. Additional production is derived from the fractional crystallization of surface and subsurface brines and, to a lesser extent, as a byproduct from nitrate deposits and the sugar beet and cement industries.

The potassium salts of marine evaporite deposits occur as beds in thick intervals of halite (NaCl). These halite deposits also contain beds of anhydrite (CaSO$_4$), and clay or shale. The potassium beds range from a few inches to several feet in thickness, but current commercial production is limited to ore intervals of at least 3 feet in thickness and mining thicknesses greater than 4 feet. The mineralogy of the beds may be simple or complex, depending on the depositional and post-depositional history of the deposit. The regularity, continuity, complexity, and ultimate value of a deposit can often be assessed by the study and proper interpretation of the mineralogy. The common ores of potassium include (i) *sylvinite*, composed essentially of halite and sylvite; (ii) *hartsalz*, composed of sylvite, halite, and kieserite or anhydrite; (iii) *langbeinite* ore (langbeinite and halite); (iv) *kainite* ore (kainite and halite); and (v) *carnallite* (carnallite and halite). Some of these ores contain appreciable amounts of other minerals, principally sulfate minerals and carnallite. Carnallite-rich beds may be mined more extensively in the future as sylvite reserves are depleted and the technology of converting carnallite to a KCl product is improved. More than 90% of the estimated world potassium reserves occur as bedded potassium deposits, principally sylvinite and carnallite. The mineral sylvite (KCl) contains the highest percent potassium of any readily available potassium compound and is in greatest demand as a fertilizer component. Sylvite is present in several important bedded evaporite deposits in the USA, Canada, Russia, and several other countries. Sylvinite ores represent the principal economically exploitable reserves for the next 50 years.

III. WORLD POTASSIUM RESERVES

The world reserves of economic and potentially economic potassium are sufficient to yield over 48 billion recoverable metric tons of K$_2$O. Enormous reserves are present in Canada which boasts approximately 5 billion tons of K$_2$O recoverable by conventional mining methods and an additional 13 billion tons of K$_2$O recoverable by solution mining methods. These reserves represent 37% of the world reserves. The greatest concentration of reserves, however, is found in the numerous evaporite deposits of Russia and represent 49% of the world reserves. The remaining known reserves are distributed between the USA, Europe, Asia, Africa, and South America as shown in Table 2. A brief discussion of the major world reserves will serve to illustrate the relative value of the various deposits and the likely shifts in world production during the new few decades.

A. North America

1. Canada

The discovery and development of the vast potassium deposits of the Devonian Prairie Evaporite Formation in Saskatchewan and adjacent Alberta and Manitoba, has overshadowed other events in the potassium industry in the past two decades. With the possible exception of recent discoveries in Russia, these deposits contain the greatest reserves of high-grade ore of any deposits in the world. The careful selection of techniques and equipment for the development, mining, and refining of the ores has assured the operations a strong competitive world position for many years.

Continous potassium production in Saskatchewan began in 1962 with the completion of the International Minerals & Chemical Corporation mine and refinery. By 1970, eight companies intend to be operating, with a total annual production capacity of between 6.1 and 6.3 million metric tons of K_2O. On the basis of world potassium production and demand estimates, this will account for a staggering 30% of world potassium production. The estimated recoverable reserves from these deposits are equally impressive, and represent 37% of the total world reserves.

The potassium deposits of the Prairie Evaporite Formation are exceptional for their high-grade (25 to 30% K_2O as sylvite), thickness (7 to 10 feet), uniformity of mineralization, and absence of structural deformation. These factors are sufficient to offset the great depth of the ore zones (below 3,000 feet) and the high cost of consructing shafts through the water-bearing overburden formations. Three main ore zones have been identified in the upper 200 feet of the Formation together with additional discontinuous potassium beds. In addition to the vast sylvinite reserves, the ore zones are composed of carnallite over additional wide areas.

Recent drilling by government agencies in Nova Scotia has located new areas of low-grade sylvinite and carnallite mineralization. In one hole 98 feet of intermittent mineralization (below a depth of 3,900 feet) averaged 5.05% K_2O, but no interval exceeding 8% K_2O was encountered. Further exploration is planned in the area, which is intensely deformed by both faulting and salt flowage.

The salt beds of the Arctic islands have been intersected in the course of oil exploration, but no potassium minerals have yet been reported. Due to the remote location, this may be the last major world evaporite basin to be explored adequately for potassium deposits.

Table 2—Potassium reserves in the world

Country	Mineralogical ore types	Grade % K$_2$O	Depth feet	Recoverable Metric K$_2$O tons (Millions) total / breakdown	Remarks
NORTH AMERICA					
Canada				18,000	
Saskatchewan	Sylvinite	14-32	3,000 - 8,000	5,000	Conventional mining
Alberta and Manitoba	and Carnallitite			13,000	Solution mining
Nova Scotia	Carnallitite	5	3,850 - 4,000	potential	Under exploration
Arctic Islands	ND			potential	
United States				225-360	
New Mexico	Sylvinite	10-22	600-2,000		
	Langbeinite	7-12	600-2,000		
Utah	Sylvinite	20-25	3,000	90-136	Paradox Basin
	Brine	0.5	Surface		Great Salt Lake
	Brine	0.7	Lake Bed		Wendover, Utah
Arizona	Sylvinite	14-16	1,800		
North Dakota	Sylvinite	NA	5,500 - 12,500		
California	Brine	3.0	Lake Bed		Searles Lake
Nevada	Brine	1.2	500-1,000		Silver Peak
Mexico	Carnallitite and Sylvinite	NA	200-2,000	potential	
SOUTH AMERICA					
Brazil	Carnallitite and Sylvinite (?)	8-16	1,500	16 / 5	Sergipe
Chile	Nitrates	1.0	Surface	NA	
Peru	Brine	2.0	Shallow	11	Sechura desert
EUROPE				5,723-9,400	
Czechoslovakia				potential	
Denmark	Sylvinite and Carnallitite	10-20	700-3,000	6	
England	Sylvinite	15-30	3,200-4,500	13-45	
	Shale	11	Surface	10	
France	Sylvinite	15-25	1,400-3,400	180-225	
East Germany	Sylvinite	13-17	1,000-3,200	3,600-5,400	
	Carnallitite	12	1,000-3,200		
	Sylvinite	15-20	1,300-3,500	1,800-3,600	
West Germany	Hartsalz	9-15	1,300-3,500		
	Carnallitite	9-15	1,300-3,500		

WORLD RESERVES

Greece		Kainite	12-18	1,000-2,600	potential
Italy		Carnallitite	11-13	1,000-2,600	23
		Sylvinite	18-19	1,000-2,600	
Netherlands					potential
Poland		Carnallitite	8	600-900	8
Rumania					potential
Spain		Sylvinite	15-25	1,000-4,000	63
		Carnallitite	14	1,000-4,000	20
RUSSIA					
West Ukraine		Hartsalz	16	250-1,000	24,000
(Stebnik-Kalush)		Sylvinite	8-30	250-1,000	
		Kainite	10	250-1,000	
		Langbeinite	NA	1,000-3,100	
Byelorussia		Sylvinite	9-30		
(Starobin-Sollgorsk)					
Urals		Sylvinite	15-30	220-700	
(Solikamsk-Brezhnikt)		Carnallitite	11-17	220-700	
Siberia					
Irkutsk		Sylvinite	NA	2,300-3,400	
Krasnoyarsk		Sylvinite,	NA	4,000-4,600	
		Kainite and Langbeinite			
Yakutsk		NA	NA	NA	
AFRICA					141-150
Angola					potential
Congo (Brazzaville)		Sylvinite	15-20	850-1,300	27
		Carnallitite	15	850-1,300	50
Egypt		Sylvinite	20-30	80-400	potential
Ethiopia		Kainite	NA		18-27
Gabon		Carnallitite and			potential
		Sylvinite			
Libya		Brine and Salts	1-4	Shallow	23
Mali		Sulfates			potential
Morocco		Sylvinite	10-11	1,600-2,600	23
		Carnallitite	11-13	1,600-2,600	
Niger and Nigeria		Brine	2.4-6.0	NA	potential
South Africa		Nitrates	NA	Surface	NA
Tunisia		Brine	NA	Surface	NA

(continued on next page)

Country	Mineralogical ore types	Grade % K$_2$O	Depth feet	Recoverable Metric K$_2$O tons (Millions) total tons	breakdown	Remarks
ASIA						
China	NA	NA	NA	544	NA	
Israel and Jordan	Brine	0.7	Surface		544	
Pakistan	Brine	4.2	4,000-8,000		potential	
	Sylvinite	NA	NA		potential	
	Carnallitite	NA	NA		potential	
	Langbeinite	NA	NA		potential	
Thailand						
OCEANIA						
Australia	Brine	3.6	Surface		potential	Lake McLeod
	Bedded salt				potential	
TOTAL				48,649 - 52,470		

NA - Not Available; ND - None Discovered.

2. United States

Bedded potassium deposits were first exploited in the USA in 1931 by the U. S. Borax and Chemical Company near Carlsbad, New Mexico. Subsequently, 6 additional companies have established operations in the district which now includes 11 mines. In 1961 the gross product revenue from these mines exceeded 1 billion dollars, making Carlsbad one of the major mining districts in North America. Of the 12 potasium horizons in the Permian Salado Formation, 4 are presently mined and 5 contain economic or potentially economic mineralization. The ores are sylvinite; sylvinite with more than 3% water insoluble material; and two mixed-ore beds containing sylvite, langbeinite, and lesser amounts of leonite, kainite, kieserite, bloedite, loewite, and polyhalite. The ore beds are between 600 and 2,000 feet in depth where reserves are present, and no major problems are encountered in shaft construction or mining. These deposits represent the principal USA potassium reserves. Due to lower grade, higher mining and freight costs, and beneficiation problems related to the abundance of clay and sulphate minerals, much of the Carlsbad sylvinite ore will be unable to compete with potassium produced in Canada and Europe except in the immediate southwestern USA. The production of langbeinite concentrates and the manufacture of potassium sulfate will continue to increase slowly, the rate of increase depending on the availability of Sicilian and German sulfate products and the success of marketing programs.

At Moab, Utah, Texas Gulf Sulphur Company operates the only potassium mine in the USA outside of the Calsbad district. The ore is present as a bed of sylvinite in the Pennsylvanian Paradox Member of the Hermosa Formation. Mining conditions are more challenging than had been anticipated due to numerous folds in the ore zone, and back stabilization and beneficiation problems. Considerable progress has been made, however, and it now appears that economic mining may be possible in this deposit.

Brines have long contributed to potassium production in the USA. The first producer, the American Potash & Chemical Company operation at Searles Lake, California, has been joined by Kaiser Aluminum & Chemical Company at Bonneville, Utah, and other smaller producers. In the future, the relative importance of potassium produced from brine will increase as Foote Mineral Company reaches full capacity as its Silver Peak, Nevada, operation and legal and production problems at the Great Salt Lake are overcome.

Potassium reserves may eventually be developed in other major evaporite basins of the USA. Considerable mineralization has been drilled in the Permian Supai Formation of northeast Arizona. Potassium may also be present in evaporite beds in Colorado, and at a considerable depth in the Devonian salts of the Williston Basin in North Dakota, which are an extension of the Saskatchewan deposits.

3. Mexico

Potassium minerals have repeatedly been reported in the Jurassic evaporites of the Isthmus of Tehuantepec and the Yucatan Peninsula. Carnallite is apparently the abundant potassium mineral, but sylvite is also reportedly present. Available information suggests that reasonably shallow (2,000 feet) moderately to strongly deformed, potentially economic potassium beds may be discovered. The deposits appear to be of considerable interest because of their proximity to the ocean, but the rate of exploration will be conditioned by the government and business climate in Mexico.

B. South America

The total known potassium resources of the South American countries are considerably less than might be expected, considering the geology and size of the continent. Bedded evaporites are known in widely separated areas, including Peru, the Amazon Basin, and the Province of Sergipe in Brazil. Potassium-bearing beds have been reported, however, only in the salt deposits near Carmopolis, Sergipe, Brazil. Here Cretaceous salt is found in a graben structure at depths in excess of 1,500 feet. The principal potassium mineral is carnallite, although sylvinite may locally be present.

The development of the phosphate deposits and potassium-bearing brine of the Sechura desert in Peru, will exploit the other major potassium reserve in South America. A method has reportedly been developed to exploit the chemically complex brine which occurs as a shallow subsurface fluid. The other currently significant source of potassium production in South America is as a byproduct from the Chilean nitrate industry. The search for potassium deposits in South America will continue because the market incentive is substantial. The early success of these programs may be deferred, however, by remote exploration areas and high exploration costs.

C. Europe

1. Denmark

A potassium-bearing evaporite bed has been intersected in the course of oil drilling in North Jutland. The potassium is reported to be present in beds up to 18 inches in thickness between 2,500 and 4,000 feet in depth. This is probably an extension of the German Zechstein deposits. The dips of the salt beds are beween 45° and 70° suggesting the area is one of salt doming. The significance of this occurrence is being evaluated by the Danish Government.

2. England

Near Whitby, in Yorkshire, potassium beds have been repeatedly intersected at depths between 3,200 and 4,500 feet. The two principal sylvinite beds in the Permian evaporite section, which exceeds 1,800 feet in thickness, averages 20 and 25 feet in thickness, 10 and 19% K_2O. The depth of these deposits may inhibit their exploitation, but one company is said to be considering conventional mining methods, and two companies are considering solution mining. The location and market incentive for a potassium operation in England is sufficient to justify the research and experimentation that will be necessary to develop a competitive solution mining process. It is likely that such an operation will reach production within the next 10 years. Recoverable reserves will approach 50 million metric tons of K_2O.

3. France

Potassium reserves are present in Triassic evaporites near Dax in southwest France and in the Oligocene evaporites of Alsace. In the Triassic deposits, erratic and irregular sylvinite beds up to 10 meters in thickness, contain between 12 and 19% K_2O. The total recoverable reserves, however, probably do not exceed 1 to 2 million tons of K_2O.

The potassium deposits of Alsace constitute the principal reserves in France. Two ore zones have been mined. The upper contains 20 to 25% K_2O as sylvite over 3 to 6 feet at depths between 1,200 and 3,400 feet. The lower bed is 6 to 15 feet in thickness, with 15 to 20% K_2O as sylvite. Six mining divisions are currently producing from the lower ore zone. The projected decrease in potassium production in France reflects the depletion of high-grade reserves. Continued production of some facilities has been possible only through the introduction of highly mechanized, low-cost mining methods.

4. East Germany

High-grade sylvinite resources are rapidly becoming depleted and the bulk of remaining reserves are reportedly carnallite mineralization. Whereas sylvinite is currently produced from more than 17 mines, future production (if the growth indicated in Table 4 is to be realized) must come increasingly from carnallite-rich ores. Additional sylvinite reserves may be discovered and exploited, such as the deposit near Zielitz scheduled for production in 1970, but additional recovery from carnallite reserves will be necessary to sustain the projected increase in production. Considering the higher production costs associated with carnallite ores, it might be anticipated that some of the growth projected for the East German potassium industry will be absorbed by an accelerated exploitation of the Russian sylvinite reserves.

5. West Germany

Long the principal world source of potassium, together with adjacent areas in East Germany, West Germany presently produces from three major districts and a few outlying deposits. In the Alsace district, sylvinite, containing 18 to 20% K_2O, is mined from structurally deformed beds at depths between 2,000 and 3,200 feet. In the Werra-Fulda district, the principal ores contain sylvite, kieserite, and carnallite with grades between 9 and 15% K_2O. The intensely deformed and irregular deposits of the Hannover district yield sylvinite, langbeinite, and kainite-bearing sylvinite, hartsalz, and sylvite-carnallite ores with K_2O grades between 15 and 18% K_2O.

The reserve picture for West Germany is less favorable than the figures in Table 2 might suggest. Proven recoverable K_2O tonnage is substantially less than the 1.8 to 3.6 billion tons of probable ore. The difference is mainly in lower grade, structurally complex reserves which will have difficulty competing in the foreseeable future with the lower production costs of Canadian and Russian deposits. The reserves are, nonetheless, sufficient to allow West Germany to retain its position as a major producer of potassium, especially for the European market.

6. Italy

The economic kainite reserves of Sicily are unique in the world and constitute the bulk of Italy's potassium reserves. The deposits are of Miocene age and the mineral is present in at least four separate horizons. In addition to kainite minor reserves of carnallite and sylvinite are reported. The potassium beds range in thickness from approximately 6 to 100 feet and dip up to 60 degrees.

Production from the Sicilian deposits will increase, but will constitute a decreasing percentage of world potassium production. If the projected increase in production capacity to 400,000 metric tons of K_2O by 1980 is realized, the life of the deposits will be at least 50 years. Modest increases in current reserves may be outlined by further exploration.

7. Netherlands

Equivalents of the Permian salts of Germany and England are present in the Netherlands. Potassium salts containing up to 15% K_2O have been encountered at a depth of 3,198 to 3,237 feet, but no commercial deposits have been identified. The potential of this area is apparent and further exploration is warranted.

8. Poland

The eastern extension of the Permian Zechstein of Germany contains bedded potassium deposits in the vicinity of Klodawa. The known deposits contain low-grade (8% K_2O) carnallite ores between 1,500 and 5,000 feet in depth. Current recoverable reserves are estimated to be only 8 million tons. The magnitude of potassium production in Poland will, however, depend on the development of methods to process carnallite ore and the exploration for and discovery of sylvinite reserves.

9. Rumania

Low-grade, sulfate-bearing sylvinite deposits of Miocene age have been reported from northern Rumania. Exposures in salt mines and data from bore holes indicate the potassium beds contain between 7 and 12% K_2O, at depths to 600 feet. Langbeinite, leonite, and kainite are associated with the sylvite. The salt masses are structurally complex and reserve data is not available.

10. Spain

The potassium deposits of Spain may be referred to the Catalonia and Navarra districts. Four mines exploit the Catalonia deposits, whereas only one operates in the Navarra district. The deposits at Catalonia have been locally deformed through faulting, folding, and flowage bringing the evaporites to the surface in places. Up to four potassium beds are locally present, but no more than three are exploited in a single mine. The ore zones contain sylvite and carnallite and range in grade from 15 to 29% K_2O. Dips up to 30° are common. The reasonably high grade, however, assures that in spite of the structural problem the majority of the potassium reserves will be mined.

Two potassium beds are present in the Navarra district between 400 and 2,000 feet in depth. The upper seam contains principally carnallite with an average grade of about 14% K_2O. The grade of the sylvinite bed is between 18 and 20% K_2O. This deposit is less deformed than those in Catalonia and the extensive high-grade, shallow reserves will be efficiently and economically exploited for many years.

11. Russia

The Soviet Union is reported to possess the most extensive reserves of high-grade potassium ores in the world, together with what may be the greatest potential for the discovery of extensive new reserves. Recent additions to the reserves of the Upper Kama River deposits in the Urals have substantially increased both the tonnage and average grade of the Russian reserves. These deposits now reportedly represent 90% of potassium reserves of the Soviet Union. The principal deposits of Russia may be divided into four geographic units.

The deposits near Stebnik and Kalush in the West Ukraine were formerly part of, and exploited by, Poland. Hartsalz, sylvinite, kainite, and langbeinite mineralization occur in at least five potassium zones. The deposits have been moderately to strongly deformed and mining in steeply dipping beds is common. The reserves are lower grade than the deposits in the Urals and the production costs are higher. The deposits represent, nonetheless, a substantial potassium sulfate production capacity.

The sylvinite deposits of Starobin and Soligorsk in Byelorussia reportedly contain between 1.5 and 3.0 billion recoverable metric tons of K_2O. The reserves are present in three sylvinite beds and an additional sylvinite-carnallite bed. The high average grade of much of the reserve (20 to 25% K_2O) is offset by a 4 to 8% water insoluble content.

The bulk of Russian potassium reserves are considered to be in the vicinity of Solikamsk and Brezhniki. Three major zones of mineralization composed of several potassium beds are present over a large area which has not been fully explored. The deposits are essentially undeformed and the combination of shallow depth, thick mineralization, and uniform high grade combine to assure vast reserves of low cost potassium. The significance of these deposits, particularly in terms of the world potassium market, has yet to be realized.

Additional, but as yet poorly documented, potassium deposits are present in east and south Russia. In Siberia, comparatively deep-bedded deposits are known for Irkutsk, Krasnoyarsk, and Yakutsk. Widespread mineralization exits north and east of the Caspian Sea and several beds of potassium minerals have been encountered in salt deposits in Turkmenistan. The potential for the development of reserves in these areas is good, as is the potential for discovering new deposits in the sparsely explored regions of Siberia. In addition, consideration is being given to the recovery of Na_2SO_4 and KCl from the brine of the Karabogas Gulf through solar evaporation. Clearly, production of potassium in the Soviet Union is destined to increase substantially.

D. AFRICA

1. Congo

Substantial carnallite and sylvinite reserves have been discovered near Pointe Noire. The carnallite averages 16% K_2O and occurs in beds 10 to 40 feet thick. The sylvinite bodies, on the other hand, are smaller and erratic in grade, distribution, and thickness. The development of these reserves has proceeded slowly, with foreign companies hesitating to invest heavily under current economic and political conditions. The deposits are ideally located close to water and will doubtless be exploited within the near future. Further exploration can be expected to add substantially to existing reserves.

2. Ethiopia

The evaporite deposits of the Danakil Depression contain substantial reserves of potassium as sylvinite, carnallite and kainite. These deposits have only been partially explored and the reserve estimate in Table 2 will undoubtedly prove to be conservative. High-grade sylvinite ore is found at a shallow depth and in spite of the reported occurrence of brine and high temperatures in the subsurface, the deposit will likely be mined in the next decade. Because of the remote and formidable location, however, the venture is destined to be costly. Exploration in other parts of Ethiopia, particularly adjacent to the Red Sea, may locate new important deposits. Further interest in this area is spurred by the ideal location with respect to world markets.

3. Morocco

The potassium reserves of the Khemisset basin are present over an area in excess of 30 square kilometers. The principal potassium mineralization is carnallite, which occurs in two beds 3 to 30 feet thick. Around the edges of the carnallite deposits are sylvinite bodies which presumably were formed by the alteration of the carnallite. The sylvinite deposits are low grade, erratic, and probably irregular. The United Nations is currently assisting in the development of sufficient reserves to warrant the construction and operation of a mine and refinery. Although high-grade deposits will probably not be located in the Khemisset basin, they may be present in other sedimentary basins of Morocco.

E. ASIA

The countries of Asia, which include those not previously discussed, except Australia and New Zealand, contain a small percentage of world potassium reserves, but considerable potential for the discovery of new deposits. The potassium resources of the Dead Sea will continue to be exploited by Israel and perhaps Jordan. The recoverable reserves are estimated at approximately 540 million tons of K_2O.

Pakistan possesses potential for potassium reserves both in bedded salts and brine. The salt bed occurrences thus far reported are sporadic zones of sylvite and langbeinite in salt mines. The brine occurrences are of more interest as they contain one of the highest potassium concentrations of any natural brine known to this writer. According to reports, the brine encountered in an oil well drilled near Dhariala, West Pakistan, contains 4.2% K_2O between 4,500 and 8,000 feet in depth. The significance of this occurrence is currently being evaluated with additional drill holes.

The known and potential potassium reserves of the remainder of Asia are difficult to estimate because of the paucity of information. The occurrences of bedded salt in such countries as Saudi Arabia, Iran, and Thailand offer the strong possibility that additional deposits will be found. With respect to potassium production and reserves in China, little is known.

F. OCEANIA

Several extensive evaporite basins have long been known in Australia. To date, the salt sections have been intersected only in the course of the oil exploration and the deposits are generally deep and strongly deformed. No potassium salts have been reported from these deposits, though their presence is certainly to be expected considering the volume of salt encountered.

Recent attention has been drawn to the potassium resources of Lake McLeod in Western Austrialia. One company is currently exploring the feasibility of producing potassium chloride from the lake brine which reportedly contains 3.6% K_2O. If successful, the operation would establish substantial reserves in the 800-square mile lake.

G. OTHER RESERVES

Egypt currently produces minor amounts of potassium from soils, nitrates, and as a by-product from a sugar refinery. Bedded potassium salts may be present, however, in the deeply buried evaporites of the Red Sea graben. Potassium is also recovered from the nitrate and molasses industries of South Africa, but there is little likelihood that bedded deposits will be found in this country. In Tunisia coastal lake beds yield some potassium, but the major potential of the country is in the buried bedded evaporites. Substantial potassium reserves reputedly exist in Libya in the form of bedded salts and interstitial brine. Little has been published about the magnitude and character of this deposit.

Potassium reserves may be developed in other African countries. Sizeable deposits of potassium and magnesium sulfate minerals have been reported from Mali, presumably in lake beds. The buried evaporite deposits of Angola, Gabon, Tanzania, and other coastal countries may prove to contain potassium-rich beds. In addition, surface and subsurface brines, such as occur in Niger and Nigeria, are probably widespread on the continent and will doubtless be exploited in some areas in the future.

IV. PROJECTED POTASSIUM RESERVES

The separate potassium reserves tabulated in Table 2 are estimates which will increase or decrease as exploration continues and new information becomes available. The total world reserves will increase, however, because (i) large areas of the continents have not been adequately explored; (ii) more widespread use will be made of surface and subsurface brines; and (iii) improvements in mining and processing methods will increase recovery in current operations and permit the exploitation of presently uneconomic deposits and brines.

New discoveries will be made as extensions of known districts and as new deposits. Extensions of known districts are most likely where exploration is in the early stages such as in the Congo (Brazzaville), Ethiopia, Siberia, and Morocco. New deposits are likely to be discovered in sparsely explored areas such as Central America, Siberia, Eastern Canada, and parts of Asia, Africa, and Australia. These discoveries might be expected to add 5 to 10 billion recoverable K_2O tons to the world reserves.

V. RECOVERY OF POTASSIUM

A. Brines

The economics of potassium production from brines are complicated by the (i) low potassium content of the brines, (ii) the common consideration that solar evaporation be utilized, (iii) the presence of components in the brine which interfere with the production of potassium chloride, and (iv) the necessity of recovering some of these components with the potassium. The technology required to surmount some of these problems is now being developed.

The recovery of potassium from brines is currently practiced at Wendover, Utah, and Searles Lake, California, in the USA, the Dead Sea in Israel, and to a lesser extent in other places. Plants designed to recover potassium from brine are under consideration or construction at the Great Salt Lake, Utah, Silver Peak, Nevada, the Salt Range, Pakistan, Lake McLeod, Western Australia; in the geothermal brine province of California and elsewhere. Potassium-bearing brines are more widely distributed in the sedimentary basins of the world than are potassium-rich evaporite deposits.

The greatest reserve of potassium is contained in the oceans. As with terrestrial brines, the economics of producing potassium from sea water depends on the recovery of associated resources such as bromine and magnesium. The ultimate magnitude of potassium recovery from sea water, however, will probably depend on the demand for water for agricultural and human consumption. It is apparent that as domestic water supplies become depleted and world food production increases, the requirements for desalinized water must increase. It seems likely that the saline-enriched effluent from desalinization plants will be processed, probably through solar evaporation, to produce chemical products for industry and agriculture. The technology of water desalinization and byproduct recovery of potassium, magnesium, bromine, and other chemicals is now experiencing rapid development.

B. Underground Deposits

Better methods of mining and refining potassium reserves are continually evolving. In certain underground mines, low height and longwall mining methods have been successful. These developments increase the portion of the in-place mineralization which can be delivered to the refinery. Developments in the field of rock mechanics have provided new knowledge for design of mine openings which permit the safe exploitation of deeper deposits. Equipment development continues to decrease mining costs, thus permitting the exploitation of lower grade ores.

A most significant development in mining methods is the solution mining venture now operating near Belle Plaine, Saskatchewan. Built by Kalium Chemicals Ltd., the operation has been producing approximately 325,000 metric tons K_2O / year since 1964. While this is by no means the first attempt at solution mining of potassium, it is the first to attain sustained production.

Prior to choosing the conventional mining method, several of the future producers in Saskatchewan gave serious consideration to solution mining. Two additional companies are presently conducting field tests of the solution method; Lynbar in Saskatchewan, and Whitby Potash Limited in Yorkshire, England. The solution mining method will continue to be refined and can be expected to produce a significant portion of world potassium production in the future.

New technology will be as important to the processing of some ores as it is to the mining. Improved beneficiation methods have been needed, for example, in the processing of mixed chloride-sulfate ores, high carnallite ores, and ores containing abundant clay. Recent process developments in Carlsbad, New Mexico will permit the recovery of both sylvite and langbeinite from beds too low in grade to be mined solely for either. Recrystallization of sylvite mines at Moab, Utah will permit the upgrading and recovery of material which otherwise could not meet product requirements.

Similar developments will continue to meet specific problems and assure that full advantage is taken of the potassium resources at hand.

C. Associated Constituents of Brine

The profitable recovery of potassium from surface and subsurface brines is often closely tied to the recovery of associated brine constituents. At Searles Lake, California, the American Potash and Chemical Corporation extracts potassium chloride, potassium sulfate, sodium sulfate, lithium, soda ash, bromine, and boron from the lake bed brines. The Dead Sea Works in Israel recovers potassium chloride, bromine and magnesium chloride from the Dead Sea brine by solar evaporation and crystallization in evaporating ponds, followed by leaching and flotation. The recovery of potassium from brines in new operations will be in locations favorable to solar evaporation, such as Western Australia and Pakistan; and in situations with unique brine compositions, such as the lithium-rich brine to be exploited by Foote Mineral Company at Silver Peak, Nevada.

D. Potassium as a Byproduct

Potassium production as the byproduct of other manufacturing processes continues to increase, but will not become a significant percentage of total world production. This includes potassium produced as a byproduct of nitrate mining in India, Chile, and South Africa; sugar refineries in Egypt and the USA; and cement plants in this country. Attention has long been directed toward the possibility of recovering potassium from alunite deposits and even potassium-rich shales and granites. If exploited at all, these reserves will attain only temporary and local importance.

VI. POTASSIUM SUPPLY AND DEMAND

No independent estimates of future world potassium demands were prepared during this summary. For a frame of reference the projections of Noyes (1965, p. 4), similar in magnitude to the projections of others, are reproduced in Table 3. For comparison, the potassium production for the period 1962-1966 (Lewis, 1967) and the projected production to 1980 (Bartley, 1966) are tabulated in Table 4. The inferred future overproduction, which decreases from 4.8 million tons K_2O in 1970, to 1.2 million tons in 1980 is not particularly significant because of inaccuracies in estimating both demands and production capacities. It is likely, nonetheless, that substantial overproduction will occur by 1968 and continue into the 1970's.

Table 3—Projected potassium consumption in the world

Region	1962-1963	1970	1975	1980
	\multicolumn{4}{c}{1,000 metric tons K_2O}			
North America				
Canada	109	204	317	453
United States	2,180	2,948	3,809	4,898
Europe	4,883	7,982	10,884	14,512
Soviet Union	826	1,814	2,540	3,628
Latin America	231	476	726	998
Africa	115	295	453	703
Asia	706	1,270	1,814	2,449
Oceania	90	295	453	612
World Total	9,140	15,284	20,996	28,253

* Adapted from Noyes (1965, p. 4).

Table 4—Potassium production and projected production in the world

	\multicolumn{5}{c}{World production of marketable potassium*}	\multicolumn{3}{c}{Estimated future world production†}
	1962	1963	1964	1965	1966	1970	1975	1980
	\multicolumn{8}{c}{1,000 metric tons K_2O}
North America								
Canada	136	569	779	1,353	1,855	6,350	8,100	10,900
United States	2,225	2,598	2,628	2,848	3,012	3,300	3,000	2,500
South America	18	19	13	14	15	20	25	30
Europe								
France	1,722	1,721	1,806	1,879	1,910	1,500	1,300	1,000
East Germany	1,752	1,845	1,857	1,926	2,000	2,200	2,500	3,000
West Germany	1,940	1,948	2,201	2,385	2,300	2,000	1,800	1,500
Italy	154	188	206	241	261	350	400	400
Spain	235	260	292	362	443	400	500	600
USSR	1,905	2,050	2,200	2,349	2,540	3,000	5,000	7,000
Asia								
Israel	91	113	255	310	371	600	800	1,000
World Total	10,178	11,311	12,237	13,667	14,707	19,720	23,425	27,930
Future Producers‡						375	875	1,500
Projected World Totals						20,095	24,300	29,430

* Adapted from Lewis (1967).
† Adapted from Bartley (1966).
‡ Some combination of Australia, Congo, England, Ethiopia, Jordan, Morocco, Peru, and possibly other countries.

At the supply and demand figures of 1980, the gross world potassium reserves (as tabulated in Table 2) would be sufficient to last 1,750 years. This is of little more significance than saying there are sufficient potassium reserves to last a long time. The world demand for potassium is certain to increase beyond the 1980 level. In addition, only a portion of the present known reserves will be mined because (i) new high-grade bedded deposits will be discovered, (ii) new sources of surface and subsurface brine will be exploited, and (iii) potassium will be recovered from the ocean. Until these developments take place, there are fully sufficient reserves which can be economically exploited at current world prices to match increased market demands. Considering the important role potassium must play in feeding the growing world population, this is a reassuring conclusion.

ACKNOWLEDGEMENTS

The preparation of this survey has been possible through the cooperation of the International Minerals and Chemical Corporation, and particular thanks are due Dr. D. L. Everhart, Chief Geologist, for permission to prepare the survey. The original manuscript has been improved through the critical review of D. L. Everhart, D. H. Freas, H. D. Strain, and A. V. Mitterer. Use has been made of numerous articles which have appeared in newspapers, trade, business journals. Whereas each reference has not been cited, the important publications are listed in the References section. The responsibility for interpretation and extrapolations of the published information rests with the writer.

REFERENCES

Bartley, E. M. 1966. Potash. *In* Canadian Minerals Yearbook 1965. Mines Branch and Min. Resources Div., Dept. Energy, Mines and Resources, Ottawa, Canada. 11 p.

Bartley, C. M. 1966. Canadian potash developments in 1965. Paper presented to Amer. Inst. Mining, Metall. and Petrol. Engineers, Feb. 27 - March 3, 1966. 10 p.

Borchert, H., and R. O. Muir 1964. Salt deposits. D. Van Nostrand Co. Ltd., London. 388 p.

Colorado School of Mines Research Foundation, Inc. 1967 Potash. Mineral Ind. Bull. Vol. 10, no. 3, Colorado School of Mines. 18 p.

Engineering and Mining Journal. McGraw-Hill, Inc., New York.

Institute of Petroleum and the Geological Society, London, March 3, 1965. Salt basins around Africa. Proceedings of joint meeting. The Institute of Petroleum, Elsevier Publ. Co., Amsterdam. 122 p.

Lewis, R. W. 1966. Potash. *In* Bureau of Mines Minerals Yearbook 1965. U.S. Bureau of Mines. 15 p.

Lewis, R. W. 1967. Potash in 1966. Mineral Industry Surveys Letter. U.S. Bureau of Mines, July 10, 1967. 4 p.

Noyes, R. 1965. Potash and potassium fertilizers. Chemical Process Monograph no. 15. Noyes Development Corp., Park Ridge, N. J. 210 p.

Pickard, F. C. 1967. Potash. Eng. Mining J., Feb. p. 173-174

The British Sulphur Corp. Ltd. Phosphorus and potassium. London.

The British Sulphur Corp. Ltd. 1966. World survey of potash. 93 p. London.

Potassium Fertilizer Technology

EDWIN C. KAPUSTA

Potash Company of America
New York, New York

I. INTRODUCTION

The element potassium occurs in abundance over widespread areas of the earth's surface as a component of various rocks, minerals and brines. Most of these potassium-bearing materials have limited value as plant nutrients and require further processing to convert them into economically useful potassium fertilizers. This discussion will examine the methods by which the principal potassium raw materials are recovered and transformed into marketable potassium fertilizers.

Primary emphasis is placed on the mining and manufacturing practices prevalent in North America; the technological principles involved may also be applied to the production of potassium fertilizers in other areas of the world.

The technology of potassium fertilizer production includes the following operations:

1) Extraction of the basic potassium-bearing raw materials from the earth's crust.

2) Separation of the desired potassium material from the undersirable portions of the original ores or brines.

3) Modification of the physical properties or transformation of the chemical composition of the beneficiated concentrates to yield varied product types and forms.

A description of these operations and processes, and the products which result from their utilization, forms the basis of this paper.

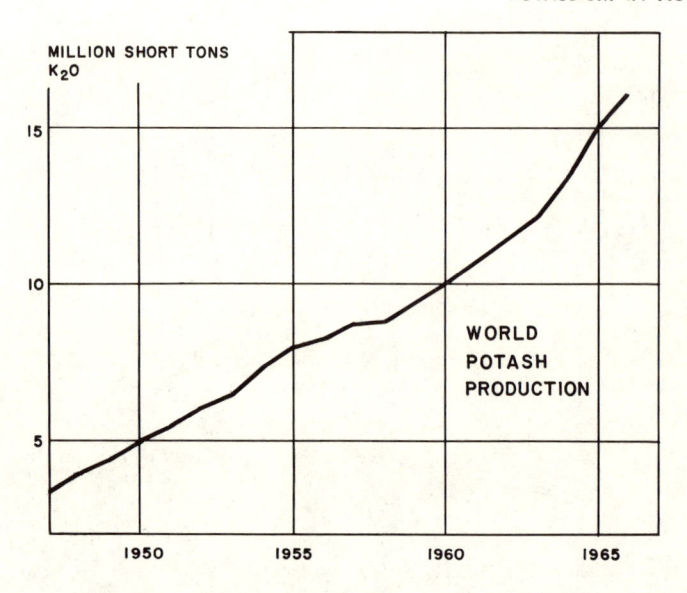

Fig. 1—World production of K$_2$O, 1947-1966.

II. THE WORLD POTASH INDUSTRY

Over 95% of the potash industry's annual output is consumed as fertilizer. The early history and development of this century-old industry have been well chronicled by many authors (Anon., 1966g; Noyes, 1966; Turrentine, 1926), as has been the critical role of potassium fertilizers in sustaining the world's burgeoning population (Anon., 1965d; Ewell, 1964). The data in Fig. 1 illustrate the remarkable growth of the potassium industry in the past 2 decades. World production of potassium materials has increased from approximately 3.3 million short tons K$_2$O in 1947 to about 16.2 million short tons of K$_2$O in 1966, a five-fold increase (Ruhlman, 1960; Lewis, 1966).

Commercially important quantities of potassium fertilizers were produced in 10 countries on 4 continents during 1966 (Lewis, 1966). The principal production centers are North America, Central Europe, and Asia (Anon., 1966a, Bartley, 1964). The leading producers are West and East Germany, the USA, Russia, Canada, and France. Significant quantities of potassium materials are also produced in Spain, Israel, and Italy.

III. POTASSIUM RAW MATERIALS

Although the earth abounds with potassium-bearing materials, relatively few constitute important sources of fertilizer potassium. The currently valuable potassium raw materials are those which are water soluble and occur in more readily accessible deposits of sufficient magnitude to justify commercial exploitation. Practically all of the potassium fertilizers produced today are derived from bedded deposits of water-soluble potassium minerals or brines.

The compositions of the commercially important potassium minerals are listed in Table 1. These ores are believed to supply more than 95% of the total fertilizer potassium produced annually.

Abundance of supply and comparative ease of extraction and beneficiation make sylvite the predominate source of fertilizer potassium in most of the producing areas. The ore which is mined is sylvinite, a mechanical mixture of sylvite (KCl) and halite (NaCl).

Carnallite, once a major potassium raw material in Europe, has given way to sylvite, but remains the major source of potassium in Israel. Kainite is the predominate material mined in Italy and is also mined to some extent in Germany. Langbeinite is an important potassium ore in the USA. The only deposit of nitre of commercial significance is located in Chile.

Natural brines have been significant sources of fertilizer potassium for many years. Currently, the major potassium brine sources include those from Searles Lake in California, the Salduro Marsh brines near Wendover, Utah, and the Dead Sea brines in Israel. These are discussed later.

IV. POTASSIUM RAW MATERIAL RETRIEVAL

The techniques employed in extracting potassium-bearing raw materials are dependent on the location and nature of the deposit as well as the character of its environment. Today, the greatest quantity of potassium is obtained from subterranean deposits of soluble potassium minerals. Smaller quantities of potassium are derived from subsurface brines. This section describes prevailing practices in mining underground potassium ore deposits and handling of brines relative to potassium fertilizer production.

A. Mining Potassium Minerals

Underground deposits of solid potassium minerals are mined by: (i) the time-honored conventional shaft mining procedure by which solid ore is removed from the ore bed and hoisted to the surface; and (ii) the solution mining technique recently applied to potassium ore in Canada. In the latter method, the potassium-bearing ore is dissolved and brought to the surface in the form of a solution.

Potassium ore is currently being mined by these methods at depths ranging from several hundred feet to over 5,000 feet (Kyle, 1964; Ruhlman, 1960). Reports (Anon., 1962; Anon., 1966d) indicate that a depth of

Table 1—Commercially important potassium minerals

Mineral	Composition	Approx. plant food content, % K$_2$O	K
Sylvite	KCl	63.17	52.45
Sylvinite	KCl, NaCl mixture	–	–
Carnallite	KCL · MgCl$_2$ · 6H$_2$O	17.0	14.1
Kainite	KCl · MgSO$_4$ · 3H$_2$O	18.9	15.7
Langbeinite	K$_2$SO$_4$ · 2MgSO$_4$	22.6	18.8
Nitre	KNO$_3$	46.5	38.6

approximately 3,500 to 4,000 feet is probably the maximum depth at which potassium may be mined by conventional shaft mining methods from both safety and cost viewpoints. The only potassium solution mine in operation in the world is in Canada. Potassium is removed from a depth of approximately 5,300 feet.

1. Shaft Mining

Shaft excavation, lining, and equipping costs constitute the largest cost item in the development of a potassium production facility. The formidable and costly shaft sinking operations entailed in the development of the extensive Canadian potassium reserves have been well documented by many authorities in this field (Scott, 1953; Walli, 1966; York, 1966).

Elaborate and expensive freezing processes are required to consolidate the several water-bearing strata and unstable quicksand-like formations encountered in these shaft sinking operations. From 4 to 5 years are involved in shaft sinking operations in Canada. Although smaller shafts are common to many of the older mines, the newer ones, as typified by those in operation and underdevelopment in Canada, are served by shafts of diameters of 16 to 18.5 feet (Anon., 1966; Kyle, 1964).

A room and pillar mining plan is usually followed in extracting the raw potassium ore. Here rectangular rooms are mined out and pillars of ore are left behind to support the overburden. In the Carlsbad mines, initial mining extracts about 50 to 60% of the ore (Kapusta, 1963). Subsequent second-stage mining or "pillar robbing" often results in removal of about 90% of the ore. In one Canadian mine, primary mining results in extraction of 35 to 40% of the ore in place (Anon., 1966c). Dimensions of entryways, rooms, and pillars vary with individual mines. In one Carlsbad mine, rooms and breakthroughs are 36 feet wide, with pillars in panel areas 36 by 82 feet, or 58 by 85 feet, depending on the nature of the salt formation above the sylvinite vein. In Canadian mines, room widths of 20 and 29 feet have been reported (Anon., 1966c; Anon., 1966f).

In the newer potassium mines, continous mining machines have largely replaced the conventional mining technique which involved undercutting,

Fig 2—Boring type continuous mining machine (Courtesy International Minerals and Chemical Corp.)

drilling, and blasting to fragment the ore. Although continuous miners of various designs are available, the essential function of each is to remove ore continuously from the potassium vein. One such machine of the boring type, illustrated in Fig. 2, is able to cut up to 5 tons ore/min. (Anon., 1966c).

The fragmented ore is gathered by mechanical loaders and loaded on conveyor belts or into shuttle cars for transport to the main line conveyor belt for delivery to the shaft. At the shaft, the ore is fed into roll crushers for primary size reduction. The crushed ore, approximately 3-to-6-inches in size, is directed into underground storage bins from which it is automatically loaded into skips for hoisting to the surface. Skips vary in capacity, some holding up to 24 tons of ore. At the surface, the crushed ore is delivered into bins for storage and subsequent processing prior to its entry into the beneficiation system.

2. Solution Mining

The practical application of the solution mining technique to potassium ore extraction is a very recent accomplishment. Although such processes have been studied for many years, the first, and to date, only successful commercial solution mining facility was brought into operation in Canada in 1964 (Anon., 1965e). The following advantages of solution mining have been cited:

1) Deeper potassium deposits may be mined. Published reports, as previously noted have indicated that conventional shaft mining is generally restricted to depths less than 3,500-to-4,000 feet because of safety and cost limitations.

2) Greater total extraction of the potassium deposit may be permitted than in the conventional room and pillar shaft mining procedure since the entire potassium deposit may be removed.

Technical details regarding the solution mining method have not been released. However, several patents and articles relating to the development and variations in the application of the method are available (Anon., 1964a; Dahms and Edmunds, 1962). As described in these literature sources, the process basically entails:

1) injection of a brine, of controlled NaCl-KCl composition and temperature, to dissolve KCl and NaCl from the deposit in the desired proportions,

2) pumping of the resultant solution from the cavity to the surface,

3) concentration and removal of the desired potassium chloride values, and

4) adjustment of the composition of remaining brine for recirculation to the mining process.

Means of establishing and controlling proper cavity formation and growth, maintenance of desired rates of KCl and NaCl dissolution and flow rates, as well as use of multiple holes to effect entry and/or withdrawal of the initial and saturated brines are described by Dahms and Edmunds 1964a, b; and Edmunds et al., 1963.

The holes reportedly contain standard 7-inch diameter casing. Temperatures in the solution cavities are reported to be about 135F. Brine injection and recovery may be accomplished through the same well or by means of individual bore holes in communication with a single cavity.

Figure 3 shows the surface facility of the Canadian solution mining operation. The possible use of solution mining for potassium extraction from other sylvinite deposits continues to receive study.

Fig. 3—Refinery associated with potassium solution mine (Courtesy Kalium Chemicals, Ltd.)

B. Extraction of Potassium Brines

Recovery of potassium products from natural brines has been practiced for many years, principally in the USA and Israel. Most brines contain a low percentage of potassium compounds in association with other minerals. The processing of brines normally necessitates an initial concentration step, usually accomplished by either forced and/or solar evaporation, followed by beneficiation operations to separate potassium compounds from the other minerals present. These methods are discussed below. The beneficiation processes used to produce the final potassium products are more fully described in later sections.

1. Searles Lake Brines

The brines from Searles Lake, located in the northern portion of the Mojave Desert, furnished the raw material for the first large-scale production of potassium chloride in the USA in 1916 (Garrett, 1960; Garrett, 1958; McDonald, 1960).

Wells sunk 70 to 130 feet to the bottom of the lake's two salt structures provide the raw brine for the two production plant cycles which yield a variety of chemicals in addition to potassium. Typical brine compositions are given in Table 2 (Kapusta, 1963). Most of the potassium is recovered from the upper salt structure brine which contains about 5% KCl. This brine is pumped several miles to the extraction plant where it is treated in the main plant process depicted in Fig. 4. The brine entering the plant is first heated in heat exchangers and then combined with recycled process liquors from a previous cycle and concentrated in large triple-effect evaporators. During the evaporation process, burkeite, a double salt of Na_2SO_4 and Na_2CO_3, and NaCl are separated from boiling liquors and withdrawn via salt traps. Burkeite is also separated from the hot concentrated liquors and crystallized NaCl by means of clarifiers, thickeners, and rotary vacuum filters. Upon removal of the sodium compounds, the hot concentrat-

Table 2—Typical Searles Lake brine compositions

Component	Formula	Upper structure	Lower structure
Potassium Chloride	KCl	5.02%	2.94%
Sodium Chloride	NaCl	16.06	15.51
Sodium Sulfate	Na_2SO_4	6.75	6.56
Sodium Carbonate	Na_2CO_3	4.80	6.78
Sodium Tetraborate	$Na_2B_4O_7$	1.63	1.96
Potassium Bromide	KBr	0.12	0.08
Sodium Sulfide	Na_2S	0.08	0.038
Phosphorus Pentoxide	P_2O_5	0.07	0.044
Lithium Oxide	Li_2O	0.015	0.006
Tungsten Oxide	WO_3	0.007	0.004

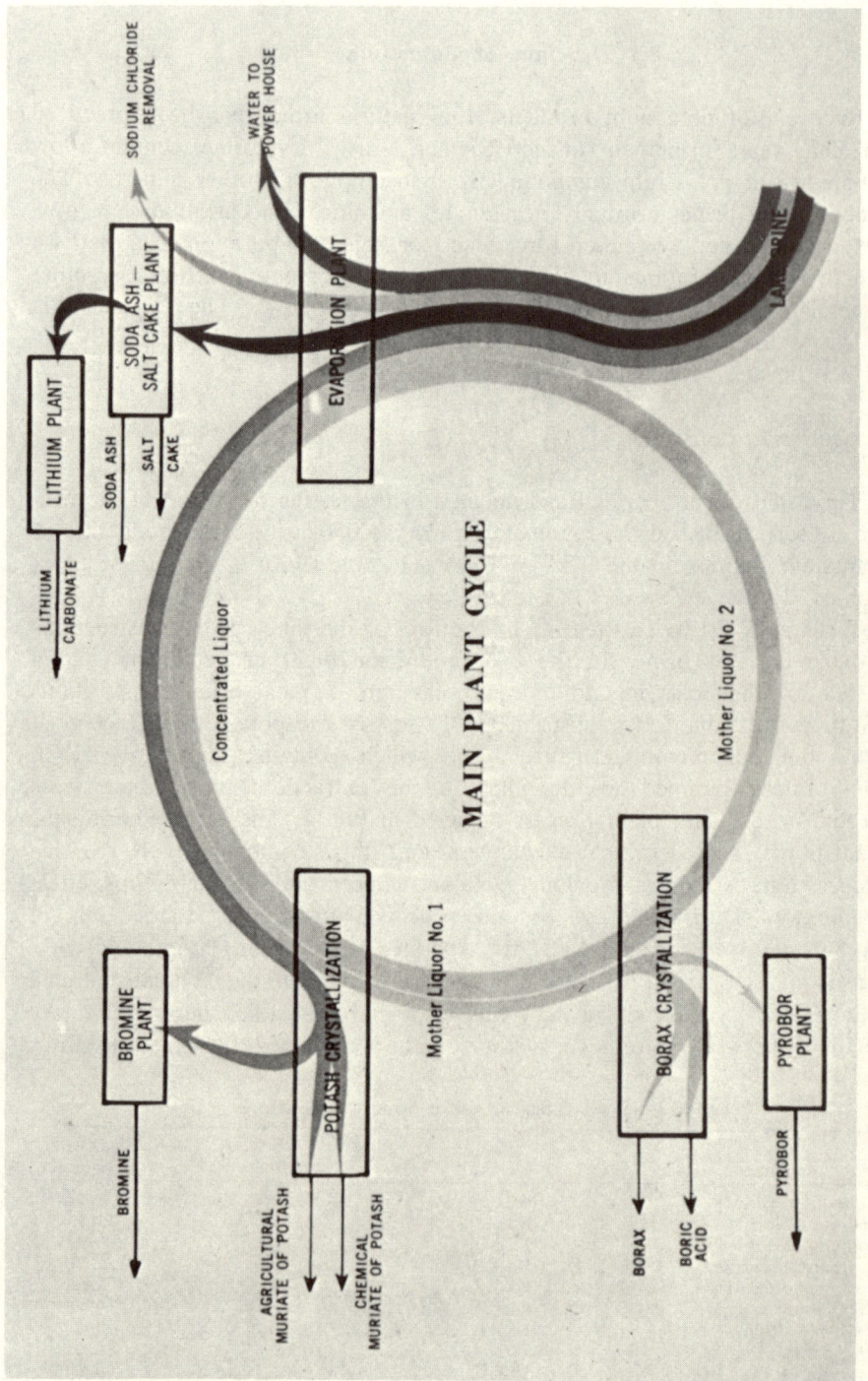

Fig. 4—Main plant cycle Trona process (Courtesy American Potash and Chemical Corp.)

ed liquor leaving the evaporators is rich in potassium and borax. The concentrated liquor is clarified and sent to the potassium recovery plant where KCl is separated by cooling in three-stage, vacuum-type coolers. The crystalline KCl is centrifuged, washed, and dried to yield potassium chloride which may be marketed as such, or converted into other physical forms by the secondary processes methods noted in subsequent paragraphs. Potassium chloride also serves as a raw material for potassium sulfate production described in a later section.

To complete the cycle, the mother liquor from the potassium crystallization step is diluted and cooled further in vacuum coolers to remove borax which is processed into various borate compounds. The resultant liquor is recycled and mixed with new incoming brine.

2. Salduro Marsh (Bonneville) Brines

The brines, containing about 1% KCl, occur in fractured clay deposits underlying the Bonneville Salt Flats in the northwestern corner of Utah (Kapusta and Wendt, 1963). The brine is collected from the entire 14-foot depth of a fissured clay layer by means of drainage canals whose bottoms extend into impervious clay. A network of 70 or more miles of main and lateral canals is employed to convey the brine to pumping stations and then to ponds for concentration by solar evaporation. Solar evaporation is made possible by the hot and dry climate prevailing from approximately mid-May to October. A typical analysis of the original brine is given in Table 3 (Hadzeriga, 1964). In the primary ponds, brine concentration, accompanied by NaCl deposition, is allowed to continue until the brine becomes saturated in respect to KCl. The KCl-laden brine is then directed to a second series of ponds where further evaporation results in the crystallization of a mixture of KCl and NaCl salts. When the $MgCl_2$ concentration reaches a level at which carnallite ($KCl \cdot MgCl_2$) begins to crystallize out of solution, the brine is removed to still other ponds.

The residual mixture of KCl and NaCl crystals is harvested by bulldozers and loaded into bottom-dump wagons for transport to the processing plant. The KCl is separated from the NaCl by means of flotation beneficiation to yield an agricultural grade potassium chloride.

Table 3—Typical Bonneville Lake brine composition

Component	Formula	Percent
Sodium Chloride	NaCl	18.0 - 24.0
Potassium Chloride	KCl	0.8 - 1.2
Magnesium Chloride	$MgCl_2$	0.9 - 1.2
Magnesium Sulfate	$MgSO_4$	0.2 - 0.3
Calcium Sulfate	$CaSO_4$	0.3 - 0.4
Lithium Chloride	LiCl	0.03 - 0.04

Table 4—Dead Sea brine composition

Component	Formula	Grams/liter
Magnesium Chloride	$MgCl_2$	130
Sodium Chloride	$NaCl$	87
Calcium Chloride	$CaCl_2$	37
Potassium Chloride	KCl	11.5
Magnesium Bromide	$MgBr_2$	5
Calcium Sulfate	$CaSO_4$	1

The $MgCl_2$ brine which remains is subjected to additional evaporation in other ponds. There NaCl and double salts of potassium and magnesium are deposited resulting in an increase in the brine's $MgCl_2$ content. At the proper $MgCl_2$ concentration, the brine is moved to a final series of evaporation ponds for recovery of the $MgCl_2$. The remaining deposit of NaCl and double salts of potassium and magnesium are washed with brackish water and the resultant brines directed back to the appropriate ponds according to their chemical compositions. This brine recycle step permits recovery of additional KCl.

3. Dead Sea Brines

The Dead Sea on the border of Israel and Jordan contains a substantial reserve of potassium and other minerals, chiefly $MgCl_2$ and NaCl. The potassium production operations, confined to the southwestern end of the lake in the vicinity of Sodom, extend over an area of apprixmately 250 sq km. The depth is reportedly less than 10 m. The feed brine analysis is given in Table 4 (Nadel, 1965).

The brine is first pumped into large earthen evaporation pans arranged to permit gravity flow from one pan to another. Concentration is effected by solar evaporation and NaCl crystallizes from solution. The NaCl-depleted brine is then directed to a series of smaller carnallite pans where further evaporation results in the crystallization of carnallite ($KCl \cdot MgCl_2 \cdot 6H_2O$) and additional NaCl. Essentially all of the potassium is deposited in the form of carnallite. The crystalline mixture containing approximately 20% KCl is harvested by floating dredges and pumped as a slurry to the two processing plants for futher purification and beneficiation.

The solid carnallite and NaCl crystals are removed from the brine by filtration and leached with water to dissolve the $MgCl_2$ in the carnallite, leaving a mixture of solid KCl and NaCl crystals. The $MgCl_2$-rich brine is returned to the evaporation ponds. In the older plant the KCl is removed from the NaCl-KCl mixture by flotation benefication. In the newer plant the KCl is leached from the mixture and the hot KCl-laden brine is sent to conventional vacuum crystallizers to produce potassium chloride products of desired particle size and purity.

V. PRODUCTION OF POTASSIUM FERTILIZERS

Potassium chloride is the most important form of potassium fertilizer. Estimates indicate that KCl accounts for over 95% of the fertilizer potassium consumed annually throughout the world. Potassium sulfate, potassium-magnesium sulfate, and potassium nitrate are the other commercially important forms of fertilizer potassium. These major potassium fertilizers, together with their approximate plant nutrient contents, are listed in Table 5. Other potassium compounds, such as potassium hydroxide(KOH), potassium carbonate (K_2CO_3) and potassium metaphosphate (KPO_3), have found but minor fertilizer usage to date.

This section deals with processes utilized in production of these potassium fertilizer products from the raw materials discussed in the preceding paragraphs.

A. Potassium Chloride

Potassium chloride is obtained from sylvinite ore and potassium brines, sylvinite ore being the major source. Recovery from sylvinite ore is accomplished by two basically different techniques: (i) flotation beneficiation and (ii) solution and recrystallization. Extraction from brines entails fractional crystallization or concentration followed by the application of flotation and/or recrystallization techniques. At the present time, in North America, the flotation beneficiation processes are the most popular. However, recrystallization processes are also used in both the USA and Canada. In Europe, and other production centers, recrystallization processes continue to enjoy widespread popularity.

1. Flotation Beneficiation

Flotation is a process by which particles of one or more minerals in a pulp, or slurry, are selectively caused to rise to surface of the slurry by the action of bubbles of air (Gaudin, 1957). The mineral particles entrained in the resultant froth on the surface of the slurry are removed by skimming off

Table 5—Principal potassium fertilizers

Material	Approximate plant food content, %	
	K_2O	K
Potassium chloride	60.0 - 62.5	49.8 - 51.8
Potassium sulfate	50.0 - 53.3	41.5 - 44.2
Potassium-magnesium sulfate	22.0 - 22.9	18.3 - 19.0
Potassium nitrate	44.0 - 44.6	36.5 - 37.0

the froth. The particles which do not rise to the surface remain in the slurry and are discharged from the bottom of the cell. Floatability is a surface phenomenon associated with the nature of the film coating the particle to be floated. The selective coating of the particles of a given mineral in a mixture of minerals by a specific reagent enhances floatability of the coated mineral particles in preference to that of the other minerals present. Reagents which will selectively film certain minerals to the exclusion of others are called collectors. Their over-all effect is to make the surface of the mineral to be floated water repellent. Air bubbles generated within the flotation cell preferentially attach themselves to the nonwetted, coated particles, thus carrying the particles to the surface.

Sylvinite ore is a physical mixture of inter-locked crystals of sylvite (KCl) and halite (NaCl) containing small quantities of dispersed clay and other impurities. The flotation beneficiation process is an important method of separating the sylvite from the halite.

Through the use of appropriate collecting (flotation) agents, either the sylvite or halite may be floated. However, industry practice favors flotation of the sylvite and depression of the sylvite.

The simplified flow diagram in Fig. 5 illustrates the general features of the flotation beneficiation process. Processes, reagents, and equipment em-

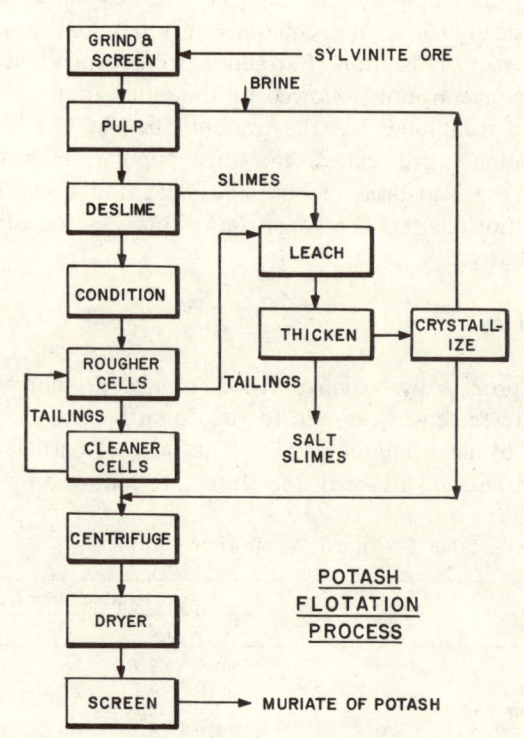

Fig. 5—Schematic diagram, potassium chloride flotation beneficiation process

ployed by different producers may vary because of differences in ore composition, types of products desired, and other individual preferences.

The sylvinite ore from surface storage bins is crushed, generally to about −4 mesh or smaller, and screened to specified particle size ranges. The ground ore is pulped in a brine saturated with respect to KCl and NaCl, to preclude ore solution, and scrubbed to disperse the clay slimes and other insolubles. Rotary tumbler mixers or high-speed agitators may be used for this purpose (Anon., 1957). The slurry is deslimed by various means, such as spiral classifiers, wet screening, hydroseparators, etc., to remove the finely divided clays and insolubles. Potassium chloride which may be carried over with the clay slimes is reclaimed in subsequent solution and recrystallization recovery operations. Effective clay removal is important because clays could consume large quantities of the costly flotation reagents added in a subsequent step.

The deslimed clay-free pulp is treated with so-called binding agents, such as guar or starch (Atwood and Bourne, 1953) or other materials. These serve to coat the remaining clay particles and prevent them from combining with flotation agents introduced in the next step. Primary aliphatic amine acetate salts derived from beef tallow are extensively used as the frothing and sylvite-collecting agent. These reagents selectively coat the sylvite particles. The treated pulp then proceeds to an initial bank of flotation cells, commonly referred to as rougher cells, where air is introduced into the pulp by mechanical agitation. The air bubbles formed attach themselves to the surface of the reagent-treated sylvite particles causing them to rise and concentrate in the surface froth. Mechanically driven paddles skim off the sylvite-rich froth. Aliphatic alcohols, such as methyl isobutyl carbinol, may be used as frothing agents. A bank of flotation cells in a potash refinery is

Fig. 6—Flotation cells in potassium refinery (Courtesy Potash Company of America)

shown in Fig. 6. The overflow, or concentrate, from the rougher cells, rich in KCl but still containing some NaCl and other impurities is fed into a second bank of "cleaner" cells for further concentration of the KCl. Overflow from the cleaner cells is centrifuged to separate the KCl from the brine. After drying and screening, the KCl is stored or subjected to other secondary processes for conversion into products of other particle sizes. The underflow from the cleaner cells is returned to the rougher cell circuit for reprocessing. The salt-laden tailings from the rougher cells are subjected to leaching and thickening. The potassium salts are recovered from the resulting solution by means of recrystallization. Waste salt and slimes are centrifuged, washed, and discarded. The final product is usually a pink tinted crystalline potassium chloride containing approximately 96 to 97% KCl. The pink cast associated with flotation beneficiated potassium chloride products is due to the presence of small amounts of residual iron and clay.

2. Solution-Recrystallization Process

Solution-recrystallization processes have been utilized for many years for the extraction of potassium chloride from sylvinite ores. The difference in the temperature-solubility relationships of KCl and NaCl forms the basis of the process (Garrett, 1963; Gaska et al., 1965). In solutions saturated in respect to both salts, the solubility of KCl increases with increase in temperature; solubility of NaCl decreases slightly as temperature is increased. This phenomenon is depicted in Fig. 7. Vacuum crystallization is, therefore, a useful technique for separting KCl from brines saturated with respect to both KCl and NaCl. This method has proved to be popular in Europe and was the process chosen for the first commerical plant built to exploit the Carlsbad deposits in the early 1930's.

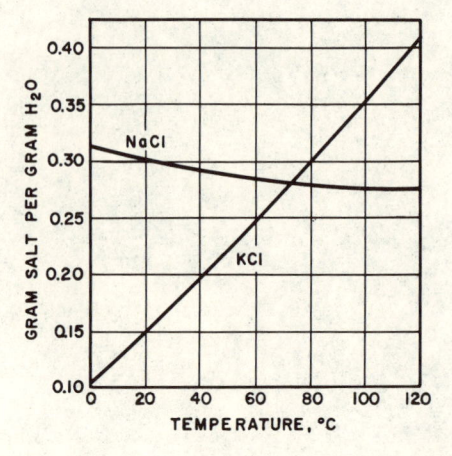

Fig. 7—Solubility of NaCl and KCl in NaCl-KCl-H_2O system

FERTILIZER TECHNOLOGY

The basic features of the solution-recrystallization process, as schematically depicted in Fig. 8, include:

1) Crushing of the sylvinite ore to expose a large surface area,

2) **Contacting** of the crushed ore with hot brine, approximately 100C, saturated with respect to NaCl but unsaturated with respect to KCl, in a counter-current leaching system, to effect dissolution of KCl from the ore,

3) Dewatering and rejection of the remaining NaCl tailings,

4) Clarification of the hot saturated KCl-NaCl brine in thickeners to remove clay slimes and insolubles.

5) Filtration and rejection of the slimes and recirculation of the recovered brine to the clarifier.

6) Pumping of the hot clear brine from the thickener to a bank of vacuum crystallizers for crystallization of KCl,

7) Separation of the KCl crystals by filtration followed by drying, screening, and storage of the product,

8) **Reheating** of the NaCl-rich, KCl-lean filtrate prior to recycling it to the head of the process.

As noted earlier, crystallization is also employed in extracting potassium chloride from the sylvite-halite solution obtained from the solution mine in operation in Canada (Anon., 1966d; Anon., 1965e; Anon., 1966e). The feed solution is fed to two parallel lines of quadruple-effect evaporators for

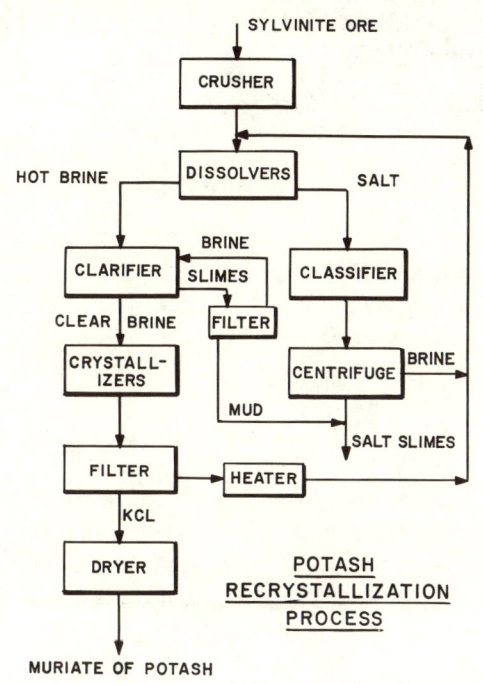

Fig. 8—Schematic diagram, KCl recrystallization process

Fig. 9—Vacuum crystallizer used in KCl plant (Courtesy of Struthers Wells Corp.)

initial concentration and NaCl removal. The crystallized salt is centrifuged, and pumped to a disposal area as a water slurry. The potassium-enriched brine from the evaporators is directed to a large thickener for clarification and then on to a series of four vacuum crystallizers for potassium crystallization. One type of crystallizer used in KCl production is shown in Fig. 9. The crystalline KCl products are centrifuged to remove the residual brine and dried in gas-fired rotary driers. The products are then screened and stored.

Crystallization processes are employed in flotation beneficiation plants to recover potassium leached from slimes and tailings. These processes are an integral and economically important part of the flotation recovery plant.

The agricultural potassium chloride products from crystallization recovery units are usually white or off-white in color, depending on the nature and quantity of the impurities present, and contain approximately 98 to 99% KCl.

3. Product Modification

Although significant quantities of potassium are applied directly to the soil, most of the potassium is used in the form of mixed fertilizers.

Changes in the types and grades of mixed fertilizers produced, and their methods of manufacture, have resulted in an increasing demand for larger quantities of potassium chloride of larger particle sizes than have been used in the recent past.

Relatively coarse particle sizes may be produced directly by the flotation-beneficiation process. However, process efficiency, as well as product grade, may be adversely affected in attempting to produce larger size products from some ores. Crystallization equipment has been developed which is capable of producing larger crystals by the use of the solution and recrystallization process described above. Again, crystallization-process efficiency and product-size limitations, as well as other factors, have dictated the desirability of utilizing other methods of producing some of the larger product sizes required by the purchaser of potassium products. Techniques involving fusion and compaction have been developed for transforming finely divided potassium chloride that has been recovered by flotation and crystallization processes, into larger particle sizes. Compaction is the newer and more widely used method.

In the compaction process (Anon., 1961; Kapusta, 1963, p. 215) finely divided potassium chloride from the flotation or crystallization recovery circuit is compressed into a thin sheet about 1/16-to 1/8-inch in thickness by means of a pair of hydraulically operated high-pressure rolls. Potassium

Fig. 10—Potassium chloride compaction unit (Courtesy Komarek-Greaves Corp.)

chloride, at temperatures of 200 to 250F, is fed between the smooth-faced compacting rolls. Brine or moisture additions may be made to the potassium material to enhance compaction. Compaction pressures range up to 300 tons. The compacted sheet is directed to a flake breaker, which reduces the sheet into large flakes. These are further reduced in size in an impact-type crusher and then screened to remove the material in the desired particle size range. The crushed oversize and fines are screened and recycled to process. The compaction process yields a granular product having an irregular though often somewhat cubical shape. Figure 10 shows a compaction unit.

In the fusion process (Kapusta, 1963, p. 214) finely divided potassium chloride is heated to the molten state, cooled, crushed, and screened to provide the desired particle size. Finely divided potassium chloride is charged into the furnace through entryways in the roof to form a series of conical piles on the furnace floor adjacent to the side walls. Heat is applied and the molten potassium material flows along the trough-shaped furnace floor forming a pool at the discharge end. The melt is continuously withdrawn from the furnace at a temperature of about 750C, and flows onto the upper surface of an internally water-cooled rotating metal disc, where it forms a layer about 1/16 to 1/8 inch in thickness. Upon cooling the mass contracts and shatters into large, thin flakes which are removed by a scraper. The flakes are cooled and then ground and screened to the desired particle size and sent to storage.

4. Product Characteristics

Potassium chloride in its pure state is a white crystalline product containing 63.17% K_2O (52.43% K). See Table 6 for some of the principal properties.

Agricultural potassium chloride products currently marketed are of a reddish tint or white crystalline materials containing approximately 60 to 62.5% K_2O (49.8 to 51.9% K). Flotation-processed materials contain

Table 6—Properties of potassium compounds

	Potassium chloride	Potassium sulfate	Potassium-magnesium sulfate	Potassium nitrate
Formula	KCl	K_2SO_4	$K_2SO_4 \cdot 2MgSO_4$	KNO_3
Color	white	white	white	white
Molecular weight	74.56	174.26	514.01	101.11
Specific gravity	1.988	2.66	2.829	2.11
Melting point, F	1,454	2,312	1,700	631
K_2O content, %	63.17	54.06	22.7	46.58
K content, %	52.44	44.88	18.8	38.67
Solubility in water, parts/100 part H_2O				
at 32F	27.6	6.85	-*	13.3
at 212F	56.7	24.1	-	246

* Unstable in water. Solids phase transformed to schoenite, $K_2SO_4 \cdot MgSO_4 \cdot 6HO$.

Table 7—Typical analysis of flotation beneficiated fertilizer grade and recrystallized fertilizer grade KCl

Chemical composition component	Flotation beneficiated fertilizer grade	Recrystallized fertilizer grade
	%	%
K_2O equivalent	60.8	62.50
KCl	96.25	98.94
Potassium, K	50.47	51.88
Chloride, Cl	47.60	47.65
Sodium, Na	1.13	0.40
Calcium, Ca	0.04	0.003
Magnesium, Mg	0.03	0.001
Bromine, Br	0.03	0.014
Sulfate, SO_4	0.10	0.004
Water Insoluble	0.60	0.08
Moisture, H_2O	0.10	0.05

from 60.0 to nearly 62.0% K_2O (See Table 7 for a typical chemical analysis). Potassium chloride made by crystallization methods is white to off-white in color, the principal impurity being NaCl, and contains approximately 62 to 62.5% K_2O (See Table 7 for a typical analysis). In some instances, the commercial materials are mixtures of potassium chloride recovered by one or more of the flotation, crystallization, compaction, or fusion processes described above.

Fertilizer-grade potassium chloride products are sold in three principal particle-size designations: standard, coarse, or granular. Other special forms are also available. These product forms are available in both flotation and recrystallized products. The following Tyler mesh size ranges (approximate) are typical of the products currently being marketed in the USA: Standard, −14 +100; Coarse, −8 +35; Granular, −6 +14.

B. Potassium Sulfate

Agricultural potassium sulfate (K_2SO_4) is produced in the USA by the following methods:
1) From langbeinite ($K_2SO_4 \cdot 2MgSO_4$) ore by an ion-exchange technique for the removal of magnesium,
2) From burkeite ($Na_2CO_3 \cdot 2NaSO_4$) and KCl in the Trona process,
3) The Hargreaves process by combining KCl and sulfur (S),
4) In Mannheim furnaces from KCl and sulfuric acid (H_2SO_4).

In Europe, potassium sulfate is derived from kainite and hartsalz.

1. Langbeinite Process

Production of potassium sulfate from langbeinite is carried out by two producers at Carlsbad. Briefly stated, the process is based on the reaction of langbeinite ($K_2SO_4 \cdot 2MgSO_4$) and potassium chloride (KCl) according to the following equation:

$$K_2SO_4 \cdot 2MgSO_4 + 4KCl = 3K_2SO_4 + 2MgCl_2$$

The actual process involves complex reactions of liquid and solid phases in the reciprocal salt pair system $MgCl_2$, K_2SO_4, and water. The process is described in detail in the literature (Kapusta, 1963; McDonald, 1960). To achieve optimum potassium recovery, mother liquor from the production of refined potassium chloride, finely ground langbeinite ore, recycled mixed salts (principally KCl and leonite — $K_2SO_4 \cdot MgSO_4 \cdot 4H_2O$), and water are combined in the proper proportions. This reaction yields K_2SO_4 as the stable solid phase and a mother liquor saturated with respect to K_2SO_4 is separateded with respect to both KCl and leonite. The crystallized K_2SO_4 is separated by centrifuging, dried, and stored. The resulting slurry containing langbenite and KCl is subjected to crystallization, converting the langbeinite to leonite and KCl and some kainite $KCl \cdot MgSO_4 \cdot 3H_2O$. These mixed salts are removed via a thickener and vacuum filter and are recycled to the process.

2. *Trona Process*

In the Trona process (McDonald, 1960), K_2SO_4 is produced from burkeite ($Na_2CO_3 \cdot 2Na_2SO_4$) and KCl in a two-stage process. Finely divided agricultural-grade KCl is first reacted with burkeite from the main brine evaporators to yield a high potassium glaserite ($Na_2SO_4 \cdot 3K_2SO_4$). The glaserite is separated and combined with a high-purity KCl brine. The K_2SO_4 forms as a solid phase and is removed from the resulting NaCl brine by filtration. The brine is recycled to the burkeite digestion step. The K_2SO_4 is washed, dried, and conveyed to storage.

3. *Hargreaves Process*

The Hargreaves process (Anon., 1954) produces potassium sulfate directly from sulfur and potassium chloride. In this process, the desired reaction is obtained by passing a gas containing sulfur dioxite (SO_2), excess air, and water vapor through beds of KCl briquettes. The gas contains more than 1 mole of water vapor and one-half mole or more of oxygen for each mole of SO_2. Process gas is obtained from a sulfur burner, and the composite gas is passed through a series of chambers or reactors charged with KCl briquettes. The gas flows first through the bed of most reacted briquettes and ends its flow through the most recently charged chamber of fresh briquettes. The reaction is endothermic and temperature control is critical because of the narrow margin between a satisfactory reaction temperature and the fusion point of the eutectic mixture. The converted K_2SO_4 briquettes are cooled and removed from the chamber with a clamshell crane. After crushing and screening, the product is ready to be marketed. Hydrogen chloride is a byproduct of the process.

4. Mannheim Process

The Mannheim furnace process is employed to produce potassium sulfate and hydrochloric acid from potassium chloride and sulfuric acid according to the following reactions:

$$KCl + H_2SO_4 = KHSO_4 + HCl$$
$$KHSO_4 + KCl = K_2SO_4 + HCl.$$

The first state is exothermic and proceeds at a relatively low temperature to yield $KHSO_4$. In the second stage, which is endothermic, the acid sulfate combines with KCl to give K_2SO_4. External heat is applied to promote the second-stage reaction. The K_2SO_4 is ground, screened, and sent to storage. The Mannheim process is also employed in France and Belgium.

5. From Kainite

Potassium sulfate is produced from kainite in Italy (Anon., 1962; Noyes, 1966) and from hartsalz and associated minerals in Germany (Anon., 1965a; Anon., 1965c).

Kainite is a double salt represented by the formula $KCl \cdot MgSO_4 \cdot 3H_2O$. Its processing into K_2SO_4 is dependent on the relative solubilities of the four salts which may be obtained from this combination of constituents. Briefly stated, the process involves two-stage leaching to effect the extraction of $MgCl_2$ from the ore.

The crushed and beneficiated kainite ore, containing about 17% K_2O is leached with brine, recycled from the second leaching stage at about 25C and yields a liquid containing $MgCl_2$ in solution and a solid phase consisting of schoenite, $K_2SO_4 \cdot MgSO_4 \cdot 6H_2O$. This conversion is illustrated by the following equation:

$$2KCl \cdot MgSO_4 \cdot 3H_2O = K_2SO_4 \cdot MgSO_4 6H_2O + MgCl_2$$

The schoenite is separated by means of classifiers and centrifuges. It is then weighed and subjected to the second leaching stage. Fresh water at about 50C is used. The more soluble $MgSO_4$ is extracted along with some of the potassium salts present. The extraction is illustrated by the following equation:

$$K_2SO_4 \cdot MgSO_4 \cdot 6H_2O = K_2SO_4 + MgSO + 6H_2O.$$

The K_2SO_4 is separated by centrifuging. It is then dried, screened, and cooled prior to storage. The mother liquor is cooled and recycled to the head of the process.

In West Germany, hartsalz is used to make K_2SO_4. Hartsalz is essentially a mixture of KCl and $MgSO_4$. Sodium chloride is present in varying quantities. The NaCl is removed by leaching the ore with a saturated brine solution. The composition of the resulting $KCl-MgSO_4$ mixture is adjusted by addition of more KCl or $MgSO_4$ as may be required. The adjusted mixture is treated with water in agitated extraction tanks at about 35C to yield potassium-magnesium sulfate and a mother liquor low in potassium. The potassium-magnesium sulfate is removed by filtration and subjected to a second-stage reaction with KCl brine to yield K_2SO_4 and $MgCl_2$. The K_2SO_4 is recovered by filtration.

In West Germany, K_2SO_4 is also reportedly made by a process involving the treatment of a KCl solution with kieserite, Epsom salt, or anhydrous $MgSO_4$ (Anon., 1965a). The magnesium salt is reacted with the KCl brine to yield potassium-magnesium sulfate and $MgCl_2$:

$$2MgSO_4 \cdot 7H_2O + 2KCl = K_2SO_4 \cdot MgSO_4 \cdot 6H_2O + MgCl_2 + 8H_2O.$$

The potassium-magnesium sulfate may be marketed as such or treated further with a pure KCl brine to give K_2SO_4:

$$K_2SO_4 \cdot MgSO_4 \cdot 6H_2O + 2KCl = 2K_2SO_4 + MgCl_2 + 6H_2O.$$

6. Product Characteristics

Pure K_2SO_4 is a white crystalline compound containing 54.06% K_2O equivalent to 44.87% K. The more important properties of potassium sulfate are indicated in Table 6.

Commercially available agricultural grades of potassium sulfate contain from 50.0 to 53.3% K_2O equivalent to 41.5 to 44.2% K. Products are white in color and are available in the standard, coarse, and granular grades described in the section dealing with potassium chloride products. The coarse grades may be produced by crystallization and granular grades by compaction of fines from any of the processes described or by selective screening in the case of the Hargreaves process.

The typical chemical analysis of the Hargreaves process product is listed in Table 8. Potassium sulfate products are guaranteed to contain no more than **2.5% Cl.**

C. Potassium-Magnesium Sulfate

Potassium-magnesium sulfate is recovered from langbeinite ore by two producers in the Carlsbad, New Mexico, area. The mineral langbeinite is a double salt of potassium and magnesium sulfate ($K_2SO_4 \cdot 2MgSO_4$) and occurs in association with sylvite and halite in the Carlsbad basin. The principal properties of langbeinite are depicted in Table 6.

FERTILIZER TECHNOLOGY

Table 8—Typical analysis of fertilizer grade potassium sulfate made by Hargreaves process

Chemical composition component	%
K_2O equivalent	52.40
K_2SO_4	96.96
Potassium, K	43.50
Sulfate, SO_4	53.03
Chloride, Cl	1.55
Sodium, Na	1.10
Calcium, Ca	0.02
Magnesium, Mg	0.08
Water Insoluble	0.50
Moisture, H_2O	0.02
Free acid, H_2SO_4	trace

The langbeinite ore is first crushed and screened and then subjected to a controlled, continuous counter-current washing process to dissolve away the undesirable chlorides, leaving the langbeinite virtually unaltered. The residual solid phase contains approximately 97% langbeinite. It is separated from the liquid phase by centrifuging. It is then dried and sent to storage. The commercial product, available in various particle sizes, is a white-to-pink crystalline solid. It contains a guaranteed analysis of 22.0% K_2O equivalent to 18.26% K, 18.0% MgO equivalent to 10.85% Mg and a maximum chloride content of 2.5% Cl. The magnesium, as well as the potassium, is water-soluble.

D. Potassium Nitrate

Potassium nitrate (KNO_3), commonly called nitre or saltpeter, is obtained as a byproduct of $NaNO_3$ production or by the reaction of KCl with $NaNO_3$ or HNO_3. These production methods are briefly described below.

1. Chilean Production

In Chile, KNO_3 is reclaimed as a byproduct of the Guggenheim process for producing $NaNO_3$ from the natural deposits of caliche. The residues from the $NaNO_3$ process are leached to yield a weak brine containing KNO_3 and other salts. The KNO_3 brine is concentrated by solar evaporation. The extraneous salts are crystallized out from solution. When the proper KNO_3 concentration is achieved, the purified solution is subjected to crystallization for recovery of potassium nitrate.

2. From KCl and $NaNO_3$

In the USA, KNO_3 has been produced for many years by the double decomposition reaction between $NaNO_3$ and KCl:

$$NaNO_3 + KCl = KNO_3 + NaCl.$$

In this process, KCl is reacted with a hot solution of $NaNO_3$. The NaCl thus formed crystallizes from solution and is filtered from the hot liquor. Solid KNO_3 is crystallized from the solution and dried. The process is readily carried out in standardized processing equipment. However, raw materials costs are high and the final product has been limited to specialty fertilizer users.

3. *From KCl and HNO_3*

The producion of KNO_3 by direct combination of KCl and HNO_3 has been studied by various investigators. The first commerical plant based on this reaction came into production in the USA in 1963 (Anon., 1965b; Noyes, 1966). The overall process is depicted by the following equation:

$$2KCl + 2HNO_3 + \tfrac{1}{2} O_2 = 2KNO_3 + Cl_2 + H_2O.$$

Actually, the process involves a series of steps leading to the formation and reclamation of various intermediates (Spealman, 1965). Potassium chloride and chilled HNO_3 are first combined in an agitated reaction tank. The partially reacted solution from this reactor is directed to a reaction column where hot HNO_3 vapors strip the HNO_3-KNO_3 solution of chlorine.

$$3KCl + 4HNO_3 = 3KNO_3 + NOCl + Cl_2 + 2H_2O.$$

The HNO_3-KNO_3 solution from the reactor tower passes to a water-stripping column, operated in conjuction with a KNO_3 saturator, to provide the more highly concentrated HNO_3 necessary to effect the nitrosyl chloride oxidation step:

$$NOCl + 2HNO_3 = 3NO_2 + \tfrac{1}{2}\text{-}Cl_2 + H_2O.$$

The nitrogen dioxide thus formed is regenerated to nitric acid in a nitric-acid-absorption unit and is recycled to process.

The KNO_3-HNO_3 solution from the water stripper is fed into three vacuum crystallizers where KNO_3 crystals are separated from the solution. The KNO_3 crystals are centrifuged, dried, and conveyed to storage or to a melter which feeds a prilling tower to produce larger sized particles. The chlorine obtained in the process is purified.

The highly corrosive nature of the various intermediates formed necessitates the use of costly materials of contsruction, such as inconel, stainless steel, and titanium metal.

4. *Product Characteristics*

Pure KNO_3 is a white crystalline solid containing 13.85% N and 46.58% K_2O (38.66% K). Table 6 summarizes some of the properties of the pure compound.

Table 9—Typical analysis of standard grade fertilizer potassium nitrate

Chemical composition component	%
K_2O equivalent	44.6
KNO_3	95.5
Sodium nitrate, $NaNO_3$	4.0
Sulfur, S	0.1
Nitrogen, N	13.8
Chloride, Cl	0.4
Water Insoluble	0.1
Moisture, H_2O	0.1

Fertilizer grades of KNO_3 contain a minimum of 13.0% N and 44.0% K_2O and a maximum of 0.1% Cl equivalent to 95.5% KNO_3. The KNO_3 is available in a standard grade and a prilled grade. The standard grade is of approximately a −20 +100 mesh size and the prilled product a −6 +14 mesh size. The typical chemical composition of commerically available standard grade potassium nitrate is given in Table 9.

E. Potassium Hydroxide

Potassium hydroxide (KOH), or caustic potash, is produced by the electrolysis of a KCl brine in either diaphragm or mercury cells of various designs (Sconce, 1962). Chlorine and hydrogen are co-products of the electrolysis. The typical diaphragm cell consists of anodes of graphite separated from iron screen cathodes by a porous asbestos diaphragm deposited on the cathodes. Saturated KCl brine, at a temperature of 60 to 70C, is fed into the anolyte which flows through the diaphragm into the catholyte where alkali is formed. A differential head maintains continous flow through the diaphragm. Chlorine gas is formed at the anode and removed through the top of the cell. Hydrogen and KOH solution are formed at the cathode. Hydrogen is removed from the top of the cathode compartment and the weak 10-15% KOH solution, containing residual KCl, is withdrawn from the bottom. The weak caustic solution is concentrated by evaporation. The KCl is removed by decantation and filtration. The resulting 50% KOH solution may be further concentrated by evaporation and fusion to produce solid and flake forms.

Mercury cells used in the manufacture of KOH differ in design and operation (Sconce, 1962). All consist of two essential components, (i) an electrolysis unit, and (ii) a decomposition unit. In the electrolysis section, the saturated KCl-feed brine is electrolyzed between a fixed graphite anode and a flowing mercury cathode. Chlorine gas is evolved at the anode, and potassium metal is deposited at the surface of the mercury cathode in which it dissolves to form an amalgam. When water is added, KOH solution and hydrogen gas are formed; both of which are withdrawn from the process.

The denuded mercury is recycled through the cell to take on more potassium metal and again passes through the decomposition unit. The mercury-cell process has the advantage over the diaphragm-cell method in that it directly produces a 50% KOH solution without resorting to an evaporation step. The 50% solution may be further concentrated into the desired solid forms.

Pure KOH is a white, odorless solid containing 83.9% K_2O (69.6% K). It is a very deliquescent compound whose water solutions are strongly alkaline. Potassium hydroxide is readily soluble in water, affording highly concentrated potassium solutions. At ordinary temperatures, KOH solutions are slightly corrosive to iron or steel. Corrosion increases greatly at elevated temperatures and is further aggravated by the presence of oxygen.

Potassium hydroxide is marketed in both solution and dry forms. Solution forms include a 50% liquid, for shipment in tank-car quantities, and a 45% KOH liquid for drum shipments. The dry forms include solid, flake, ground, and other specialty types. The solid form is produced by evaporating KOH solution to its maximum concentration and pouring it into drums for solidification. The solid and other dry forms, also shipped in steel drums, containing 88-92% KOH.

The fertilizer use of KOH has been limited chiefly to specialty liquid mixed fertilizers. Interest has been evidenced in its use in commerical liquid fertilizers, but its high cost relative to other potassium fertilizers has restricted such utilization. Caustic potash is employed in the production of potassium phosphates which have also found some utility as specialty fertilizers, notably in water-soluble concentrates.

F. Potassium Carbonate and Bicarbonate

In the USA, potassium carbonate (K_2CO_3) is made by the neutralization of a strong caustic potash solution with CO_2. The resultant solution is purified and filtered to yield a 48 to 52% K_2CO_3 liquid. Evaporation and subsequent crystallization of this liquid leads to a solid sesquihydrate ($K_2CO_3 \cdot 1\frac{1}{2} H_2O$) form containing 83 to 85% K_2CO_3. This granular product is centrifuged, washed, dried, and screened prior to packaging. Calcination of the sesquihydrate to remove water of crystallization yields a 99 to 100% K_2CO_3 which is marketed as a calcined grade. The 48 to 50% liquid is shipped in drums and tank cars. The calcined and hydrated solid K_2CO_3 are packaged in barrels, bags, and fiber kegs. Calcined K_2CO_3 is also shipped in bulk form in hopper-bottom cars.

In its pure form, K_2CO_3 is a white, odorless solid containing 68.16% K_2O (56.57% K). It is readily soluble in water. Its solubility increases from 105.5 parts per 100 parts of water at 32F to 146 parts per 100 parts of water at 212F. Like caustic potash, K_2CO_3 readily absorbs moisture from the atmosphere. The calcined product may absorb several percent moisture and still remain dry and free flowing. The hydrated grade becomes wet quickly upon exposure to normal atmospheric conditions. Potassium carbonate hydrolyzes in water to form moderately alkaline solutions. At 68F a 0.25% solution has a pH of 11.1.

Potassium carbonate has not found wide usage as a potassium-bearing fertilizer. It has been used in small quantities as a source of nonchloride potassium in some tobacco fertilizers. Its comparatively high price and limited availability, compared to the other more common potassium fertilizers, probably limited its use.

Potassium bicarbonate ($KHCO_3$) is produced by the carbonation of K_2CO_3. In its pure state, $KHCO_3$ is a white solid containing approximately 47% K_2O (39% K). It has a density of 2.17 and is less hygoscopic than K_2CO_3. Its solubility ranges from 22.4 to 60 parts per 100 parts of water at temperatures of 32F and 140F, respectively. Upon calcination, the material decomposes to yield K_2CO_3 and CO_2. Reports indicate that $KHCO_3$ is produced in France by a nonelectrolytic process. This process features the direct carbonation of potassium chloride in an aqueous solution containing an organic amine. Although considered as a desirable source of potassium for liquid fertilizer manufacture, $KHCO_3$ has found little if any commerical use as a fertilizer material in this country.

G. Potassium Phosphates

The development of economical methods of producing various potassium phosphates has been the goal of continuing investigations for many years primarily because of their high plant food content, excellent physical condition, and desirable solubility characteristics. Ortho and pyrophosphates, produced by the neutralization of phosphoric acid with KOH or K_2CO_3, are common industrial chemicals. However, their high cost has restricted their use as potassium fertilizers to the realm of specialty materials. Reviews of processes for the production of various potassium phosphates are available in the literature (Noyes, 1966; Waggaman, 1955).

Potassium metaphosphate (KPO_3), is a highly concentrated plantfood source, which, when pure, contains 39.87% K_2O (33.09% K) and 60.13% P_2O_5 (26.27% P). This compound has attracted considerable study since its possibilities as a fertilizer material were first noted several decades ago.

Its unique feature is its low solubility in water, and its possibilities as a slow-release fertilizer have been considered by agronomists. Several methods of manufacture have been proposed and pilot-plant quantities have been produced.

In an early TVA study (Noyes, 1966), KPO_3 was made by burning elemental phosphorus in a combustion chamber at about 1,000C and reacting the resultant P$_2$O$_5$ gas with finely divided potassium chloride to yield KPO$_3$ in a molten state. The molten product was tapped from the furnace, cooled, and ground to the desired size. In other TVA studies, KPO$_3$ was produced by heating a mixture of phosphoric acid with potassium chloride in a fuel-fired furnace. The pilot-plant products contained about 58% P$_2$O$_5$ and 39% K$_2$O.

Other investigators have also prepared KPO$_3$ by reacting KCl with orthophosphoric acid at elevated temperatures (Waggaman, 1955). The reaction proceeds according to the following equations:

$$2KCl + 2H_3PO_4 = 2KH_2PO_4 + 2HCl$$
$$3KH_2PO_4 + heat = K_2H_2P_2O_7 + H_2O.$$
$$K_2H_2P_2O_7 + heat = 2KPO_3 + H_2O.$$

The use of relatively expensive raw materials coupled with high-temperature processing, high fuel costs and aggravated corrosion are notable deficiencies of the processes.

More recently, a lower temperature process utilizing wet process phosphoric acid and potassium chloride has been described (Darden and Strelzoff, 1963; Raistrick et al., 1960).

Reports indicate that the process is being operated on a one-ton/day pilot-plant scale in Israel.

The process has been described (Raistrick et. al., 1960) as follows: A slurry of wet process phosphoric acid (50% P$_2$O$_5$) and fertilizer-grade potassium chloride is added to the feed end of an insulated rotary drum reactor with the simultaneous addition of hot recycle material at a temperature of about 500C. The heat in the hot recycle brings about reaction of and evaporation from the slurry. The mass leaves the reactor at a temperature of approximately 200C in a relatively free-flowing condition. The material then enters a counter-current heated rotary kiln from which it leaves fully reacted at about 500C. A portion of the reacted material is cooled and screened and withdrawn, as product. A portion of the reacted material from the kiln is recycled to the process along with reheated fines from the screening operation. The hydrogen chloride evolved in the reactor and kiln may be recovered as hydrochloric acid. The product is a fertilizer grade KPO$_3$ containing approximately 58% P$_2$O$_5$ and 37% K$_2$O in a granular form.

VI. SHIPPING POTASSIUM FERTILIZERS

Potassium fertilizers are normally solid compounds and, except for the small amounts of specialty materials which may be transported in liquid form, are available to consumers in bulk and packaged forms. The major tonnage is shipped in bulk form in railcars. Boxcars and hopper-bottom cars are used. The greatest portion is shipped in hopper-bottom cars having capacities of 60 to 100 tons. Packaged shipments are usually made in multi-wall paper bags which hold 80 or 100 lb. Bagged and bulk products are also shipped by truck to areas where such shipments can be made economically. Trailer tank trucks equipped with blower systems for pneumatic discharge of the contents have found increased use in recent years.

REFERENCES

Anonymous. 1954. For verteran Hargreaves process — a new job in the potash industry. Chem Eng. 61 (5):132- 134.

Anonymous. 1957. Newest Carlsbad producer is shipping potash. Eng. Mining J. 158 (3):70-76.

Anonymous. 1961. More pressure-fewer fines. Chem. Week. 88(16):51-52.

Anonymous. 1962. Mining kainite in Sicily. Phosphorus and Potassium. British Sulpher Corp., London. no. 1, p. 41-54.

Anonymous. 1964a, June. Cavity control — key to solution mining. Can. Chem. Process. 48(6):63-66.

Anonymous. 1964b, Oct. The French potash industry. Phosphorous and Potassium. British Sulpher Corp., London. no. 13, p. 34-36.

Anonymous. 1964c. Kalium's potash solution mining plant will go on-stream this fall in Canada. Eng. Mining J. 165:110-113.

Anonymous. 1965a, Feb. Sul-Po-Mag and patent Kali. Phosphorus and Potassium. British Sulpher Corp., London. no. 15, p. 41-43.

Anonymous. 1965b, July. Synthetic saltpeter scores. Chem. Week. 97(4):35-38.

Anonymous. 1965c, Oct. The potash mines of the Wintershall group. Phosphorus and Potassium. British Sulpher Corp., London. no. 19, p. 27-30.

Anonymous. 1965d, Dec. Estimated world fertilizer production capacity as related to future needs. A report to A.I.D., U.S. Dept. of State. Tennessee Vally Authority, Muscle Shoals, Ala.

Anonymous. 1966a, April. Potash and fertilizers for growing world markets. Monthly review of the Bank of Novia Scotia. Toronto, Canada.

Anonymous. 1966b, July 14. Design specialized equipment. The Northern Miner. Mag. Suppl. (Toronto, Can.) p. 68-69.

Anonymous. 1966c, July 14. International Minerals has world's largest plant. The Northern Miner. Mag. Suppl. (Toronto, Can.) p. 40-42.

Anonymous. 1966d, July 14. Kalium Chemicals initiates solution mining. The Northern Miner. Mag. Suppl. (Toronto, Can.) p. 48-49.

Anonymous. 1966e, July 14. New concept of mining at depth may open more potash deposits. The Northern Miner. Mag. Suppl. (Toronto, Can.) p. 67, 90.

Anonymous. 1966f, July 14. Potash Company of America first to go into production. The Northern Miner. Mag. Suppl. (Toronto, Can.) p. 43-45.

Anonymous. 1966g,. World survey of potash. British Sulphur Corp. Ltd., London.

Atwood, G. E., and D. J. Bourne. 1953. Process development and practice of the Potash Division of Duval Sulfur and Potash Company. Mining Eng. 5:1099-1104

Bartley, C. M. 1964, Feb. Potash. Can. Mining J. 85(2):130-133.

Dahms, J. B., and B. P. Edmonds. 1962. Solution mining method. U.S. Pat. 3,058,729.

Dahms, J. B., and B. P. Edmonds. 1964a. In situ potassium chloride recovery by selective solution. U.S. Pat. 3,135,501.

Dahms, J. B., and B. P. Edmonds. 1964b. Solution mining of potassium chloride. U.S. Pat. 3,148,000.

Darden, T., and S. Strelzoff. 1963. Sept. Relative economics of potassium metaphosphate as a fertilizer. Presented at Amer. Chem. Soc. meeting, New York.

Edmonds, B. P. et. al. 1963. Recovery of potassium chloride. U.S. Pat. 3,096,969.

Ewell, Raymond. 1964. Famine and fertilizer. Chem. Eng. News 42(50):106-116.

Garrett, D. E. 1960. Borax processing at Searles Lake, p. 119-122. In Industrial minerals and rocks, 3rd ed. Amer. Inst. Min., Met., and Petrol. Engrs., New York.

Garett, D. E. 1963. Crystallization of potash. Chem. Eng. Progr. 59(10):59-64.

Garrett, D. E. 1958. Industrial crystallization at Trona. Chem. Eng. Progr. 54(12):65-69.

Gaska, R. A., R. D. Goodenough, and G. A. Stuart. 1965. Ammonia as a solvent. Chem. Eng. Progr. 61(1):139-144.

Gaudin, A. M. 1957. Flotation. 2nd ed. McGraw-Hill Book Co., New York. p. 500-520.

Goudie, M. A. 1957. Middle Devonian potash beds of Central Saskatchewan. Dept. of Min. Resources, Sask., Canada. Report 31.

Hadzeriga, P. 1964, June. Some aspects of the physical chemistry of potash recovery by solar evaporation of brines. Trans. Soc. Mining Eng. p. 169-174.

Kapusta, E. C. and N. E. Wendt. 1963. Advances in fertilizer potash production, p. 189-230. In M. H. McVickar et al. (ed.) Fertilizer technology and usage. Soil Science Society of Amer., Madison, Wis.

Kyle, A. J. 1964. Mining methods and equipment used at IMC's Esterhazy operations. Can. Inst. Mining Met. Trans. LXVII:83-91

Lewis, R. W. 1966. Potash. U.S. Dept. Interior, Bur. Mines Minerals Yearbook.

MacDonald R. A. 1960. Potash: Occurences, processes, production, p. 367-402. In The chemistry and technology of fertilizers. ACS mono. no. 148. Reinhold Publishing Corp., New York.

Nadel, Shlomo. 1965. Harvesting more Israel potash. Engr. Mining J. 166(10):84-90.

Noyes, Robert. 1966. Potash and potassium fertilizers. Chem. Proc. Mono. no. 15. Noyes Development Corp., Park Ridge, N. J.

Raistrick, Bernard, and John Stewart Raitt. 1960. Granular potassium metaphosphate. Gr. Britain Pat. 832,011.

Ruhlman, E. R. 1960. Potash, p. 669-680. In Industrial minerals and rocks. 3rd ed. Amer. Inst. Mineral., Met., and Petrol. Eng. New York.

Ruhlman, E. R., and G. E. Tucker. 1952. Potash. U.S. Dept. Interior Bur. Mines Minerals Yearbook. 1:825-843.

Sconce, James S. 1962. Chlorine — its manufacture, properties and uses. ACS Monograph 154. Reinhold Publishing Corp., New York. p. 81-199.

Scott, S. A. 1963. Shaft sinking through Blairmore sands and paleozoic water-bearing limestones. Can. Mining Met. Bull. 56(2):94-103

Spealman, M. L. 1965. New route to chlorine and saltpeter. Chem. Eng. 72(23):198-200.

Turrentine, J. W. 1926. Potash. John Wiley & Sons, Inc. New York.

Waggaman, W. H. 1955. Calcium and potassium metaphosphates and miscellaneous products, p. 411-415. In Phosphoric acid, phosphates, and phosphatic fertilizers. 2nd ed., Reinhold Publishing Corp., New York.

Walli, J. R. O. 1966. The application of European shaft-sinking techniques to the Blairmore formation. In R. Noyes (ed.) Potash and potassium fertilizers. Chem. Proc. Mono. no. 15. Noyes Development Corp., Park Ridge, N. J.

York, L. A. 1966. Grouting the 'prairie sediments'. Can. Mining Met. Bull. In R. Noyes (ed.) Potash and potassium fertilizers. Chem. Proc. Mono. no. 15. Noyes Development Corp., Park Ridge, N. J.

3

Preparation of Finished Fertilizers Containing Potassium

RONALD D. YOUNG

Tennessee Valley Authority
Muscle Shoals, Alabama

I. INTRODUCTION

Recent years have brought about rapid strides in nitrogen fertilizer technology, and much publicity has been given to the very large, economical ammonia and urea plants and their effect on cost of fertilizer production. Likewise, improvements in phosphate fertilizer technology, including larger and more efficient wet-process acid plants, the granular diammonium phosphate boom, and development and introduction of polyphosphate fertilizers, have been highlighted. The present sulfur situation with tighter supply and higher price is being emphasized, and alternatives such as nitric phosphate processes and production of elemental phosphorus are receiving deserved attention.

During this period of rapid advances in technology and change in the nitrogen and phosphate industry, the potash industry too has continuously made steady and impressive progress, though it has been generally unheralded. Actually, very significant accomplishments have been largely overlooked. Among these have been (i) improvements in the physical and chemical quality of potassium materials provided in plentiful supply at lower cost, (ii) preparation of the materials in several ranges of particle sizes that are desired, and (iii) development of deep mining technology and of solution mining that previously had been considered impratical. Fertilizer manufacturers have benefited from a competent and competitive industry by being able to rely on good quality, low-cost, potassium materials at the time needed for inclusion in their finished fertilizers.

The purpose of this paper is to discuss the preparation of mixed or compound fertilizers containing potassium. There is not a large amount of detailed or intricate technology involved; the appropriate potassium materials can be readily included in formulations for granular, liquid, or suspension fertilizers, as well as bulk blends. Some of the procedures that have been found to be effective and some considerations in type and particle size of materials to be used will be pointed out. Also, some of the present and potential products and processes for two-component fertilizers, such as potassium nitrate, potassium metaphosphate, and potassium polyphosphate, will be described.

II. TYPES OF MATERIALS USED

A. General Considerations

The least expensive and most widely used material is potassium chloride (often called muriate of potash). Potassium sulfate, a more expensive form, is desired for some types of fertilizers, and potassium nitrate is available as a material to supply both nitrogen and potassium. Potassium hydroxide and potassium carbonates are more readily soluble and allow production of higher analysis liquid fertilizers. These materials are available in a variety of particle sizes, each developed for a particular use.

B. K_2O Equivalent of Materials

Potassium chloride is the least expensive and by far the most widely used material to supply potassium in granular fertilizers and bulk blends. This material is suitable agronomically except for a few specific crops such as tobacco. The more expensive potassium material, potassium sulfate, is generally used in formulating tobacco fertilizers to avoid the harmful chloride. Potassium nitrate can be used to provide the potassium and part of the nitrogen in formulating fertilizers, and is suitable for tobacco. Some advantages in granulation from use of potassium nitrate have been reported by Smith (1964).

The following tabulation gives the K_2O equivalent of the potassium materials used as fertilizers.

	K_2O equivalent, %
Potassium chloride (KCl)	60-62
Potassium sulfate (K_2SO_4)	50-52
Potassium nitrate (KNO_3)	44
Potassium hydroxide (KOH) (45% solution)	37.7
Potassium carbonate K_2CO_3	59-63
Potassium bicarbonate ($KHCO_3$)	45

III. PREPARATION OF FERTILIZERS CONTAINING POTASSIUM

A. Granular Fertilizers

In the earlier days of fertilizer production, potassium materials were used in nongranular mixed fertilizers of low analysis such as 4-10-7 and 6-8-4 grades. These were usually prepared by batch mixing and ammoniation of formulations based on ordinary superphosphate. Often the potassium material was crudely handmixed with other fertilizers by the farmer at the time of application. With the advent of granulation, which started in the early 1950's in the USA and somewhat earlier in Europe, potassium became a major component in the production of higher analysis fertilizers of better quality.

Formulations for the usual 1:1:1 or 1:2:2 $N:P_2O_5:K_2O$ ratios, such as 10-10-10 and 6-12-12 grades, imposed no particular problems. The potassium material, usually potassium chloride, was fed in the proper proportion and conditions of granulation were adjusted to incorporate it uniformly in the granules in the work reported by Hein et al. (1956) and by Yates, Nielsson, and Hicks (1954). When higher potassium ratios such as the 1:4:4 $N:P_2O_5:K_2O$ ratio for grades like 5-20-20 and 6-24-24 became popular, the problem of uniform incorporation of the larger proportion of potassium materials arose. Phillips et al. (1958) report that the use of potassium materials of larger particle size, referred to as "granular," was advantageous in formulations for these high potassium grades. The coarse potassium chloride particles provided nuclei that were coated with the other materials in the formulation during the ammoniation-granulation process. The industry provided this granular material by variations in flotation technique and later by compaction and other methods. Various sizes of coarse and granular potassium materials became popular for use in certain types of granular fertilizer formulations.

Experience has led some manufacturers to prefer the regular size potassium material for the formulation of conventional granular fertilizers. Some prefer coarse, and still others prefer granular or a combination of sizes. Determination of the optimum size is not clear cut or consistent. Consequently, differences in opinion and practice exist.

B. Addition of Potassium Materials in Slurry-Type Processes

Processes for making granular fertilizers by using slurries or concentrated solutions as the primary feed material for the granulation step have come into prominence in recent years. These processes include those for producing ammonium phosphate, nitric phosphate, and ammonium phosphate nitrate. Processes of this general type are being used to make newer products, such as urea-ammonium phosphate and ammonium polyphosphate, which are creating fairly widespread interest. The most common practice in the U S A has been to produce grades, such as 18-46-0 or 11-55-0 ammonium phosphate, 20-20-0 or 26-13-0 nitric phosphate, and 25-25-0 or 30-10-0 ammonium phosphate nitrate. However, grades that contain potassium can readily be produced by these processes if logistics favor this approach. These include grades, such as 15-15-15, 17-17-17, 18-18-18, and 22-11-11.

Some producers of fertilizers of these types prefer to add the potassium material to the slurry or concentrated solution before it is fed to the granulation step. This method of operation has the potential advantages of more intimate mixing of the material with the other components, and of possibly accelerating any reactions of the potassium salt and other materials. With some slurries, however, this method has one disadvantage in that extra water is needed to thin the slurry. Premixing of potassium materials in the slurry or solution, of course, is necessary in a prilling operation and is desirable in a Spherodizer process.

Most of TVA's granulation work has been done with rotary drum and inclined pan granulators, and a lesser amount with a pug mill or blunger. In practically all of this development work the potassium material was added with the recycled solids to the granulator. No advantage for mixing of this material in the slurries or solutions was apparent, and operation with solids was simpler. Young, Hicks, and Davis (1962) report that the addition of solid potassium material was actually beneficial in supplementing recycle for control of granulation. With proper control, uniform incorporation was obtained. No advantage in storage properties was indicated for products prepared by mixing the material in the slurry or solution. When potassium chloride is mixed with ammonium nitrate, they react to form potassium nitrate and ammonium chloride. Petrographic examination of products prepared by mixing the material in the fluids or by adding it separately shows essentially complete reaction of this type. For a slurry-type operation, the very high potassium ratios are troublesome in the same manner as in conventional granulation. The 7-28-28 high-analysis grade by ammonium phosphate processes has been an example of this type according to Surber (1967).

C. Addition of Potassium Materials in Bulk Blends

Most bulk-blended fertilizers contain potassium salts. The only major concerns in blending (a physical mixing operation) are compatibility with other materials and proper matching of particle size. Granular potassium salts are compatible with all the materials normally used in blends. The addition of potassium materials usually increases to some extent the caking tendency of blends, but this has not been a significant problem. Conditioning of blends of this type that are to be bagged is recommended; blends handled in bulk are suitable without conditioning. The granular potassium materials are now available in a particle size range that blends well with other granular or

FINISHED FERTILIZERS

prilled components. Hoffmeister, Watkins, and Silverberg (1964) state that the proper matching of particle sizes is the single most important factor in satisfactory bulk blending. In the earlier days of bulk blending, granular potassium materials of suitable size were not available in plentiful supply; it sometimes was necessary to use material that was not well suited for this purpose. However, the industry soon developed procedures for producing larger crystals and for agglomerating particles by granulation, compaction, or melting and casting. Typical particle size ranges of present day commercial granular potassium chloride are given below.

		Particle size, Tyler mesh			%
+6	-6+8	-8 +10	-10+14	-14+20	-20
0-1	25-30	50-55	15-20	1-2	1
1-2	40-50	34-40	7-20	1-3	1

Other materials such as the sulfate and nitrate now are available in suitable granular size. Granular potassium materials with particle sizes such as those shown above are suitable for bulk blending with other granular materials that have about the same size distribution. The importance of good matching of particle sizes is shown in Fig. 1. This figure from TVA experimental data reported by Hoffmeister, Watkins, and Silverberg (1964) shows that the screen analyses of several samples from a pile of 0-27-27 grade were quite uniform when the granular components for the blend were well matched in particle size. With unmatched sizes, very wide variations in analyses of the samples were evident.

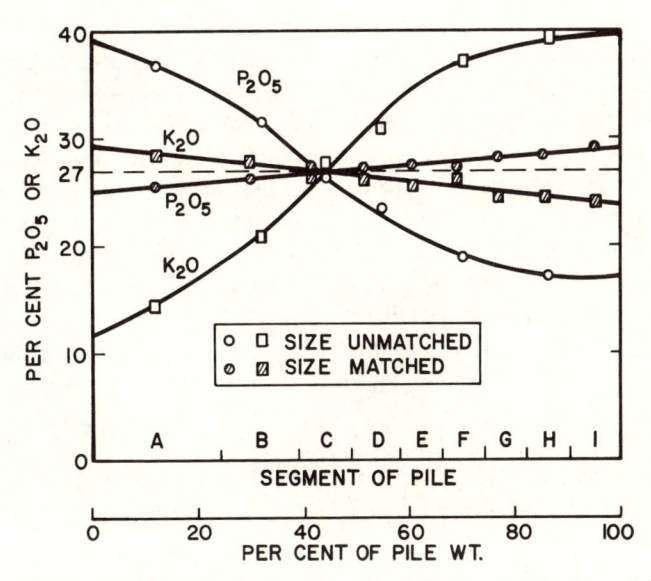

Fig. 1—Control of segregation by matching particle-size distribution of ingredients of 0-27-27 grade, dry blend.

D. Potassium Salts in Fluid Fertilizers

Potassium is included as a nutrient in most grades of liquid and suspension fertilizers. In clear liquids, solubility of the potassium salt is the primary limiting factor; the following tabulation shows some of the grades that can be produced with various salts and stored and handled satisfactorily at 32F.

	Practical maximum liquid grades at 32F	
Source of potassium	1:1:1 ratio †	1:2:2 ratio †
KCl	8-8-8	5-10-10
KNO_3	5-5-5	3-6-6
K_2SO_4	5-5-5	3-6-6
KOH	11-11-11*	5-20-20
K_2CO_3 or $KHCO_3$	11-11-11*	5-20-20

* With urea as supplemental N.
† $N:P_2O_5:K_2O$.

The less expensive potassium chloride is more soluble and can be used in preparation of higher grades of clear liquids than are practical with the sulfate or nitrate. The purer and slightly higher grade of recrystallized white potassium chloride is preferred and most widely used. Potts, Elder, and Scott (1961) state that potassium hydroxide allows preparation of substantially higher grades, but this material is more costly than the other materials and is not widely used at present. Potassium carbonates also are more soluble, but foaming can be a problem in their use.

Suspension fertilizers offer a main advantage in largely avoiding the problem of the solubility of potassium salts. Scott and Wilbanks (1967) report that these fertilizers, with a large part of the plant food as small crystals in suspension, can be produced in grades such as 15-15-15 and 7-21-21 that are almost double the analysis of liquid grades. For use in suspensions the potassium materials should be all minus 20 mesh in size to minimize plugging of spray nozzles. The regular red potassium chloride is suitable.

The sequence of addition of potassium material is important in preparation of liquid and suspension fertilizers. The TVA's development work and experience shows that adding this material last, after all of the fluid components have been fed to the mixing tank, is preferred.

IV. SENSITIVITY OF NITRATE FERTILIZERS CONTAINING POTASSIUM CHLORIDE

Fertilizers that contain a substantial amount of ammonium nitrate are subject to thermal decomposition. The presence of chloride can sensitize this decomposition and cause it to occur at a lower temperature according to Parker and Watchorn (1965). This must be taken into account in production, storage, and shipment of fertilizers of this type that contain potassium chloride. Overheating of the fertilizers in the dryer or even limited contact in storage piles with a source of heat such as a submerged light bulb or hot metal fragments from welding have resulted in a few severe instances of decomposition.

Proper precautions in operation and handling and selection of proportions of components that fall in a less sensitive range are appropriate safeguards, according to Huygen and Perbal (1965).

V. TWO-COMPONENT FERTILIZERS CONTAINING POTASSIUM

A. Potassium Nitrate

The previous discussion has been related entirely to materials and methods used in adding potassium nutrient during preparation or production of mixed fertilizers. There are processes for production of fertilizer materials, such as potassium nitrate, potassium metaphosphate, and potassium polyphosphate that supply two of the major nutrients. These products are not widely used at present but have some merits.

Jacob (1963) gives information about a large plant for production of potassium nitrate from potassium chloride and nitric acid which has been in operation in Mississippi for about 4 years. This fertilizer contains both nitrogen and potassium in a single compound and has a total nutrient content of about 60%. Its use in preparation of granular fertilizers has been evaluated in studies at the Engineering Experiment Station of Iowa State University reported by Boyland and Karmit (1964).

B. Potassium Metaphosphate

Potassium metaphosphate is a very high-analysis fertilizer compound that contains the equivalent of 90 to 100% plant food. Grades up to about 0-60-40 are possible. In addition to the very high analysis, there had been interest in this type of fertilizer to provide a more slowly soluble source of potassium. However, agronomic tests by TVA and others have not established definite advantages for the more slowly available potassium. Harris (1963) reports one comparatively new process utilizing a lower temperature technique in production. Although the very high analysis would be desirable, the economics of production do not appear to be attractive except at a very few locations near sources of both the potassium material and phosphate rock. The TVA has done some limited development work on potassium polyphosphate fertilizers (1964).

VI. ANALYTICAL METHODS

Potassium in mixed fertilizers is usually determined by the volumetric sodium tetraphenylboron method recommended by the AOAC (1965) after extraction of the sample with ammonium oxalate. The AOAC method of flame photometry (1965) continues to be used for determining potassium, and an increasing number of people use atomic absorption although this is not yet an official procedure. Also, a few laboratories still use the time-honored gravimetric chloroplantinate method. Potassium salts, such as the chloride, nitrate, or sulfate, are analyzed by any of the above-mentioned procedures after the salts have been dissolved in water.

The determination of potassium in water or ammonium oxalate extracts offers few problems, whereas phosphorus and nitrogen must be converted to the orthophosphate and ammonium forms before they can be determined. The oxalate extraction procedure was devised many years ago for calcium-based fertilizers and may not be suited for newer materials, such as potassium polyphosphate. Oxalate is a precipitant for calcium and decomposes such compounds as syngenite, $CaK_2(SO_4)_2 \cdot H_2O$, releasing the potassium, but it has little effect on potassium solubility when calcium is absent.

REFERENCES

Association of Official Agricultural Chemists. 1965. Official methods of analysis, 10th ed. p. 23, Sec. 2.083, 2.084, 2.085.

Ibid. pp. 22-23, Sec. 2.077, 2.078, 2.079, 2.080, 2.081, 2.082.

Boyland, D. R., and D. V. Karmit. 1964. Granulation characteristics of a 5-4-12 (5-10-15) fertilizer containing potassium nitrate. J. Agr. Food Chem. 12:423-8.

Harris, F. J. 1963. Potassium metaphosphate: a novel method of manufacture and a summary of its behavior as a fertilizer. Proc. Fertilizer Soc. Anal. Chem. 76:1-48.

Hein, L. B., G. C. Hicks, Julius Silverberg, and L. F. Seatz. 1956. Fertilizer technology: granulation of high-analysis fertilizer. J. Agr. Food Chem. 4:318-30.

Hoffmeister, G., S. C. Watkins, and Julius Silverberg. 1964. Bulk blending of fertilizer materials: effect of size, shape, and density on segregation. J. Agr. Food Chem. 12:64-9.

Huygen, D. C., and G. Perbal. 1965. ISMA Tech. Conf. LEE/65/XIII. Edinburg, Scotland.

Jacob, K. D. 1963. Nitrate of potash—a new product for the industry in production at Vicksburg plant. Com. Fertilizer 107 (5):23-5.

Parker, A. B., and N. Watchorn. 1965. Self-propagating decomposition of inorganic fertilizers containing ammonium nitrate. J. Sci. Food Agr. 16:355-68.

Phillips, A. B., G. C. Hicks, J. E. Jordan and T. P. Hignett. 1958. Fertilizer granulation: effect of particle size of raw materials on granulation of fertilizers. J. Agr. Food Chem. 6:449-53.

Potts, J. M., H. W. Elder, and W. C. Scott. 1961. Liquid fertilizers from superphosphoric acid and potassium hydroxide. J. Agr. Food Chem. 9:178-81.

Scott, W. C., and J. A. Wilbanks. 1967. Fluid fertilizer production. Chem. Eng. Prog. 63:58-66.

Smith, R. C. 1964. Nitrate of potash in quality mixes. Farm Chem. 127 (6):74-6.

Surber, J. 1967. Pilot-plant production of 7-28-28 granular fertilizer. Proc. Fertilizer Ind. Roundtable, Washington, D. C.

Tennessee Valley Authority. 1964. New developments in fertilizer technology: 5th demonstration. p. 23-4.

Yates, L. D., F. T. Nielsson, and G. C. Hicks. 1954. TVA continuous ammoniator for superphosphates and fertilizer mixtures. Farm Chem. Part I 117:38, 41, 43, 45, 47-8. Part II Ibid. 117:34, 36-8, 40-41.

Young, R. D., G. C. Hicks, and C. H. Davis. 1962. Application of slurry-type processes in the TVA ammoniator-granulator. J. Agr. Food Chem. 10:68-72.

Agronomic Evaluation of Potassium Polyphosphate and Potassium Calcium Pyrophosphates as Sources of Potassium

ORVIS P. ENGELSTAD

Tennessee Valley Authority
Muscle Shoals, Alabama

I. INTRODUCTION

Sources of potassium in common use are completely water-soluble. In most cases this poses no serious agronomic problems. However, there are situations in which a slower release or dissolution rate may be advantageous. The objectives in typical situations might be to reduce leaching losses of potassium in sandy soils, reduce luxury consumption of potassium, reduce salt injury to germinating plants, and supply potassium to perennial plants or trees with only infrequent application.

In the search for potassium sources that would provide such effects, potassium salts of condensed phosphates and certain potassium calcium pyrophosphates have been tested by several investigators. It is the purpose of this paper to review the agronomic research conducted with these various sources and to discuss their relative merits with respect to delayed release of potassium.

II. CHEMICAL CHARACTERISTICS

A. Potassium Polyphosphate (KPP)

Pfanstiel and Iler (1952) concluded that KPP is a polymer and that the molecular weight may be as high as 120,000. According to Lehr et al. (1967), KPP is a very long-chain potassium polyphosphate of formula

(KPO$_3$)$_n$. This necessitated a change of name from potassium metaphosphate (which implies a ring structure) to the more accurate one of potassium polyphosphate as proposed by Jost (1962). According to Van Wazer (1953), the confusion in nomenclature stems from the fact that when n in the general formula for polyphosphate [K$_{n+2}$ P$_n$O$_{3n+1}$] becomes very large, this form is analytically indistinguishable from the metaphosphate (KPO$_3$)$_n$.

Potassium polyphosphate, also known as potassium Kurrol's salt, is generally produced by the reaction of KCl and H$_3$PO$_4$, and the pure form contains 33% K and 26% P. In pure crystalline form, KPP is a water-soluble compound. However, because of its slow rate of solution, the official method of the Association of Official Analytical Chemists (AOAC) indicates this salt to be only slightly soluble. Madorsky and Clark (1940) reported the solubility of KPP to be less than 0.004%. This slow rate of solution results from the formation of a viscous surface layer. Volkerding concluded that while KPP showed very little initial water solubility, the amount that dissolves is dependent on the amount of solid phase present and on the time of standing in aqueous solution (C. C. Volkerding, 1942. A study of some chemical and biological properties of metaphosphates and their utilization as soil fertilizers. Ph.D. Thesis, Cornell University). Volkerding and Bradfield (1944) found that the solubility of KPP in aqueous solution is markedly increased by the addition of small amounts of various sodium and calcium salts. Soklakov et al. (1966) reported that when CaO, MgO, Fe$_2$O$_3$, and B$_2$O$_3$ were added to the melt, very little effect on the structure of KPP was noted. They concluded that the increase in water solubility upon addition of such additives resulted from changes in crystallization conditions.

B. Potassium Calcium Pyrophosphates

Using laboratory methods, Brown et al. (1963) of TVA prepared 10 potassium calcium pyrophosphates for chemical and crystallographic characterization. These compounds are listed in Table 1 along with the stoichiometric contents of potassium and phosphorus and relative water solubilities of the compounds. For methods of laboratory preparation and for chemical and structural characterization, see Brown et al. (1963) or Lehr et al. (1967).

C. TVA Plant Products

Potassium polyphosphate and K$_2$CaP$_2$O$_7$ (KCP), were selected by TVA for pilot plant production, largely on the basis of their apparent low water solubility according to the AOAC procedure and to relatively high contents of potassium.

Table 1—Potassium calcium pyrophosphates

Compounds	% K	% P	Relative rate of solution in water
1. $K_2CaP_2O_7$	26.7	21.1	Extremely slow
2. $K_2CaP_2O_7 \cdot 4H_2O$	21.5	17.0	Slow
3. $K_2Ca_3(P_2O_7)_2 \cdot 2H_2O$	13.4	21.3	Slow
4. $K_2Ca_5(P_2O_7)_3 \cdot 6H_2O$	8.6	20.5	Extremely slow
5. $K_4CaH_2(P_2O_7)_2$	28.6	22.7	Moderate
6. $K_2CaH_4(P_2O_7)_2$	16.6	26.4	Rapid
7. $KCaHP_2O_7$	15.4	24.5	Extremely slow
8. $KCaHP_2O_7 \cdot 2H_2O$	13.5	21.4	Moderate
9. $KCa_2H_3(P_2O_7)_2 \cdot 3H_2O$	7.4	23.7	Slow
10. $KCa_3H(P_2O_7)_2 \cdot 4H_2O$	6.7	21.4	Extremely slow

* From Brown et al. (1963) and from Lehr et al. (1967).

Pilot-plant KPP was produced by burning elemental phosphorus in the presence of KCl (Walthall, 1953). This product varied in K content from 29 to 32%, in Cl content from 1 to 4%, and in P content from 24 to 25%. This product was also made by reacting wet-process H_3PO_4 with KCl (New Developments in Fertilizer Technology. p. 23-24. TVA 5th Dem. Oct. 6-7, 1964). These experimental products consisted primarily of crystals of KPP with minor amounts of KCl, beta-$Ca_2P_2O_7$, and KH_2PO_4.

The presence of impurities in the melt and rate of cooling greatly affect the extent of crystallization and hence the solubility of KPP. Stanford and Hignett (1957) and Potts, Scott, and Anderson (1958) reported that the presence of as little as 2 to 3% Al_2O_3 or Fe_2O_3 as impurities results in a glassy or vitreous product that dissolves fairly readily in water. The results of these tests, plotted in Fig. 1, show that the K water solubility of KPP increased from 5 to 95% when the Al_2O_3, Fe_2O_3, or Al_2O_3 plus Fe_2O_3 content was increased from 0 to 3%. Potassium water solubility also increased over the same range when the SiO_2 content of KPP was increased from zero to 6%. The potassium water solubility decreased, however, when the Al_2O_3 or Fe_2O_3 content was increased above 3%. By such results it is apparent that KPP can be prepared with any desired level of potassium water solubility, merely by manipulating the content of impurities.

The TVA produced pilot plant quantities of melts approaching the composition of KCP by burning elemental phosphorus in the presence of finely ground phosphate rock and KCl. Rapid quenching of the melt tended to produce low potassium solubility and high citrate solubility of the phosphorus; these conditions favor the formation of crystalline KCP. Slow cooling increased the content of citrate-insoluble phosphorus and water-soluble potassium; under these conditions crystals of citrate-insoluble beta - $Ca_2P_2O_7$ form, leaving a potassium-rich glass. While pure KCP has a composition of 0-27-21 (N-P-K), these pilot-plant products varied in analysis from 0-17-22 to 0-22-21 (N-P-K), depending upon the content of KPP and unreacted KCl.

Because of the manner in which these products were produced in early

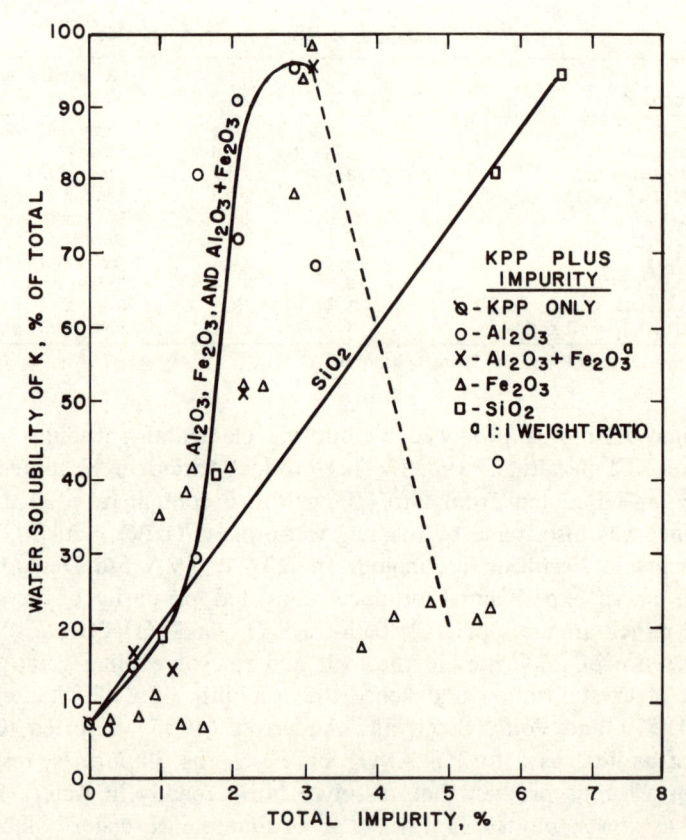

Fig. 1—Effect of alumina, iron, and silica on the water solubility of potassium in KPP prepared at 1650F. (From Potts et al., 1958).

stages of development, they were often referred to as "fused potassium phosphates." For further discussion of the TVA processes, see Copson, Pole, and Baskervill (1942) and Potts et al. (1958).

These pilot-plant products of KPP and KCP varied quite widely in degree of crystallinity and were seldom composed of a single pure crystalline species. This variation resulted in differences in water-solubility. This was especially true for products made from wet-process phosphoric acid, suggesting that some of the impurities contained in the acid affected the solubility of the product. Accordingly, products containing varying amounts of impurities were synthesized in the laboratory and their solubility determined over a 6-month period. The results are illustrated in Fig. 2 (Potts et al., 1958). The material designated number 1 is a vitreous glass which contained occluded KPP, $Ca_2P_2O_7$, and quartz crystals; number 2 had partially devitrified and contained crystals of KPP and KCl; number 3 was primarily crystalline KCP; and number 4 was primarily crystalline KPP. The solubility test was performed by placing 4 grams of the experi-

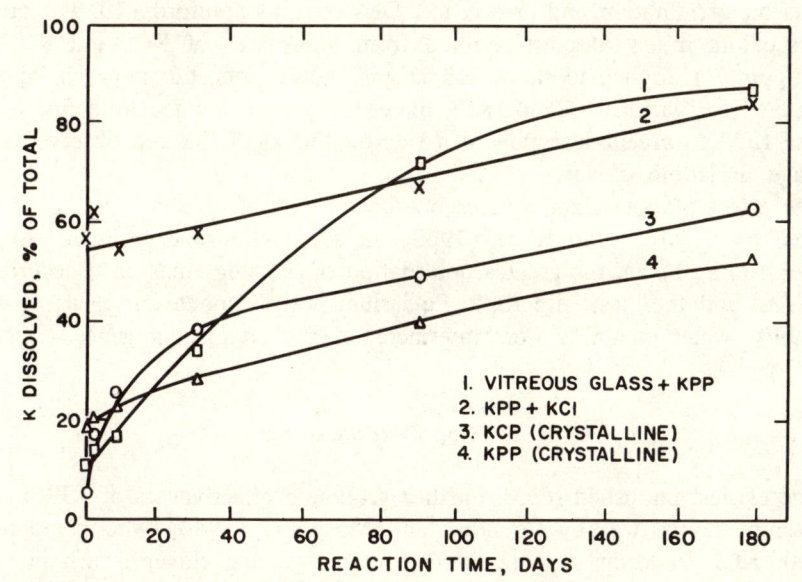

[Fig. 2—Solubility of potassium phosphates in water (4.0 g. of solid in 1 liter of water) at 70 to 80F. (From Potts et al., 1958).

1. VITREOUS GLASS + KPP
2. KPP + KCl
3. KCP (CRYSTALLINE)
4. KPP (CRYSTALLINE)]

mental materials in 1 liter of water and agitating intermittently at room temperature. The data plotted in Fig. 2 show that materials varied in initial solubility but all dissolved slowly in water over a period of several months.

D. Other Plant Products

Scottish Agricultural Industries Ltd. produced KPP by reacting wet-process phosphoric acid and fertilizer-grade KCl in a rotary reactor (Harris, 1963). The resultant product contains P high in citrate solubility, low in water solubility (5%), and contains approximately 31% K. Using the same process, Chemical and Phosphates Ltd., at Haifa, Israel, is also producing KPP of presumably similar characteristics (Hagin and Scherzer, 1967).

III. AGRONOMIC RESULTS

A. Germination Effects

It has been hypothesized that KPP and KCP should have less depressive effect on germinating seeds than KCl or K_2SO_4. Presumably this hypothesis was based on lower salt effects and a lower rate of dissolution for the former materials. However, very few studies have been made to study their effects on germination and emergence. DeMent and Stanford (1959) noted that seedling injury to corn resulted from high rates of KCl and K_2SO_4 (200 ppm K) applied to sandy soil in greenhouse pots, but no such injury was noted for various KCP or KPP materials. Hsiao[1] applied both fine and coarse KCP to greenhouse pots at a rate of 930 kg K/ha and observed no damage to ladino clover.

The effect of potassium sources on the emergence of flax seedlings was studied by Caldwell and Kline (1963) in a greenhouse experiment. At a rate of 460 kg K/ha, the greatest retardation of seedling emergence occurred with KCl and the least with KCP. Potassium polyphosphate materials of 45 and 99% water solubility were intermediate in delaying emergence.

B. Crop Response

The earliest published report on the agronomic effectiveness of KPP as a potassium source was by Chandler and Musgrave (1944) who compared it with KCl. In terms of potassium content of ladino clover grown in the greenhouse, KCl was more effective than KPP. No differences in yield of four crops were noted. Yields of sudangrass, grown on field plots on Mardin silt loam were higher with KPP than with KCl by a fairly wide margin.

DeMent and Stanford (1959) tested several finely divided (-35 mesh) KPP plant products varying in potassium water solubility from <4% to 100%. Two KCP products of low water solubility (< 4 and 4% were also included. As shown in Table 2, no clear relationship was found between

---

[1]Hsiao, T. C. 1960. Potassium absorption by ladino clover as related to solubilities of potassium carriers and lime levels of the soil. M. S. Thesis. University of Connecticut.

Table 2—Relative effectiveness of potassium added as −35 mesh sources varying in water solubility. (From DeMent and Stanford, 1959)

Source	H_2O-Sol. K, % of total	Relative effectiveness†
KCP	< 4*	88
	4	93
KPP	< 4*	85
	16	98
	52	77
	100	93
K_2SO_4	100	100
KCl	100	92

* Pure products.
† Based on uptake of K by a first crop of corn.

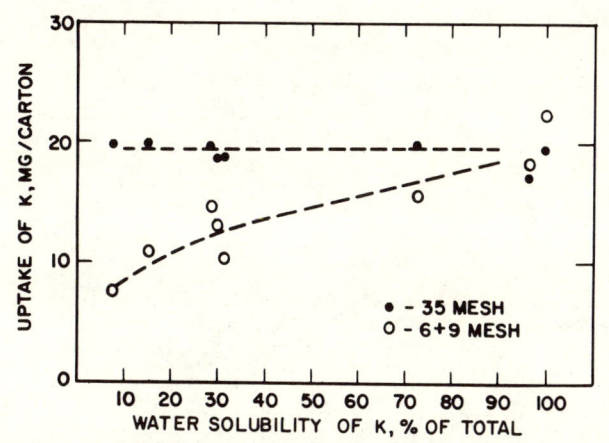

Fig. 3—Uptake of K (minus uptake from control of 8.1) as affected by water solubility and particle size of various potassium sources. A constant rate of 50 mg K were applied to 200 g of soil. (From DeMent and Stanford, 1959).

water solubility of these various sources and uptake of potassium by a first crop of corn.

DeMent and Stanford (1959) also found that uptake of potassium by German millet over a 1-week period following application showed that all of the fine (-35 mesh) KPP and KCP sources were equal in effectiveness to KCl. However, the KPP and KCP sources of low water solubility (7 to 31%) applied as −6+9 mesh particles were much lower in availability during this period (see Fig. 3). The effect of particle size was still evident after a 2-week reaction period prior to plant uptake.

Subsequent TVA experiments reported by DeMent, Terman, and Bradford (1963) also showed that the immediate effectiveness of KCP and KPP declined noticeably with increase in particle size. These results are summarized in terms of relative effectiveness values in Table 3. Again, no significant differences were noted when these sources and KCl were added in −35 mesh particle size.

Table 3—Relative effectiveness of potassium sources as affected by water solubility and particle size. (From DeMent, Terman, and Bradford, 1963)

Particle size	(Experiment 1)*			(Experiment 2)†		
	KCl	KPP	KCP	KCl	KPP	KCP
−35 mesh	100	102	93	100	80	98
−14 + 20 mesh		97	97		105	66
−4 + 6 mesh		68	77		71	65
1/4 inch		52	70		51	48
3/8 inch		–	–		47	42
H_2O Sol. K, % of total	100	38	35	100	31	34

* Based on K uptake by corn from mixed treatments.
† Based on yield of dry matter by corn from mixed treatments.

Table 4—Relative effectiveness of pure potassium sources in two greenhouse experiments

Source	H$_2$O-Sol. K, % of total	Relative effectiveness Lehr et al. (1964)* −100 mesh	−6+14 mesh	1967 Experiment† −100 mesh	−6+14 mesh
KCP	5	63	23	73	22
HKCP	3	44	10	54	0
KPP	22	-	-	83	90
K$_2$SO$_4$	100	100	115	-	-
KCl	100	-	-	100	89

* Based on K uptake by first crop of corn.
† Based on first clipping yield of sorghum-sudangrass hybrid. (Unpublished TVA data).

The results of two other TVA greenhouse experiments are summarized in Table 4. The results reported by Lehr, Engelstad, and Brown (1964) included comparisons of KCP, KCaHP$_2$O$_7$ (HKCP), and K$_2$SO$_4$ in −100 and −6+14 mesh particle sizes. The 1967 TVA experiment included the two particle sizes of KCP, HKCP, KPP and KCl. Relative effectiveness values were calculated using first crop uptake data.

The results of the 1967 experiment (Table 4) indicate that KPP was only slightly less effective than fine KCl as a source of potassium for the first clipping of sorghum-sudangrass, regardless of particle size. The relative effectiveness values for KPP (fine KCl = 100) were 83 and 90 for fine and −6+14 mesh particles, respectively. The potassium calcium pyrophosphates were definitely lower in effectiveness, especially in the coarse particle size. Relative effectiveness based on first-crop potassium uptake in both experiments decreased as follows for both particle sizes: K$_2$SO$_4$ or KCl > KCP > HKCP. In each case the HKCP compound exhibited the lowest effectiveness, indicating the lowest rate of dissolution. The KCP member of the series was intermediate in effectiveness, and like HKCP, exhibited a substantial particle-size effect.

In greenhouse experiments with ladino clover, Hsiao[2] found no differences in dry matter yield among coarse (−6+14 mesh) and fine (−35 mesh) KPP and KCP, and fine KCl, but uptake of potassium from coarse KCP was lower than from the other sources.

Caldwell and Kline (1963) reported that KPP of 99 and 45% water solubility and KCP of 25% water solubility were equal to KCl for corn and alfalfa. This was apparently true for both −35 and −6+14 mesh sizes. While they did find that KCP was less effective for corn in the −6+14 mesh size, they concluded that these experimental potassuim sources did not restrict luxury consumption by plants at normal application rates.

[2]Ibid.

In greenhouse experiments, Pritchett and Nolan (1960) also found that the availability of potassium in finely divided KPP and KCP was unrelated to water solubility. Increasing the particle size to −6+14 and −3+4 mesh markedly reduced the availability and leaching of these potassium sources in sandy soils.

Adams et al. (1967) reported that KCP (in which 7.7% of K was water-soluble) produced more Coastal bermudagrass forage with a lower potassium content than either KCl or K_2SO_4. This was attributed to the lower rate of dissolution of KCP.

In field experiments with corn, Younts and Musgrave (1958) reported that banded KPP, K_2SO_4, and KCl were equal as potassium sources in terms of potassium uptake.

Harris (1963) summarized evaluation results in Britain comparing KPP with mixtures of superphosphate and KCl. The results indicated that KPP was a satisfactory fertilizer, but potassium effects were confounded with phosphorus effects in the data presented.

In greenhouse experiments in Israel, Lachover and Feldhay (1966) and Hagin (1966) found that KPP was as effective as KCl and K_2SO_4, regardless of particle size.

C. Alterations in Soil

In laboratory studies, Menon (1960), found that particles of KPP softened and began to disintegrate after leaching with water, whereas KCP particles remained largely unchanged. When particles of various sizes were incubated in moist organic soil, 46 and 94% of the water-soluble K was released in a 24-hour period from the −4+14 mesh particles of KCP and KPP respectively. In a field study in which particles of KPP, KCP and KCl were left exposed on the soil surface, no KCl or KPP particles could be found after 40 days' exposure. Potassium calcium pyrophosphate particles were recoverable in unaltered form after this same period.

DeMent et al. (1963) reported that the effectiveness of large particles of KPP was greater with surface placement than when mixed with the soil. In fact, at the termination of the experiment, all surface-applied KPP particles had completely dissolved, presumably by the frequent surface watering; residues of KPP mixed with the soil were still evident. They speculated that rate of dissolution of KPP was more important than immediate solubility in water in explaining availability to plants.

After completion of the experiments reported by Lehr et al. (1964) and after the 1967 experiment, the soil from pots receiving coarse particles was examined for residues. In each of these experiments, residues of the pyrophosphates were easily found, but not of KPP. This is not surprising in view of the relative effectiveness of this material for the first crop (See

Table 4). Petrographic examination of the residues from the pyrophosphates showed that no appreciable alterations had occurred in the soil during a period of 4 months after application.

D. Recovery of Added Potassium

The recovery of added potassium was studied by a number of investigators. These studies involved recovery by cropping, by leaching, or by cropping plus leaching. As an example of early results, Andrews (1947) reported that when banded in Ruston soil in laboratory systems, only 1% of the K in KPP was leached, as compared with 39% of the K in KCl.

DeMent and Stanford (1959) determined the recovery of added K from −35 mesh sources of KCP and KPP. These data are presented in Table 5 for the first crop and as totals for three crops. Recoveries were similar for all sources. About 50% of the added K was recovered in the first crop and a total of 60-70% by the three crops. Percentage recoveries of added potassium by individual clippings in the 1967 experiment are presented in Fig. 4. In all cases where uptake of added potassium was significant in the first clipping, uptake was still important in the third. The pyrophosphates in general supplied potassium more uniformly to the respective clippings than did KPP or KCl. Marked decreases in recovery resulted with increase in particle size of the pyrophosphates. No such decreases were found for KPP or KCl.

After one experiment involving leaching and plant uptake, Pritchett and Nolan (1960) found that as much as 50% of the K applied as KCP remained in the soil as undissolved particles, as compared to 20-40% for KPP. In a greenhouse lysimeter experiment involving frequent leachings, losses of potassium from KCl were about 3.5 times that from KPP or KCP. Leaching losses of potassium from the latter sources were small, regardless of particle size. The percentage recovery of added potassium in this experiment is plotted in Fig. 5. A higher recovery of potassium was

Table 5—Recovery by corn of added potassium from −35 mesh sources varying in water solubility. (From DeMent and Stanford, 1959)

Source	H_2O-Sol. K % of total	Recovery First crop	Total 3 crops
KCP	< 4*	50	64
	4	53	68
KPP	< 4*	48	60
	16	56	66
	52	44	55
	100	53	65
K_2SO_4	100	56	70
KCl	100	52	67

* Pure products

obtained from the large particles of KCl and KPP; however, potassium recovery from large particles of KCP was less than from −35 mesh particles. There was also a delay in potassium recovery from the coarse particles of KPP and KCP by the first and second crops.

DeMent and Stanford (1959) reported that leaching of soil systems immediately after application of potassium sources removed potassium from KPP in relation to initial potassium water solubility. An interaction of particle size and potassium water solubility was evident; that is, low solubility and coarser particle size reduced loss of potassium. With higher water solubility, approximately twice as much potassium was leached from the larger particle size (−6+9 mesh) than from the smaller (−35 mesh). Increased soil-fertilizer reaction time caused an increase in potassium recovery from the materials of low water solubility, but a reduction from materials with a higher content of water-soluble potassium.

Using leaching tubes in the laboratory, Hsiao[3] found that K losses decreased in the following order: KCl>KCP(fine)>KCP(coarse). These recoveries of K in the leachate from soil of pH 5.1 were 27, 18, and 13%, respectively, of the added K. Leaching of potassium was lower from soil limed to pH 6.5.

---

[3]Ibid.

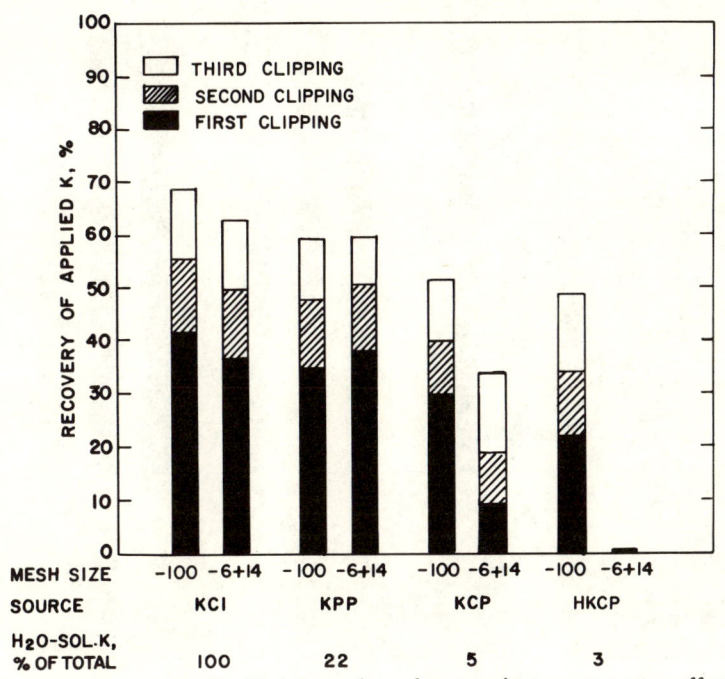

Fig. 4—Percent recovery of added potassium from various sources as affected by particle size over 3 successive clippings of a sorghum-sudangrass hybrid. (unpublished TVA data, 1967 experiment).

Ayres and Hagihara (1953) found that after heavy leaching, recovery of potassium by sudangrass from potassium-containing compounds in latosolic Hawaiian soils decreased in the following order: $KPP > KH_2PO_4 > K_2SO_4 > KCl$. It was implied that potassium in the leachate was in the reverse order. They concluded, however, that potassium in KPP was subject to more leaching than is indicated by its initial low water solubility.

Losses of potassium from KPP were also quite low in a leaching experiment reported by Thorup and Mehlich (1961). The recovery of potassium in the leachate decreased as follows: $KNO_3 = KH_2PO_4 > K_2HPO_4 > KPP$. The KPP used in this experiment was -10 mesh in particle size.

MacIntire et al. (1954), on the other hand, found that recovery of K from lysimeter studies was similar for KPP and K_2SO_4 at annual rates of 166 lb. K/acre. At a single rate of 1042 lb. K/acre, twice as much K was recovered in the leachate from KPP as from the K_2SO_4. In the latter, half of the applied potassium was not recovered and assumed to be fixed in nonreplaceable form.

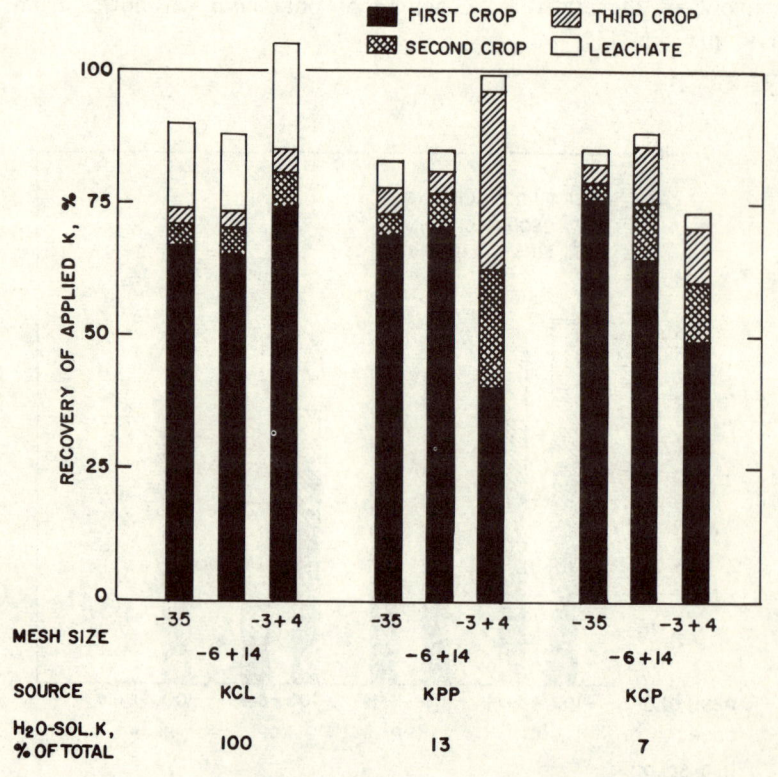

Fig. 5—Percent recovery of added potassium from various sources, as affected by particle size over 3 successive grain crops with frequent leaching. (From Pritchett and Nolan, 1960).

After application to a sandy soil, Lutrick (1958) found that potassium from KCP moved very little with percolating water, whereas potassium from KCl moved substantially more. This apparently reflects the low solubility of KCP.

For a review of the factors influencing movement of potassium in soils, the reader is referred to Munson and Nelson (1963).

IV. DISCUSSION AND CONCLUSIONS

The data available on germination and emergence effects indicate that KPP and KCP are much less injurious than are KCl or K_2SO_4. This would be expected in view of the relatively slower rates of dissolution and lower salt effects. Fairly heavy rates of KPP or KCP can apparently be placed near germinating seeds without causing injury.

In most of the experiments reviewed, the first-crop effectiveness of finely divided KPP and KCP was similar to that of KCl or K_2SO_4 and largely unrelated to water solubility level. Apparently, the rate of dissolution of these compounds is sufficiently rapid in soil to supply ample quantities of potassium to plants, when added in finely divided form.

The effect of increasing the particle size is apparently different for KCP and KPP, however. While immediate effectiveness generally decreased with increase in particle size, this decrease in effectiveness was often much more marked for KCP than KPP. In fact, in certain experiments, no change in effectiveness of KPP was noted with change in particle size. Also, even coarse particles of KPP tended to disappear in soil fairly rapidly. In view of its higher immediate effectiveness and apparent rate of dissolution in soil, KPP does not have great potential as a "slow-release" source of potassium unless an unusually large particle size is used.

The KCP compound, on the other hand, does exhibit a fairly low rate of dissolution when added as coarse particles. The other potassium-containing calcuim pyrophosphate for which data are presented, HKCP, dissolved too slowly from particle sizes typical of commercial fertilizer to be of serious interest. Its content of K is also quite low (15%).

The recovery of added potassium by cropping or removal by leaching is fairly consistent with the data on agronomic effectiveness. Recoveries of potassium from KPP and KCP were similar to that from KCl or K_2SO_4 when added in finely divided form. When added as coarse particles, the crop recovery decreased for KPP and KCP relative to that from the soluble sources.

Both KPP and KCP were consistently less subject to leaching losses than the conventional sources.

REFERENCES

Adams, W. E., A. W. White, R. A. McCreery, and R. N. Dawson. 1967. Coastal bermudagrass forage production and chemical composition as influenced by potassium source, rate, and frequency of application. Agron. J. 59:247-250.

Andrews, W. B. 1947. The response of crops and soils to fertilizers and manures. W. B. Andrews, State College, Mississippi.

Ayers, A. S., and H. H. Hagihara. 1953. Effect of the anion on the absorption of potassium by some humic and hydrol humic latosols. Soil Sci. 75:1-17.

Brown, E. H., J. R. Lehr, J. P. Smith, and A. W. Frazier. 1963. Preparation and characterization of some calcium pyrophosphates. J. Agr. Food Chem. 11:214-22.

Caldwell, A. C., and J. R. Kline. 1963. Effect of variously soluble potassium fertilizers on yield and composition of some crop plants. Agron. J. 55:542-5.

Chandler, R. F., Jr., and R. B. Musgrave. 1944. A comparison of potassium chloride and potassium metaphosphate as sources of potassium for plants. Soil Sci. Soc. Amer. Proc. 9:151-153.

Copson, R. L., G. R. Pole, and W. H. Baskervill. 1942. Development of processes for metaphosphate productions. Ind. Eng. Chem. 34:26-32.

DeMent, J. D. and G. Stanford. 1959. Potassium availability of fused potassium phosphates. Agron. J. 51:282-285.

DeMent, J. D., G. L. Terman, and B. N. Bradford. 1963. Crop response to phosphorus and potassium phosphates varying widely in particle size. J. Agr. Food Chem. 11:207-212.

Hagin, J. 1966. Evaluation of potassium metaphosphate as fertilizer. Soil Sci. 102:373-379.

Hagin, J., and S. Sherzer. 1967. Potassium metaphosphate as a fertilizer, p. 175-185. In G. V. Jacks (ed.) Soil chemistry and fertility. Trans. of Comm. II, IV, Int. Soc. Soil Sci. (Aberdeen, Scotland.) 1966.

Harris, F. J. 1963. Potassium metaphosphate: a novel method of manufacture and a summary of its behaviour as a fertilizer. Proc. no. 76, The Fertilizer Soc., London.

Jost, K. H. 1962. The structure of potassium polyphosphates. Naturwissenschaften 49:229-230.

Lachover, D. and H. Feldhay. 1966. Evaluation of potassium metaphosphate as a source of potassium for potatoes with special reference to foliar diagnosis, yield, and quality of tubers. J. Agr. Sci. 67:281-285.

Lehr, J. R., O. P. Engelstad, and E. H. Brown. 1964. Evaluation of calcium ammonium and calcium potassium pyrophosphates as fertilizers. Soil Sci. Soc. Amer. Proc. 28:396-400.

Lehr, J. R., E. H. Brown, A. W. Frazier, J. P. Smith, and R. D. Thrasher. 1967. Crystallographic properties of fertilizer compounds. Chem. Eng. Bull. No. 6, TVA, Muscle Shoals, Ala.

Lutrick, M. C. 1958. The downward movement of potassium in Eustis loamy fine sand. Soil and Crop Sci. Soc. of Florida Proc. 18:198-202.

MacIntire, W. H., W. M. Shaw, B. Robinson, and C. Veal. 1954. Potassium fixation: differential fixation of potassium from incorporations of metaphosphate and sulfate in two soils. J. Agr. Food Chem. 2:85-91.

Madorsky, S. L., and K. G. Clark. 1940. Potassium metaphosphate—a potential high-analysis fertilizer material. Ind. Eng. Chem. 32:244-248.

Menon, R. G. 1960. A study of the behavior of fused potassium phosphates in soil. Ph. D. Thesis, Michigan State Univ. microfilms. Ann Arbor. (Diss. Abstr. 21:3212).

Munson, R. D., and W. L. Nelson. 1963. Movement of applied potassium in soils. J. Ag. Food Chem. 11:193-201.

Pfanstiel, R., and R. K. Iler. 1952. Potassium metaphosphate: molecular weight, viscosity behavior and rate hydrolysis of non-cross-linked polymer. J. Amer. Chem. Soc. 74:6059-6064.

Potts, J. M., W. C. Scott, and J. F. Anderson. 1958. Fused potassium phosphates. Mimeo. TVA, Muscle Shoals, Ala.

Pritchett, W. L., and C. N. Nolan. 1960. The effects of particle size and rate of solution on the availability of potassium materials. Soil and Crop Sci. Soc. of Florida Proc. 20:146-153.

Soklakov, A. I., A. S. Cherepanova, I. A. Grishina, and E. M. Popova. 1966. Potassium metaphosphate (Russ) CA 67. 63388.

Stanford, G., and T. P. Hignett. 1957. Fertilizer development in the TVA: new processes and agronomic implications. Soil and Crop Sci. Soc. of Florida, 17:161-185.

Thorup, R. M., and A. Mehlich. 1961. Retention of potassium meta- and orthophosphates by soils and minerals. Soil Sci. 91:38-43.

Van Wazer, J. R. 1953. Phosphoric acids and phosphates. *In* Kirk-Othmer Encyclopedia of Chemical Technology, Vol. 10.

Volkerding, C. C., and R. Bradfield. 1943. The solubility and reversion of calcium and potassium metaphosphates. Soil Sci. Soc. Amer. Proc. 8:159-166.

Walthall, J. H. 1953. Chemistry and technology of new phosphate materials, p. 205-255. *In* K. D. Jacob (ed.) Fertilizer technology and resources in the United States. Academic Press, Inc., New York.

Younts, S. E., and R. B. Musgrave. 1958. Chemical composition, nutrient absorption, and stalk rot incidence of corn as affected by chloride in potassium fertilizer. Agron. J. 50:426-429.

Mineralogy of Soil Potassium

CHARLES I. RICH

Virginia Polytechnic Institute
Blacksburg, Virginia

I. INTRODUCTION

Of the many plant nutrient-soil mineral relationships, those involving potassium are of major if not prime significance. A major component of the earth's crust and plants, potassium is also present in substantial quantities in most soils. Its availability to plants is varied and related in many ways to the crystal chemistry and structure of soil minerals.

The lithosphere contains an average of about 2.3% K_2O (Ahrens, 1965), an abundant element compared to phosphorus or sulfur. In the soil this percentage decreases to about 1.4 on the average. In the weathering of igneous rocks, some of the potassium is lost by leaching, but as water containing this potassium moves to the sea, a very large part of the potassium is incorporated in minerals which form sedimentary rocks. Thus, salts of sea water contain 1.1% K compared to 30% Na. Sodium is present in about the same concentration as potassium in the lithosphere, but once brought into solution by weathering a large part remains in solution. Sedimentary rocks, particularly shales, consequently contain considerable potassium. In the earth's crust, potassium silicates are abundant. Orthoclase and its polymorphs make up 16% of the total crust, whereas biotite and closely associated trioctahedral micas make up 3.8% and muscovite contributes another 1.4% (Ahrens, 1965, p. 29).

Potassium, among the mineral cations required by plants, is the largest in size (r = 1.33A, Table 1). Thus, the number of O^{2-} ions surrounding it in mineral structures is high, 8-14, and consequently the strength of each K-O bond is relatively weak.

Potassium, relative to Ca^{2+}, Mg^{2+}, Li^+, and Na^+ ions, has a higher polarizability but NH_4^+, Rb^+, Cs^+, and Ba^{2+} are some of the cations of interest which have higher values (Table 1). Other things being equal, the higher the polarizability the more highly the cation is attracted in exchange reactions (Helfferich, 1962, p. 162) since the distance of closest approach is less.

In regard to hydration, however, potassium has a low hydration energy compared to Mg^{2+}, Al^{3+}, Ca^{2+}, Na^+, and Li^+ (Table 1). This means that in the interlayer space, K^+ ions would cause little swelling and have less effect on other groups, including organic molecules in the interlayer space, than the other ions commonly found in these positions.

II. STRUCTURES OF SOIL MINERALS INVOLVED IN POTASSIUM REACTIONS

A. General Significance of Mineral Structures

Most of the potassium that is readily available to plants exists as exchangeable ions principally on clay mineral surfaces. As this potassium is removed, it is replaced at various rates depending on the amount and characteristics of the potassium-bearing minerals present in the soil. The potassium-micas and the potassium-feldspars constitute the principal potassium-bearing minerals in soils.

The type of bonding within the crystal structure, the extent and type of disorder within the crystal or at crystal terminations, and crystal size are mineral properties that determine the rate at which weathering occurs and potassium is released. Disorder exists at crystal cleavage and fracture surfaces and cracks, twinning planes, dislocations, and vacant sites. Interstratification, partial opening of interlayer space, and warping of silicate layers are also types of disorder. These crystal properties also determine,

Table 1—Radius, polarizability, and hydration energy of certain cations

Cation	Radius*	Polarizability†	Hydration energy‡ First shell	Total shells
	Å	Å^3	kcal/g ion	
Li^+	0.60	0.02	71	74
Na^+	0.95	0.21	51	51
K^+	1.33	0.97	34	34
Rb^+	1.48	1.47	27	27
Cs^+	1.69	2.37	21	21
Mg^{2+}	0.65	±0.2	300	411
Ca^{2+}	0.99	0.44	227	311
Sr^{2+}	1.13	0.84	189	263
Ba^{2+}	1.35	1.63	163	227
NH_4^+	1.43	1.60	30	30
H^+	Very small	Negative	---	---

* Pauling.
† Böttcher (quoted from De Bruyn and Marel, 1954).
‡ MacKenzie, 1964.

to a great extent, the exchange properties, including potassium-selectivity and fixation.

B. Brief Description of Mineral Structures

1. Micas

The micas are phyllosilicates consisting of unit layers each composed of two Si, Al-O tetrahedral sheets between which is a M-O, OH octahedral sheet, where M consists of Al^{3+}, Fe^{2+}, Fe^{3+}, Mg^{2+}, and some other cations. Potassium ions occupy positions between the unit layers in facing ditrigonal holes. Micas are classified in two major groups, dioctahedral and trioctahedral. In the dioctahedral group, two out of three octahedral cation positions are occupied, whereas in trioctahedral micas, all three positions are occupied. In octahedral coordination in dioctahedral minerals Al^{3+} are the principal cations, whereas the divalent ions Mg^{2+} and Fe^{2+} are the principal cations in the trioctahedral minerals.

2. Vermiculites

Expansible layer silicates, the vermiculite and montmorillonite-saponite groups, are particularly important in exchange reactions, particularly those involving potassium, because of the very extensive internal exchange surface of these minerals. The surface charge density is 0.5-1.0 per $O_{10}(OH)_2$, usually somewhat less than that of the micas. In soils, particularly acid soils, dioctahedral vermiculite is much more common than the trioctahedral variety.

3. Montmorillonite-Saponites (Smectites)

The montmorillonite-saponite (smectite) group is similar to the vermiculites except that the charge density in less [0.25-0.5 per $O_{10}(OH)_2$]. The dioctahedral varieties, including those which have aluminum for silicon substitution in the tetrahedral layer (beidellite), are much more common in soils than trioctahedral varieties.

4. Illite

Illite is considered to be a mixed-layer mica-montmorillonite (or vermiculite). It is suggested that this mixing may be in the X Y plane as well as as in the Z direction. The latter type of interstratification is usually recognized as the normal structure of mixed layer minerals. In addition to the question concerning the type of mixing, there is also uncertainty as to the presence of H_3O^+ in the normal potassium positions of the mica portions.

5. Kaolinite and Halloysite

These are 1:1 layer silicates composed of single Si-O tetrahedral groups superimposed on Al-O,OH octahedral groups. In kaolinite the 1:1 layers occur in parallel stacks with no interlayer material. In halloysite water occurs between the unit layers and a tubular morphology is common. In contrast to the 2:1 layer silicates, cation exchange in kaolinite and halloysite is at external surfaces only.

6. Potassium Feldspars

These are framework silicates consisting of SiO_4 and AlO_4 tetrahedra linked in all directions through the oxygen of the tetrahedra. Potassium in the potassium-feldspars and sodium, calcium, and other cations in other feldspars, are held in the interstices of the Si,Al-O framework. The negative charge produced by aluminum in tetrahedral coordination is balanced by the positive charge of the cations in the interstices. The formula for the various potassium feldspars is $KAlSi_3O_8$, showing that only one out of four of the tetrahedra contain aluminum. There are several minerals, namely, sanadine, orthoclase, and microcline, which have the same chemical formula.

7. Zeolites

Zeolites are framework silicates with channels that incorporate water and selectively exchangeable cations. They are "cation sieves" due to the critical size of the channels. Although the calcium-zeolite, analcine, has been found in soils (Shultz et al., 1964), as far as the writer is aware, a potassium-zeolite has not been identified in soils. Since acid treatments destroy zeolites, they may have been missed in many past analyses in which acids are used to remove carbonates.

8. Amorphous Material

Amorphous material that contains Si-O groups as well as Al, Fe-O,OH groups often has a high capacity for retention of both cations and anions. Although allophane is amorphous in that it gives no X-ray diffraction pattern, there is apparently some organization, probably a loose linkage of Si, Al-O tetrahedra with some surrounding loosely bound positively-charged Al-OH, H_2O groups in octahedral coordination (de Villiers and Jackson, 1967). The role of these minerals in potassium retention needs further study.

C. Recent Structure Determinations and their Implications Concerning Soil Potassium Reactions

1. Layer Silicates

Refined structural analyses of the micas and other layer silicates show that the dimensions of the octahedral and tetrahedral sheets vary with occupancy and non-occupancy of octahedral cation sites and the size and charge of those cations occupying these sites (Bailey, 1966). When the unrestrained tetrahedral (Si, Al-O) sheet is oversize compared to the octahedral sheet, the tetrahedra rotate and, in the case of dioctahedral minerals, tilt around vacant octahedra. Thus, in the case of the micas, the anion coordination about potassium (or other interlayer cations) is not 12 equidistant oxygens but 6 near-by oxygens plus 6 oxygens at some further distance. Furthermore, the surface oxygens are not in a hexagonal pattern but, because of the twisting of the tetrahedra, form a ditrigonal pattern (Fig. 1). Tilting also causes the oxygens to be located in different planes rather than in one plane as in the ideal undistorted configuration.

Radoslovich (1962) determined regression equations relating b-dimensions of layer silicates with cation composition. The regression equation for micas, where ions are substituted in the paragonite composition, $NaAl_2(Si_3Al)O_{10}(OH)_2$ is as follows:

$$b = (8.925 + 0.099\ K - 0.069\ Ca + 0.062 + 0.116\ Fe + 0.098\ Fe^{3+} + 0.166\ Ti) \pm 0.03 A.$$

Thus, an increase in any of the cations indicated, with the exception of Ca^{2+}, would tend to increase the b-dimensions of the micas. Only K^+ and Ca^{2+} of this group are interlayer cations. As the mica takes in more Mg^{2+} and Fe^{2+} in the octahedral layer and becomes more trioctahedral, the b-dimension increases and the rotation of the tetrahedra becomes less.

Since the degree of tetrahedral twisting and tilting varies, the interlayer cation-oxygen distance also varies. Thus, the bonding of interlayer cations in micas would be expected to be very different.

Radoslovich (1960, 1962) showed that in muscovite and other dioctahedral micas, the potassium ion is too large for the six coordinating oxygens. This causes less tetrahedral rotation and more stretching of the octahedral sheet than otherwise would occur. When potassium was removed, shrinking in the b-dimension was observed by Burns and White (1963), and Leonard and Weed (1967) have shown that the b-dimension of micas varies with the kind of interlayer cation present. In trioctahedral micas with larger octahedral sheets, such as biotite, there is less tetrahedral twisting and a longer K-O bond.

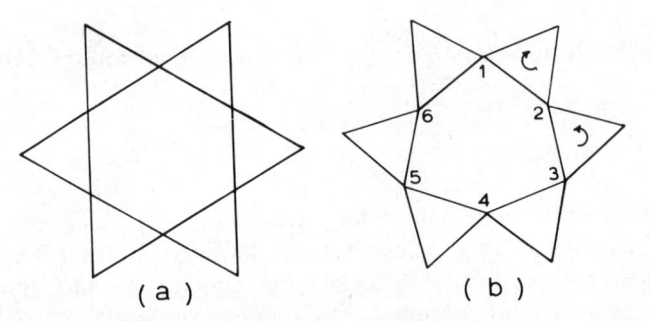

Fig. 1—Ditrigonal arrangement of tetrahedral groups in layer silicates (b) in contrast to "ideal" hexagonal arrangement (a).

These observations suggest that there is considerable flexibility of the structural units in layer silicates in response to interatomic forces within the crystal.

Cowley and Goswani (1961) found that after treatment with organic bases the b-dimension of some crystals of montmorillonite increased by 2%. In those crystals showing this alteration, the new b-dimension was that which would be expected for montmorillonite with the same composition in which the tetrahedra were unrestrained. Thus, one cannot assume that the layers of montmorillonite and vermiculite are structurally constant when different cations and different amounts of water are present in the interlayer space.

Another feature of layer silicates that has implications for soil chemistry is the orientation of the O-H bond in the 2:1 layer silicates. In the dioctahedral micas the O-H bond is considerably inclined from an axis perpendicular to the (001), whereas in trioctahedral micas it is nearly perpendicular. It is not clear whether the interaction between the weak positive charge of hydrogen and potassium is significant enough to affect the equilibration position of potassium (Takéuchi, 1965). Its affect, if significant, would appear to promote the release of potassium more readily from trioctahedral micas than from dioctahedral micas. Thus, the direction of the O-H bond as well as the tetrahedral rotation and tilting may favor a more rapid release of potassium from trioctahedral micas such as biotite or phlogopite than from dioctahedral micas such as muscovite.

These factors may help to explain the preponderance of dioctahedral micas in acid soils. The oxidation of Fe^{2+}, to Fe^{3+} in biotite cannot explain the faster rate at which it releases potassium, compared to muscovite (Newman and Brown, 1966). The rate is much greater even when iron is not oxidized. As weathering proceeds, oxidation of iron probably becomes more important.

Based on single crystal studies, there is ordering among the octahedral cations. These studies indicate that the vacant cation sites in dioctahedral micas do not occur at random but are located in a regular fashion so as to place the cations in occupied octahedral sites as far from each other as possible. In the tetrahedral layer there also is evidence of ordering in the aluminum-for-silicon substitution in muscovite (Brown, 1966). In this case substitution appears to be in rows favoring [10] or [11] or [1$\bar{1}$] directions in the XY plane of the layer. The arrangement of substituted and unsubstituted rows leads to local charge balance around each potassium ion.

These recent findings are significant in considering theories on interlayer diffusion of ions involving hopping distances, local bonding, selectivity, etc.

2. Potassium Feldspars

Sanadine is a polymorph formed at high temperature. The aluminum positions are located at random among the tetrahedra. Following sanadine, orthoclase, intermediate microcline, and maximum microcline represent increasingly stable forms. In maximum microcline there is nearly complete ordering of aluminum ions in the tetrahedra. This mineral is formed at low temperature. Thus, in soils, authigenic microcline is frequently found (Jeffries and White, 1938; Jeffries 1956; Jeffries et al., 1956).

Granitic and authigenic micoclines have two important differences (Finney and Bailey, 1964). Granitic microclines usually are monoclinic and are twinned in the familiar cross-hatch pattern. This cross-hatch twinning has been attributed to the inversion of disordered phases, orginally formed at high temperature, by a diffuse Si/Al ordering transformation on cooling. Authigenic microclines, on the other hand, are triclinic and lack cross-hatch twinning since they form at low temperature under conditions where the phase formed is stable.

The second major difference is that granitic microclines are perthitic. During the same transformation that produced Si/Al ordering, the original K/Na solid solution unmixed to form albite-microcline perthitic structures. Authigenic microclines are never perthitic and are essentially free of sodium.

Sanadine and orthoclase are monoclinic, whereas microline is triclinic. Sodium is frequently present, partly as albite, in fine, often microscopic, lamellar twinning. In granitic microcline, twinning is seen as cross-hatching. Since the albite weathers before the potassium-feldspar, the loss of albite can markedly affect the weathering rate of the potassium feldspar by exposing more surface to the soil solution (Fig. 2).

III. RELEASE OF SOIL POTASSIUM BY WEATHERING

A. Availability of Potassium in Soil Minerals to Plants

A large supply of potassium is usually present in soils. This potassium is present in the clay fraction, as well as in the silt and sand, and thus its availability is of considerable interest. An example is the residual soil, Tatum

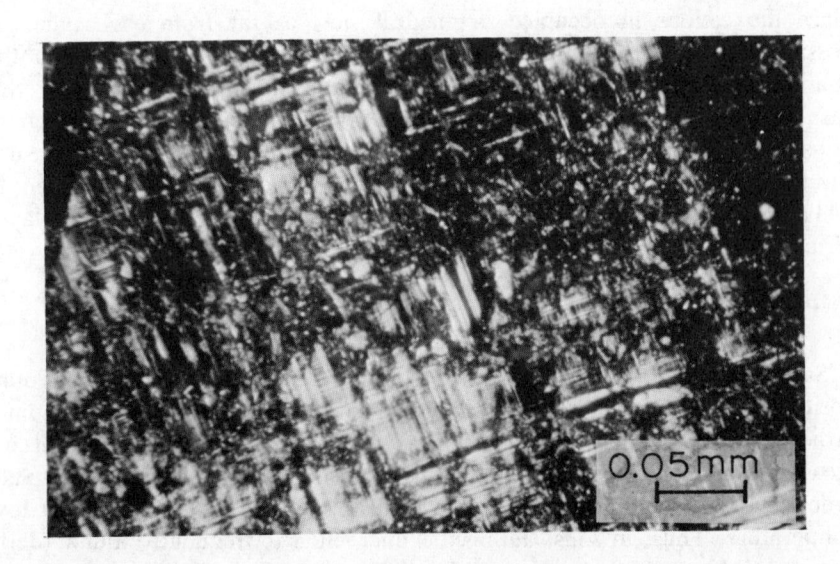

Fig. 2—Weathered microline (particle with cross-hatching). Preferential weathering of mineral (albite ?) at twinning planes indicated. From "border conglomerate," Chesterfield Co., Virginia. (S. B. Cotton and C. I. Rich, 1964, unpublished report).

silt loam. Data on the total potassium content of a profile sampled in Virginia is given in Table 2. In the first 127 cm (50 inches) of this soil there is 368,000 kg/ha (164 tons/acre) of K. Although this soil probably contains more potassium than the average, the large supply of potassium in soils has been well established for many years, and these data are meant to emphasize this large supply.

The availability to plants of the large potassium supply in soils has been of interest ever since chemical analyses revealed this supply. Much of the research work on potassium availability in silicates was conducted at an early stage in soils research. Plummer (1918) reviewed much of this work and presented the results of comprehensive cultural and chemical tests of his own on the availability of potassium in the principal potassium silicates found in soils. A small portion of Plummer's data is presented in Table 3.

Table 2—Percentage of potassium in whole soil and fractions; clay content and pH of Tatum, silt loam, Orange County, Virginia (Rich, 1956)

Horizon	Depth, cm	Whole Soil	>50µ	50-20µ	20-5µ	5-2µ	2-0.2µ	0.2-.08µ	Clay (<2µ), %	pH
A_1	0-5	1.12	0.19	0.34	1.88	2.38	2.02	0.99	11.7	4.26
A_2	5-18	1.38	0.24	0.40	1.73	2.37	2.19	1.35	14.1	4.61
B_1	18-31	1.42	0.26	0.85	2.07	2.46	2.21	1.43	26.1	4.74
B_2	31-66	1.58	0.99	1.32	2.65	2.56	1.95	1.05	46.2	5.31
B_3	66-97	3.09	4.16	3.70	4.69	4.50	2.63	1.10	35.8	4.97
C_1	97-127	3.47	3.41	3.39	4.66	4.56	3.09	0.91	19.4	4.94

Table 3—Availability of K in freshly ground minerals (Plumner, 1918)

Source of K	K removed by oats (gains over check)	K removed by 5 extractions*	
		H$_2$O	Carbonated H$_2$O
	g K$_2$O/pot	ppm K$_2$O	
K$_2$SO$_4$	0.305	----	-----
Biotite	0.244	26.3	261.3
Muscovite	0.213	24.2	169.4
Orthoclase	0.075	20.4	94.0
Microcline	0.016	18.1	61.4

* 30 g in 200 ml, shaken for 96 hours.

Each mineral was "ground to an impalpable powder and sifted through the finest grade of bolting cloth." These data show that potassium in freshly ground potassium-silicates is appreciably available to plants and that carbonated water is much more effective than water in its release. Furthermore, the relative availability of potassium was in the following order: biotite > muscovite > orthoclase > microcline. Other data presented by Plummer showed that the amount of potassium removed decreased markedly with time. This is particularly important in attempting to understand the release of potassium from potassium silicate minerals occurring in soils.

Reitemeier et al. (1951) made a comprehensive study and review of the release of native and fixed potassium in soils. The mineralogy of the various fractions of the soils of widely different origins was included in this study. There was no obvious relationship between the extent of potassium release and the content of hydrous mica. It was pointed out that the role of potassium minerals must depend not only on their total abundance but also on their present potassium content and stages of weathering or formation. It is of interest that the two soils having the highest rate of potassium release also had the highest montmorillonite content. Since the clay fraction of all the soils except one had more than 60% hydrous mica, perhaps a combination of a good source of potassium, that is, fine-grained mica and a mineral that has a high cation exchange capacity but a low fixing capacity, may be one that would promote the maintenance of an adequate exchangeable potassium supply. A high exchangeable potassium level may not be necessary if the supply is renewed rapidly.

B. General Mechanisms of Weathering of Potassium-Bearing

The agents of chemical weathering-water, H$^+$ ions, and electrons (Jackson et al., 1948) do not simply dissolve away the outer fringes of silicate. Because of the different diffusion rates and reactivity of the elemental and group components of the silicate minerals, together with the simultaneous reactions of water, protons, and electrons, one would not expect such reactions to be simple and to follow the same course in all soil environments. In general, the most vulnerable groups in layer silicates are the K^+ or Na^+ ions in the micas. Octahedral groups, particularly the divalent cations, are next in susceptibility. Micas weather differentially by opening up of the interlayer region with the exchange of potassium for hydrated ions, but also by a slower dissolution of the silicate layer, principally at the edges. Thus, the "frayed edges" of mica have been proposed by Bray (1937) and many other workers. As will be pointed out, the progress of potassium exchange may not be uniform; some layers may be skipped or opened laterally less than others, leading to regular or random interstratification of layers and to wedge zones between "open" and "closed" mica layers. The relative rates of potassium removal plus expansion, compared to dissolution, probably depends on the composition, microstructure, and bonding. Also, in different soils the kind of weathering, as well as the rate of weathering of the same mica, may be different, depending on the rate of leaching and the solution composition as determined by the total soil environment.

Weathering of the framework silicates containing different cations leads to the development of external layers or "skin" on individual particles different from the unweathered portion within the particle (Correns, 1963). The skin is deficient in the least stable cation(s) of the crystal. In the case of the potassium-feldspars the skin is deficient in potassium. Thus, the rate of potassium removal from freshly exposed crystals is rapid at first but decreases to a much slower rate as weathering proceeds. Particles in soils then would be expected to release potassium at a lower rate than fresh particles of the same size.

C. Weathering of Mica

1. Effect of Kind of Mica

As has been implied in the last section, the micas have very different weathering rates. There is considerable evidence that biotite weathers more rapidly than muscovite (Jackson and Sherman, 1953). On the other hand, Dennison, Fry, and Gile ('1929) concluded that muscovite and biotite do not differ appreciably in the rates at which they are decomposed in the soil. These last authors found that biotite altered rather uniformly to about 4% K_2O, whereas weathered muscovite contained from 1 to 9% K_2O. They further suggest that secondary muscovite weathers more rapidly than the primary variety.

The weathering of biotite in soils is described by Walker (1949), McAleese and Mitchell (1958), Wilson (1966), and many others. Phlogopite, the magnesium-rich trioctahedral mica, apparently weathers as does biotite to form vermiculite (Aitken, 1965).

The principal dioctahedral mica, muscovite, apparently weathers at varying rates, depending on a number of factors. In general, however, muscovite is the most resistant of the micas. As Dennison, Fry, and Gile (1929) point out, secondary muscovite may weather more readily than primary muscovite. Perhaps this is due to a particle size affect, but there also may be differences in disorder, occlusions, etc. (Schwartz, 1958).

Large, well-crystallized muscovite flakes may weather more by dissolution and kaolinite-composed pseudomorphs of muscovite may form, whereas the fine-grained muscovite in schists may form dioctahedral vermiculite by a more rapid replacement of potassium because of greater loss in negative charge in these micas which often contain appreciable ferrous iron.

Paragonite, the sodium dioctahedral mica, often contains considerable potassium. One study of weathering of such a mica in soil (Cook and Rich, 1962) indicated that the sodium and potassium were removed by weathering at about the same rate.

Illite, usually dioctahedral, is considered an interstratification of mica and montmorillonite. Because of this and other types of disorder, the mica component may lose its potassium more rapidly than non-interstratified mica.

Glauconite, a dioctahedral mica, high in Fe^{2+}, apparently of marine origin and organic association, and possessing considerable disorder (Burst, 1958), would be expected to weather rapidly in the environment of a well-drained soil. It may be somewhat analogous to illite except that illite is aluminous whereas glauconite is ferriferous. Glauconite would be expected to release potassium more rapidly than illite.

2. Effect of Particle Size on K Removal

Mortland and Lawton (1961) studied the rate of potassium removal from different size fractions of biotite weathered in one lot. At an early stage of weathering, the fine particles released potassium at the most rapid rate, but with time the coarse particles lost more potassium. At the final stage of weathering there was little difference in potassium content of the different sand, silt, and clay size fractions. In fact, the two finest fractions contained slightly more potassium than the coarser fractions.

Reichenbach and Rich studied the removal of potassium from different size fractions of muscovite by using $BaCl_2$ solutions (Reichenbach, H. G. v., and C. I. Rich. *Unpublished report.* 1967). The results presented in Fig. 3 show that with time more potassium is removed from the coarser fractions. X-ray diffraction shows 10A reflections remaining in the clay size fractions but nearly all of the silt fractions were converted to a dioctahedral vermiculite. These results are supported by analyses of many soil clays which contain appreciable potassium, even in the finest fractions. The cause of the slowing down of potassium exchange in the fine clay needs further study. Only 1 or 2 ppm of K in solution is necessary to bring the system to equili-

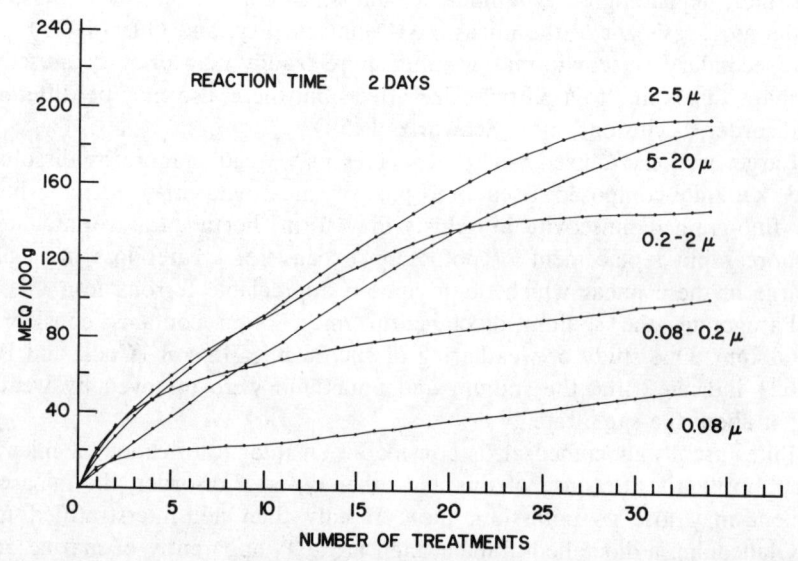

Fig. 3—Effect of particle size on K exchange from musovite by 0.1NBaCl$_2$ treatments (1 g, 3,000 ml, 120C, each for 2 days).

brium so that the mica no longer loses K. In the biotite system, on the other hand, nearly all of the potassium can be removed in 2 or 3 treatments with BaCl$_2$ under these conditions, demonstrating the difference in potassium exchange from the two minerals.

3. *Loss of Negative Charge*

The natural vermiculites and montmorillonites derived from micas have a negative charge somewhat lower than that of the original micas. The classification of the product depends indirectly on the negative charge, since its swelling properties with (e.g., glycerol-solvated magnesium-saturated 2:1 layer silicate) depend on the charge density. The fixation and release of potassium also depends in part on the charge density. Thus, the mechanisms by which the negative charge is changed are of significance.

Gruner (1934) proposed that trioctahedral micas lost charge by oxidation of Fe^{2+} to Fe^{3+}. That such oxidation may not be accompanied by an equal loss in negative charge was suggested by Bradley and Serratosa (1960), who proposed the reactions:

$$Fe^{2+} + \text{structural (OH}^-) \rightarrow Fe^{3+} + \text{structural (O}^{2-}) + H$$
$$4H + O_2 \rightarrow 2H_2O$$

Newman and Brown (1966) incorporated these reactions in proposing the following sequence of reactions:

1) K^+ in interlayer sites at the edge of the mica flakes exchange for solution cations.

2) A few of the structural hydroxyls that are exposed by removing potassium are released to solution and the loss of negative charge allows local expansion of the layers.

3) Expansion of the structure allows more cations from solution to enter the interlayer space to exchange for K^+, and the expansion thus continues.

4) Atmospheric oxygen enters the interlayer space and oxidizes the ferrous iron thus:

$$O_2 + 4Fe^2 + 4 \text{ structural } (OH^-) \rightarrow 4 Fe^{3+} + 4 \text{ structural } (O^{2-}) + 2H_2O.$$

5) Divalent octahedral ions are released, possibly from the sites exposed by loss of structural (OH^-).

The loss of charge by dioctahedral micas is less pronounced and in laboratory expansion of "pure" muscovites by Reichenbach and Rich no loss was detected (Reichenbach, H. G. v., and C. I. Rich, 1968. Preparation of dioctahedral vermiculite from muscovite and subsequent exchange properties. Int. Congr. Soil Sci., Trans. 8th Adelaide, Austr. 1968. Accepted for publication). Similar results with soil derived micas were obtained by Cook and Rich (1963). Dioctahedral micas may contain a small amount of ferrous iron, and its oxidation could account for some loss in charge. An additional mechanism is the incorporation of protons in the original structure.

McConnell (1950) suggested that substitution of OH for O in the tetrahedral layer may account for the extra water found by analysis of montmorillonite. Tetrahedrally arranged hydroxyl, $(OH)_4$, groups as well as statistical O-OH substitution were visualized (Jackson et al., 1952). Rosenqvist (1963) postulated that protons may be associated with the apical oxygens of the tetrahedral group occupied by Al^{3+} ions.

The charge density in even "pure" layer silicates is difficult to evaluate because of the blocking of exchange sites by hydroxy-aluminum groups and the many side reactions that may take place. The location of protons also adds to the difficulty.

D. Weathering of Potassium Feldspars

Nash and Marshall (1957) studied the reactivity of seven different feldspars to water, acids, and salt solutions. Ammonium chloride released from 2.74 to 9.06 meq cations/100 g of ball-milled specimens. The ammonium ion was found to be much more effective in displacing cations than were Mg^{2+} and Sr^{2+} ions. Furthermore, NH_4^+ replaced more cations than could be accounted for by the NH_4^+ subsequently found in the feldspars. Much of the NH_4^+ in the feldspar could not be replaced by Mg^{2+} ions. The authors concluded that feldspar surfaces do not have fixed cation exchange capacities and that the capacity indicated is very much a function of the cation present in the feldspar and the exchanging cation. The authors further concluded that the surface-reactive layer is thin, probably no more than 1 or 2 unit cells in depth. Two types of exchange sites were proposed. The exposed pores would provide sites of high bonding energy for cations of the proper dimensions. Thus, a mechanism for cation selectivity at feldspar surfaces can be visualized. The authors further suggest that where breakage of Si-O-Si bonds occurs, such sites would be freely accessible to cation exchange. However, the probable formation of very stable SiOH bonds at these sites would make them available at only very high pH.

Minerals ground extensively may lose large proportions of their constituents to reacting solutions and exhibit large apparent exchange capacities (Jackson and Truog, 1940). Furthermore, in soils, mineral surfaces may have external layers quite different from fresh surfaces. Correns (1963) reports on extensive experiments on weathering of feldspars and micas. He visualized a complex reaction and suggested that Al^{3+} is the ion which counters the negative charge produced by loss of K^+ to the solution phase. Water and weak acids initially released potassium from a potassium feldspar at a more rapid rate than other constituents, but in the course of weathering a Si-Al-O residue layer developed about the grain and reduced the rate of potassium loss to that of the decomposition rate of the Si-Al-O layer. The thickness of the residue layer formed on K feldspar of $< 1\mu$ radius was calculated to be about 300A, or 30 elementary cells. Thus, a clay particle of fine clay size would be nearly "all skin." As weathering continued, the thickness did not increase.

The residue layers produced by this artificial weathering at room temperature gave no X-ray or electron diffraction lines and thus by these criteria are considered amorphous. The SiO_2/Al_2O_3 ratio of the residual layer was approximately 5.5:5.8.

It is seen that feldspars and other minerals in soils do not react as rapidly as one might conclude from studies of minerals with fresh cleavage surfaces. Nevertheless, crystal structure does have a significant bearing on relative weathering rates.

IV. RELEASE OF NONEXCHANGEABLE POTASSIUM BY DRYING

Attoe (1947) found that on air drying certain soils the level of exchangeable potassium was increased. Previously Bray and DeTurk (1939), by heating certain Illinois soils at 200C, increased the exchangeable potassium level in those soils containing small amounts of exchangeable potassium and decreased it in those soils containing large amounts. Also, Fine et al. (1942) increased the potassium level by freezing moist soils.

Subsequently, it has been learned that this release process is slowly reversible when the water content is restored (Scott et al., 1957) and that the addition of polar liquids with a low vapor pressure inhibit or prevent potassium release on drying (Bates and Scott, 1964).

This potassium-release phenomenon may not be important under field conditions, since soils, except for the few centimeters at the surface, do not normally dry to the extent necessary to cause release (Luebs et al., 1956). Hanway et al. (1961) also found that subsoils showed the principal response to drying. However, the reaction is of considerable significance in a soil testing program and of considerable interest in trying to understand the chemistry of soil potassium.

As Scott, et al. (1957) point out, the release of K on drying at 110C or below probably is due to an exchange of other ions for the fixed K during the drying process. Cook and Hutcheson (1960) showed that the response of Kentucky soils depended on the potassium level. Those high in exchangeable potassium fixed this element, whereas those low in potassium released it on drying. It is reasonable to assume that the principal source of potassium is illite and other 2:1 minerals containing mica-like zones, and vermiculite is chiefly responsible for fixation.

Accompanying hydration of interlayer cations of partially altered micas there probably is an increase of warping of the silicate layers. Warping opens up the layers for more ready access of water and ions; and therefore, exchange reactions between the layers should also increase. At the same time there may be a closer association of oxygen ions about potassium ions at some point away from the place of bending due to the strain on the silicate layer, thus inhibiting potassium release. The difficulty in removing all the potassium from mica, especially the fine fraction, suggests that excessive warping may inhibit potassium release (III C 2). Warping and stress about remaining potassium ions would be expected to increase as the potassium is depleted and as the particle size is decreased. Thus, drying may relieve this excessive warping and permit exchange.

Another possible explanation of potassium release on drying is that the relative mobility of ions may change with water content. Oster and Low (1963) showed that the relative activation energy of three ions at high water content was $Na^+ > Li^+ > K^+$, but at medium and low water contents the order was $K^+ > Li^+ > Na^+$. The highest activation energies for the three ions was not at the lowest water content as might be expected but at the medium level. (The low level produced one layer of H_2O between the clay lamina and the medium level produced two water layers.) Oster and Low explain this on the basis that more water molecules had to be moved at the "medium" level. Thus, on drying there may be a redistribution of ions due to change in the relative mobilities of K^+ and the competing ions such as Ca^{2+}. The effect of different cation saturation on potassium-release caused by soil drying needs further study.

Another facet of the problem is the possible role of H_3O^+ and protons in potassium chemistry. The effect of drying on loss of H_2O from H_3O^+ and the fate of the proton is of interest. There is a possibility that the proton becomes associated with OH groups in the silicate structure, thereby lowering the negative charge.

Fixation and release may be occurring simultaneously (DeMumbrum and and Hoover, 1958; Bates and Scott, 1964). Mica-like zones may release K^+ and the expansible layers of vermiculite in mixed-layer minerals, and vermiculite alone, may be "sinks" by exchange of other ions for K^+, thereby keeping the potassium activity low yet exchangeable.

V. POTASSIUM FIXATION

Potassium fixation, as a crystal-chemical process, is certainly more complex, but as a soil fertility problem is less forboding than once visualized. The term "fixation" is a relative term, since fixed fertilizer potassium may be more available to plants than is native potassium, but this availability probably varies widely. Volk (1934) cited evidence that potassium added in fertilizer caused an increase in the mica content and proposed that potassium and soil components such as kaolinite and amorphous material reacted to form mica. This process is visualized as a reversal of the weathering process by which potassium is entrapped in the facing ditrigonal holes of expansible 2:1 layer silicates. That plants themselves could recycle potassium to the soil surface in sufficient quantity to form mica has been proposed by a number of workers, including Swindale and Uehara (1966).

Bray and DeTurk (1939) proposed that soils have an equilibrium value for exchangeable potassium. If cropping lowers the exchangeable potassium level, potassium is released with time to the orginal level; whereas if an excess is applied, potassium is fixed so that the level tends to be maintained.

The degree of potassium fixation depends on the charge density of the mineral, the degree of interlayering, the moisture content, the concentration of potassium ions, as well as the kind and concentration of competing cations.

Some minerals, such as weathered micas and vermiculites, fix potassium under moist as well as dry conditions, whereas montmorillonites fix potassium only under dry conditions. In the case of montmorillonite the amount of fixation is very small unless the charge density is high (Weir, 1965). A low charge montmorillonite (Wyoming) tends to maintain a 15A spacing when K saturated unless heated (Laffer et al., 1966). Some soil montmorillonites have a greater capacity to fix potassium than do many specimen type montmorillonites (Schwertmann, 1962a, b). These soil montmorillonites have a higher charge density and probably have wedge positions near mica like zones where selectivity of potassium and fixation can take place.

In acid soils the principal mineral responsible for K fixation probably is dioctahedral vermiculite (Brown, 1953). This mineral seems to be widespread if such minerals as "clay vermiculite" (Fieldes and Swindale, 1954) and "ammersooite" Marel (1954) are included in this group.

Expansion and contraction of clays depend on a balance between competing forces. Tending to expand the clay is the swelling caused by hydration of the interlayers cation. In the case of potassium this is small (Table 1). The forces of attraction (and contraction) are electrostatic cation-clay surface forces and van der Waals forces between the clay platelets. Kittrick (1966) concluded in a study of hydration of vermiculite, saturated with various ions, that the first force, electrostatic attraction, is more important in the interlayer space considered. The mathematical analysis by Hurst and Jordine (1964) would also bear out this conclusion.

It is usually assumed that when potassium is fixed water molecules are excluded from the interlayer space. Thus, the K^+ ions must fit into the ditrigonal holes of the Si, Al–O sheet.

Countering the effect of high charge density on potassium-fixation in many acid soils is the presence of hydroxyl–Al, Fe^{3+} interlayer groups (Rich, 1968). These groups act as "props" between the unit silicate layers and inhibit or prevent collapse of the layers about the K^+ ions.

In Table 4 it is noted that the introduction of hydroxy–Al groups into vermiculite increased the Gapon coefficient k K/Ca from 5.7 to 11.1 liters /mmol)$^{1/2}$ × 10^{-2}. However, potassium fixation was reduced markedly. When the vermiculite saturated with the mixed calcium-potassium solution was dried, 2.8 meq/100 g K was removed by $1N$ NH_4OAc, but in the case the aluminum-treated sample 23.2 meq/100 g was removed. In the two cases it is noted that, prior to drying, 17.8 and 28.9 meq/100 g, respectively, had been removed.

In summary, it appears that in most soils potassium fixation, in the sense that potassium is made extremely unavailable to plants, is low. Potassium frequently takes positions in the layer silicates in which it has a high selectivity with respect to divalent ions, but in an acid soil these K^+ ions are apparently available to plants. Depending on the cation and the pH of the extracting solution, these ions may be termed "fixed." Thus, a Mg $(OAc)_2$ solution at pH 7.0 would not extract K^+ ions from wedge zones or where the large hydrated Mg^{2+} ion could not enter. In an acid soil, however, H_3O^+ ions could exchange with this K^+ (Rich and Black, 1964).

Weir (1965) compared the fixation of potassium by montmorillonites having marked differences in location and density of charge. There was an increase in fixation with layer charge, but there was no relationship between potassium fixation and site of charge.

Barshad (1954) also concluded that the magnitude of the interlayer charge rather than origin of the charge determines potassium or NH_4 fixation. It is seen that a relationship, which would seem to be present between potassium

Table 4—Exchange Ca and K and Gapon exchange coefficient for clays saturated with a mixed $CaCl_2$-KCl (each 0.005N) solution*

Clay mineral or soil clay	Origin	Size (μ)	Ca	K	kK/Ca†
			meq/100g		liters/$_{1}$ (mmol)$^{\overline{2}}$ × 10^{-2}
Clay Minerals					
Illite	Morris, Illinois	2-0.2	21.0	4.3	6.5
Muscovite	Ontario, Canada	<0.2	18.0	18.6	32.8
Montmorillonite	Otay, California	<2	87.5	22.8	8.2
Montmorillonite	Wyoming	<2	71.5	7.8	3.4
Vermiculite	Libby, Montana	<0.2	99.6	17.8	5.7
Vermiculite-treated ‡	Libby, Montana	<0.2	82.0	28.9	11.1
Vermiculite-Biotite	Libby, Montana	20-5	57.0	3.6	14.7
Soil Clays					
Nason (A)	Virginia	2-0.2	22.0	16.8	24.3
Nason (C)	Virginia	2-0.2	20.3	20.9	32.6
Berks (Ap)	Virginia	2-0.2	8.6	7.5	27.6
Carrington (Ap)	Iowa	2-0.2	18.8	11.0	18.5
Putnam (B22)	Missouri	<2	42.0	21.5	16.2

* From Rich and Black (1964).
† Based on Ca and K extracted from the wet samples by 1 N NH_4OAc, except for the untreated vermiculite, in which case the values for 1N $Mg(OAc)_2$ were used (NH_4OAc only removed 15.8 meq).
‡ Al^{3+} saturated and boiled in H_2O for 2 hours prior to K^+ selectivity determination.

or NH_4 fixation and charge location, is difficult to demonstrate. Ions such as potassium that may occur in the ditrigonal holes of 2:1 layer silicates apparently are retained with different energies depending in part on the rotation and tilting of the tetrahedra about the ditrigonal holes. These distortions of the ideal hexagonal arrangement are determined in turn by the size and number of cations in the silicate sheets. In addition to charge density, the configuration of the oxygen about exchange sites probably determines, partially at least, the observed differences in replaceability of fixed or native potassium in micas of the same layer charge.

VI. POTASSIUM SELECTIVITY OF MICA-VERMICULITES AS AFFECTED BY MINERAL PROPERTIES AND HYDROGEN IONS

A. Mineral Properties

1. *Negative Charge*

The site of negative charge as well as the charge density of minerals may affect potassium selectivity. Schwertmann (1962a) suggested that the high potassium selectivity of clay-size muscovite is due to the tetrahedral site of the negative charge. However, the negative sites of vermiculite, a mineral which has a low potassium selectivity compared to muscovite (Schwertmann, 1962a; Rich and Black, 1964), are also located largely in the tetrahedral layer. In fact, there is evidence (Foster, 1960) that many trioctahedral micas (and presumably also the vermiculites derived from them) have a tetrahedral charge exceeding that of muscovite and this excess charge is balanced by an excess of positive charge in the octahedral layer.

Barshad (1954), Schwertmann (1962a,b), and Weir (1965) found that the selectivity of K^+ was not well correlated with charge density of clay minerals. Moreover, when Schwertmann (1962a) treated muscovite with $LiNO_3$, so that some of the vacant octahedral sites were occupied, the potassium selectivity was not altered significantly, although the measured cation exchange capacity increased from 24.1 to 80.4 meq/100g due to extensive opening of the interlamellar space and substituting readily exchangeable hydrated Li^+ for "nonexchangeable," nonhydrated K^+. Thus, it is apparent that potassium selectivity cannot be attributed entirely to charge density or location of charge.

2. Surface Features

The rotation and tilting of the tetrahedra, changes in the b-dimension, and the orientation of the OH bond in micas has been discussed (II C). These features could affect the retention and selection of potassium.

Within a single particle there may be many expressions of disorder that affect ion exchange. Raman and Jackson (1964) demonstrated in electron microscope studies that mica surfaces have cracks and that the silicate layers at these cracks roll back in scroll-like fashion when the exchange sites are saturated with large hydrated cations and return to a flat morphology when again potassium saturated. Fleischer et al. (1964) reports that, over geological time, the spontaneous fission of uranium, present as an "impurity," may damage minerals so as to induce the formation of pits and cracks. Aside from providing "wedge zones" for high selectivity of potassium-size ions, these cracks open up the interior of the particle to a more rapid exchange of cations.

3. Interlayer Features

Since most of the exchange of vermiculite and mica-vermiculite minerals is in the interlamellar space, the characteristics of this space probably affect the exchange markedly. The distance between the silicate sheets, the continuity of this distance, the lateral extent of individual interlayer spaces, and the stability of this space, may be some of the important features. The interlamellar distance can be altered readily by cations of differing hydration (Barshad, 1959) and can be stabilized by hydroxy–Al groups (Rich, 1960) between the silicate sheets.

Keay and Wild (1961) found that the activation energy for the NH_4^+ ion on vermiculite varied with the degree of NH_4^+ saturation. At 0.3, 0.5, and 0.7 saturation, the activation energies were 3.6, 7.3, and 22.4 kcal mole^{-1}, respectively. A similar relationship would be expected for potassium exchange in vermiculite. This may be interpreted as due to gradual closing of interlamellar space and closer coordination with the 12 surrounding oxygens in the holes of the silicate sheets.

Closing of the silicate layers at the edge of the particle also affects the exchange of larger hydrated cations in the system. For instance, Frysinger and Thomas (1961) noted that 12% of the exchange sites in initially Sr^{2+} saturated vermiculite were still occupied by Sr^{2+} after subsequent saturation of the mineral with Cs^+, an ion with properties similar to those of K^+. This amount of Sr^{2+} could not be exchanged by further Cs^+ treatment.

4. Nonuniform Weathering of Mica

One can visualize the replacement of potassium by other ions during weathering as a diffusion of potassium out of the interlayer space and a diffusion of other ions to the vacated sites. The rate of potassium loss would decrease as the center of particle is approached. Such a model was proposed by Reed and Scott (1962) and the experimental results obtained by these authors would support such a model.

However, many mineralogical studies of soils and weathered rocks demonstrate that weathering is generally not this regular. Bassett (1959) reported that the weathering of biotite commonly formed a regular interstratification of biotite and vermiculite in a Libby, Montana location and suggested that as potassium was removed from one layer, the bonds about the K^+ ions in the next layer were strengthened.

It is reasonable to assume that removal of potassium from mica is a slow diffusion-controlled process but that the rate varies with individual layers. Such differences in diffusion rate could result from different degrees of disorder, local variation in charge distribution, and perhaps local substitution of Na^+ or other ions for K^+. Rich and Cook (1963) and Cook and Rich (1963) demonstrated marked differences between the rate of potassium exchange by a well-ordered muscovite and other muscovite-like, dioctahedral-type micas. Marshall and McDowell (1965) noted marked differences in micas.

Even the crystallographic direction affects the rate of potassium release. In muscovite, the exchange rate in the direction of [110] was about 3.5 times as high as in the direction [100] (Weiss et. al., 1956).

In Fig. 4 and 5 are photographs made by S. B. Cotton of thin sections made across the XY plane of highly weathered muscovite from a border conglomerate in Chesterfield County, Virginia. These photographs demonstrate types of weathering, partial opening, and wedge zones that may occur on a scale in which unit silicate layers are involved. Recent electron microscope studies by John Brown of the Georgia Institute of Technology and the

MINERALOGY OF SOIL POTASSIUM 99

Fig. 4—Weathered muscovite, with occlusions, from "border conglomerate," Chesterfield Co., Virginia. (S. B. Cotton and C. I. Rich, 1964, unpublished report).

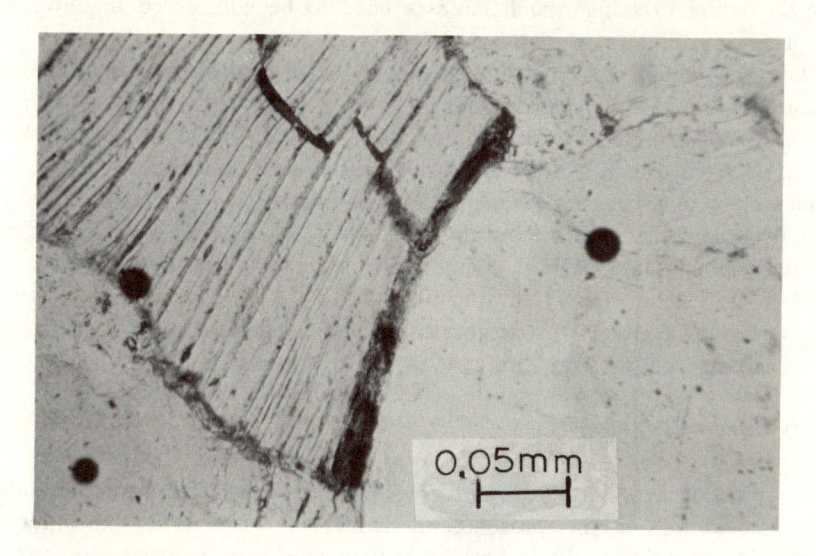

Fig. 5—Weathered muscovite showing "wedge zones." From "border conglomerate," Chesterfield Co., Virginia. (S. B. Cotton and C. I. Rich, 1964, unpublished report).

author, of ultramicrotome sections of weathered muscovite support these suggestions. Note that in Fig. 4 the openings in the mica particle do not match across the large crack normal to XY plane. One possible explanation is that major weathering occurred subsequent to the formation of the large crack and

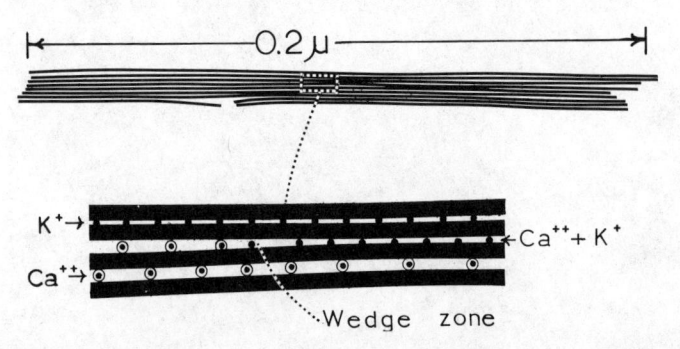

Fig. 6—Proposed model of an edge section of a medium clay-size vermiculite particle with mica-like zones. Expanded section shows wedge zone and ion selectivity.

that opening of individual silicate layers is a somewhat random process. Figure 5, on the other hand, suggests weathering about occlusions. Thus, it is apparent that the mode of weathering of micas is complicated and in view of the probable important effects of wedge zones and different degrees of opening on cation exchange, such features need to be considered in developing theories on exchange mechanisms.

Jackson (1963) has pointed out that the structural alignment of the 10A mica cores in partially opened micas favors re-entry of potassium from the open 14A "vermiculite" zones along the same XY plane of the weathered mica. Rich and Black (1964) and Rich (1964) proposed that steric effects in such wedge zones favor the entry of K^+ and cations of similar size, and this would explain the high selectivity of K^+ in the presence of large hydrated cations such as Ca^{2+} or Mg^{2+}. Figure 6 is a diagram of the proposed model.

Rich and Black (1964) showed further that, in the case of $<0.2\mu$ muscovite and micaceous soils, some potassium, exchangeable by NH_4^+ ions, was not exchangeable by Mg^{2+} ions.

5. Hydroxy–Al Interlayers

Another well-established feature of weathered clay-size mica, particularly that in acid soils, is the presence of hydroxy–Al islands in the interlayer space. It was pointed out in Section V that interlayers may increase potassium selectivity. The Gapon coefficient for a number of soil clays and specimen clay minerals is indicated in Table 4. Apparently K^+ ions can move more easily in "propped open" interlayer space than in interlayer spaces where the silicate layers tend to collapse about the K^+ ions (Fig. 7). Thus, in the first instance, K^+ can reach the wedge zones and can be selected preferentially. The $<0.2\mu$ vermiculite fraction still contained 1.5% K and thus mica zones and wedge zones were also probably present.

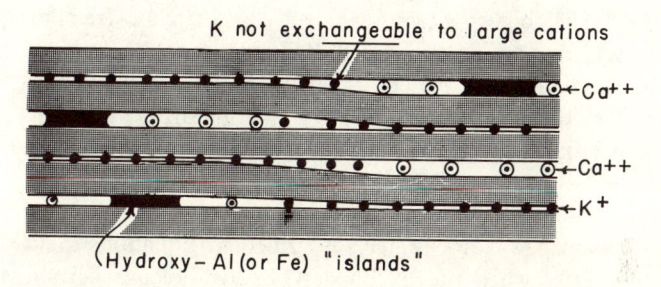

Fig. 7—Proposed model of an expansible layer silicate with interlayers indicating effect on potassium-fixation.

6. Initial Cation Saturation

Dennis and Ellis (1962) noted that sodium-vermiculite (Libby, Montana) collapsed when 40% of the Na$^+$ ions were replaced by K; calcium-vermiculite collapsed when only 20% of the Ca^{2+} ions were replaced by K$^+$.

Jacobs and Tamura (1960) found that as increasing amounts of potassium were added to sodium-vermiculite there was a continuous drop in CEC. X-ray diffraction also indicated that structure collapse occured in a continuous fashion, rather than in an abrupt single collapse at a critical concentration of sorbed potassium. Although X-ray diffraction indicated complete collapse of the structure at approximately 60% K saturation, the sorption results show that even higher degrees of potassium saturation do not so completely collapse the structure as to prevent additional interlayer fixation or cesium.

7. Specific Sorption Sites for Potassium

Jacobs and Tamura (1960) and the same authors in separate reports (1963) distinguish between "edge" fixation and "interlayer" fixation of Cs. They also point out that all interlayer positions of vermiculite are not equivalent in their exchange properties and suggest that interlayer charge density plays an important role in determining which interlayers are most easily collapsed.

The importance of "edge" sites in selectivity measurements is limited by the number of these sites at the edge and whether interlamellar sites near the edge are included. According to Gaines (1957), there is an exchange capacity of 3.48 meq/sq m at cleavage surfaces on muscovite. Assuming even this density of exchange capacity on true edges and also assuming disk-shaped particles of 2μ diameter, this is equivalent to about 0.25 meq/100 g of exchange capacity at edges or 2.5 meq/100 g for particles of 0.2μ diameter. It is also significant that potassium-selectivity *increases* with particle size in some soils (Schwertmann, 1962a; Rich and Black,

1964). At later stages of weathering selectivity may decrease (IIIC2) with particle size. The high potassium-selectivity of clay-size muscovite and many soil clays cannot be explained by selection of potassium at specific sites of such low exchange capacity. Much of the potassium-selectivity most likely occurs in the interlamellar region. An important question is whether all of the specific sites occur near the edge or whether many are deeper within the crystal.

Beckett (1964) distinguished between two types of potassium exchange. One type of exchange is adequately described by equations of the Gapon or Eriksson type only when the exchanger already holds a certain amount of exchangeable potassium. The exchange of the second quantity of exchangeable potassium appears to be held at sites possessing a specific binding power for potassium. The entropy of exchange of this potassium has an anomalously high value.

Bolt et al. (1963) proposed that the specific sites for potassium in illite are intra-lattice positions near the edges of the clay crystal.

B. Hydrogen Ions

The H^+ ion is particularly effective in displacing other cations in clay-size, partially weathered muscovite (Schachtschabel, 1940). In muscovite the series $H > K > Ca > Mg > Na$ was found, but in montmorillonite the series was $Ca > Mg > H > K > Na$. The H^+ ion differs in many respects from other cations in exchange reactions. Krishnamoorthy and Overstreet (1950) found satisfactory exchange constants for all ion pairs except those involving H^+ ions.

Although the H^+ ion is usually associated with one or more water molecules, reaction rates involving this ion are not restricted by diffusion rates of the hydrated ion (e.g. H_3O^+) (Bernal and Fowler, 1933). The lifetime of a proton–H_2O association is extremely brief, and very rapid reactions are possible through hydrogen bond or proton transfer (Eigen, 1959).

Furthermore, it is well known that H–clays decompose spontaneously and rapidly to form $Al^{3+}-$, $Mg^{2+}-$ (or other cation-clay) systems, depending on the decomposition of the minerals. Even so there is evidence that there are stable 2:1 layer silicates that contain H_3O^+ or H^+ ions (Burns and White, 1963). Studies of exchange rates using isotopes of hydrogen (Rosenqvist, 1963), show that there are three components in clays with which hydrogen exchange occurs:

1) Hydrogen of interlayer and external surface water which exchanges in a matter of seconds or minutes.

2) H^+ and H_2O associated with exchange sites of closed mica layers, which exchanges several percent in a week at 115C.

3) Hydrogen in OH groups of the octahedral layer which exchanges at a very slow rate.

These results suggest that proton diffusion through the silicate mineral is very rapid in the presence of a continuous film of water, as in the interlayer space of expanded 2:1 layer silicates. Thus, proton transfer may be rapid to the wedge zone at the junction of "open" and "closed" mica. Here the rate of H_3O^+–K^+ exchange may be limited by the rate of K^+ diffusion.

Regarding the hydrogen associated with the exchange sites of the closed mica layers, a further complication is the possibility that the proton is not associated with the H_2O molecule at the exchange site as H_3O^+ but with a nearby apical oxygen of the tetrahedral layer (Rosenqvist, 1963). The stability of these H^+ + H_2O groups (or H_3O^+) may be due to their isolation and the need for the immediate presence of another cation so that cation exchange can take place. Isolated cations also offer greater problems of maintaining neutrality when exchange of ions of differing valence are involved. There is evidence that the exchangeable potassium, for which there are specific sites, competes with hydrogen ions for these sites. An acid Virginia soil, in which the dominant mineral is dioctahedral vermiculite-mica, was able to select potassium from a $1N$ $Ca(OAc)_2$ solution containing only a small amount of potassium (K/Ca ratio was 1.35:10,000) as long as the pH was above 4.35. The amount of potassium removed increased with pH, whereas below pH 4.35, K^+ ions were removed from the soil by the same solution. At this low concentration, H^+ and K^+ ions appeared to be competing for the same sites, whereas Ca^{2+} ions were excluded (Rich, 1964).

In weathering of mica, H^+ or H_3O^+ ions may be precursors to other ions such as Mg^{2+} or Al^{3+}. The interaction of large ions with hydrogen ions in the exchange of potassium is also indicated. An acidified Mg^{2+} solution is more effective in exchanging K^+ than either acid or Mg^{2+} solution alone (Rich, 1964).

The changes in the mineralogy of sediments, as they move along an increasing salinity gradient may be affected by the same selectivity-pH relationships. Nelson (1963) found an increase in illite and a decrease in dioctahedral vermiculite in the Rappahannock estuary as the salinity and pH increased. A higher pH would favor the greater selectivity of K^+ ions and the formation of illite from vermiculite in the presence of a considerable concentration of Ca^{2+} and Mg^{2+} ions.

C. Proposed Theory on the Effect of the Wedge Zone on Potassium Selectivity

The position of the wedge zone is important to the extent of the "catalytic" effect of this configuration in potassium selectivity. Potassium is selected from among large hydrated ions (e.g. Ca^{2+} or Mg^{2+}) because of the space limitations for diffusion in the wedge zone. If the wedge zone is near the edge of the particle, it can be effective in only a small amount of potassium

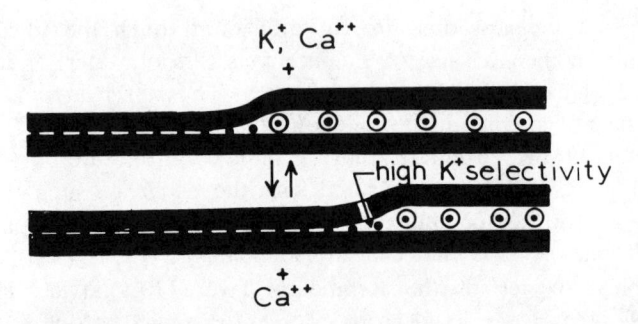

Fig. 8—Proposed model of "wedge" zone in weathered mica showing continuity of site of high potassium-selectivity.

selection; but if it is deep within the particle, then the "catalytic" effect can result in considerably more potassium being selected. As the silicate layers close about the selected K^+ ions, a new wedge zone is formed (Fig. 8). This progression depends on the concentration of K^+ ions, structural alignment, charge density of the mineral, concentration of large, hydrated ions, and the presence of hydroxy-Al or other "prop"-like interlayer features.

At very low concentrations of potassium, there is selection of potassium at the wedge site and as the potassium concentration increases, first there is collapse of the silicate layer at the wedge site. With further increase in potassium concentration, there is a tendency for collapse of the silicate layers first where exchange is initiated—at the edge of the particles—and an entrapment of ions deeper within the crystal.

In the absence of wedge zones (true vermiculite), a large concentration of K^+ (or similar ions) is necessary to initiate collapse of the vermiculite. Potassium selectivity would then be expected to be low until collapse was initiated.

According to the proposed idea on K^+–H^+ exchange at the wedge position, small amounts of K^+ may be selected from a solution also having a high concentration of Ca^{2+}, Mg^{2+}, or other large, highly-hydrated cations. Acid conditions would favor H_3O^+ ions at the wedge position, whereas as the pH increased, selection of K^+ ions would be more likely. Protons can diffuse easily in the wedge zone and can compete with K^+ ions at the wedge site.

REFERENCES

Ahrens, L. H. 1965. Distribution of elements in our planet. McGraw-Hill, New York. 110 p.

Aitken, W. W. S. 1965. An occurrence of phlogopite and its transformation to vermiculite by weathering. Mineral. Mag. 35:151-164.

Attoe, O. J. 1947. Potassium fixation and release in soils occurring under moist and drying conditions. Soil Sci. Soc. Amer. Proc. (1946) 11:145-149.

Bailey, S. W. 1966. The status of clay mineral structures. Clays and clay minerals. 14:1-23. (Pergamon Press, N. Y.)

Barshad, I. 1954. Cation exchange in micaceous minerals II. Replaceability of ammonium and potassium from vermiculite, biotite, and montmorillonite. Soil Sci. 78:57-76.

Barshad, I. 1959. Vermiculite and its relation to biotite as revealed by base-exchange reactions, X-ray analyses, differential thermal curves, and water content. Amer. Mineral 33:655-678.

Bates, T. E., and A. D. Scott. 1964. Changes in exchangeable potassium observed on drying soils after treatment with organic compounds I. Release. Soil Sci. Soc. Amer. Proc. 28:769-772.

Beckett, P. 1964. Potassium-calcium exchange equilibria in soils: specific sorption sites for potassium. Soil Sci. 97:376-383.

Bernal, J. D., and R. H. Fowler. 1933. A theory of water and ionic solution, with particular reference to hydrogen and hydroxyl ions. Chem. Physics 1:515-548.

Bolt, G. W., M. E. Sumner, and A. Kamphorst. 1963. A study of three categories of potassium in an illitic soil. Soil Sci. Soc. Amer. Pro. 27:294-299.

Bradley, W. F., and J. M. Serratosa. 1960. A discussion of the water content of vermiculite. Clays and clay minerals. 7:260-270, (Pergamon Press, New York.)

Bray, R. H. 1937. Chemical and physical changes in soil colloids with advancing development in Illonois soils. Soil Sci. 43:1-14.

Bray, R. H., and E. E. DeTurk. 1939. The release of potassium from non-replaceable forms in Illinois soils. Soil Sci. Soc. Amer. Proc. (1938) 3:101-106.

Brown, G. 1953. The dioctahedral analogue of vermiculite. Clay Minerals Bull. 2:64-69.

Burns, A. F., and J. L. White. 1963. The effect of potassium removal in the b-dimension of muscovite and dioctahedral soil micas. Proc. Int. Clay Conf., Stockholm, Sweden. Macmillan Press, New York. p. 9-16.

Burst, J. F. 1958. Mineral heterogeneity in "glauconite" pellets. Amer. Mineral 43:481-497.

Cook, M. G., and T. B. Hutcheson, Jr. 1960. Soil potassium reactions as related to clay mineralogy of selected Kentucky soils. Soil Sci. Soc. Amer, Proc. 24:252-256.

Cook, M. G., and C. I. Rich. 1962. Weathering of sodium-potassium mica in soils of the Virginia Piedmont. Soil Sci. Soc. Amer. Proc. 26:591-595.

Cook, M. G., and C. I. Rich. 1963. Negative charge of dioctahedral micas as related to weathering. Clays and clay minerals. 11:47-64. (Pergamon Press, New York.)

Correns, C. W. 1963. Experiments on the decomposition of silicates and discussion of chemical weathering. Clays and clay minerals. 10:443-459. (Pergamon Press, New York.)

Cowley, J. M., and A. Goswami. 1961. Electron diffraction patterns from montmorillonite. Acta Cryst. 14:1071-1079.

De Bruyn, C. M. A., and H. W. van der Marel. 1954. Mineralogical analyses of soil clays. Geol. en Mijnbouw Nw, Ser. 10:407-428.

DeMumbrum, L. E., and C. D. Hoover. 1958. Potassium release and fixation related to illite and vermiculite as single mixtures and in mixtures. Soil Sci. Soc. Amer. Proc. 22:222-225.

Denison, I. A., W. H. Fry, and P. L. Gile. 1929. Alteration of muscovite and biotite in the soil. USDA Tech. Bull. 128. 32 p.

Dennis, E. J., and R. Ellis, Jr. 1962. Potassium ion fixation, equilibria, and lattice changes in vermiculite. Soil Sci. Soc. Amer. Proc. 26:230-233.

de Villiers, J. M., and M. L. Jackson. 1967. Cation exchange capacity variations with pH in soil clays. Soil Sci. Soc. Amer. Proc. 31:473-476.

Eigen, M. 1959. The protonic charge transfer in hydrogen bonded systems, p. 429-432. In D. Hadzi (ed.) Hydrogen bonding. Pergamon Press, N. Y.

Fieldes, M., and L. D. Swindale. 1954. Chemical weathering of silicates in soil formation. New Zealand J. Scil. Tech., B, 36:140-154.

Fine, L. O., T. A. Bailey, and E. Truog. 1942. Availability of fixed potassium as influenced by freezing and thawing. Soil Sci. Amer. Proc. (1941) 6:183-186.

Finney, J. J., and S. W. Bailey. 1964. Crystal structure of an authigenic maximum microline. Z. Kristallogr. 119:413-436.

Fleischer, R. L., P. B. Price, and E. M. Symes. 1964. On the origin of anamalous etch pits in minerals. Amer. Mineral 49:794-800.

Foster, Margaret D. 1960. Layer charge relations in the dioctahedral and trioctahedral micas. Amer. Mineral 45:383-398.

Frysinger, G. K., and H. C. Thomas. 1961. The ion exchange behaviour of vermiculite-biotite mixtures. Soil Sci. 91:400-405.

Gaines, G. L., Jr. 1957. The ion exchange properties of muscovite mica. J. Phys. Chem. 61:1408-1413.

Gruner, J. W. 1934. Vermiculite and hydrobiotite structures. Amer. Mineral. 19:557-575.

Hanway, J. J., S. A. Barber, R. H. Bray, A. C. Caldwell, L. E. Engelbert, R. L. Fox, M. Fried, D. Hovland, J. W. Ketcheson, W. M. Laughlin, K. Lawton, R. C. Lipps, R. A. Olson, J. T. Pesek, K. Pretty, F. W. Smith, and E. M. Stockney. 1961. North Central regional potassium studies: Field studies with alfalfa. Iowa State Univ., Ames, Iowa. Res. Bull. 494 p. 161-187.

Helferrich, F. 1962. Ion exchange. McGraw Hill, New York. 624 p.

Hurst, C. A., and E. St. A. Jordine. 1964. Role of electrostatic energy barriers in expansion of lamellar crystals. J. Phys. Chem. 41:2735-2745.

Jackson, M. L. 1963. Interlayering of expansible layer silicates in soils by chemical weathering. Clays and clay minerals. 11:29-46 (Pergamon Press, New York.)

Jackson, M. L., Y. Hseung, R. B. Corey, E. J. Evans, and R. C. Vanden Heuvel. 1952. Weathering sequence of clay-size minerals in soils and sediments: II. Chemical weathering of layer silicates. Soil Sci. Soc. Amer. Proc. 16:3-6.

Jackson, J. L., and G. D. Sherman, 1953. Chemical weathering in soils. Advance. Agron. 5:219-318.

Jackson, M. L., and E. Truog. 1940. Influence of grinding soil minerals to near molecular size on their solubility and base exchange properties. Soil Sci. Soc. Amer. Proc. (1939) 4:136-143.

Jackson, J. L., S. A. Tyler, A. L. Willis, G. A. Bourbeau, and R. P. Pennington. 1948. Weathering sequence of clay-size minerals in soils and sediments. J. Phys. Colloid Chem. 52:1237-1260.

Jacobs, D. G. 1963. The effect of collapse-inducing cations on the cesium sorption properties of hydrobiotite. Proc. Int. Clay Conf. Stockholm. Macmillan Press, New York. p. 239-248.

Jacobs, D. G., and T. Tamura. 1960. The mechanisms of ion fixation using radioisotope techniques. Int. Congr. Soil Sci. Madison, Wisc. II: 206-214.

Jeffries, C. D. 1956. Mineralogical studies of Jordon Plots and other Pennsylvania soils. In 75th Anniversary of the Jordon Fertility Plots. Penn. State Agr. Exp. Sta. Bull. 613:26-36.

Jeffries, C. D., E. Grissinger, and L. Johnson. 1956. The distribution of important soil forming minerals in Pennsylvania soils. Soil Sci. Soc. Amer. Proc. 20:400-403.

Jeffries, C. D., and J. W. White. 1938. Variation in the composition of the feldspar from a Hagerstown soil profile. Soil Sci. Soc. Amer. Proc. 2:133-141.

Keay, J., and Wild, A. 1961. The kinetics of cation exchange in vermiculite. Soil Sci. 92:54-60.

Kittrick, J. A. 1966. Forces involved in ion fixation by vermiculite. Soil Sci. Soc. Amer. Proc. 30:801-803.

Krishnamoorthy, C., and R. Overstreet. 1950. An experimental evaluation of ion exchange relationships. Soil Sci. 69:41-53.

Laffer, B. G., A. M. Posner, and J. P. Quirk. 1966. Hysteresis in the crystal swelling of montmorillonite. Clay Minerals 6:311-321.

Leonard, R. A., and S. B. Weed. 1967. Influence of exchange ions on the dimension of dioctahedral vermiculites. Clays and Clay Mineral. Proc. 15th Natl. Conf. Pergamon Press, N. Y. p. 149-161.

Luebs, R. E., G. Stanford, and A. D. Scott. 1956. Relation of available potassium to soil moisture. Soil Sci. Soc. Amer. Proc. 20:45-50.

Mackenzie, R. C. 1954. Hydrationseigenschaften von Montmorillonit. Ber. Dtsch. Keram. Ges. 41:696-708.

Marel, H. W. van. 1954. Potassium fixation in Dutch soils: mineralogical analyses. Soil Sci. 78:163-179.

Marshall, C. E., and L. L. McDowell. 1965. The surface reactivity of the micas. Soil Sci. 99:115-131.

McAleese, D. M., and W. A. Mitchell. 1958. Studies on the basaltic soils of Northern Ireland V. Cation-exchange capacities and mineralogy of the silt separates (2-20u). J. Soil Sci. 9:81-88.

McConnell, D. 1950. The crystal chemistry of montmorillonite. Amer. Mineral 35:166-172.

Mortland, M. M., and K. Lawton. 1961. Relationships between particle size and potassium release from biotite and its analogues. Soil Sci. Soc. Amer. Proc. 25:473-476.

Nash, V. E., and C. E. Marshall. 1957. Cationic reactions of feldspar surfaces. Soil Sci. Soc. Amer. Proc. 21:149-153.

Nelson, B. W. 1963. Clay mineral diagenesis in the Rappahannock estuary; an explanation. Clays and clay minerals. 11:210. (Pergamon Press, N. Y.)

Newman, A. C. D., and G. Brown. 1966. Chemical changes during alteration of micas. Clay Minerals 6:297-309.

Oster, D., Jr., and P. F. Low. 1963. Activation energy for ion movement in thin films on montmorillonite. Soil Sci. Soc. Amer. Proc. 27:369-373.

Plummer, J. K. 1918. Availability of potassium in some common soil forming minerals. J. Agr. Res. 14:297-315.

Radoslovich, E. W. 1960. The structure of muscovite, $K AL_2(Si_3Al) O_{10}(OH)_2$. Acta Cryst. 13:919-932.

Radoslovich, E. W. 1962. The cell dimensions and symmetry of layer-lattice sillicates: II. Regression relations. Amer. Mineral 46:617-636.

Raman, K. V., and M. L. Jackson. 1964. Vermiculite surface morphology. Clays and Clay Minerals, 12:423-429 (Pergamon Press, New York.)

Reed, M. G., and R. D. Scott. 1962. Kinetics of potassium release from biotite and muscovite in sodium tetraphenylboron solutions. Soil Sci. Soc. Amer. Proc. 26:437-440.

Reitemeier, R. F., I. C. Brown, and R. S. Holmes. 1951. Release of native and fixed nonexchangeable potassium of soils containing hydrous mica. USDA Tech. Bull. 1049.

Rich, C. I. 1958. Muscovite weathering in a soil developed in the Virginia Piedmont. Clays and clay minerals. 5:203-212 (Nat. Acad. Sci.-Nat. Res. Council Pub. 566).

Rich, C. I. 1960. Aluminum in interlayers of vermiculite. Soil Sci. Soc. Amer. Proc. 24:26-32.

Rich, C. I. 1964. Effect of cation size and pH on potassium exchange in Nason soil. Soil Sci. 98:100-106.

Rich, C. I. 1968. Hydroxy interlayers in expansible layer silicates. Clays and clay minerals Vol. 16. (In Press).

Rich, C. I., and W. R. Black. 1964. Potassium exchange as affected by cation size, pH, and mineral structure. Soil Sci. 97:384-390.

Rich, C. I., and M. G. Cook. 1963. Formation of dioctahedral vermiculite in Virginia soils. Clays and clay minerals. 10:96-106. (Pergamon Press, New York.)

Rosenquist, I. Th. 1963. Studies in position and mobility of the H atoms in hydrous micas. Clays and clay minerals. 11:117-135. [Pergamon Press, New York.]

Schachtschabel, P. 1940. Untersuchunger über die Sorption der Tonmineralien und organishen Bodenkolloide, und die Bestimmung des Antiels dieser Kolloid an der Sorption im Boden. Kolloidbeih. 51:199-276.

Schultz, R. K., Ray Overstreet, and I. Barshad. 1964. Some unusual ionic exchange properties of sodium in certain salt-affected soils. Soil Sci. 99:161-165.

Schwartz, G. M. 1958. Alteration of biotite under mesothermal conditions. Econ. Geol. 53:164-177.

Schwertmann, U. 1962a. Die selecktive Kationensorption der Tonfraktion einiger Boden aus Sedimenten. Z. Pflan. Düng., Bodenkundle 97:9-25.

Schwertmann, U. 1962b. Eigenschaften und beldung aufweitbarer (quellbarer) Dreischicht-tonminerale in Böden aus Sedimenten. Beiträge zur Miner. und Petrog. 8:199-209.

Scott, A. D., and T. E. Bates. 1967. Changes in exchangeable potassium observed on drying soils after treatment with organic compounds II. Reversion. Soil Sci. Soc. Amer. Proc. 31:481-485.

Scott, A. D., J. J. Hanway, and E. M. Stickney. 1957. Soil-potassium relations I. Potassium release observed on drying Iowa soils with added salts or HCl. Soil Sci. Soc. Amer. Proc. 21:498-501.

Swindale, L. D., and G. Uehara. 1966. Ionic relationships in the pedogenesis of Hawaiian soils. Soil Sci. Soc. Amer. Proc. 30:726-730.

Takéuchi, Y. 1966. Structures of brittle micas. Clays and clay minerals. 13:1-25. (Pergamon Press,, N. Y.)

Tamura, T. 1963. Cesium sorption reactions as indicator of clay mineral structures. First Int. Clay Con., Stockholm. Pergamon Press, New York. 229-237.

Volk, N. J. 1934. The fixation of potash in difficultly available forms in soils. Soil Sci. 37:267-287.

Walker, G. F. 1949. The decomposition of biotite in soils. Mineral Mag. 28:693-703.

Weir, A. H. 1965. Potassium retention in montmorillonite. Clay Minerals 6:17-22.

Weiss, A., A. Mehler, and U. Hoffman. 1956. Kationenaustausch und innerkrystallines Quellungsvermögen bei den Mineralen der Glimergruppe Z. Naturforschg 16:435-438.

Wilson, M. J. 1966. The weathering of biotite in some Aberdeenshire soils. Mineral Mag. 35:1080-1093.

Role of Potassium in Photosynthesis and Respiration[1]

WILLIAM A. JACKSON and RICHARD J. VOLK

North Carolina State University
Raleigh, North Carolina

I. INTRODUCTION

Growth responses of two different kinds have indicated that potassium may have an essential role in photosynthesis and respiration of higher plants. First, the potassium requirement for optimal growth of starch-storing species of plants appears to be greater than that of species grown primarily for protein. Root crops especially have a high potassium requirement, and it is commonly observed that root or tuber enlargement is depressed relatively more than leaf development when potassium is in short supply. Figure 1 illustrates this point with sweet potatoes grown in sand culture at adequate and limiting levels of potassium. Leaf growth was rather similar in both situations, but enlarged root development was severely restricted in the low potassium treatment.

The second general observation suggesting a close relationship between potassium and photosynthetic and/or respiratory events is the tendency for potassium fertilization to be relatively more effective under growing conditions in which light is restricted (Russell, 1927; Noguchi and Sugawara, 1966). An example of the influence of shading on response of two varieties of rice to applied potassium is shown in Table 1. Although dry weights were depressed by shading with two layers of cheesecloth, the percentage response to the applied potassium was larger with the shading treatment.

Neither of the general observations cited above indicates a direct causal relationship between potassium nutrition and the photosynthetic or respir-

---

[1] Paper no. 2551 of the Journal Series of the North Carolina State Univ. Agr. Exp. Sta., Raleigh, N. C.

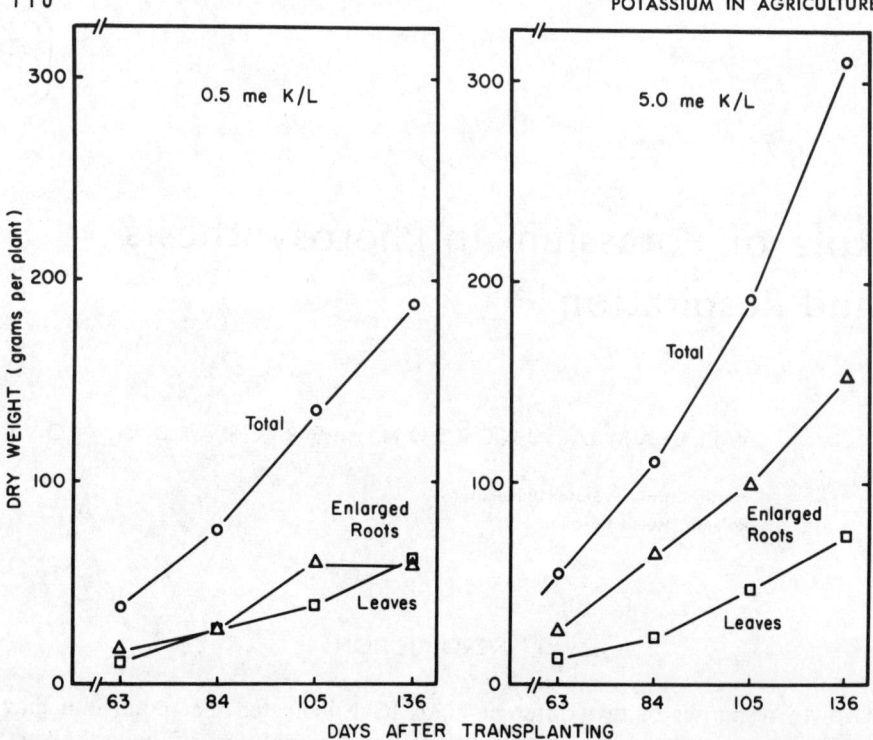

Fig. 1—Influence of potassium on relative growth of leaves and enlarged roots of sweet potatoes. Plants grown in sand culture. Irrigated daily with 1 liter/plant of complete nutrient solutions. Total values include stems, petioles and fibrous roots as well as leaves and enlarged roots. (Jackson, W. A., Unpublished data. North Carolina Agr. Exp. Sta., 1960).

atory processes. Other influences on plant behavior could explain the observations. Among these are restricted translocation of photosynthate in the first case and alteration in tillering and stem thickness in the second. Both such suggestions have in fact been made and no doubt are of importance in the responses noted. Nevertheless, as we shall show, there is now abundant evidence indicating that various facets of the photosynthetic process are affected by the potassium status of the plant. Moreover, respiratory processes are definitely altered as potassium comes into short supply. Effects are seen in respiration occurring in leaves in the light as well as in darkness and in non-chlorophyllous tissue.

II. MEASUREMENT OF GAS EXCHANGE COMPONENTS

Estimates of the overall rates of photosynthesis and respiration of intact tissue are usually based on the net flux of either O_2 or CO_2; fluxes of both gases are seldom measured. Such estimates may create confusion and

Table 1—Influence of shading on response of two rice varieties to K (from Table 31, Noguchi and Sugawara, 1966)

Variety	Cheesecloth covering*	Fertilization†	Total yield‡	Increase due to K	Percentage increase
	layers		g dry wt/pot		%
Rikuu No. 132	None	NP	65.8		
		NPK$_1$	142.7	77	115
		NPK$_2$	156.6	91	138
	One	NP	62.2		
		NPK$_1$	155.7	94	151
		NPK$_2$	184.7	123	195
	Two	NP	44.1		
		NPK$_1$	108.4	64	145
		NPK$_2$	130.4	86	197
Mubo-aikoku	None	NP	43.7		
		NPK$_1$	139.0	95	216
		NPK$_2$	151.4	107	243
	One	NP	48.3		
		NPK$_1$	135.9	88	183
		NPK$_2$	172.9	125	261
	Two	NP	34.7		
		NPK$_1$	112.7	78	222
		NPK$_2$	131.8	97	277

* Light intensities for one and two layers of cheesecloth were 69.4% and 47.2% respectively of natural intensity.
† 12kg soil, NP = 1g N, 1g P$_2$O$_5$; NPK$_1$ = 1g N, 1g P$_2$O$_5$, 1g K$_2$O; NPK$_2$ = 1g N, 1g P$_2$O$_5$, 2g K$_2$O. Supplied as ammonium sulfate, monocalcium phosphate and potassium sulfate.
‡ Includes grain and straw.

uncertainty because a particular environmental or nutritional condition most likely will not alter the exchange of CO_2 and O_2 to the same extent. Moreover, release of CO_2 and uptake of O_2 do occur upon illumination in spite of the fact that the dominant processes are uptake of CO_2 and release of O_2 (Ozbun et al., 1964). Similarly, in darkness or in non-chlorophyllous tissue the rate of CO_2 release dominates although sizeable uptake of CO_2 may occur. This dark CO_2 fixation may be appreciable at times (Barker et al., 1965), and probably larger in leaves than in roots (Hartt and Kortschak, 1967). Release of O_2 by tissue in darkness, however, has never been reported. The concomitant uptake and release of gases may complicate the interpretation of experimentally induced variables, especially at low light intensities.

As an example of this behavior, uptake and release of oxygen by a soybean leaf is shown in Fig. 2. The leaf was enclosed in a 1 liter plexiglass compartment (Volk and Jackson, 1964). When the lights (500 ft-c at leaf surface) were turned on, the chamber contained 92 μmoles $^{16}O_2$ and 631 μmoles $^{18}O_2$. Carbon dioxide was present at a compensation concentration (1.7 μmoles), and 346 μmoles Ar were present as an internal standard. Nitrogen was used as a ballast to maintain the chamber at atmospheric pressure, and the gases were circulated at 1.5 liters/min. During the period (0 to 90 min) in which the ambient CO_2 concentration remained at compensation

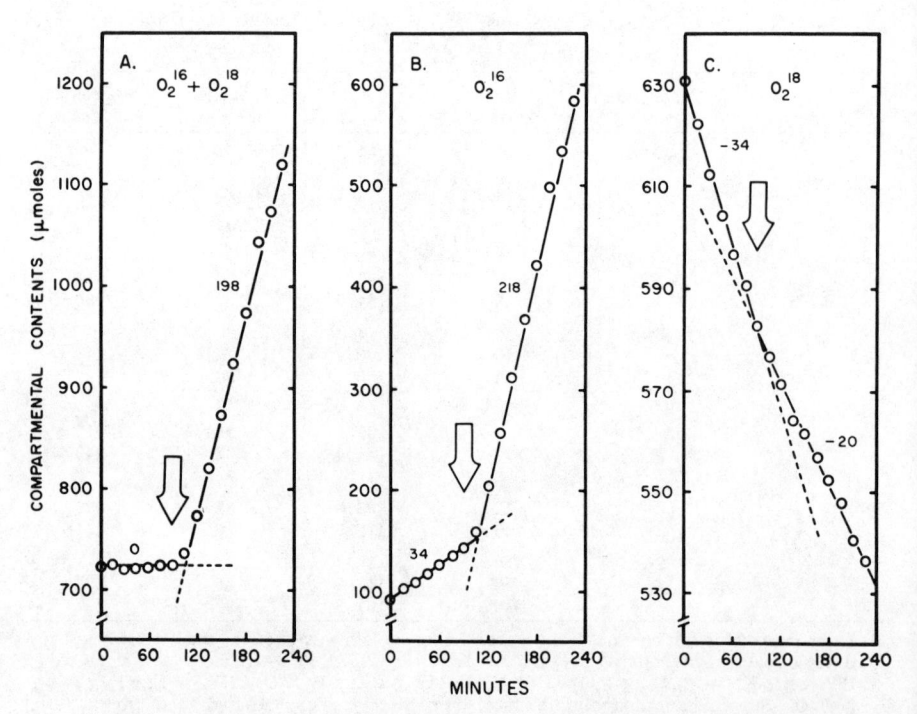

Fig. 2—Rates of appearance of $^{16}O_2$, and disappearance of $^{18}O_2$, in the atmosphere surrounding a mature soybean leaf at two rates of net CO_2 uptake. The light intensity was 500 ft-c. At zero time the closed 1-liter compartment contained 92 μmoles $^{16}O_2$, 631 μmoles $^{18}O_2$, 1.7 μmoles CO_2, 346 μmoles Ar, and 37,129 μmoles N_2. During the first 90 min., CO_2 remained at the compensation level of 1.7 μmoles. At 91 min. (indicated by the arrow) CO_2 was injected into the chamber at a constant rate of 174 μmoles/hr. The numbers adjacent to the lines indicate the rates of change of oxygen isotopes in μmoles/hr. Note that for $^{18}O_2$(C) the ordinate is expanded 5-fold. (Mulchi, C. M., R. J. Volk, and W. A. Jackson, Unpublished data, North Carolina State Univ. 1967).

levels (Fig. 2A), mass spectrometric analysis of periodic samples revealed no change in total O_2 content of the chamber. The lack of net O_2 release and net CO_2 uptake would lead to the conclusion that photosynthesis was nil. However, examination of the behavior of the two isotopic species of oxygen clearly leads to quite another conclusion. Appearance of $^{16}O_2$ in the chamber was occurring at the rate of 34 μmoles/hr (Fig. 2B) and, in spite of this release of $^{16}O_2$ from the leaf, $^{18}O_2$ was disappearing from the chamber at an equal rate (Fig. 2C). A considerable O_2 exchange therefore was occurring when there was no net uptake of CO_2. Photosynthesis, viewed as capacity of the leaf to convert light energy into biochemical energy via photolysis of H_2O was occurring at a significant rate as shown by the release of $^{16}O_2$. Simultaneously, a substantial uptake of oxygen was taking place. The latter may be considered as a measure of photorespiration, although one cannot clearly conclude what proportion of the total oxygen uptake was a result of

(i) a glycolic acid oxidase reaction, (ii) a mitochondrial terminal electron transport reaction, or (iii) a chloroplast reaction in which substances of high reducing potential (e.g., ferredoxin) derived from photosynthetic electron transport were oxidized more or less directly by oxygen.

When CO_2 was injected into the chamber at a constant rate of 174 μmoles hr^{-1} (91 to 225 min) net O_2 release increased to 198 μmoles hr^{-1} indicating substantial net photosynthesis (Fig. 2). This was a result of a larger increase in release of $^{16}O_2$ from the leaf (218 μmoles hr^{-1}) and a depressed uptake of $^{18}O_2$ (20 μmoles hr^{-1}). Calculation of the actual rates of oxygen release and uptake were made from the isotopic concentrations assuming that each molecule of $^{16}O_2$ and $^{18}O_2$ in the chamber have an equal opportunity for being taken up, and that negligible amounts of $^{18}O_2$ are released (Volk and Jackson, 1964). It must also be realized that such calculations reveal only minimal rates; the calculated O_2 release rate is less than the actual rate by the extent to which released O_2 is consumed internally by light respiratory reactions before it appears in the external atmosphere. The calculated O_2 uptake rate also underestimates the actual O_2 uptake rate to the same extent. Nevertheless, it is quite clear that net O_2 changes in the atmosphere surrounding illuminated leaves are the result of both an influx and efflux component. The same has been shown for CO_2 (Ozbun et al., 1964). The influence of a given nutritional or environmental variable on the net rates of gas exchange therefore may be through an effect on the influx component, on the efflux component, or on both. For the purposes of this report it is of special interest that one of the influences of potassium deficiency on photosynthesis is to increase the rate of O_2 uptake and CO_2 release, thus indicating a distinct effect on photorespiration (Ozbun et al., 1965 a). This and other effects will be discussed later.

Since most experimental procedures do not permit simultaneous estimates of the uptake and release rates, we will in this report continue to refer to the net gas measurements as either "photosynthesis" or "respiration" regardless of whether O_2 or CO_2 measurements were made. In those cases where attempts were made to separate them, the directional fluxes will be specified (e.g., "O_2 uptake in the light, O_2 release" etc).

Oxygen exchange information of the sort shown in Fig. 2, as well as other types of measurements (Forrester et al., 1966 a, b; Zelitch, 1966; Moss, 1966; Goldsworthy, 1967) indicates (i) that "respiration" occurs in the light, (ii) that it may be of substantial magnitude, and (iii) that it may in fact be the result of quite a different sequence of events than occurs in dark respiration. Therefore we will have to distinguish between dark respiration (and/or respiration of non-chlorophyllous tissues), and respiration of chlorophyllous tissues in the light. For purposes of convenience the latter is referred to as "photorespiration," the former simply as "respiration." Since photorespiration is so closely associated with photosynthesis, and because there have been so few studies of how photorespiration is affected during potassium deficiency, the two general processes are discussed together in the following section.

III. PHOTOSYNTHESIS AND PHOTORESPIRATION

The manner in which some other essential elements are involved in photosynthetic processes is much more clearly established than is the case for potassium. It is evident, for example, that manganese and chloride are essential in the O_2 evolving processes, whereas iron (a component of ferredoxin and cytochromes) and copper (a component of plastocyanin) play important roles in the transport of electrons and generation of photosynthetic reductant. But we cannot now assign such a specific role to potassium in the photosynthetic scheme. The supply of potassium can, and probably does, influence many individual processes in the sequence of events, and influences them to differing extents. Delineating the primary influence of potassium is difficult because very little is known about the relative distribution of potassium and other metabolites in, and mobility between, individual organelles of an intact cell. Moreover, the movement and distribution of potassium among individual organs of intact higher plants beclouds attempts to specify primary and secondary influences, even for the gross processes of photosynthesis and respiration. In spite of these complications, however, there is a sizeable literature on the photosynthetic behavior of plants subjected to variable potassium nutrition from which some important conclusions can be drawn. Attention in this paper will be focused primarily on studies with intact plant tissue, although some observations made with chloroplast and mitochondrial preparations will be mentioned.

A. Historical Perspective

In 1922 Briggs published results of experiments with bean leaves illustrating a depressed net release of O_2 when the plants had been grown at low potassium levels. The depression was observed under conditions where light intensity was limiting photosynthesis (Fig. 3). Subsequently, Gregory and Richards (1929) determined CO_2 uptake by strips of tissue from the youngest fully developed leaves of barley at weekly intervals. At the 4th and 5th weeks after planting, lower net CO_2 uptake was observed with the low potassium plants as compared to the normal plants. Thereafter only small differences were noted. The detrimental effect during the 4th and 5th week occurred both at 5,000 metre-candles and at a light intensity which just compensated CO_2 release. Subsequent studies (Richards, 1932) illustrated the relationship between the potassium concentration of leaves and their photosynthetic capacities (Fig. 4). Measurements were made at weekly

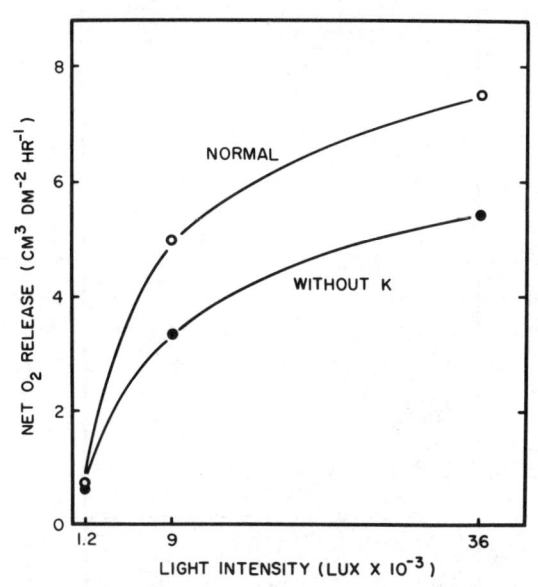

Fig. 3—Influence of potassium and light intensity on net O_2 evolution of *Phaseolus vulgaris*. Measurements made on leaflets of first pinate leaves 32 and 35 days after planting; 5% CO_2, 15-19C. (From Table V, Briggs, 1922).

intervals following seeding using the youngest fully-expanded leaves from plants exposed to three potassium treatments. The data are quite scattered which is not surprising since tissue age was somewhat variable. Redistribution of potassium was clearly occurring (cf. Gregory and Richards, 1929), and measurements of both parameters were rather inexact. Nevertheless, the general tendency was for positive relationship between leaf potassium and CO_2 uptake for K_2O values less than about 1%.

Current methodologies have established more precise relationships between leaf potassium concentration and CO_2 uptake rates for corn (Peaslee and Moss, 1966) and alfalfa (Cooper et al., 1967) at high light intensities, although even these studies show rather substantial deviations from the general relationship. In both reports it nonetheless is clear that photosynthesis was restricted when leaves contained less than 1.5 or 2% potassium.

Eckstein's (1939) experiments with wheat revealed three additional features of importance: (i) the detrimental influence of potassium deficiency was evident before degradation of chlorophyll was apparent, (ii) transpiration as well as CO_2 uptake was affected, and (iii) the beneficial action of added potassium depended on the supply of other nutrients. Figures 5A and 5B show CO_2 uptake and transpiration of excised leaves of 25-day-old wheat plants. The three lower potassium treatments resulted in a depression in the chlorophyll content only in the first leaves. Although photosynthesis and transpiration were similarly affected by potassium supply, a direct relationship could not be established because the second leaf took up CO_2 at a

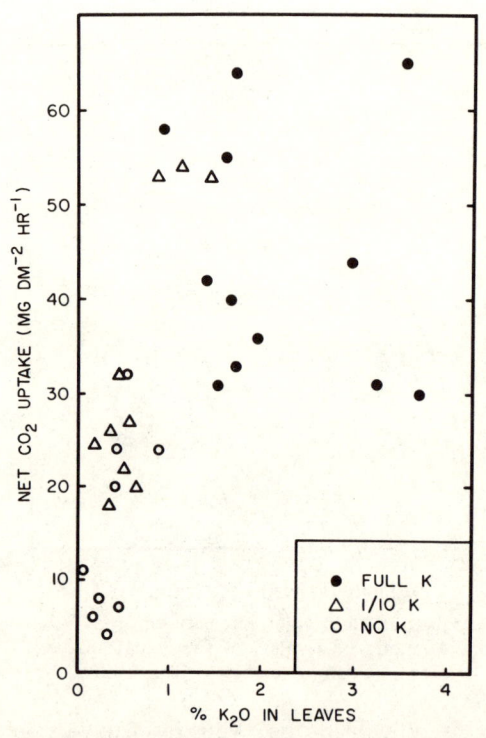

Fig. 4—Relationship between potassium concentration in leaves and CO_2 uptake by young leaves of barley. Measurements were with excess CO_2 at 15,250 metre-candles and 24C. Symbols refer to measurements made at weekly intervals. Sand culture, plants grown outside (From Richards, 1932).

greater rate than the first leaf, but transpired less. Lower transpiration rates in potassium depleted plants had been shown previously by Snow (1936). In an examination of stomata on both upper and lower surfaces Eckstein (1939) was unable to detect differences in opening due to the potassium treatments. We shall see that more recent work has shown an influence of potassium on guard cell behavior (Fujino 1959a,b,c; 1967) and stomatal openings (Cooper et al., 1967).

Figure 5C presents the third significant observation by Eckstein (1939). The nature of the photosynthetic response to applied potassium was substantially influenced by the nutritional status of the plants. When the potassium supply was low, the higher rate of nitrogen and phosphorus was detrimental to photosynthesis (Fig. 5C). Maximal photosynthetic rates with low nitrogen and phosphorus were obtained at 25 mg K_2O/pot with depressions at either side. But 100 mg K_2O/pot was required for maximal activity when the nitrogen and phosporus levels were high, and under these conditions there was no evidence of detrimental effects at even higher potassium applications. Alten et al. (1937) had shown previously that under limiting nitrogen supply there was first an increase in photosynthesis as the potassium

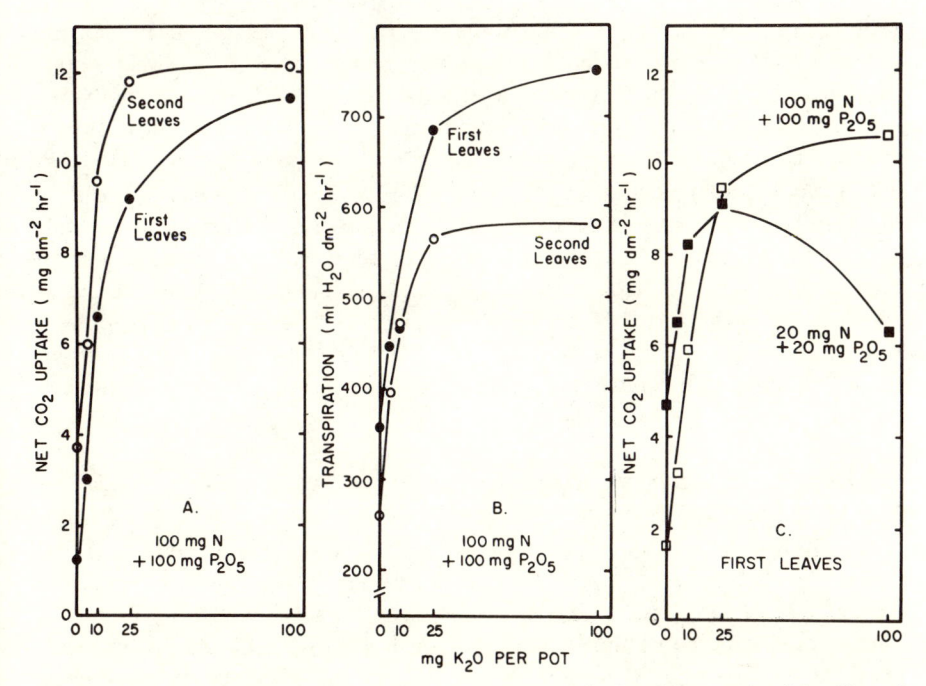

Fig. 5—Influence of potassium supply on photosynthetic CO_2 uptake (A, C) and transpiration (B) by first and second excised leaves from 25-day-old wheat plants. (From Eckstein, 1939).

supply was increased, but at progressively higher rates of potassium the photosynthetic rates were depressed. When the nitrogen supply was high, the depression at high potassium was not observed.

By 1940, therefore, a number of facts regarding the relationship between photosynthesis and potassium nutrition had been established. Photosynthesis was restricted under potassium deficient conditions even at non-saturating light intensities (Briggs, 1922; Gregory and Richards, 1929); a gross correlation was evident between potassium content and photosynthetic activity (Richards, 1932); the response to potassium was influenced by the supply of other nutrient elements (Alten et al., 1937; Eckstein, 1939); and transpiration was restricted in a manner similar to the restriction in CO_2 uptake (Eckstein, 1939). Since that time a number of other studies have expanded these original observations.

B. Recent Concepts of the Role of Potassium in Photosynthesis and Photorespiration

Since the original investigations in the 1920's and 1930's the beneficial effects of potassium on photosynthesis have been shown with a wide range of higher plants including tung (Loustalot et al., 1950); *Lemna minor*

(Bierhuizen, 1954), spinach and tomatoes (Schmidt, 1959), bean (Ozbun et al., 1965a, 1965b), rice (Noguchi and Sugawara, 1966), sweet potato (Tsuno and Fujise, 1965), corn (Moss and Peaslee, 1965; Peaslee and Moss, 1966), alfalfa (Cooper et al., 1967) and sugar cane (Hartt and Burr, 1967). Moreover, the influence of potassium has also been clearly shown with various algae including *Chlorella* (Pirson, 1939), *Ankistrodesmus* (Pirson et al., 1952), *Hydrodactyon* (Neeb, 1952) and *Nostoc* (Clendenning et al., 1956). It therefore seems clear that the requirement of potassium for normal photosynthesis is a universal one, but the manner in which the potassium influence is exerted is not nearly so evident. In this section we present various suggestions which have been advanced and also point out the regulating influence other nutritional factors or the environment exert on the potassium effect.

1. *Primary and Secondary Effects*

By culturing *Ankistrodesmus* for varying periods of time in potassium-free media and then measuring photosynthesis with and without added potassium Pirson et al., (1952) were able to show a remarkable dual effect (Table 2). Under moderate deficiencies, the depressed photosynthetic rates returned nearly to normal upon the addition of potassium to the medium; substantial restoration was effected within one-half hour. However, when potassium deficiency was more intense, only a slight recovery was observed. It was also shown that rubidium was effective in the short time recovery phenomenon although sodium was not (Pirson, 1939). The long term growth of algae (Kellner, 1955, Pirson and Kellner, 1952) and higher plants (Richards, 1941) is, of course, never as good when rubidium is substituted for potassium. The rapidity with which photosynthesis was stimulated by potassium or rubidium additions under moderate potassium deficiency mitigates against an effect on alteration of protein or chlorophyll synthesis. Pirson et al. (1952) suggested that it was a result of an enzyme activation phenomenon (cf. Evans and Sorger, 1966). That an additional effect of potassium was exerted was evident from the failure of stimulation to occur within reasonably short periods un-

Table 2—Recovery of photosynthesis in K− deficient *Ankistrodesmus* upon addition of K (values taken from Fig. 4, Pirson et al., 1952)*

Hours after adding K	Days of potassium deficiency		
	9	11	14
	—— mm^3 O$_2$ evolved/hr/mg dry wt. ——		
0	58	53	40
0.5	110	58	--
1	117	62	38
2	122	68	41
3	132	67	40

* Rates with normal cultures were around 160 mm^3 O$_2$ evolved/hr/mg dry wt.

der severe potassium-deficiency (Table 2). Under these more extreme conditions, failure of specific protein or chlorophyll maintenance must be invoked. The data strongly suggest two separate influences of potassium nutrition on the photosynthetic process. One was specifically due to potassium and was exerted through an alteration in general protein metabolism while the other was due to an effect of potassium or rubidium ions on the activity of the system.

That the two effects may not be so clearly separated is indicated by the report of Clendenning, et al. (1956) who noted that blue green algae rapidly lost photosynthetic ability when washed with potassium-free solutions. Adding potassium to the wash solution prevented most of the damage but, once initiated, the washing injury could not be repaired by potassium addition. However, it was not shown that the tissue had the ability to reabsorb potassium. No pigment changes were involved and capacity of the tissue to photoreduce quinone was not influenced, suggesting an effect on the CO_2-fixing reactions rather than on photosynthetic electron transport. *Chlorella* was not damaged by the washing procedure and the evidence suggests that the damage was of a different sort than Pirson et al. (1952) had noted with *Chlorella* and *Ankistrodesmus*. In any event the specific way in which the lack of potassium influences photosynthetic events in algae still remains obscure.

It is of course experimentally more difficult to undertake critical depletion and resupply studies with leaves of higher plants. Nevertheless, Peaslee and Moss (1966) have reported recovery of photosynthesis by potassium-deficient corn by placing excised leaves in KNO_3 solutions. Leaf porosity, as measured by a porometer, was less in the potassium deficient leaves, and also increased upon exposure to the KNO_3 solutions. The parallel stimulation of photosynthesis and leaf porosity strongly suggests that a reversible stomatal closure occurred in the potassium-deficient leaves (cf. Cooper et al. 1967). Using bean leaves, Ozbun (Ozbun, J. L., R. J. Volk, and W. A. Jackson, 1963. *Unpublished data. North Carolina Agr. Exp. Sta.*) was unable to induce recovery of impaired photosynthesis by adding KCl via the petiole, but the experiments were carried out at lower light intensities and at much higher CO_2 concentrations than used by Peaslee and Moss (1966). The reversible effect on leaf porosity is of course an effect quite different from the short term recovery observed by Pirson et al. (1952) with algae. The possible involvement of potassium movement into and out of the guard cells in stomatal opening and closing will be discussed further in section II B 5.

2. *Influence of Other Nutrients*

The combined effects of nitrogen and potassium on net photosynthesis are evident from experiments on mature leaves of sweet potato grown under a variety of nutritional treatments (Tsuno and Fujise, 1965). Photosynthesis increased with nitrogen concentration of the leaves in the range from 2 to 3%, but the leaves clearly could be separated into two groups (Fig. 6). Those leaves with more than 4% K_2O had substantially greater photosynthetic rates than those with lower concentrations. Even at quite high nitrogen concentrations, there was no indication of photosynthesis falling off, which is in contrast to the work of Eckstein (1939) presented in Fig. 5.

We have mentioned that rubidium could substitute for potassium in the short term recovery of photosynthesis in moderately potassium-deficient algae. However, it was not effective in overcoming the secondary and more deep seated effects caused by prolonged potassium absence during growth (Pirson et al., 1952). In those studies, sodium was not effective in replacing potassium for the direct recovery of photosynthesis. In the "washing damage" reported by Clendenning et al. (1956) with *Nostoc*, sodium in fact accentuated the decline in photosynthesis by increasing the loss of potassium from the tissue. *Lemna minor* grown with sodium in the absence of potassium was limited in photosynthesis over a wide range of light intensities, and adding potassium exerted only a slight beneficial effect (Bierhuizen, 1954). Nevertheless Schmidt (1959) reported that spinach and tomato plants grown with sodium but without potassium showed fully as much photosynthetic capacity per unit leaf area as plants grown with potassium, even though the growth of the former was restricted. Likewise in halophytes, where NaCl is as effective as KCl in growth, the photosynthetic rates were not different (Baumeister and Schmidt, 1962). It was concluded that sodium could replace potassi-

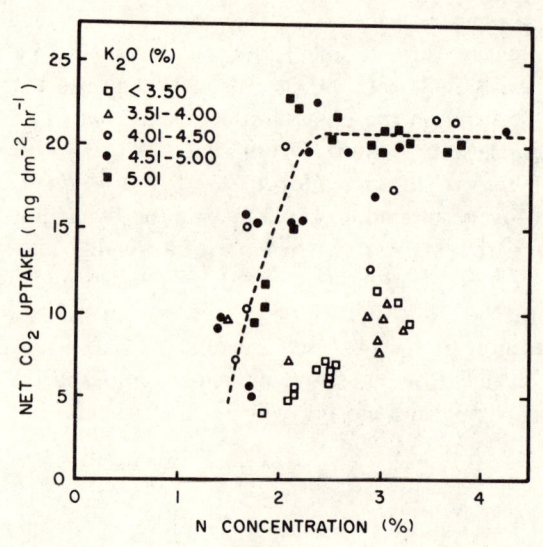

Fig. 6—Net CO_2 uptake by intact mature leaves of sweet potato plants grown with various nutritional conditions. Light intensity of 30 Klux, 0.03% CO_2 (From Tsuno and Fujise, 1965).

um in CO_2 assimilation processes of higher plants, but that it could not assume the role of potassium in protein synthesis. However, in corn (a plant not studied by Baumeister and Schmidt) Peaslee and Moss (1966) reported that the relationship between photosynthesis and leaf potassium concentration was not altered by inclusion of sodium in the nutrient medium. Here again we see a nutritional influence whose pattern of response is not all clear. Further descriptive investigation is required to delineate the extent to which the effect of potassium on photosynthetic and photorespiratory events is modified by other nutritional influences.

3. Light Intensity

The fact that potassium deficiency restricts net photosynthesis at light-limiting as well as light-saturating conditions is well established (Briggs, 1922; Gregory and Richards, 1929; Bierhuizen, 1954; Noguchi and Sugawara, 1966). A part of the restriction at low light intensities is due to increased photorespiration in potassium deficient leaves as revealed by Ozbun et al. (1965a, b). They employed isotopic procedures which permitted separation of the net gas exchange rates into the uptake and release components. The studies were conducted at 2-4% O_2 with nonlimiting CO_2 concentrations, conditions which would minimize photorespiration. Multiple effects of inadequate potassium nutrition were revealed. When the potassium supply was curtailed while the leaves were still expanding, net O_2 release and net CO_2 uptake were both repressed. The manner in which the depressions in net rates came about were a function both of the light intensity and of the state of development of potassium deficiency. Figures 7 and 8 present the gas exchange rates at 300 ft-c of first trifoliate leaves of 27-day-old bean plants, the potassium deficient leaves having been deprived of potassium fifteen days. By this time chlorophyll degradation had occurred and the concentration of K in the deficient leaf was down to 0.12%. Both net O_2 release (Fig. 7) and net CO_2 uptake (Fig. 8) were depressed in the potassium-deficient leaf. The latter resulted from a concomitant depression of CO_2 uptake and stimulation of CO_2 release. The decrease in net O_2 release also resulted from effects on both uptake and release of O_2. The oxygen uptake rate was essentially doubled by the potassium deficiency while O_2 release was depressed about 37%. At a higher light intensity (1,500 ft-c) net O_2 release and net CO_2 uptake were also curtailed in the potassium deficient leaf (Ozbun et al., 1965a). This was largely a consequence of O_2 release and CO_2 uptake being depressed because neither O_2 uptake nor CO_2 release was affected significantly by the increase in light intensity from 300 to 1,500 ft-c.

Substantial effects of limited potassium supply on photosynthesis and photorespiration were also noted prior to the appearance of visual abnormality of the leaf tissue. Table 3 shows results for leaves 10 days after potassium

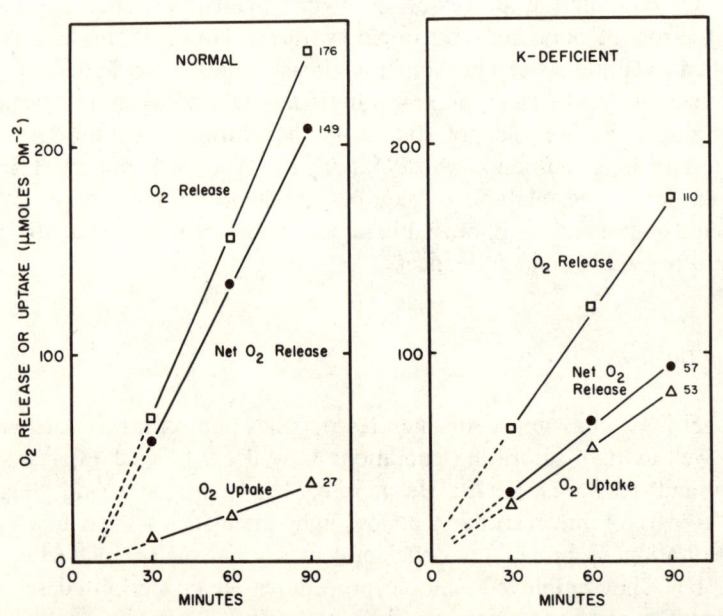

Fig. 7—Net O_2 release, and separation of the rate into O_2 release and O_2 uptake at 300 ft-c in normal and potassium deficient first trifoliate leaves of 27-day-old bean plants. The plant with the potassium deficient leaf had not received potassium since the 12th day. (From Table 2, Ozbun et al., 1965a).

removal. At 300 ft-c, net O_2 release and CO_2 uptake were depressed in the potassium-deficient leaf. In contrast to the severely deficient leaves, however, the effect on net O_2 release was entirely due to the much higher O_2 uptake rate, for the capacity to carry out those processes leading to O_2 release was in fact slightly greater with the deficient leaf. On the other hand, both an impairment in CO_2 uptake and an enhancement in CO_2 release were responsible for the depressed net CO_2 uptake due to potassium deficiency. At 1,500 ft-c, lower O_2 release and increased O_2 uptake both contributed to the depressed net O_2 uptake. As at 300 ft-c, lower CO_2 uptake and enhanced CO_2 release contributed to the lower net CO_2 uptake.

Table 3 also shows the rates of O_2 release and O_2 uptake at 1,500 ft-c after the CO_2 concentration had been depleted to the compensation level. Here it is seen that O_2 uptake rates increased greatly compared to their rates when abundant CO_2 was present. Moreover, the O_2 uptake rate of the potassium deficient leaf nearly equaled the O_2 release rate, whereas in the normal leaf O_2 uptake was considerably less than O_2 release. Photorespiration was therefore much larger at limiting CO_2 concentrations, and was accelerated in the potassium deficient leaf. It is instructive to note that O_2 release at 1,500 ft-c with compensation CO_2 concentrations was greater than O_2 release at 300 ft-c when CO_2 was being taken up. These data emphasize the magnitude of O_2 exchange phenomona in leaf tissue and the influence of po-

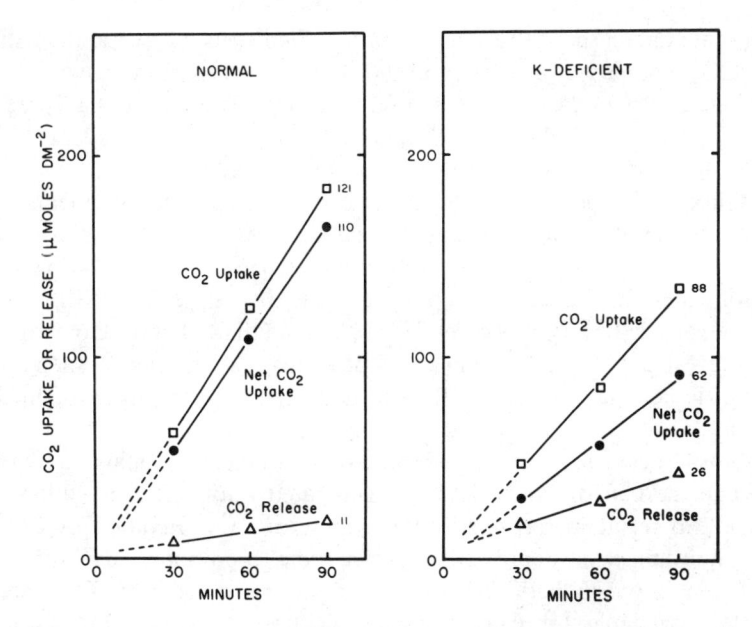

Fig. 8—Net CO_2 uptake, and separation of the rate into CO_2 uptake and CO_2 release, at 300 ft-c in normal and K deficient first trifoliate leaves of 27-day-old bean plants. Experimental conditions as in Fig. 7.

tassium deficiency in increasing the O_2 exchange. It also is of interest that the O_2 uptake rate of the potassium deficient leaf at 300 ft-c actually was greater than the rate of CO_2 uptake (Table 3). This did not occur in the normal leaf where more than twice as much CO_2 as O_2 was taken up.

The enhanced O_2 uptake and CO_2 release rates of illuminated potassium deficient leaves (Ozbun et al. 1965 a, 1965 b) show that photorespiratory

Table 3—Influence of K nutrition on photosynthesis and photorespiration of the first trifoliate leaves of pole beans 22 days after seeding (Ozbun et al., 1965a, Table 2)*

	300 ft-c†		1500 ft-c†		1500 ft-c, Compensation CO_2‡	
	Normal	-K	Normal	-K	Normal	-K
	—————— μmoles dm^{-2} hr^{-1} ——————					
O_2 Release	169	182	793	745	217	217
O_2 Uptake	42	91	42	91	152	197
Net O_2 Release	127	91	751	654	65	20
CO_2 Uptake	91	73	564	515		
CO_2 Release	11	29	11	29		
Net CO_2 Uptake	80	44	553	486		

* The potassium supply was removed from the -K plants on the 12th day.
† Measurements at 1-2% CO_2.
‡ Measurements at 0.01% CO_2.

processes are increased substantially under conditions of potassium stress. Higher CO_2 compensation concentrations with potassium-deficient alfalfa (Cooper et al. 1967) lead to the same conclusion. The chloroplast preparations of Fujiwara and Iida (1967) absorbed O_2 and also evolved previously fixed $^{14}CO_2$ in darkness. Although both processes were increased in the preparations from potassium deficient leaves, these measurements may not reflect the photorespiratory activity of intact tissue. The fact that photorespiration has a number of characteristics strikingly different than dark respiration, and that these are rapidly evident during light-dark transitions (Ozbun, et al., 1965; Forrester et al., 1966a; Zetlich, 1966; Goldsworthy, 1967), suggest that the enhancement of photorespiration, due to shortage of potassium, is not via the same influence that results in the increase in dark respiration.

In photosynthesis, it usually is assumed that sufficient reducing equivalents (reduced ferrodoxin or $NADPH_2$) are generated during the evolution of 1 mole of O_2 to result in the fixation of 1 mole of CO_2. Fixation of CO_2 by isolated chloroplasts is depressed under aerobic conditions, suggesting that some of the photosynthetic reductant can be reoxidized by O_2, thereby leaving less available for CO_2 reduction (Gibbs et al., 1967). Since, in the studies of Ozbun et al. (1964, 1965a, 1965b), there was not a stoichiometric quantity of CO_2 taken up for each mole of O_2 evolved, some of the reducing equivalents must have been used to reduce oxidants other than CO_2. For the normal leaf at 300 ft-c in Fig. 7 and 8, the utilization of photosynthetic reductant power was 55% for CO_2 uptake and 45% for other oxidants, one of which no doubt was O_2. With the potassium deficient leaf, the proportion was 40% for CO_2 and 60% for other oxidants, and a much larger proportion of the latter was O_2. It seems clear that a major influence of potassium deficiency under these conditions was to diminish the capacity of the tissues to fix CO_2 and simultaneously to enhance the capacity to take up O_2. Ozbun et al. (1965a) have advanced three postulates to explain the increased O_2 uptake. These are (i) a restriction in the photosynthetic phosphorylation mechanism, (ii) an increase in ATPase activity, or (iii) an increased use of ATP relative to reducing equivalents for reactions other than CO_2 reduction. All three possibilities involve the idea that the relative amount of ATP is low compared to the amounts of reducing equivalents generated so that the ATP requirement for CO_2 fixation is insufficient. The excess reductant is thus free to reduce other oxidants (e.g., O_2) which in normal leaves are generally less able to compete with CO_2. Because potassium deficiency restricts protein synthesis and alters its structural configurations (Evans and Sorger, 1966), one would expect that anabolic reactions which require photosynthetic reductant would not be able to function as well in potassium deficient tissue. Thus the use of the photosynthetically generated components ($NADPH_2$ and ATP) for reactions other than CO_2 fixation would be expected to be less, and more ready reduction of O_2 may be the consequence if ATP synthesis is less than $NADPH_2$ synthesis.

Although O_2 release was impaired in the moderately deficient leaves at 1,500 ft-c, (Table 3), accelerated O_2 uptake and depressed CO_2 uptake resulting from the lack of potassium also were noted. Thus the general tendency for altered use of photosynthetic reductant as discussed in the preceeding paragraph still prevailed prior to chlorophyll degradation. When plants remained on potassium deficient media until chlorophyll degradation was advanced, O_2 evolution was impaired even at 300 ft-c (Fig. 7). Nevertheless, increased O_2 uptake as well as depressed CO_2 uptake (Fig. 8) were characteristic of these tissues as well.

It must be emphasized that the foregoing comments relate to experimental observations in which the plants were deprived of potassium while the leaves, subsequently used for the photosynthetic measurements, were still expanding. Other experiments, in which the potassium removal was delayed until the first trifoliates reached complete enlargement, revealed that the potassium deficient leaves were not materially affected, either in O_2 release or CO_2 uptake, until chlorophyll loss was advanced. Illuminated, mature leaves had lower O_2 uptake rates than young expanding leaves (Ozbun et al., 1964) and a modest increase was noted after five days without potassium. The effect was not large, however, and photosynthesis was not materially affected even at 14 days after potassium removal. By this time, a threefold increase in soluble carbohydrates and marked changes in amounts and proportions of soluble nitrogenous constituents of the leaves indicated substantial alterations in other metabolic events.

Moss and Peaslee (1965) measured photosynthesis of corn leaves of various ages in potassium depleted plants. A definite relationship was found between potassium concentration in the leaf and net CO_2 uptake at 9,000 ft-c regardless of leaf age. When plants were supplied varying amounts of potassium and measurements were made on the topmost leaf, the data fell on the same curve (Peaslee and Moss, 1966). The authors thus concluded that the potassium concentration of the leaf, rather than leaf age, regulated the photosynthetic rates. Their studies were conducted at normal atmospheric CO_2 and O_2 concentrations at which leaf porosity and stomatal apertures may be regulatory.

4. Subcellular Effects

The suggestion presented by Ozbun et al., (1965a) that potassium-deficient leaves have a wider ratio of photosynthetic reductant to ATP than normal leaves is supported by the studies of Latzko and Mechsner (1958) and Mechsner (1959) with potassium-depleted *Chlorella*. These investigators followed the changes in inorganic phosphate immediately after transferring from darkness to light. When potassium was added to the cells, sub-

Table 4—Influence of K on light-induced photophosphorylation in K—depleted *Chlorella* (calculated from Fig. 3, Latzko and Mechsner, 1958)

Time	Decrease in inorganic phosphate*	
	−KCl	$+10^{-4}M$ KCl
sec	μg PO_4/mg dry wt.	
5	1.44	5.40
25	2.16	5.40
35	--	4.32
50	1.44	3.60
65	0	5.40
125	1.80	6.48
305	2.16	5.04

* Indicates the amount of inorganic phosphorus removed from tissue and solution after the lights were turned on.

stantial inorganic phosphorus was converted to organic phosphorus upon exposure to illumination (Table 4). The conversion did not occur in linear fashion with time but rather a biphasic pattern was noted. When potassium was not added, the cells were unable to convert inorganic to organic phosphorus as rapidly although the biphasic pattern was retained. A phosphorus fraction containing nucleotides showed changes inversely related to those of inorganic phosphorus, confirming that phosphorylation was responsible for the observations (Mechsner, 1959). A rather small amount of potassium ($10^{-5}M$) was adequate for complete recovery, and rubidium was about 80% as effective as potassium. Other monovalent cations were without effect. Failure to observe these patterns in the presence of a Hill reaction inhibitor, and the lack of any influence of inhibitors presumed to influence only the CO_2 fixation or oxidative phosphorylation reactions, seems to establish that photophosphorylation was impaired in the absence of potassium (Mechsner, 1959).

Other studies confirm the concept that the photophosphorylation reactions are more sensitive to lack of potassium than are the reactions involved in electron transport and generation of photosynthetic reductant. Experiments with isolated chloroplasts have indicated that potassium does not participate directly in the Hill reaction (Gorham and Clendenning, 1952). However, Spencer and Possingham (1960) showed that the Hill reaction of chloroplasts isolated from potassium deficient tomato leaves was less than those from control plants. In most instances the inhibition was not as evident at low light as at high light intensities. Adding potassium to the deficient chloroplasts was of no benefit. The data thus indicate that the site of the effect was a dark reaction in the sequence of electron transport to the indophenol dye employed in the assay.

The organization of the photosynthetic system is disrupted as chlorophyllous tissue becomes potassium deficient, and electron miscroscopic studies indicate a number of similarities between chloroplasts from moderately potassium-deficient and phosphorus-deficient leaves (Thomson and Weier, 1962). This observation supports the concept that potassium has an important influence upon phosphorus metabolism. Thomson and Weier (1962) also noted that chloroplast structure was modified in nearly fully developed plastids as the potassium migrated to newer tissue. It would be expected that a considerable degree of macro-molecular organization is required for the proper functioning of the total photosynthetic process. Nevertheless, Ozbun et al. (1965b) have shown that once leaves are fully expanded under adequate potassium nutrition, they can suffer a remarkable depletion of potassium and still carry out both the O_2 evolving and the CO_2 reducing reactions very effectively. These data together with those of Thomson and Weier (1962) suggest that some structural alterations in chloroplasts can take place in intact potassium deficient leaves without untoward effects on photosynthetic functions.

Chloroplasts normally contain a considerable proportion of the total potassium in leaves. Although the original observations of Neish (1939) indicated that potassium was concentrated outside the chloroplasts, non-aqueous extraction techniques (which depress the loss of soluble components during the extraction) clearly reveal that a substantial amount of potassium is present in the chloroplasts. Stocking and Ongun (1962) report that about half of the total leaf potassium was found in the chloroplasts of tobacco and bean when non-aqueous extraction procedures were used. Saltman et al. (1963) also noted that the potassium concentration in chloroplasts of *Nitella opaca* was slightly higher than in the surrounding cytoplasm. Nevertheless, the distribution of potassium between chloroplasts and cytoplasm probably does not remain constant. A substantial efflux of potassium from isolated chloroplasts can occur in the light (Dilley, 1964; Dilley and Vernon 1965). This efflux and light-induced chloroplast shrinkage are closely related kinetically, and both reactions are coupled to photosynthetic electron transport reaction. It was further suggested that inward proton movement in light was partially countered by potassium efflux. However, Crofts et al. (1967) have shown that the fluxes of potassium into and out of isolated chloroplasts are highly complex, the direction, kinetics, and magnitude being strongly influenced by experimental conditions. The situation obtaining in the intact leaf remains unknown, but the fact that older leaves rapidly become depleted of potassium upon removal from the growth medium implies that considerable mobility of potassium into and out of chloroplasts is possible. This mobility probably is a first without severe deleterious effects on photosynthesis (cf. Ozbun et al. 1965a, b).

5. Carbon-14 Labeling Patterns

Determining the kinetics of incorporation of ^{14}C into metabolic intermediates has been enormously successful in characterizing the sequence of events in the CO_2 reduction phase of photosynthesis. It therefore is surprising to note that this technique has seldom been used to delineate the ways in which photosynthesis is modified by potassium nutrition. That significant metabolic alterations in the fate of the administered $^{14}CO_2$ do take place under potassium deficiency is shown by experiments with tomato plants (Jones, 1966). After 5 min exposure in the light, the potassium-deficient leaves contained substantially less of the fixed carbon in the ethanol-insoluble fraction than did the normal leaves. There was less incorporation of ^{14}C into glucose and fructose in the potassium deficient leaves, and the specific activity of the starch in these leaves was less than half that in normal leaves. During the five-minute period, ^{14}C accumulated in 3-phosphoglyceric acid but little was found in the phosphoenolpyruvate or glyceric acid of deficient leaves. The normal leaves, in contrast, contained substantially more label in phosphoenolpyruvate and in pyruvate as well. Although much further experimentation is required to delineate the pathways affected, these data reveal that the metabolic fate of the reduced carbon can be significantly altered by potassium deficiency.

6. Stomatal Regulation

Since the concentration of CO_2 in the atmosphere surrounding the leaves is rather low (.03%), diffusion into the leaves may be quite sensitive to the size of the stomatal openings, especially where turbulence is low. It seems quite evident that potassium nutrition may be of considerable importance in regulating the actions of guard cells. Measurements of net water loss per unit leaf area usually reveal that potassium-deficient plants transpire less than control plants. This has been observed with sunflower, tobacco and bean (Snow, 1936) and wheat plants (Eckstein, 1939; cf. Fig. 5B) although Noguchi and Sugawara (1966) report an exception with rice. There is evidence (Snow, 1936) to indicate that sodium may be partially effective in off-setting the lower transpiration caused by lack of potassium. More recently, Peaslee and Moss (1966) observed a similar depression of transpiration in potassium-deficient corn leaves. When the cut leaves were exposed to solutions of KNO_3, leaf porosity increased slowly, finally approaching that of the control leaves. Snow (1936), however, noted that when potassium was resupplied to deficient, intact beans, the recovery in transpiration did not take place until 5 days had passed. Under the light saturating, low CO_2 conditions employed by Peaslee and Moss (1966), there was a clear relationship between net CO_2 uptake and leaf porosity. Moreover, Cooper et al. (1967) found that the stomatal openings of potassium deficient alfalfa leaves usually were much smaller than those of normal leaves. It therefore appears that the influence of potassium on photosynthesis under natural conditions may in part be exerted through a regulation of stomatal openings.

In 1959 Fujino proposed a mechanism to account for these observations. It was noted that the stomata of epidermal leaf strips placed in water failed to open whereas those placed in solutions of KCl opened readily (Fujino, 1959a, b). Histochemical studies revealed that the degree of opening was related to the potassium content of the guard cells (Fujino, 1959c). The effects of metabolic restrictions of various sorts lead to the suggestion (Fujino, 1959c) that active accumulation of potassium into the guard cells in light resulted in increased osmotic values thereby inducing swelling with consequent opening of stomata. Subsequent studies (Fujino, 1967) further confirmed the relationship between stomatal opening and migration of potassium into and out of guard cells and, moreover, a mechanism was proposed to account for the potassium movement. Immersing leaf strips in solutions of ATP increased the potassium content of guard cells and increased the rate of stomatal opening. They remained open and the guard cells maintained their potassium when kept in the light. Thus the ATP (which in natural conditions would be produced in photophosphorylation and perhaps to a lesser extent by respiratory phosphorylation) was suggested to be responsible for potassium accumulation in the light by the guard cells. Open stomata of epidermal strips in ATP solutions closed when the strips were placed in darkness and the guard cells simultaneously lost potassium. It was further noted that an active guard cell ATPase developed in darkness. Inhibition of this enzyme by p-chloromercuric benzoate increased potassium accumulation in guard cells and accelerated stomatal opening in both light and darkness. The suggestion therefore was made that ATPase activity was involved in pumping potassium from the guard cells. This would result in lowering of the osmotic value of the guard cells and stomatal closure. It remains to be worked out how the influence of CO_2 levels and other regulators of stomatal behavior can be mechanistically related to the intriguing concept developed by Fujino. It is perhaps sufficient at present to note that high CO_2 levels lead to stomatal closure and it would be important to know if ATPase is stimulated and potassium lost from guard cells under these conditions. Fujino's (1967) scheme provides an explanation for the lack of stomatal opening during potassium deficiency because of the small amount of potassium present in the mesophyll cells. This, together with the possibility of lowered photophosphorylation abilities of potassium deficient tissue (Latzko and Meschner, 1958), would lead to failure of accumulation of potassium in the guard cells and restricted stomatal opening. Under conditions where potassium is resupplied and a fair amount reaches the mesophyll cells adjacent to the guard cells, stomatal opening would be expected to occur again, provided that the guard cells retained ability to synthesize ATP.

7. *End Product Removal*

A common feature of potassium deficient leaves is the accumulation of soluble sugars and a lowered concentration of the more complex carbohydrates (Evans and Sorger, 1966). Direct $^{14}CO_2$ labeling experiments indicate that translocation of recent photosynthetic products from potassium-deficient leaves is impaired even more than the rate of CO_2 fixation (Baver et al., 1964). Since it is conceivable that an accumulation of sugars in leaves may restrict photosynthesis (Hartt and Burr, 1967), translocation from the leaf may exert an internal regulatory influence on the photosynthetic process. Investigations with sweet potatoes revealed that cultural conditions which decreased the size of the sink (the enlarged roots in this case) serving as a depository for the translocated carbohydrate decreased the leaf photosynthetic rates (Tsuno and Fujise, 1965). Low potassium treatments were very effective in this regard, leading to the postulate that the initial influence of potassium deficiency was to restrict root enlargement. The decrease in sink size limited translocation of carbohydrates from the leaves thereby lowering the photosynthetic rate. Although impaired translocation is no doubt an important consequence of potassium deficiency, stomatal regulation, gas exchange, and distribution of applied ^{14}C are also modified as mentioned previously. Thus, a large number of metabolic processes are affected by potassium deficiency and the dominant influence thereof is undoubtedly related to the environmental conditions under which the deficiency develops.

IV. RESPIRATION

The detrimental influence exerted by potassium deficiency on photosynthesis and the concomitant enhancement of photorespiration is paralled during the initial deficiency stages by an increased dark respiration rate in both chlorophyllous and non-chlorophyllous tissue. Moreover, the respiratory pattern is also modified. The multiplicity of metabolic reactions which require potassium for maximal activity (Evans and Sorger, 1966), prevents assigning a specific cause for the stimulation of respiration induced by potassium deficiency. Nevertheless, some of the physiological responses are clear and they seem to suggest the crucial involvement of potassium in the energy conserving reactions of respiration.

A. Historical Perspective

An increase in the respiratory rate of leaves of moderately potassium-deficient barley was shown by Gregory and Richards (1929). Subsequent experiments revealed that severe deficiency conditions often depressed respiration (Richards, 1932). As the severity of the deficiency increased, the respiration rate usually passed through a maximum. This trend generally held for all fully expanded leaves after the third, although the magnitude of the increase and subsequent decline showed a substantial degree of variation (Fig. 9). The trends were confirmed in further experiments by Gregory and Sen (1937) who suggested that the low respiratory rates under severe deficiency

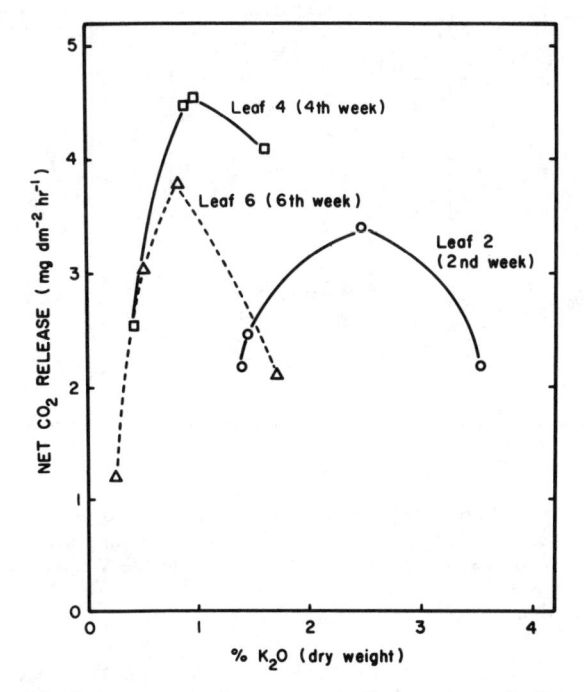

Fig. 9—Relationship between potassium concentration and respiration rate of barley leaves. Measurements made on each leaf as it reached full expansion after seeding. (From Richards, 1932).

were a consequence of lack of oxidizable substrate. These authors quote experiments by Said who apparently demonstrated that adding sugar to the severely potassium-deficient leaves gave an immediate increase in respiration whereas adding potassium without sugar was without effect.

B. Recent Concepts of the Role of Potassium in Respiration

The general pattern of respiration described above has subsequently been noted in leaves of other species (Fujiwara and Iida, 1967; Yamashita and Fujiwara, 1966). Multiple consequences of potassium deficiency are evident also in root tissue. Figure 10 shows respiratory rates for the 2nd cm sections from root tips of bean plants at various times after removal of potassium from the nutrient solution. Within 5 days after potassium removal, O_2 uptake per unit fresh weight was increased substantially, although fresh weight of the segments was not yet affected. By the 12th day there was no difference between the segments from potassium-deficient and control plants when expressed per unit weight, but the potassium-deficient segments had a lower respiratory rate when expressed per unit of nitrogen.

The majority of studies have been concerned with the enhanced respiration which occurs under moderate potassium-deficiency. This enhancement has

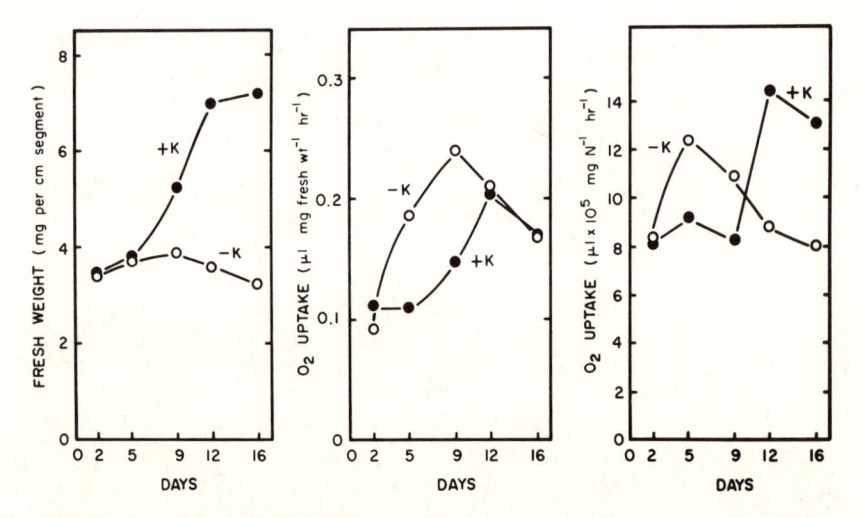

Fig. 10—Fresh weight and O_2 uptake of the second 1-cm segments of bean roots. At 14 days after germination (day zero on the abscissa), one series of plants (-K) were transferred to solutions without potassium while the second series (+K) continue to receive K. (Volk, R. J. and W. A. Jackson, Unpublishd data, North Carolina Agr. Exp. Sta., 1966).

been observed in tissues from all parts of higher plants as well as in algae. It, rather than the depressed rate due to severe deficiency, is the more normal influence of inadequate potassium under natural conditions and will have a strong influence on the net carbon balance of plants. Most of the following discussion is therefore related to the stimulation in respiration induced by moderate potassium deficiency.

1. Respiratory Control

Regulation of overall respiration at the cellular level is enormously complex. Evans and Sorger (1966) have listed in detail enzymes and enzyme systems, the activities of which *in vitro* are affected by potassium. These include enzymes in glycolysis, starch synthesis and protein synthesis, as well as in phosphorylation and respiration. The respiratory control mechanisms for intact tissue may consequently be delineated only in a general way. A supply of respiratory substrate at the reaction loci is of course essential. The extent to which oxidative phosphorylation is coupled to electron transport, the amounts and proportions of ADP, ATP, and inorganic phosphate, and the relative concentrations of the oxidized and reduced forms of NAD and NADP all may exert strong regulatory effects. The role of these key transient metabolites may be modified by the extent and activity of shuttle systems which influence their subcellular distribution as for example between cytoplasm and mitochondria, and between mitochondrial compartments (Lehninger, 1964). Complex fluxes of inorganic ions including potassium occur between mitochondria and cytoplasm (Rottenberg and Solomon, 1966) as well as between chloroplasts and cytoplasm (Dilley, 1966). The influx and efflux patterns are closely associated with metabolic activity of the organelles. The inorganic ion movements may in turn influence transport of organic substrate and product molecules into and out of the organelles (Harris et al., 1967) thereby affecting reaction rates in the different compartments. Regulation also may be imposed by the availability of electron acceptors such as O_2, or by presence and activity of enzymes permitting alternative electron acceptors, such as nitrate, to be reduced. Finally, respiratory substrate may be oxidized to various extents through the pentose phosphate pathway rather than by classical glycolysis and the tricarboxylic acid cycle, the reactions involved being considerably different in the two cases. Accordingly, a curtailed potassium supply may influence respiration in very many ways at the cellular level. In addition, some plant organs (e.g., roots) are dependent for their respiratory substrate on long distance transport processes which may themselves be affected by potassium deficiency. Many different manifestations of respiratory response to potassium deficient conditions have been observed. It is therefore not surprising that a specific metabolic role for potassium in respiration has not yet been clearly established.

2. CO_2 Release and O_2 Uptake

The accelerated respiration resulting from moderate potassium deficiency is characterized by O_2 uptake being increased more than CO_2 release. Thus, normal cultures of *Ankistrodesmus* growing in nitrate medium had a respiratory quotient (RQ = CO_2 Release/O_2 Uptake) of 1.38, while the moderately potassium deficient cultures with substantially higher respiration had an RQ of 1.13 (Table 5). First trifoliate leaves from bean plants that were deprived

Table 5—Respiration of moderately K–deficient *Ankistrodesmus* as influenced by glucose and KCl additions (from Table 3, Pirson et al., 1952)

	O_2 Uptake	CO_2 Release	Extra CO_2	RQ
	— mm^3 mg^{-1} hr^{-1} —			
I. Grown with Nitrate-Nitrogen				
Normal culture	2.4	3.3	0.9	1.38
Potassium-deficient	5.5	6.2	0.7	1.13
Potassium-deficient + KCl*	3.2	4.4	1.2	1.38
Normal Culture + Glucose	12.5	18.7	5.2	1.50
Potassium-deficient + Glucose	8.9	12.3	3.4	1.38
Potassium-deficient + Glucose + KCl	2.7	4.2	1.5	1.56
II. Grown with Ammonium-Nitrogen				
Normal culture	2.3	2.5	0.2	1.09
Potassium-deficient	5.9	4.8	−1.1	0.81
Potassium-deficient + KCl*	4.1	4.2	0.1	1.02

* Measured after 4 hrs exposure to KCl during which the rates were steadily declining to the reported values (cf. Fig. 12).

of potassium while the leaves were still expanding also showed decreased RQ values resulting from the more rapid increase in O_2 uptake than in CO_2 release (Fig. 11). In these experiments, CO_2 release was corrected for dark CO_2 uptake which was slightly larger in the potassium-deficient leaves. When algae were grown with ammonium nitrogen, O_2 uptake and CO_2 release values were more nearly equal than when grown with nitrate (Table 5; cf. Barker et al., 1965), but potassium deficiency still resulted in a greater stimulation in O_2 uptake than CO_2 release.

Respiratory quotients in excess of unity are interpreted as indicating partial utilization of respiratory reductant by oxidants other than molecular oxygen (e.g., nitrate). The low RQ values associated with potassium deficiency suggest that respiratory reductant is diverted from endogenous oxidants to molecular oxygen. This postulate is supported by the observation that potassium is required for maximal synthesis of nitrate reductase (Nitos and Evans, 1960). This may well be the dominant reason for the relatively larger increase in O_2 uptake than CO_2 release resulting from moderate potassium deficiency. However, the changes in tissue concentration of various metabolites (cf. Evans and Sorger, 1966) indicate that the rates of many reactions are affected, some of which undoubtedly modify the relative rates of exchange of the two gases.

Since dark CO_2 uptake may occur, measurements of net CO_2 release underestimate the intensity of decarboxylation reactions. In young bean leaves, dark CO_2 uptake rates have been noted to approach 25% of the corrected CO_2 release rates (Ozbun et al., 1965a, Barker et al., 1965). Accordingly, respiratory quotients calculated from net CO_2 release values may be some-

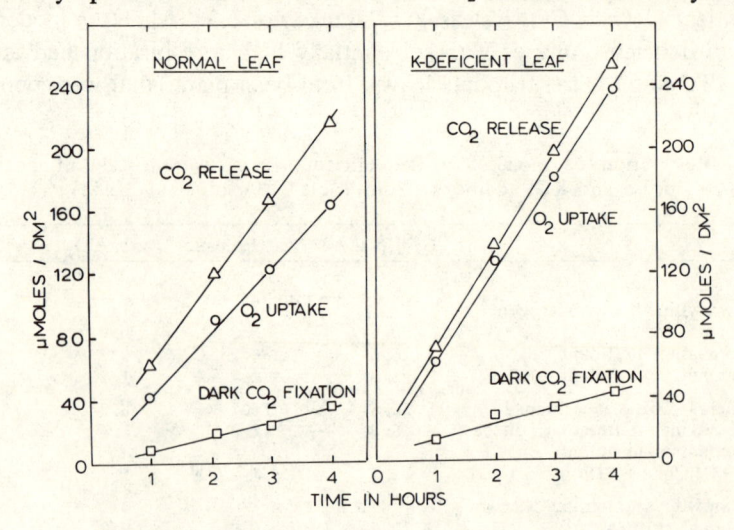

Fig. 11—Dark respiration and CO_2 uptake by terminal leaflets of first trifoliates of normal and K-deficient 20-day-old bean plants. The plant with the K-deficient leaf had not received K since the 11th day (Fig. 1, Ozbun et al., 1965a).

what lower than the actual values, thereby underestimating the extent to which other oxidants consume respiratory reductant. The evidence suggests that dark CO_2 fixation of bean leaves may be slightly enhanced when the plants are deprived of potassium while the leaves are still expanding (Ozbun, 1965a). The extent to which this increase occurs serves to accentuate the lower RQ of potassium-deficient tissue when net CO_2 release measurements are made.

3. Short-Term Recovery

Table 5 and Fig. 12 illustrate another important feature of the enhanced respiration associated with moderate potassium deficiency. Addition of KCl to the deficient cultures caused the respiration rate to approach the slower rate of normal cultures. Since uptake of O_2 was depressed by the added KCl more than CO_2 release, the RQ values also tended to approach those of normal cultures. These effects were rapid, being noted within 1 hr. after adding potassium. This indicates that extensive protein synthesis was not a prerequisite of the reaction. Rubidium was also effective in depressing the stimulated respiration although sodium was not (cf. Bierhuizen, 1954).

4. Respiratory Substrate Utilization

Moderately potassium deficient tissues usually accumulate reducing sugars. Results of experiments in which respiratory substrates are added to potassium deficient algae indicate that the deficiency has very complex effects on substrate utilization. When glucose was added to normal *Ankistrodesmus* or *Chlorella* cultures, respiration was increased considerably, net CO_2 release being accelerated more than O_2 uptake (Table 5; Fig. 12). The greater "extra CO_2" values and the increase in RQ values suggest that assimilation of nitrate was concomitantly enhanced. Respiration of the potassium deficient cultures also was enhanced upon adding glucose, but the increase over "ground respiration" was not nearly as large as with normal cultures. Thus the potassium-deficient cultures were inefficient in carrying on glucose respiration although their "ground respiration" was supra-normal. In addition, the lower RQ and "extra CO_2" values of the potassium deficient cultures indicate less nitrate assimilation during "glucose respiration" than occurred in the normal cultures. Adding KCl with glucose to potassium-deficient *Ankistrodesmus* cultures (Table 5) resulted in a most curious effect. The rates were similar to those where no glucose was added. It thus appears that "glucose respiration" was curtailed by potassium deficiency but eliminated completely by potassium addition. Subsequent experiments with *Chlorella* also revealed that potassium limited "glucose respiration" (Fig. 12). However, a somewhat more complex pattern was evident upon adding KCl. "Glucose respiration" at first decreased, but after 7 hr. with KCl it was

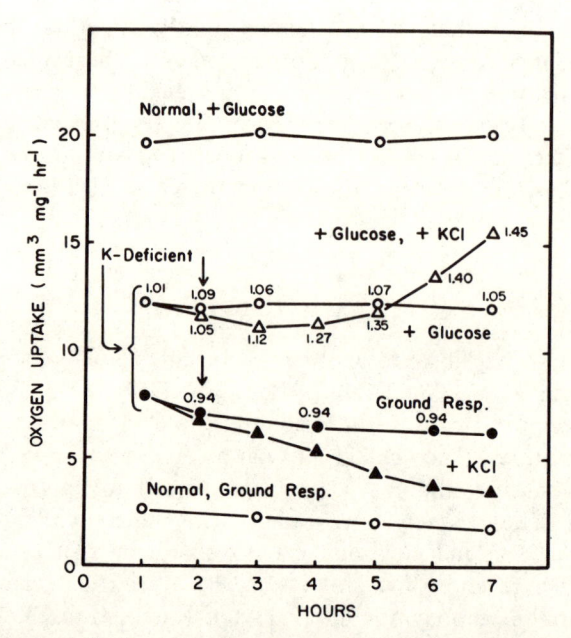

Fig. 12—Ground respiration and glucose respiration of moderately K-deficient *Chlorella* and the influence of added KCl. Glucose (0.25%) was added prior to the experiment and time of KCl additions ($8 \times 10^{-5}M$) is indicated by the arrow. Numbers represent respiratory quotient values at the indicated time. The uppermost line is glucose respiration, and the lowest line is ground respiration, of normal cells. The four middle lines are for the K-deficient cultures. (From Fig. 1, Daniel, 1956).

approaching that of the normal cultures. It may be inferred that adding KCl to the potassium-deficient cells exposed to glucose induced an immediate decrease in the "ground respiration" similar to that observed in absence of glucose. This was followed by a slow increase in "glucose respiration" which after a few hours overcame the decline in "ground respiration" (Daniel, 1956). Other experiments by Daniel (1956) revealed that a greater proportion of the glucose was respired, and less glucose was assimilated, into the cells of the potassium-deficient cultures compared to normal cultures. Addition of KCl to the former increased the proportion of glucose converted to cell material indicating greater efficiency in synthetic reactions.

An effect of potassium-deficiency, and its reversal by added potassium, on respiratory substrate utilization by leaves of higher plants has been reported by Sugiyama and Goto (1966). When potassium-deficient sugar beet leaves were infiltrated with uniformly labeled ^{14}C-glucose or ^{14}C-fructose, disappearance of ^{14}C from the hexoses was more rapid when potassium was simultaneously infiltrated. Some of the ^{14}C was recovered in sucrose, and added potassium increased this conversion, especially when the leaves had been starved to lower the sucrose concentration. Even in this case, however, net loss of ^{14}C was substantial and was increased by potassium additions. No respiratory data were given so it is not known what proportion of the ^{14}C from the hexoses not converted to sucrose was respired as CO_2. It would be of interest to learn if "glucose respiration" in potassium-deficient leaves is subnormal and whether the high "ground respiration" falls away upon potassium additions as seems to be the case with algae.

5. Energy Conversion and Transformation

The accelerated respiratory rates accompanying potassium deficiency suggested to Daniel (1956) that the metabolism might be similar to that in which the restriction in electron transport imposed by oxidative phosphorylation had been removed by uncoupling agents. If the deficiency did cause uncoupling, less efficient phosphorus transformations should result. Experiments with isolated mammalian mitochondria suggest that, under certain conditions, potassium is required for maximal ATP synthesis per mole of O_2 taken up (Pressman and Lardy, 1955; Gamble, 1957). Rivenbark and Hanson (1962) have suggested that potassium may prevent a detrimental influence of endogenous basic proteins. There is some evidence that potassium deficiency does indeed result in inefficient ATP synthesis in tissues of higher plants. Roots of 35-day-old potassium-deficient pumpkin plants, compared to controls, contained lower amounts of phosphorus in the nucleotide, nucleic acid, and protein fractions as well as in sugar phosphates (Vyskrebentseva, 1963a). A higher proportion of the total phosphorus was in the inorganic form in the deficient roots. When excised roots were exposed to $H_2^{32}PO_4$ solutions, less ^{32}P was incorporated into nucleotides and sugar phosphates of the potassium-deficient roots. Concomitant addition of potassium was partially effective in overcoming the lower rates of incorporation. Very similar observations have been made on potassium-deficient roots of sugarcane (Hartt and Kortschak, 1967).

Kursanov et al. (1965) have suggested that the pentose phosphate pathway is curtailed in potassium deficient roots. Their evidence was a much lower glucose-6-phosphate dehydrogenase activity in the root extracts from deficient plants. The low activity would result in limited synthesis of ribulose-5-phosphate, thereby depressing nucleotide synthesis (Table 6; Vyskrebentseva, 1963a). It would also result in lessened formation of NADPH, which is required in many reductive synthetic reactions, thereby accounting in part for the failure of potassium deficient tissues to elaborate complex molecules from low molecular weight compounds. Since O_2 uptake was increased, not depressed, in the deficient pumpkin roots, the postulated curtailment of the pentose phosphate pathway means that reactions of the glycolysis and tricarboxylic acid cycle must have been increased. This is supported by the fact that NaF (an inhibitor of glycolysis) and malonate (an inhibitor of the tricarboxylic acid cycle) gave only moderate depressions in O_2 uptake of the normal roots whilst that of potassium-deficient roots

Table 6—Oxygen uptake, its inhibition by malonate and fluoride, and concentrations of nucleotides and ester phosphates in 35-day-old pumpkin roots (from Kursanaov et al., 1965)

	μliters O_2/g crude wt	.033 M Malonate	.05 M NaF	Nucleotides	P-Esters
	O_2 Uptake	—— Percentage inhibition ——		— γP/g dry wt —	
Normal	97.6	16	12.4	185.7	960
Potassium-deficient	150.9	33.4	44.5	70	760

was restricted much more. Thus, in the deficient condition, more O_2 uptake occurred by reactions which were inhibited by these reagents (Table 6).

The fact that pyruvate, hydroxypyruvate and glycolate tended to accumulate in potassium-deficient roots whereas a-ketoglutarate was depleted (Vyskrebentseva, 1963b) suggests that the tricarboxylic acid cycle, compared to glycolysis, was restricted by lack of potassium. Moreover, amino acids closely associated with pyruvate tended to accumulate while glutamate was depressed. Other studies, however, have shown a-ketoglutarate concentrations to be increased in potassium deficient tissue (Jones, 1961; Yamashita and Fujiwara, 1966). It is of course perilous to predict metabolic pathways from gross chemical composition because of the many ways in which the composition of a given metabolite may be affected. The amino acids and tricarboxylic acids concentrations in the deficient tissue for example may be largely a consequence of proteolysis and deamination.

Interference in respiratory energy conversions due to potassium deficiency is also indicated by observations on mitochondrial structure. Electron miscroscopic studies of sections taken 1 cm from root tips of pumpkin plants showed substantial and progressive structural abnormalities as the plants became increasingly deficient in potassium (Kursanov et al. 1965). Three days after removing potassium from the nutrient solution, swelling of the mitochondria was noticed. By the 10th day the outer membranes appeared to be dissolving. Increasing enlargement and distortion was evident by the 17th day and the number of cristae (an indication of physiological activity) was depressed. Some fusion of the mitochondria was noticed also. By the 34th day the entire contents appeared to be lost. As Kursanov et al. (1965) point out, these structural changes may be causally related to physiological changes in the tissue. The imperfection of the outer membrane on the 10th day suggests an altered permeability of reactants and end products. Such conditions may permit reactions to occur more readily than in normal roots but a structural arrangement necessary for coupled oxidative phosphorylation may no longer exist.

In spite of the fact that many biochemical and physiological studies suggest that the efficiency of oxidative phosphorylation is lessened as a result of potassium deficiency, Latzko (1965) has shown that the ATP concentra tion in barley roots was higher in 7-day-old seedlings grown without potassium than in roots of control seedlings (Table 7). The typical increase in respiration was observed, and it follows that use of ATP in synthetic reactions must have been restricted greatly. This observation is similar to that reported previously for yeast deprived of potassium (Latzko, 1961). Adding potassium decreased the ATP concentration, indicating an accelerated utilization thereof. It thus appears that impairment of both oxidative phosphorylation and ATP utilization are important consequences of growth in potassium-limited media. Nevertheless the buildup of ATP is not sufficient to overcome the accelerated electron transport to oxygen.

Latzko's work (1961, 1965) emphasizes an important anomaly regarding potassium nutrition and respiration in plants. When some organisms are rapidly depleted of potassium, their respiration rate is curtailed, and adding potassium enhances the respiration. This phenomenon contrasts with the stimulation in respiration which develops during growth and development at low potassium supplies. The decrease in respiration caused by rapid potassium removal has been shown in both yeast (Latzko, 1961) and in red and brown algae(Bergquist, 1959; Eppley, 1960). From its specificity for potassium and rubidium and from differences in time course of response (Fig. 13), this phenomenon appears to be independent of the less specific "salt respiration" noted in many tissues. Results with various metabolic inhibitors (Eppley, 1960) and the fact that ATP concentration was decreased upon readdition of potassium (Latzko, 1961) suggest that the accelerated respiratory phase was a consequence of enhanced ATP utilization, perhaps via a potassium activated ATPase (Blond and Whittam, 1964). The stimulation in respiration induced by potassium additions to tissue previously subjected to rapid potassium depletion has not been observed in higher plants, perhaps because rapid depletion of potassium is difficult in these tissues. The affect may, however, be restricted to the lower plant forms.

Table 7—Influence of K nutrition on various parameters of metabolic activity of roots of barley seedlings (from Table 3, Latzko, 1965)

	With K*	Without K*
Oxygen Uptake (min^{-1})†	16 ± 0.3	24 ± 0.6
ATP†	17	31
ADP†	16	18
Inorganic P‡	10.0	9.0
Oxalacetate†	0.7	1.3
Malate†	62	34
K_2O‡	15.0	12.8

* Seven days after transferring to nutrient solutions
† µMoles/100g fresh wt.
‡ mMoles/100g fresh wt.

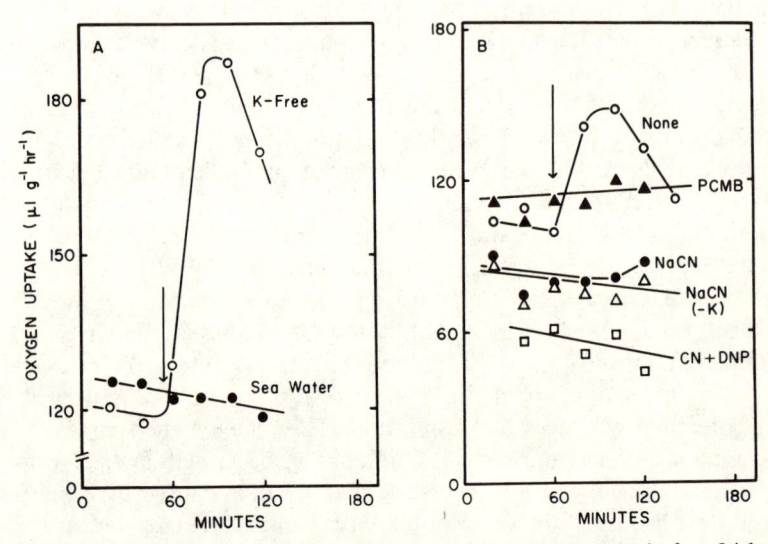

Fig. 13—Oxygen uptake of *Porphyra perforata*. (*A*) In sea water and after 24 hours in K-free sea water; KCl added to 0.01*M* at the arrow. (*B*) After 24 hours in K-free sea sea water in presence of inhibitors; KCl added to 0.01*M* at the arrow; p-chloromercuribenzoate (PCMB) 0.05m*M*; 2. 4-dinitrophenol (DNP) 0.2m*M;* NaCN 0.1m*M;* -K indicates no KCl added. (From Eppley, 1960).

V. SUMMARY

Photosynthesis, photorespiration, and dark respiration all are affected when plants are exposed to an inadequate potassium supply. These physiological alterations occur before visual abnormalities of the tissue appear. During the initial stages of deficiency photosynthetic CO_2 fixation is repressed whereas photorespiration and dark respiration are stimulated. The net result is an unfavorable rate of CO_2 retention and a concomitant limitation of dry matter accumulation.

Photosynthesis may be restricted by stomatal closure induced by potassium deficiency. This would occur primarily at normal ambient CO_2 concentrations where stomatal aperature limits diffusive entry of CO_2. The theory advanced by Fujino (1967), involving light-regulated movement of potassium between guard and mesophyll cells, offers a plausible explanation for this phenomenon. Accumulation of sugars, a characteristic of potassium deficient leaves, may also limit photosynthesis through a mass action effect, although this has not been demonstrated specifically.

Additional, and complex, effects of potassium on photosynthetic processes have been shown with algae or leaves exposed to nonlimiting CO_2

concentrations. The few data available on ^{14}C labeling patterns reveal substantial alterations in the metabolic transformations following initial CO_2 fixation. Two separate effects of lack of potassium have been shown with algae. One is observed with severe potassium deficiency and is apparently concerned with fundamental changes in protein synthesis which cannot be readily overcome by potassium additions. The second is a reversible activating effect with occurs in moderately deficient algae. The addition of potassium rapidly restores optimal photosynthetic rates.

Isotopic procedures at non-limiting CO_2 concentrations reveal that the depression in net photosynthetic CO_2 uptake during potassium deficiency is a result of an increase in photorespiratory CO_2 release as well as a decrease in CO_2 uptake. Oxygen uptake as well as CO_2 release are accelerated in potassium-deficient illuminated leaves. Increases in ambient CO_2 compensation levels also illustrate that potassium deficiency increases photorespiration. The observation that CO_2 uptake is depressed more than O_2 release together with the demonstration of lower rates of light-induced inorganic phosphorus incorporation in algae, suggest that a major effect of potassium deficiency is to depress the synthesis of high energy phosphates (or to increase their nonproductive hydrolysis to inorganic phosphate). The transfer of electrons and production of photochemical reductant, as indicated by O_2 release, is restricted to a lesser extent than is CO_2 uptake. The increase in O_2 uptake suggests diversion to O_2 of photosynthetic reductant not consumed in reducing CO_2 or endogenous oxidants (e.g., nitrate). At non-limiting CO_2 concentrations, the effects of potassium removal after full expansion of leaves is relatively minor until chlorophyll degradation is advanced. If the leaf is still developing when the potassium supply becomes inadequate the effects are much more rapid and pronounced.

Dark respiration of leaves, and respiration of nonchlorophyll tissue, are increased during the early stages when plants are grown with low potassium supplies. After prolonged and severe deficiency, the respiratory rate is often depressed. For the more general situation of stimulated respiration, O_2 uptake is increased more than CO_2 release so that the RQ is lowered. It is probable that reduction of nitrate and other endogenous oxidants declines, oxygen thereupon serving to oxidize a larger proportion of respiratory reductant. High respiration rates of moderately potassium-deficient algae are rapidly restored to normal after the addition of potassium. The RQ is increased simultaneously. Structural changes in mitochondria and slower rates of phosphorus turnover among various metabolites suggest that inefficient oxidative phosphorylation is one of the consequences of potassium deficiency. The ATP which is formed also appears to be used less rapidly but its accumulation under potassium deficient conditions does not result in decreased rates of electron transport and terminal oxidation.

ACKNOWLEDGMENTS

Support from the American Potash Institute and from the Crops Research Division, Agricultural Research Service for some of our studies reported herein is gratefully appreciated. We should also like to thank Mr. C. L. Mulchi who conducted the experiment shown in Fig. 2 and Mrs. Clara Blackwood, Mrs. Ann Matrone, and Mrs. Jacqueline Jackson for their assistance at various stages in preparation of this manuscript.

REFERENCES

Alten, F. G., Goeze, and H. Fischer. 1937. Kohlensaureassimilation und Stickstoffhaushalt bei gestaffelter Kaligabe. Bodenkunde und Pflanzenernahrung (N. S.) 5:259-289.
Barker, A. V., R. J. Volk, and W. A. Jackson. 1965. Effects of ammonium and nitrate nitrogen on dark respiration of excised bean leaves. Crop Sci. 5:439-444.
Baumeister W. and L. Schmidt. 1962. Uber die Rolle des Natriums in pfanzlichen Stoffweschsel. Flora 152:24-56.
Baver, I. D., A. S. Ayres, and T. Tanimoto. 1964. The interrelation of soil chemistry and plant biochemistry in the study of fertilizer problems. Symposium: Biochemical problems of plant nutrition as related to their metabolism. Trans. 8th Int. Congr. Soil Sci. IV:1225-1235.
Bergquist, P. L. 1959. The effect of cations and anions on the respiratory rate of the brown alga, *Homosira banksii*. Plant Physiol. 12:30-36.
Bierhuizen, J. F. 1954. Observations on potassium deficiency in *Lemna minor* L. Mededel. Landbou. Wageningen 54:311-319.
Blond, D. M., and R. Whittam. 1964. Effects of Na and K on oxidative phosphorylation in relation to respiratory control by a cell membrane ATPase. Biochem. Biophys. Res. Comm. 17:120-124.
Briggs, G. E. 1922. Experimental researches on vegetable assimilation and respiration XVI. The characteristics of subnormal photosynthetic activity resulting from deficiency of nutrient-salts. Proc. Roy. Soc. Lond. Ser. B. 94:20-35.
Clendenning, K. A., T. E. Brown, and H. C. Eyster. 1956. Comparative studies on photosynthesis in *Nostoc muscorum* and *Chlorella pyrenoidosa*. Can. J. Bot. 34:943-966.
Cooper, R. B., R. E. Blaser, and R. H. Brown. 1967. Potassium nutrition effects on net photosynthesis and morphology of alfalfa. Amer. Soil Sci. Soc. Proc. 31:231-235.
Crofts, A. R., D. W. Deamer, and L. Packer. 1967. Mechanisms of light induced structural change in chloroplasts. II. The role of ion movements in volume changes. Biochem. Biophys. Acta. 131:97-118.
Daniel, A. L., 1956. Stoffwechsel und Mineralsalzernahrung einzeller Grunalgen. Flora 143:31-66.
Dilley, R. A. 1964. Light-induced potassium efflux from spinach chloroplasts. Biochem. Biophys. Res. Comm. 17:716-722.
Dilley, R. A. 1966. Ion and water transport processes in spinach chloroplasts. Brookhaven Symposia in Biology 19:258-280.

Dilley, R. A., and L. P. Vernon. 1965. Ion and water transport processes related to the light-dependent shrinkage of spinach chloroplasts. Arch. Biochem. Biophys. 111:365-375.

Eckstein, O. 1939. Effect of potash manuring on the production of organic matter. Plant Physiol. 14:113-128.

Eppley, R. W. 1960. Respiratory responses to cations in a red alga and their relationships to ion transport. Plant Physiol. 35:637-644.

Evans, H. J., and G. J. Sorger. 1966. Role of mineral elements with emphasis on the univalent cations. Ann. Rev. Plant Physiol. 17:47-76.

Forrester, M. L., G. Krotkov, and C. D. Nelson. 1966a. Effect of oxygen on photosynthesis, photorespiration and respiration in detached leaves. I. Soybean. Plant Physiol. 41:422-427.

Forrester, M. L., G. Krotkov, and C. D. Nelson. 1966b. Effect of oxygen on photosynthesis, photorespiration and respiration in detached leaves. II. Corn and other monocotyledons. Plant Physiol. 41:428-431.

Fujino, M. 1959a. The relation of pH, salts, and starch to stomatal movement. Kagaku (Tokyo) 29:147-148.

Fujino, M. 1959b. Effect of the potassium ion and pH on the stomatal movement of the onion. Kagaku (Tokyo) 29:424-425.

Fujino, M. 1959c. Stomatal movement and active transport of potassium ion. Kagaku (Tokyo) 29:660-661.

Fujino, M. 1967. Role of adenosinetriphosphate and adenosinetriphosphatase in stomatal movement. Sci. Bull. Faculty of Education, Nagasaki Univ. no. 18 pp. 1-47.

Fujiwara, A. and S. Iida. 1967. Effect of potassium on the respiration of higher plants. English edition. International Potash Institute. Berne.

Gamble, J. L. Jr. 1957. Potassium binding and oxidative phosphorylation in mitochondrial fragments. J. Biol. Chem. 228:955-971.

Gibbs, M., E. S. Bamberger, P. W. Ellyard, and R. G. Everson. 1967. Assimilation of carbon dioxide by chloroplast preparations. In T. W. Goodwin (ed.) Biochemistry of chloroplasts, Vol. II. pp. 3-38.

Goldsworthy, A. 1967. Experiments on the origin of CO_2 released by tobacco leaf segments in the light. Phytochem. 5:1013-1019.

Gorham, P. R., and K. A. Clendenning. 1952. Anionic stimulation of the Hill reaction in isolated chloroplasts. Arch. Biochem. Biophys. 37:199-223.

Gregory, F. G., and F. J. Richards. 1929. Physiological studies in plant nutrition. I. The effect of manurial deficiency on the respiration and assimilation rate in barley. Ann. Bot. 43:119-161.

Gregory, F. G., and P. K. Sen. 1937. Physiological studies in plant nutrition. VI. The relation of respiration rate to the carbohydrate and nitrogen metabolism of the barley leaf as determined by nitrogen and potassium deficiency. Ann. Bot. (NS) 1:521-562.

Harris, E. J., K. van Dam, and B. C. Pressman. 1967. Dependence of uptake of succinate by mitochondria on energy and its relation to potassium retention. Nature 213:1126-1127.

Hartt, C. E., and G. O. Burr. 1967. Factors affecting photosynthesis in sugar cane. Proceed. 12th I. S. S. C. T. Cong. (Puerto Rico, 1965). Elsevier Publishing Co. Amsterdam. pp. 590-608.

Hartt, C. E., and H. P. Kortschak. 1967. Radioactive isotopes in sugar cane physiology. Proceed. 12th I. S. S. C. T. Cong. (Puerto Rico, 1965) Elsevier Publishing Co., Amsterdam. pp. 647-662.

Jones, L. H. 1961. Some effects of potassium deficiency on the metabolism of the tomato plant. Can. J. Bot. 39:593-606.

Jones, L. H. 1966. Carbon-14 studies of intermediary metabolism in potassium-deficient tomato plants. Can. J. Bot. 44:297-307.

Kellner, K. 1955. Die Adaptation von *Ankistrodesmus braunii* on Rubidium und Kupfer. Biologisches Entralblatt 74:662-691.

Kursanov, A., Vyskrebentseva, I. Sveshnikova, and M. Krasavina. 1965. Disorganization of energy metabolism in the roots during potassium starvation. Doklady Akademil Nauk SSSR. 162:211-214. (Eng. Transl. pp. 77-81).

Latzko, E. 1961. Zur Rolle des K bei der oxydativen Phosphorylierung. Biochem. Biophys. Acta 47:350-357.

Latzko, E. 1965. Uber die Wirking des Kaliums in der pflanzlichen Atmugn. Zeitschur. f. Pflanz.—Ernahrung, Dung., Boden. 111:114-122.

Latzko, E. and K. Meschner. 1958. Bedeutung der alkali-ionen fur die Intensitat der Lichtphosphorylierung bei. *Chlorella vulgaris*. Naturwiss. 45:247-248.

Lehninger, A. L. 1964. The Mitochondrion. W. A. Benjamin, Inc. New York.

Loustalot, A., S. A. Gilbert, and M. Drosdorff. 1950. The effect of N and K levels in tung seedlings on growth, apparent photosynthesis, and carbohydrate composition. Plant Physiol. 25:394-412.

Meschner, K. 1959. Untersuchungen an *Chlorella vulgaris* uber den Einfluss de Alkaliionen auf die Lichtphosphorylierung, Biochem. Biophys. Acta. 33:150-158.

Moss, D. N. 1966. Respiration of leaves in light and darkness. Crop Sci. 6:351-354.

Moss, D. N.,and D. E. Peaslee. 1965. Potosynthesis of maize leaves as affected by age and nutrient status. Amer. Soil Sci. Soc. Proc. 5:280-281.

Neeb, O. 1952. *Hydroactyon* als Objekteiner vergleichenden Untersuchung physiologischer Grossen. Flora 139:39-95.

Neish, A. C. 1939. Studies on chloroplasts. II. Their chemical composition and the distribution of certain metabolites between the chloroplasts and the remainder of the leaf Biochem. J. 33:300-308.

Nitsos, R. E., and H. J. Evans. 1966. Effects of inorganic cations on the inductive formation of nitrate reductase. Plant Physiol. 41:1499-1504.

Noguchi, V., and T. Sugawara. 1966. Potassium and Japonica rice. International Potash Institute, Berne.

Ozbun, J. L., R. J. Volk, and W. A. Jackson. 1964. Effects of light and darkness on gaseous exchange of bean leaves. Plant Physiol. 39:523-527.

Ozbun, J. L., R. J. Volk, and W. A. Jackson. 1965a. Effects of potassium deficiency on photosynthesis, respiration and the utilization of photosynthetic reductant by immature bean leaves. Crop Sci. 5:69-75.

Ozbun, J. L., R. J. Volk and W. A. Jackson. 1965b. Effects of potassium deficiency on photosynthesis, respiration and the utilization of photosynthetic reductant by mature bean leaves. Crop Sci. 5:497-500.

Peaslee, D. E., and D. N. Moss. 1966. Photosynthesis in K-and Mg-deficient maize (*Zea mays* L.) leaves. Amer. Soil Sci. Soc. Proc. 30:220-223.

Pirson, A. 1939. Uber die Wirkung von Alkali-Tonen auf Wachstrum und stoffweschel von *Chlorella*. Planta (Berl.) 29:231-261.

Pirson, A. and K. Kellner. 1952. Physiologische Wirkungen des Rubidiums. Deutsche Botanische Gesellschaft, Berichte 65:276-286.

Pirson, A., C. Tichy, and G. Wilhelmi. 1952. Stoffweschel und Mineral-salznahrung einzelliger Grunalgen I. Vergleichebde. Untersuchungen an Mangelkulturen. Planta (Berl.) 40:199-253.

Pressman, B. C., and H. A. Lardy. 1955. Further studies on the potassium requirements of mitochondria. Biochem. Biophys. Acta 18:482-487.

Richards, F. J. 1932. Physiological studies in plant nutrition. III. Further studies of the effect of potash deficiency on the rate of respiration in leaves of barley. Ann. Bot. 46:367-388.

Richards, F. J. 1941. Physiological studies in plant nutrition. XI. The effect on growth of rubidium with low potassium supply, and modification of this effect by other nutrients. Ann. Bot. (N. S.) 5:263-296.

Rivenbark, W. L., and J. B. Hanson. 1962. The uncoupling of oxidative phosphorylation by basic proteins, and its reversal with potassium. Biochem. Biophys. Res. Comm. 7:318-321.

Rottenberg, H., and A. K. Solomon. 1966. Energy pathways for potassium accumulation in mitochondria. Ann. New York Acad. Sci. 137:685-699.

Russell, E. J. 1927. Crop experiments. Rothamsted Annual Report. pp. 22-23 (1927/1928).

Saltman, P., J. G. Forte, and G. M. Forte. 1963. Permeability studies on chloroplasts from *Chlorella*. Expt'l. Cell Res. 29:504-514.

Schmidt, L. 1959. Unterschungen uber den Einfluss des Natriums bei Spinat und Tomaten. Flora 143:1-22.

Snow, A. G. Jr. 1936. Transpiration as modified by potassium. Plant Physiol. 11:583-594.

Spencer, D., and J. V. Possingham. 1960. The effect of nutrient deficiencies on the Hill reaction of tomato. Austr. J. Biol. Sci. 13:441-455.

Stocking, C. R., and A. Ongun. 1962. The intracellular distribution of some metallic elements in leaves. Amer. J. Bot. 49:284-289.

Sugiyama, T., and Y. Goto. 1966. Physiological role of potassium in the carbohydrate metabolism of plants (Part II). Effect of potassium on the conversion and degredation of sugar. Soil Sci. and Plant Nutr. 12:225-229.

Thomson, W. W., and T. E. Weier. 1962. The fine structure of chloroplasts from mineral deficient leaves of *Phaseolus vulgaris*. Amer. J. Bot. 49:1047-1055.

Tsuno, V., and K. Fujise. 1965. Studies on the dry matter production of sweet potato. Bull. Nat'l. Inst. Agr. Sci. (Japan) Series D. No. 13 pp. 1-131.

Volk, R J., and W. A. Jackson, 1964. Mass spectrometric measurement of photosynthesis and respiration in leaves. Crop Sci. 4:45-48.

Vyskrebentseva, E. I. 1963a. The effect of potassium on use of phosphate in root metabolism. Fiziol Rast. 10:40-47. (Eng. Transl. pp. 31-35).

Vyskrebentseva, E. I. 1963b. Oxidative and nitrogen metabolism in pumpkin roots during potassium deficiency. Fiziol. Rast. 10:307-312. (Eng. Transl. pp. 253-257).

Yamashita, T., and A. Fujiwara. 1966. Respiration and organic acid metabolism in potassium deficient rice plants. Plant Cell Physiol. 7:527-532.

Zelitch, I. 1966. Increased rate of net photosynthetic carbon dioxide uptake caused by the inhibition of glycolate oxidase. Plant Physiol. 41:1623-1631.

7

Effect of Potassium on Carbohydrate Metabolism and Translocation

W. C. LIEBHARDT

Standard Fruit Company
La Ceiba, Honduras

I. INTRODUCTION

Metabolism may be thought of as the manufacture of vital cellular constituents. Potassium's place on this assembly line is still unclear however. Research on the subject has been extremely difficult because potassium is soluble in the cell sap. It is neither a permanent constituent of a plant part nor is it laid down as part of any specific compound in the plant. Despite the difficulties of potassium research, many hypotheses dealing with possible functions have been put forth. According to the literature the potassium ion plays a role in most metabolic processes of the plant and much has been learned about this element by observing plants under potassium deficiency. The effects of potassium deficiency are expressed in a number of ways; some of these symptoms being identical to other elemental deficiencies since metabolism is such an integrated process.

As metabolism is the synthesis of products necessary to cell life, translocation is the movement of these materials within the plant. Like an industry, the ability of the plant to function well is undoubtedly strongly related to the ability of the translocation system to supply the raw materials for synthesis of new products are cellular maintenance. Mineral deficiencies may affect the functioning of this system considerably and therefore, many abnormalities of poor nutrition may be a result of a poorly functioning translocation system.

II. CARBOHYDRATE METABOLISM

Carbohydrates in the plant perform many functions. Ribulose diphosphate is the acceptor compound of the CO_2 fixed in photosynthesis resulting in the production of a triose phosphate. From this phosphorylated triose all other carbon compounds originate. Non-nitrogen compounds are translocated as a sugar, generally sucrose or a compound containing sucrose.

The respiration of glucose via glycolysis and the Krebs cycle provides carbon skeletons for synthesis of most cellular compounds plus energy to do the work of the cell. Cellulose is the material which encloses the cytoplasm and therefore, aids in maintaining a sterile environment. It also helps hold the plant upright.

Starch and other complex forms of carbohydrates make up much of the storage material of plants. For example, most seeds contain starch which is hydrolized upon germination thus giving food and energy to the young seedling.

This paper will not consider all the various aspects of carbohydrate metabolism but will look at the effect of potassium deficiency on the sugars and complex carbohydrates.

III. THE CARBOHYDRATE FRACTION

Working with potassium deficient guayule, Cooil and Slattery (1948) found that the reducing sugar concentration was consistently higher when compared to normal plants. In mature stems and roots, levulins were consistently lower in the low potassium plants, indicating that potassium is necessary for condensation of fructose to fructosan of the inulin type. Condensation of fructose in leaves of low potassium plants appeared to be related to the ratio of K/soluble calcium in leaf tissue. In all tissues, reducing sugars when expressed as a percent of total carbohydrates, were higher in potassium deficient plants than in those having adequate potassium. As further support, Eaton (1952) working with sunflowers, reported that total and reducing sugars and sucrose were highest in potassium deficient plants throughout the experiment.

Gregory and Babtiste (1936) found that under adequate fertility, the oldest barley leaves lost sugar with advancing age; whereas the leaves developed at a latter stage of growth, increased in sugar content. However, in potassium deficient leaves, sugar concentration consistently fell as the leaves matured, regardless of the relative age of the leaves on the plant. This might indicate an insufficient supply of sugars for the developing grain, hence more exhaustive translocation from leaves to grain in potassium deficient plants. Ward (1960) stated that the amount of starch in potato leaves was a direct function of the potassium level of the leaves while there appeared to be no relationship between these factors in the tuber. This might be expected since reproductive tissue tends to be less affected by a deficiency than the vegetative portions of the plant.

Buchanan, Hastings and Nesbett (1949) in research on liver slices found that the potassium ion did not affect the amount of total carbohydrates formed, however, it did affect the nature of the carbohydrate synthesized. When potassium was absent, the main product was a non-glycogen carbohydrate, but in the presence of potassium, a considerably greater portion of glycogen was formed.

Therefore, it could be generalized, that the condensation of simple forms of carbohydrates (glucose, fructose, sucrose) to more complex forms (starch, levulin) is impeded by a lack of potassium. The increase in simple sugars at the expense of the more complex types of carbohydrates could be a result of many factors.

IV. ENZYME SYSTEMS

Evans and Sorger (1966) made a comprehensive study of the literature and found that over 40 enzyme systems from plants, animals, and microbes require potassium. These enzymes were involved in photophosphorylation glycolysis, oxidative phosphorylation, respiration, protein synthesis, and glycogen or starch synthesis.

Much research has been done on the enzyme, pyruvate kinase. Miller and Evans (1957) showed that a concentration of $0.05M$ K resulted in maximum activity of this enzyme in several species of plants. This level of potassium is about equal to the amount plants with adequate potassium have in their tissue. This enzyme catalyzes the following reaction:

$$\text{PEP} + \text{ADP} \xrightarrow{\text{Mg} \quad \text{K}} \text{Pyruvate} + \text{ATP}$$

Lardy (1954) in a review of phosphorylating reactions stated that the fructokinase reaction was enhanced four-to-fivefold by potassium additions to the liver. The reaction is:

$$\text{Fructose} + \text{ATP} \xrightarrow{\text{K}} \text{Fructose 1-P or 6-P} + \text{ADP}.$$

McElroy and Nason (1954) also list this enzyme as requiring potassium in plants.

Many enzymes which affect protein or nitrogen metabolism are strongly activated by potassium. Webster and Varner (1955) isolated an enzyme from wheat germ extracts which catalyzed the following: Glutamylcysteine + glycine + ATP → glutathione + ADP + Pi. The synthesis has an absolute requirement for both magnesium and potassium, plus free SH groups on the enzyme. Many enzyme systems require SH groups for stability and it may be that a lack of potassium affects some enzymes in this manner. High levels of sugars in plant tissue could be explained as result of a reduced rate of protein synthesis. Scheck (1959) found an increased activity (or quantity) of carbohydrates, β amylases, invertase, β fructosidase, and β glucosidase in several crop plants under potassium deficiency. Buzover (1950) found an increasing amount of reducing sugars due to enhancement of invertase activity in potassium deficient potatoes.

Thomas, Coleman, and Jackson (1959) in research on crude leaf homogenates of sweet potatoes found that CO_2 fixation by the PEP dependant pathway was stimulated by magnesium but not potassium. However, homogenates of potassium sufficient plants were considerably more active than homogenates of potassium deficient plants. They theorized that magnesium worked as an enzyme activator but that potassium must be involved in the synthesis of the enzyme.

Animal tissue and yeast require potassium for glycogen and starch synthesis; therefore, it may be possible that some enzyme or cofactor involved in starch synthesis requires potassium. No literature was found, however, to support the potassium requirement in any known enzyme for starch synthesis.

Hiatt (1965) found that formlytetrahydrofolate synthetase from spinach leaves required an univalent cation for activity. The univalent cation requirement was satisfied by K, NH_4, Rb, or Na and maximum activity was obtained with 0.2-$0.3M$ concentrations of these cations. Potassium deficient plants were restored to near normal levels by adding $0.2M$ K to the spinach extract. In comparison, other enzyme systems requiring K are usually activated maximally by K concentrations of $0.05M$ or less.

Therefore, the formic acid activating system would probably be one of the first enzyme systems to be affected under conditions of potassium deficiency. The importance of the role of tetrahydrofolic acid in purine and histidine synthesis, has been discussed by Buchanan and Hartman (1959) and Rabinowitz (1960). Thus the level of potassium in the tissue might influence the rate of synthesis of such compounds as ATP, GTP, nucleic acids, NAD and NADP. Insufficient synthesis of such compounds might explain why carbohydrate condensation may proceed at a slower rate since at least some of the above compounds are necessary in starch synthesis.

It would appear that any of the above-mentioned mechanisms may be responsible for greater concentrations of simple carbohydrates. It would also seem probable that there are enough different mechanisms to explain the symptoms which occur when potassium is deficient. Therefore, solely on the basis of the role of potassium in enzyme systems, one could conclude abnormal carbohydrate metabolism in deficient plants.

V. TRANSLOCATION

Translocation is the movement of water, inorganic ions, and organic products of photosynthesis in the xylem and phloem of the plant. Essentially then, the xylem and phloem are the circulatory system of the plant. Much work has been done on the movement of materials from one place in a plant to another and yet the basic process is still not understood. This paper will not discuss the various theoretical mechanisms, but will concentrate on the effects of this process on growth or yield of plants. Nor will water movement be discussed. Excellent reviews on translocation have been presented by Nelson (1963) and Kursanov (1963).

A. General Patterns of Translocation

Biddulph and Cory (1965) found in beans that the destination of metabolites from a given leaf depends upon its position. Lower leaves export primarily to the roots with a small fraction ascending toward the apex. From the uppermost exporting leaves the metabolite flow is primarily toward the stem apex. Intermediate leaves export in both directions. It is highly probable that the metabolite from newly maturing leaves supplies the stem with various compounds other than the bulk material, sucrose, and that these are largely consumed within the rapidly growing areas of the stem. Very little of the export from young leaves reaches the roots. However, the organic nutrient requirement of roots consists of sucrose, thiamine, and possibly nicotinic acid and pyridoxine which cannot be synthesized in the roots. The bulk of this export is from the more mature leaves where sucrose export predominates. Jones, Martin, Porter (1959) found a preferential movement of assimilates to the root in early stages of tobacco growth and this decreased as the plants matured.

Nelson and Gorham in research with soybeans (1959) found that as plants exhibit a predominance of reproductive processes, the original transport scheme of assimilate directed rootwards from the leaves changes more and more, acquiring a zonal character. It seems probable that less of these materials would be reaching roots in potassium deficient plants, since lower leaves are generally most affected by potassium deficiency.

As a general phenomena, it appears that different leaves in different positions on the plant are responsible for supplying different areas.

B. Mobilizing Forces In The Plant

The compound or group of compounds responsible for mobilization of material from one plant part to the other is not known. However, the effects of plant hormones on the mobilizaion of materials is an area of intense study and appear to have a role. Leopold (1961) in his discussion of senescence stated that Mothes (1960) and Mothes, Engebrecht and Schutte (1961) found that excised leaves of tobacco lose their color and senesce quickly. However, leaves which have roots escape this fate. Mothes extrapolated this data to a whole plant and suggested that the main mobilizing areas in the plant compete for organic factors which are synthesized only in the roots. This would explain the rooted leaves' ability to live whereas an excised leaf dies. This may be overstating the case, as a continued supply of water and nutrients would appear to be quite important also.

Work by Kende (1964) showed that compounds from two regions of a chromatogram of root exudate retarded chlorophyll degradation of detached sunflower leaves. Activity of the two areas was attributed to specific compounds which are translocated from the root to the shoot and which play a regulatory role in the metabolism of the leaves. One fraction has induced cell division in soybean callus. This experiment would therefore support Mothes' interpretation.

Later data will show that when plants are deficient in potassium, the lower leaves are a very poor source of photosynthate and that potassium deficiency retards the movement of compounds in plants. It would then seem logical that the roots of potassium deficient plants would be much smaller than normal, since the lower leaves which probably supply them with their organic requirements would be considerably less efficient in photosynthesis and translocation. Therefore, since the upper portion of the plant may receive some compound(s) from the roots which delays senescence, the phenomenum of premature senescence in potassium deficient plants could be explained.

Mothes' group established that kinetin can actually produce a mobilizing effect by applying organic nitrogenous materials to the treated area of the leaf. In order to test this hypothesis, they used an amino acid which would not be incorporated into proteins: alphaminoisobutyric acid. This amino acid, too, was mobilized into the kinetin treated areas. They concluded that kinetin produced a mobilizing effect first, and that protein synthesis might follow this mobilization. They suggested then that kinetin may be a model of the forces that cause mobilization of food materials in plants.

Research by Davies, Seth, and Waring (1966) led them to suggest that increased apical dominance observed when IAA is used in combination with kinetin and gibberellic acid may be due to the mobilization of metabolites into the decapitated stump and away from the auxiliary buds, rather than a direct inhibition of buds by optimum auxin concentration.

Witham and Miller (1963) reported that a substance chemically similar to kinetin is naturally occurring in corn. Trace amounts were found in young shoots but none was detected in roots. No pronounced accumulation was found during further vegetative growth but high concentrations were found during the onset and development of reproductive organs. Some of the substance was found in tassels at anthesis and young ears contained relatively small amounts of the substance, but as they matured, there was an increasing concentration of the material.

There was a high concentration in both cob and kernels and the kernels were higher on a dry weight basis. As the kernels dried and matured the amount of the substance decreased. This would further support a mobilization theory i. e., a compound similar to kinetin was found in high concentration in the kernels and cob when translocation of materials to the ear would be proceeding vigorously. The decline of the material when translocation is reduced would be expected.

C. Experiments On The Movement Of Labelled Compounds

The use of radioactive elements has greatly contributed to our knowledge of translocation. It has positively confirmed previously hypothesized phenomena which heretofore was undemonstratable and opened new, highly reliable techniques in nutrition studies.

The effect of the environment on the sugarcane plant has been extensively studied by Hartt (1963). She has found that the translocation process was hindered by a deficiency of potassium. In this experiment a part of a leaf was enclosed in a transparent chamber and then $^{14}CO_2$ was injected into the chamber and the leaf was allowed to photosynthesize for 5 min. The chamber was removed and the plant was then allowed to translocate the photosynthate for 90 min. Table 1 shows the results of this study.

Both the velocity and amount of movement of ^{14}C photosynthate was restricted by potassium deficiency. When potassium was adequate the leading edge of the ^{14}C had moved nine joints from the joint which was attached to the fed leaf. When K was lacking the ^{14}C had moved only two joints which indicates the velocity is reduced considerably when potassium is not sufficient. The quantity of material exported was also severely reduced when potassium was not adequate. Seventy seven percent of the total activity was still in the fed part of leaf six when potassium was deficient, whereas only

Table 1—The effect of K deficiency upon translocation in sugar cane after 90 min.

Part of plant	% of total radioactivity in entire plant	
	Control	-K
Leaves and joints 1-5	1.8	0.1
Fed blade 6, above fed part	0.03	0.2
Fed blade 6, fed part	41	77
Fed blade 6, below fed part	14	18
Sheath 6	14	4
Joint 6	10	0.6
Joint 7	7	0.4
Joint 8	5	0
Joint 9	3	
Joint 10	2	
Joint 11	1	
Joint 12	1	
Joint 13	0.4	
Joint 14	0.15	
Joint 15	0.11	
Joints 16-21	0	

41% of the activity was present in the fed part of leaf six when potassium was adequate.

Anismov (1953) found that ^{14}C and ^{32}P movement from leaves to roots of wheat seedlings was reduced by potassium deficiency in wheat seedlings. No direct dependance was found between rate of translocation from the leaves and the dry weight of the roots.

It is obvious that with potassium deficiency there is a reduced rate of synthesis and the movement of materials is hindered.

D. Effect of Potassium on Dry Matter and Translocation

When a crop is growing in the field, a movement or circulation of compounds over a period of time can be seen by dividing the plant into various parts and observing them during the growing season. By this procedure, it is possible to determine translocation of material from a particular part of the plant to the developing fruit. Research by this author (Liebhardt, W. C. 1964. The relationship of soil fertility to characteristics of the corn plant which affect lodging. *M.S. Thesis. Univ. of Wisconsin, Madison, Wis.*) showed this movement of compounds in growing corn (*see* Table 2). As can be seen in Table 2, dry matter production was substantially greater when potassium was adequate. Approximately 11% more of the dry matter produced was translocated to the ear when potassium was adequate.

In another experiment with genotype and potassium as variables, (Liebhardt, W. C. 1966. The effect of potassium on the carbohydrate and nitrogenous fractions of corn. *Ph.D. Thesis. Univ. of Wisconsin, Madison, Wis.*) it was found that variety as well as potassium affected the percent of total dry matter in an ear of corn. The range among varieties was 47.3% to 64.8% when potassium was adequate and 37.6% to 56.8% when potassium was deficient. This demonstrates the effect of potassium on the movement of photosynthate and storage compounds to the ear.

In the 1964 experiment, plants were divided at nodes four and seven, the leaves remaining with the attached node, resulting in a lower, middle, and upper sample of the stover. Maximum dry matter accumulation in all plant sections occurred at the early dent stage and translocation effects could be evaluated in the various plant sections by following dry matter

Table 2—The effect of K on dry matter production of field corn

	N-P-K	N-P
	lb/acre	
Maximum dry matter accumulation	11880	8587
Dry matter in the ear	6956	4087
% of total dry matter accumulation in ear	58.5	47.6

trends (*see* Table 3). It is obvious here that the loss from the lower portion of the plant when potassium was deficient was substantially greater than the control, whereas the middle and upper portions were almost identical on a percentage basis. It would appear that the lower portion of the plant was under much more stress than the middle and upper portions. This fact was even more apparent when the effects of nitrogen, phosphorus, and potassium deficiency on concentrations in various plant sections was studied.

Whether nitrogen or phosphorus was sufficient or not, there was a higher concentration of nitrogen and phosphorus at the top of the plant. This same concentration gradient was not apparent when potassium was sufficient, as the potassium concentration was much higher in the lower portion of the plant (Table 4). When potassium was lacking, however, the concentration gradient was reversed, i.e., a higher potassium concentration was found at the top of the plant. The lower amount of potassium in the bottom of the potassium deficient plant was most likely a result of movement of potassium from the bottom to the upper leaves in an attempt to offset potassium deficiency.

Table 3—The effect of K on dry matter accumulation in field corn

Treatment	Date	Node 1-4	Node 4-7	Node 7 & above	All Stover
			lb/acre		
N-P-K	Sept. 5	1962	2248	2567	6777
N-P-K	Oct. 11	1301	1513	1458	4274
N-P	Sept. 5	1520	1479	1831	4830
N-P	Oct. 11	856	1008	1025	2889
			% Loss		
N-P-K		34	33	43	37
N-P		44	32	44	40

Table 4—The effect of K, plant section and harvest time on the K concentration in field corn

Treatment	Harvest date	Stage of growth	Node 1-4	Node 4-7	Node 7 & above
				Potassium content %	
N-P-K	July 26		3.4	2.4	2.2
	Aug. 15	Silk	2.3	1.8	1.5
	Sept. 5	Early Dent	1.9	1.4	1.4
	Oct. 14	Mature	1.9	1.2	0.9
N-P	July 26		0.5	1.0	1.3
	Aug. 15	Silk	0.5	0.6	0.6
	Sept. 5	Early Dent	0.4	0.5	0.4
	Oct. 14	Mature	0.3	0.5	0.4

Table 5—The effect of K on nutrient loss from various sections of stover from the time of maximum accumulation until final harvest

Treatment	Node 1-4	Node 4-7	Node 7+
		N % Loss	
N-P-K	60	57	66
N-P	53	56	61
		P % Loss	
N-P-K	67	73	70
N-P	61	67	70
		K % Loss	
N-P-K	39	50	60
N-P	62	56	62

Since dry matter loss was much greater in the bottom portion of the plant and the potassium concentration gradient up and down the stalk was reversed when potassium was deficient it seemed logical that the lower part of the stover would export or lose most of its K. Data to support this is shown in Table 5.

The loss of nitrogen and phosphorus was slightly higher from the nitrogen-phosphorus-potassium (N-P-K) treatment in all sections studied; however, the loss of potassium from the lower portion of the stover was greater from the nitrogen-phosphorus (N-P) treatment. Between silk and early dent the N-P-K treatment was able to obtain all the potassium it needed for the developing ear from the soil; however, the soil in the N-P treatment was so limited in potassium supplying ability that the ear obtained 68% of its K from the already K deficient stover. After the early dent stage potassium moved out of the ear and the plant in both treatments.

The corn growing in the potassium deficient plots was badly lodged and the parenchyma tissue of the roots and lower portion of the stalk had broken down severely by October 3 (Liebhardt and Murdock, 1965). In addition the root system was considerably smaller where potassium was deficient.

VI. PARENCHYMA BREAKDOWN

Research was then initiated to find a possible mechanism for the breakdown of parenchyma in corn stalks (Liebhardt, W. C. 1966. The effect of potassium on the carbohydrate and nitrogenous fractions of corn. Ph. D. Thesis. Univ. of Wisconsin, Madison, Wis.). Three possible mechanisms were postulated. These were: cell wall deterioration, excessive translocation of carbohydrate from storage tissue due to the polarizing effect of the ear, or inhibition of protein synthesis.

Table 6—The effect of K and harvest date on the cellulose, sugar and K concentration of the lower three nodes and internodes of corn stalks

Sample date	Treatment						
	N-P-K			N-P		
	Sugar	Cellulose	K	Sugar	Cellulose	K
	%			%		
7/20 (presilk)	29.8	24.8	2.4	31.6	20.3	0.4
8/4 (silk)	22.1	29.1	1.5	26.0	28.7	0.3
8/18	26.7	26.1	1.8	26.2	25.2	0.3
9/4 (early dent)	34.2	29.1	1.2	22.9	31.4	0.2
9/18 (late dent)	25.9	25.2	1.3	19.4	31.6	0.3
10/1	25.4	26.2	1.5	11.6	38.2	0.2

* Wisconsin hybrid 575, average of 3 replicates, tissue freeze dried. Nodes and internodes 1 to 3.

The materials used were adequately fertilized and potassium deficient corn plants harvested six times during the growing season starting at presilk. The leaves and sheaths were removed from the stalk. Counting from the bottom of the stalk, internodes and nodes one to three made one sample and the three internodes and nodes above it (four to six) made the other sample. The stalks were harvested, chopped, and quickly frozen in powdered dry ice and kept in a freezer at $-20C$ until dried in a commercial freeze dryer. Once dried the samples were ground and stored at $-20C$ until analyzed.

The effects of maturity and potassium fertilization on the lower portion of the stalk are demonstrated in Table 6. The K concentration when K was adequate ranged from 2.4% at the presilk stage to 1.5% at maturity. When K was lacking, the K concentration was 0.4% at presilk and only 0.2% at maturity.

A. Carbohydrate Fraction

If cell wall deterioration is the mechanism by which parenchyma breakdown proceeds, it seems logical that the percent of cellulose would decrease with maturity, especially in the lower portion of the potassium deficient stalk. In the lower portion of the stalk, where potassium was inadequate, the concentration of cellulose increased from 20.3% at presilk to 38.2% at maturity (Table 6). The increase in cellulose was most noticeable between late dent and maturity, when breakdown was proceeding rapidly. The same trends were observed in the upper portion of the stalk (Table 7).

When potassium was adequate, the concentration of cellulose remained essentially constant with time in both parts studied (Tables 6 and 7). This is contrary to what would be expected, if cell wall breakdown was the mechanism for parenchyma breakdown. Therefore, it is doubtful that cell wall deterioration is the primary cause for parenchyma breakdown. However,

Table 7—The effect of K and harvest date on the cellulose, sugar and K concentration of upper nodes and internodes

Sample date	Treatment					
	N-P-K			N-P		
	Sugar	Cellulose	K	Sugar	Cellulose	K
	%			%		
7/20 (presilk)	29.5	24.7	2.3	34.1	19.0	0.5
8/4 (silk)	30.1	27.8	1.1	26.5	26.6	0.3
8/18	32.8	24.4	1.1	30.2	24.2	0.2
9/4 (early dent)	28.3	25.7	0.8	28.7	29.7	0.3
9/18 (late dent)	27.9	25.9	1.0	25.5	27.8	0.3
10/1	29.8	27.1	1.2	17.2	34.9	0.4

* Wisconsin hybrid 575, average 3 replicates, tissue freeze dried, nodes and internodes four to six.

this does not preclude the possibility that the cell wall lost its primary function through reorientation and thereby initiated breakdown.

Cellulose is generally considered to be linear chains of β-D glucose residue linked together through carbons one and four. These chains derive some of their strength from mutual hydrogen bonding. The effect structurally is discussed by Preston (1964). It could be possible that hydrogen bonding is disrupted in potassium deficiency. If this were the case, the method used in this study would not detect this change in structure.

Research by Loomis (1945) showed that the translocation pattern in the plant is polarized by the ear and the photosynthate moves along the following path: leaves → sheath → stalk → ear shank → developing fruit. Hartt (1963) reported that the amount of photosynthate movement and velocity was limited when deficiencies of either nitrogen, phosphorus or potassium persisted in sugarcane. Moss and Peaslee (1965) demonstrated that the lower leaves of potassium deficient plants were a relatively insignificant source of photosynthate. In view of these findings, it seems logical that a greater portion of the carbohydrates stored in the lower portion of a deficient stalk would move to the ear because of its polarizing effect and also because of the insufficient supply of photosynthate in the entire plant.

Evidence to support the latter hypothesis is shown in Tables 6 and 7. Sugar levels declined rapidly in the lower portion of stalks suffering from K deficiency, i.e. from a high of 31.6% at presilk to a low of 11.6% at maturity. The decline was less in the upper portion of the stalk where potassium was limiting and changed very little in either portion where potassium was adequate. It was shown by Ozbun, Volk, and Jackson (1965) that potassium deficiency increased the rate of respiration and on this basis, part of the decline in sugar concentration might be explained. However, it seems unlikely that the large decline in sugar reported in this experiment could be explained solely on the basis of an increase in respiration rate. It is more likely attributable to the polarization effects of the ear.

Data from a previous experiment also supports this view (Liebhardt, W. C., and J. T. Murdock. 1963. *Unpublished data.*). Potassium deficient plants had the ears removed as they were formed, thus eliminating the mobilization of materials to the grain. Therefore, the rest of the plant would not be influenced by the mobilizing of material to the ear. Plants adequately fertilized and normally not exhibiting this breakdown, had the six bottom leaves removed July 26 (tassel emergence stage). It was reasoned that such a treatment would remove a large part of the plant's photosynthesizing surface, therefore, with less photosynthate available, storage compounds would be more critical, especially in the lower part of the stalk as in potassium deficient plants.

Potassium deficient plants with the ears removed at silk showed no parenchyma breakdown and were green, healthy, and succulent on October 1. Removal of the lower six leaves of the potassium sufficient plants resulted in parenchyma breakdown of the stalk and roots and the resulting ear looked typically potassium deficient, i.e., it was not filled out to the end and many kernels were smaller than normal. This would indicate a deficiency of carbohydrates to fill out the ear as in potassium deficient plants. The polar translocation of storage products from the stalk to the ear appears to be necessary for parenchyma breakdown to proceed. It also indicates that other mechanisms are of secondary importance.

Analysis of cellulose and sugar has established that sugar is moved out of the stalk when the ear is filling-out to a greater degree in potassium deficient plants than adequately fertilized plants. Since cellulose is resistant to the polarizing effect of the ear due to its chemical and physical nature it increases on a percentage basis; however, it is doubtful that there is any new synthesis of cellulose at this time.

Inhibition of protein synthesis was proposed as a mechanism for parenchyma breakdown. Therefore, nitrogen was fractionated as follows: insoluble protein, that portion insoluble in 80% ethanol (55C); soluble protein, that portion soluble in 80% ethanol (55C) and insoluble in 1% pricric acid; and non-protein nitrogen (NPN), that portion soluble in 80% ethanol (55C) and 1% picric acid (Stein and Moore, 1954).

B. Nitrogen Fraction

Inspection of the N data (Tables 8 and 9) revealed that definite changes in insoluble, soluble, and NPN took place and were a function of maturity and potassium availability. The nitrogen concentration greatly decreased between presilk and early dent in both cases and increased slightly thereafter. When total nitrogen was fractionated some large differences appeared. Where potassium was adequate, the percent of nitrogen as soluble, insoluble, and to a lesser extent NPN, remained essentially constant with time. How-

Table 8— The effect of K and harvest date on the N fraction of the lower three nodes and internodes

	Treatment						
	N-P-K				N-P		
Sample date	Total N	Insoluble† protein	Soluble‡ protein	Soluble NPN§	Total N	Insoluble† protein	Soluble‡ protein	Soluble NPN§
	%	—% of Total N—			%	— % of Total N —		
7/20	2.06	25	62	14	1.53	44	37	20
9/4	0.99	23	67	9	0.93	36	50	15
10/1	1.12	24	66	10	1.10	28	42	31

* Wisconsin hybrid 575, tissue was freeze dried, average of 3 replicates.
† Insoluble in 80% EtOH.
‡ Soluble in 80% EtOH, insoluble in 1% aqueous picrate.
§ Soluble in 80% EtOH and 1% aqueous picrate.

Table 9—The effect of K and harvest date on the N fraction of the upper nodes and internodes of field corn

	Treatment						
	N-P-K				N-P		
Sample date	Total N	Insoluble† protein	Soluble‡ protein	Soluble NPN§	Total N	Insoluble† protein	Soluble‡ protein	Soluble NPN§
	%	—— % of Total N ——			%	——% of Total N——		
7/20	1.89	36	40	23	1.96	37	43	20
9/4	0.98	34	45	21	1.05	35	53	11
10/1	0.79	40	35	24	0.93	33	52	15

* Wisconsin hybrid 575, tissue was freeze dried, average of 3 replicates, nodes and internodes 4 to 6.
† Insoluble in 80% EtOH
‡ Soluble in 80% EtOH, insoluble in 1% aqueous picrate.
§ Soluble in 80% EtOH and 1% aqueous picrate.

ever, when potassium was inadequate the amount of nitrogen as insoluble protein decreased with maturity. The soluble protein on the other hand, increased between presilk and early dent and then decreased. During the period from early dent to maturity, when breakdown was proceeding rapidly and soluble protein was decreasing, the NPN increased from 15% to 31% of total N. In general, where potassium was not applied, there was a trend of insoluble protein → soluble protein → NPN as the corn matured. This suggests internal enzymatic degradation of protein to amino acids, a result of premature senescence.

The outstanding difference between the crop receiving adequate potassium versus that not receiving potassium was the amount of soluble protein at presilk. Soluble protein was calculated using the following method: (% total N) × (% N as soluble N) × (6.25). The corn receiving adequate K contained 8% soluble protein while that which was deficient contained only 3.56%. These differences, however, were observed only in the lower portion of the stalk. Where potassium was inadequate, therefore, corn stalks had less than half the soluble protein found in normal corn stalks. It is suggested that some critical level of soluble protein or other compound(s) is necessary to maintain the cell in a functional synthesizing state and that when the level gets too low, the cell becomes structurally and metabolically disorganized and senescence takes place. It would appear that this event precedes breakdown and is not detectable under a microscope.

C. Senescence.

Pappelis and Smith (1963) reported that corn stalks with a high pith density had a high percent of living cells, whereas dry fluffy pith of low density was composed mostly of dead cells. It seems logical that a cell in a high state of metabolic activity would have a high water content whereas a cell in a low state of metabolic activity would not have as much water. Cells were stained in neutral, red plasmolyzing solution. Cells which were well hydrated were found to have stained, plazmolyzed protoplasts. Fluffy, nonhydrated cells did not have any protoplasts, nor did hydrated, autoclaved tissue.

Martens (Martens, J. W. 1965. Stalk and root rot of maize: The influence of potassium and chloride on host physiology and disease reaction. *Ph.D. Thesis. Univ. of Wisconsin, Madison, Wis.*) and Martens and Arny (1967) working with the same variety used in our study (Wisconsin 575), in the same research area, planted the same day, did pith density studies as related to potassium fertility. Fertility did not significantly affect the pith density of the second internode above the roots when sampled 10 weeks after planting. However, after the 10th week, significant drops in pith density were noted, especially when potassium was deficient. These differences in pith density continued throughout the remainder of the growing season and pith density was consistently higher where potassium was adequate.

It seems evident from the data reported by Martens and Arny that a large portion of the cells were already in a low state of physiological activity early in August, about the time of our second sampling which corresponds to the silking stage. Measuring by pith density, it is also evident that senescence preceeds a large decrease in sugar concentration.

Results of research by Craig and Hooker (1961) seem to support this phenomenon; however, they theorized that the decrease in sugar in the stalk *causes* senescence of pith tissue. They theorized that sugar should correlate with density; however, they state that differences in density may be a result of some other factor, which would seem to be the case for Wisconsin 575 corn.

Exactly what comprises the soluble protein fraction, is uncertain, but it seems that a large portion of this material would be enzymatic in nature, since it would seem to be in the cellular solution or in suspension within the cytoplasm. The insoluble material would be the structural component of the protein, possibly bound to other structural material. If this assumption is true, the lower portion of the potassium deficient stalk may lack sufficient amounts of enzymatic material needed for normal metabolism. This means that biosynthesis in this region would probably proceed at a slower rate, which may hasten senescence. It is postulated that these differences in soluble protein may increase the rate of parenchyma breakdown, once the ear starts filling out.

On the basis of the evidence presented it is postulated that the following takes place:

1) Stalk and root parenchyma senesces prematurely when potassium is deficient, possibly a result of abnormal nitrogen metabolism in the plant. At this stage (silking), however, there is no detectable parenchyma breakdown of the stalk. There are changes in tissue density which indicates less physiological activity and that cells are senescing more rapidly when potassium is deficient.

2) When the ear is filling, large amounts of carbohydrates and other materials are mobilized to the ear from the roots and stalks. This loss of material is particularly great when potassium is deficient due to a lower rate of carbohydrate production and is especially evident in the lower leaves.

3) As a result of the exhaustive translocation of material from the stalk and roots to the ear the parenchyma disintegrates. This results in a weak plant which lodges, and therefore much of the crop is lost.

Breakdown only takes place when an ear is left on the stalk thus indicating that the polarized mobilization of compounds from the stalk and roots is the primary cause of parenchyma breakdown. This is also supported by the fact that an adequately fertilized plant with six leaves removed at the tassel stage will show parenchyma breakdown also. Removal of six leaves removes much of the plant's photosynthesizing surface, thus producing a shortage of carbohydrates. The stalk and roots carbohydrate supply is then depleted to an abnormally low level resulting in parenchyma disintegration.

ACKNOWLEDGEMENTS

This work was supported in part with funds provided by the American Potash Institute, Washington, D.C. The author would like to acknowledge the assistance of Dr. Paul J. Stangel with whom much of the afore mentioned work was done and the help of Mrs. W. C. Liebhardt. The help of Srita. Mirna Ramirez in the preparation of this manuscript is sincerely appreciated.

REFERENCES

Ansinov, A. A. 1953. Translocation of assimilates in wheat seedlings in connection with root nutrition conditions. Fiziol. Rastenii 6:143. English Summary.
Biddulph, O., and R. Cory. 1965. Translocation of 14C metabolites in the phloem of the bean plant. Plant Physiol. 40:119-129.
Buchanan, J. M., and S. C. Hartman. 1959. Enzymic reactions in the synthesis of the purines. Advance. Enzymol. 21:199-261.
Buchanan, J. M., A. B. Hastings, and F. B. Nesbett. 1949. The effect of the ionic environment on the synthesis of glycogen and total carbohydrate from pyruvate in liver slices. J. Biol. Chem. 180:447-455.
Buzover, F. Y. 1950. Daklady Okad. Naulk. USSR 73:1291-1293.
Cooil, B. J. and M. C. Slattery. 1948. Effect of potassium deficiency and excess upon certain carbohydrates and nitrogenous constituents in guayule. Plant Physiol. 23:424-440.
Craig, J., and A. L. Hooker. 1961. Relation of sugar trends and pith density to diplodia stalk rot in dent corn. Phytopathology 51:376-382.
Davies, C. R., A. K. Seth, and P. F. Waring. 1966. Auxin and kinetin interaction in apical dominance. Science. 151:468-469.
Eaton, S. V. 1952. Effects of potassium deficiency on growth and metabolism of sunflower plants. Bot. Gaz. 114:165-180.
Evans, Harold J., and George J. Sorger. 1966. Role of mineral elements with emphasis on univalent cations. Ann. Rev. Plant Physiol. 17:47-76.
Gregory, F. G., and E. C. D. Babtiste. 1936. Physiological studies in plant nutrition. Ann. Bot. 50:579-619.
Hiatt, A. J. 1965. Formic acid activation II. Activation of formyltetrahydrofolate synthetase by magnesium, potassium and other univalent cations. Plant Physiol. 40:189-193.
Hartt, C. E. 1963. Translocation of sugar in the cane plant from 1963 reports. Hawaiian sugar technologists. p. 151-167.
Jones, H. E., R. V. Martin and H. K. Porter. 1959. Translocation of carbon -14 in tobacco following assimilation of carbon -14 dioxide by a single leaf. Ann. Bot. (London) (N. S.) 23:493-508.
Kende, H. 1964. Preservation of chlorophyll in leaf sections by substances obtained from root exudates. Science. 145:1066-1067.
Kursanow, A. L. 1963. Metabolism and the transport of organic substances in the phloem. Advance. Bot. Res. 1:209-278.
Lardy, H. A. 1954. Influence of inorganic ions on phosphorus reactions. Phosphorus Metabolism. 1:477-499.
Leopold, A. C. 1961. Senescence in plant development. Science. 134:1727-1732.
Liebhardt, W. C., and J. T. Murdock. 1965. Effect of potassium on morphology and lodging of corn. Agron. J. 57:325-328.
Loomis, W. E. 1945. Translocation of carbohydrates in maize. Science. 101:398-400.
Martens, J. W., and D. C. Arny. 1967. Nitrogen and sugar levels of pith tissue in corn as influenced by plant age and by potassium and chloride ion fertilization. Agron. J. 59:332-335.
McElroy, W. D., and A. Nason. 1954. A mechanism of action of micronutrient elements in enzyme systems. Ann. Rev. Plant Physiol. 5:1-30.
Miller, G., and Harold J. Evans. 1957. The influence of salts of pyruvate kinase from tissues of higher plants. Plant. Physiol. 32:346-354.

Moss, D. N., and D. E. Peaslee. 1965. Photosynthesis of maize leaves as affected by age and nutrient status. Crop Sci. 5:280-281.

Mothes, K. 1960. Naturwissenschaften. 47:337.

Mothes, K., L. Engebrecht and H. R. Schutte. 1961. Physiol. Planatarium 14:72.

Nelson, C. D. 1963. Effect of climate on the distribution and translocation of assimilates. In L. L. Evans (ed.) Environmental control of plant growth. Academic Press, New York. p. 150-172.

Nelson, C. D., and P. R. Gorham. 1959. Can. J. Bot. 37:439-447.

Ozbun, J. L., R. S. Volk, and W. A. Jackson. 1965. Effect of potassium deficiency on photosynthesis, respiration and the utilization of photosynthetic reductant by immature bean leaves. Crop Sci. 5:69-74.

Pappelis, A. J., and F. G. Smith. 1963. Relationship of water content and living cells to spread of Diplodia zeae in corn stalks. Phytopathology 53:1100-1105.

Preston, R. D. 1964. Structural plant polysaccharides. Endeavour 23:154-159.

Rabinowitz, J. C. 1960. Folic acid In P. D. Boyer, H. Lardy, and K. Myrback (ed.) The enzymes. Academic Press, New York. Vol. 2. p. 185-252.

Scheck, H. 1959. Z. Pflanzenahr Dung Bodenk. 60:209-220.

Stein, W. N., and S. Moore. 1954. The free amino acids of human blood plasma. J. Biol. Chem. 211:915-926.

Thomas, G. W., N. T. Coleman, and W. A. Jackson. 1959. Influence of magnesium, potassium and nitrogen nutrition on phosphoenolpyruvate-stimulated carbon dioxide fixation. Agron. J. 51:591-593.

Ward, G. M. 1960. Potassium in plant nutrition. Can. J. Plant. Sci. 40:729-735.

Webster, G. C., and J. E. Varner. 1955. Arch. Biochem. and Biophys. 55:95-103.

Witham, F. H., and C. O. Miller. 1963. Levels of a kinetin like factor in the developing maize plant. Plant. Physiol. 38 XXVIII.

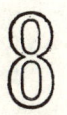

The Effect of Potassium on the Organic Acid and Nonprotein Nitrogen Content of Plant Tissue

MERLE R. TEEL

American Farm Research Association
West Lafayette, Indiana

I. INTRODUCTION

Comprehensive reviews of research on the role (s) of mineral elements in plant nutrition have been published (Barber and Humbert, 1963; DeWit et al., 1963; Evans and Sorger, 1966; Hendricks, 1966; Hewit, 1963; Nason and McElroy, 1963). Literature cited in this chapter is restricted primarily to studies which show a connection between potassium content and fluctuations in the concentration of organic acids and nonprotein nitrogen compounds in plant tissue. The discussion is projected to readers with limited understanding of plant metabolism. For fuller understanding of the reasoning of this paper, the reader is referred to publications on organic acids by Vickery and Pucher (1940), Ranson (1965), Zelitch (1964), and Beevers et al. (1966). Reviews on the metabolism of nitrogenous compounds were published by Webster (1956), Cowie (1962), Steward and Bidwell (1962), Steward and Durzan (1965), and Steward and Pollard (1961). Current concepts of some important biochemical pathways in higher plants are discussed in a series of papers edited by Pridham and Swain (1965).

A. The Agronomist's Challenge

The challenge for the investigator who works with the entire plant, the growth of which is an expression of many interactions of climate and soil, is to develop sufficient understanding of many interdependent metabolic events in plant cells to permit a synthesis of new knowledge. Two focal points of interest are: (i) organic acids and (ii) nonprotein nitrogen compounds — the latter being soluble in alcohol but precipitated by trichloracetic acid. Hereafter these nitrogenous compounds will be referred to collectively in the abbreviated form as NPN.

The synthesis of new knowledge, gained by studying fluctuations in organic acid and NPN pools, will often be based on circumstantial evidence. Steward and Durzan (1965) warn that when studying the impact of nutrient elements on nitrogen metabolism, it is not sufficient to merely describe the substances which accumulate when a given element is deficient. We should establish the symptoms of the deficiency at a reversible level, and then study changes that ensue with time when the deficient element is resupplied. Evans and Sorger (1966) add that this type of investigation sometimes reveals sufficient information to suggest metabolic blocks, and thus specific sites where the element in question may play its essential role(s).

B. Organic Acids and Ionic Balance

Organic acids have been recognized as important constituents in cytoplasm for many years (Buch, 1960; Jacobson and Ordin, 1954; Pierce and Applemen, 1943; Ulrich, 1941). Attention has been given to their relationship with cation-exchange capacity of roots (Crooke and Knight, 1962; Drake and Campbell, 1956; Drake and White, 1961; Gray et al., 1953; Huffaker and Wallace, 1959; Knight et al., 1961; McLean et al., 1956; Mehlich, 1953). There is a close relationship between organic acids and cation ratios (DeKock, 1964; Latzko, 1954; Shear et al., 1946; Ulrich, 1941; Van Itallie, 1938), and cation-anion balance (DeWit, et al., 1963; Dikjshoorn, 1963; Hiatt, 1967; Jackson and Coleman, 1958; Jacobson and Ordin, 1954; Kirkby and Mengel, 1967; Kirkby and DeKock, 1965; Ting and Dugger, 1967; Ulrich, 1941; Wallace et al., 1949).

DeWit et al. (1963), stress the importance of ionic balance as a parameter in plant nutrition. Experiments with excised roots show that cations and anions from neutral salts are not necessarily taken up in equivalent amounts. Electroneutrality is theoretically maintained by exchange of either HCO_3^- or H^+. Due to a concurrent change in the organic anion content, the pH of the plant material remains at approximately 6.

The total cation content minus the total inorganic anion content (phosphorus included at H_2PO_4) is numerically equal to the organic anion content (DeWit et al., 1963; Dikjshoorn, 1963; Hiatt, 1967; Pierce and Applemen, 1943). This value remains remarkably constant under a wide set of cation ratios provided the inorganic anion content of the nutrient solution remains constant. According to Kirkby and DeKock (1965), plant age lends another modification. Gross cation-anion ratios in Brussel sprout leaves varied from 0.2 to 1.0 depending on leaf age. This increase with age resulted essentially from the increase in calcium and decrease in nitrogen in older leaves. The calcium accumulation was in close association with increased levels of malate. According to Kirkby and DeKock (1965) the mechanism which accounts for this accumulation is not clear.

C. Cation-Anion Ratios

Much attention has been given to cation ratios since the early reports of Van Itallie (1938), Ulrich (1941), and Bear (1950). Hiatt (1967) discusses this relationship from the standpoint of organic acid changes during ion uptake. The pH of expressed root sap increased in his study when cations were absorbed in excess of anion, and decreased when anions were absorbed in excess of cations. Organic acid changes in the excised roots were proportional to expressed sap pH changes induced by unbalanced ion uptake.

Walker (1960) claims that there is considerable evidence that cation absorption is linked with the metabolic accumulation of anions, however, the instances of reports where cations have been taken up in excess of anions is rare indeed. Walker emphasizes that as long as cation uptake does not exceed anion uptake there seems no need whatsoever to postulate a theory of contact exchange to account for cation uptake. The removal of an anion by a plant from the soil solution means that a cation must also be accounted for. Walker found that the removal of anions from the soil solution by plants entailed an equivalent reduction in cations within experimental error.

Kirkby and Mengel (1967) found good evidence in tomato in support of the concept of a cation-anion balance in different plant tissues, maintained by the diffusible organic and inorganic cations and anions. Different forms of nitrogen resulted in dramatically different growth habits and dry matter yield, suggesting a close interaction between the uptake of ions by the plant and yield. This observed difference due to nitrogen source has also been reported by Barker and Bradfield (1963) and MacLeod (1966). Whereas Barker and Bradfield found greater growth of corn when 75% of the N was supplied as NH_4, MacLeod found the opposite effect with bromegrass, orchardgrass, and timothy. In Kirkby and Mengel's experiment (1967) plants fed nitrate nitrogen contained a higher concentration of cations and organic acid anions while the content of inorganic anions was lower. This phenomenon is explained by the necessity for all plants tissues to maintain equilibrium. When ammonium ions are taken up, higher amounts of inorganic anions are necessary to maintain ionic balance.

Excess cations are balanced by organic anions (Hiatt, 1967; Jacobson, and Ordin, 1954; Pierce and Applemen, 1943; Ulrich, 1941). These organic acids are present as salts of the inorganic cations, consequently they contribute to the inorganic equilibrium. Kirkby and Mengel (1967) showed that nitrate-fed tomato plants contain a much higher concentration of organic acids than ammonium-fed plants. Urea-fed plants occupied an intermediate position. With each nitrogen source, the organic acid concentration of the plant material decreased from the top leaves to the root tissue.

Hiatt (1967) proposes that carboxylation of phosphoenolpyruvate (PEP) to form oxalacetate (Fig. 1) may be causally related to excess cation absorption. In his study, the pH of the cell sap shifted in response to imbalances in cation and anion uptake. The changes were detectable within 15 min. after roots were bathed in 10^{-3} N K_2SO_4. Organic acid changes in roots were proportional to expressed sap pH changes induced by unbalanced ion uptake.

D. Organic Acids and the Krebs Cycle

Ranson (1965) concludes that the cyclic (clockwise) enzymic sequence of reactions shown in the Krebs cycle (Krebs and Kornberg, 1957); (Fig. 1) operating in the mitochondria of cells, provides for a considerable amount of fluctuation in the organic acid level in cell sap. Most growing cells are adequately equipped, although in varying degrees, to synthesize, consume, and interconvert the major acids shown in the cycle.

It is clear by mere inspection that the operation of the cycle cannot possibly account for acid accumulation. In one complete turn, the two-carbon acetyl-CoA which condenses with oxalacetate to produce citrate (Hiatt, 1962) is offset with the liberation of two molecules of CO_2. Therefore, accumulation of any acid of the cycle would soon depress the supply of oxalacetate and consequently would reduce the rate at which acetyl-CoA is consumed and energy released. Nevertheless, the cycle would allow one acid to accumulate at the expense of another; furthermore it would allow the accumulation of an acid at the expense of a reserve food supply which in the course of its break down gave rise to one or more of the acids (Beevers et al., 1966; Ranson, 1965).

E. Fluctuations in Organic Acid Pools

Bold changes in organic acid levels which accompany climatic and/or edaphic stress suggests that organic acids serve as reservoirs of potentially useful metabolites. Certain of these changes reflect unfavorable biochemical events in the cytoplasm, e. g., enzyme failure, and can be related to mineral nutrition (Cooil and Slattery, 1948; DeKock, 1964; Fowler, 1963; Kirkby and DeKock, 1965; Latzko, 1954).

Enzyme failure due to nutrient deficit or imbalance leads to the accumulation of acids, malate in particular, which appear to be metabolically inert. Their relative inertness is believed to be due in part to localization of the

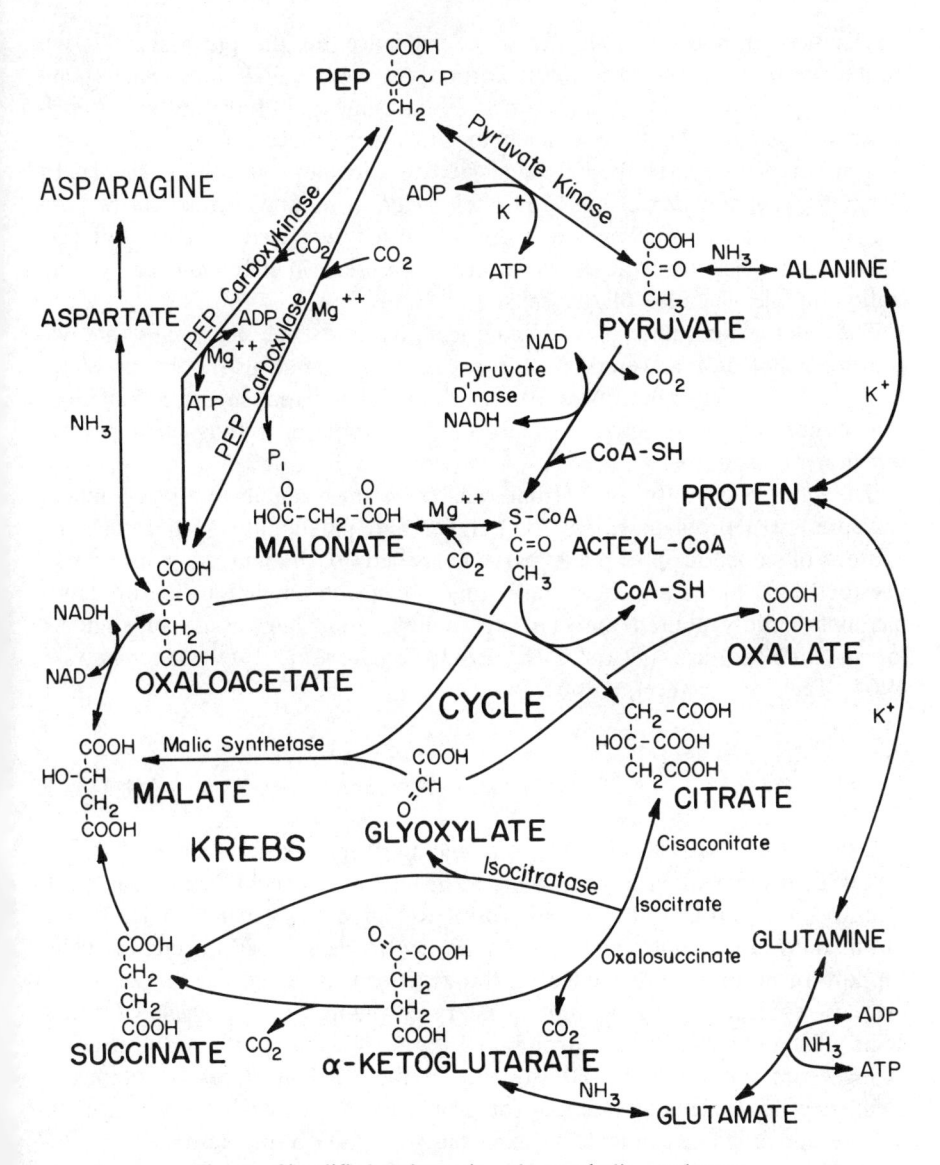

ORGANIC ACID AND NONPROTEIN NITROGEN IN PLANTS 169

Fig. 1—Simplified schematic of metabolic pathways.

acids in vacuoles where they are kept from disrupting key events in the cytoplasm. Steer and Beevers (1967) found that citrate, pyruvate, succinate, glutamate, and aspartate supplied exogenously in variously labeled forms, were used by mitochondria as rapidly as endogeneous forms of these acids. Apparently these members of the Krebs cycle penetrate readily into mitochondria and do not enter cytoplasmic pools which are not in ready equilibrium with acids in the mitochondria. Beevers et al. (1966) point out that localization in pools is a matter of convenience for the plant since it has limited capacity to excrete excess carbon or waste metabolites. Such pools enable the plant to maintain proper cytoplasmic pH, thus protecting pH-sensitive systems. That pool acids are relatively inaccessible to the cytoplasmic organelles after being transferred to vacuoles, was demonstrated by Bennett-Clark and Bexon (1943). They found a marked increase in respiration when certain vacuolar acids were added to the external medium of beet slices, even when these acids were already present in much higher concentration in the vacuoles of the cells.

The relative inertness of vacuolar acids supports the reasoning that vacuoles act not only as reservoirs for excess carbon, but also as a depository for products which might have a toxic effect. The spatial separation of toxic compounds for the sensitive enzymes of the cytoplasm is thus an important function of vacuoles.

The slow back diffusion of materials from the vacuole to the cytoplasm may present a problem where food flavor and palatability to livestock are matters of concern. The plant parts harvested for human consumption or livestock feed may be gathered at a time when nutrient imbalance or deficit has unfavorably altered the cell sap, making the plant tissue substandard for its intended use (Dikjshoorn, 1963; Jasiorowski, 1960; Nowakowski, 1964; Teel, 1962; Teel, 1966).

F. Nonprotein Nitrogen

Steward and Bidwell (1962) point out that organic acid metabolism becomes involved with nitrogen metabolism where keto acids serve as "ports of entry" for nitrogen into organic form. Reciprocally, when proteins breakdown, the reuse of nitrogen from amino acids releases carbon chains which support respiration. Knowledge of the reactions associated with the Krebs cycle is requisite to any attempt to interpret changes in the levels of amino acids which may exist in uncombined form. This is particularly true for changes brought about by adjustments in the nutrient status of plants. A deficiency of an element essential for protein synthesis, or for any key metabolic event could reasonably be expected to cause an increase in the NPN level (Cowie, 1962; Steward and Bidwell, 1962).

The crucial event in protein synthesis occurs when carbon from sugar and nitrogen in a suitable form are condensed at a template site of synthesis where amino acids are not only made but also are immediately incorporated into protein (Steward and Durzan, 1965). Thus amino acids which become the final building blocks of protein may never mingle with those that occur free and often in large amounts in cells.

As with organic acids there appears to be some mechanical barrier which permits spatial separation of soluble nitrogen compounds from the cytoplasm. Cowie (1962) mentions an internal pool of amino acids which has a closer connection with protein synthesis than exogenous amino acids. With yeast cells, when higher concentrations of exogenous amino acids are available, the quantity of amino acids contained in the NPN fraction increases. When these accumulations are compared with the steady-state amino acid concentrations in the internal pool, it is apparent that accumulation in the "expandable pool" may exceed the level of the internal pool. Cowie states that the internal pool remains fixed in size and composition. However, when exogenous amino acids are present, the cell accumulates these acids to levels exceeding their external concentration. The amino acids accumulated in this "expandable pool" have characteristics which permit their easy distinction from acids in the internal pool. The degree of accumulation varies with the type of amino acid supplied, its external concentration and the presence of other exogenous acids. The size and concentration of the pool thus varies greatly. No interconversion of acids takes place in the "expandable pool", and unlike the internal pool, the acids are sensitive to osmotic shock, extractable with cold water, and readily exchange with exogenous amino acids.

Steward and Pollard (1961) conclude that the significance of the nitrogen pool is not simply to provide prefabricated compounds for condensation into protein. Many compounds not known to be used directly in protein synthesis exist, and they occur in large amounts. A compound may accumulate in a local situation by virtue of a metabolic block, brought about by a genetic, nutritional, or environmental means, or often it may be due to circumstances inherent in normal development (Aronoff, 1962).

It is beyond the scope of this paper to discuss mechanisms which might account for the accumulation of a particular organic acid or soluble nitrogen compound. Rather, the purpose of this writing is to focus attention on a limited number of works which show that potassum-starved plants are altered biochemically in a fashion which should interest both agronomists and animal scientists.

II. VARIATION IN ORGANIC ACID COMPOSITION

Having given recognition in the introduction to some fundamental studies which help explain fluctuations in organic acid pools, let us now examine the phenomenon from a more applied viewpoint.

A. Mechanisms for Acid Accumulation

Perhaps the most convenient mechanism by which CO_2 gains entry into organic combination nonphotosynthetically is via carboxylation of PEP (phosphoenolpyruvate) (Fig. 1) which produces the key intermediate, oxalacetate (Bandurski, 1955; Bandurski and Greiner, 1953; Mazelis and Vennesland, 1957; McCollum et al., 1959).

Danner and Ting (1967), and Ting and Dugger (1967) found good evidence that PEP carboxylase is the chief enzyme mediating this reaction. They stress that PEP carboxylase lies extramitochondrial, along with a soluble form of malic dehydrogenase which converts oxalacetate to malate. Working in sequence, this pair of enzymes allows for malate accumulation independent of the Krebs cycle.

Malic enzyme, also extramitochondrial in corn root tips, provides for decarboxylation of malate to yield pyruvate (Fig. 1). Why doesn't the reverse of this reaction account for malate accumulation? According to Mazelis and Vennesland (1957) the reverse reaction is unlikely because of the unfavorable free energy exchange. Some device for supplying extra energy would be required for the reverse step to play a quantitative role in intermediary metabolism.

Danner and Ting (1967) found that malate disappeared at 45% of its production rate in this three-enzyme system. The rate of $^{14}CO_2$ turnover (measured as $^{14}CO_2$ release), after a period of $^{14}CO_2$ uptake, seems to be on the order of 50% or more of the uptake rate. They caution that these data reflect the synthesis and subsequent turnover of all products. The rate of turnover of malic $^{14}CO_2$ from $^{14}CO_2$ fixation seems to be much the same. Therefore, these authors feel that the sequence, PEP carboxylase → soluble malic dehydrogenase → malic enzyme, could account for much of the CO_2 fixation and turnover observations. They conclude further that reactions permitting nonphotosynthetic CO_2 fixation are not directly related to mitochondrial metabolism.

Danner and Ting (1967) report that CO_2 is apparently essential for growth in corn root tips. They point to research which demonstrates a relationship between protein synthesis and CO_2 fixation in corn root tissue. Under certain conditions the products (chiefly malate) can be transferred to mitochondria, however, under most conditions malate produced nonphotosynthetically is transferred to a pool where it is relatively inaccessible to cytoplasmic organelles.

Research by Jackson and Coleman (1959) provides good evidence for the operation of this system. In their study with snapbean roots, total CO_2 fixation was depressed only a slight amount when roots were bathed in 1% NH_4OH. However, the percentage of labeled carbon found in amino acids (aspartate plus glutamate) rose from 14% without treatment to 29% with treatment. Reciprocally, malate contained only 11.5% of the labeled carbon with treatment, but 64.2% without the treatment. Phosphoenolpyruvate carboxylase activity was found to be very pronounced in 12 economic plant species, consequently, these workers conclude that this enzyme occupies an important position in the metabolic scheme.

B. Potassium Relationships

If the foregoing is valid, potassium is involved in an indirect, but nevertheless important manner. As an essential cation for pyruvic kinase (Fig. 1) it exerts an influence on the availability of PEP.

Thomas et al. (1959) reasoned that CO_2 fixation should be lower in plants adequately supplied with potassium because this cation would not only enhance conversion of PEP to pyruvate but also promote related steps leading to protein synthesis. Further, they suggest that potassium levels needed to insure activation of pyruvic kinase might depress the magnesium level, thereby lowering the rate of CO_2 fixation by PEP carboxylase. This does not seem tenable however, because, as McCollum et al. (1958) point out, for maximum activity, pyruvic kinase requires magnesium as well as potassium. In addition to the transformation of carbohydrate intermediates, this enzyme also catalyzes a key step in intermediary metabolism viz., the transfer of "high energy" phosphate from PEP to ADP. It is generally accepted that Mg^{2+} is essential for maximum activity of transphosphorylating enzymes (McCollum et al. 1958).

With potassium deficiency, we might expect elevated calcium levels. This would tend to reduce pyruvic kinase activity (Broyer et al., 1942; Cooil and Slattery, 1948; Miller and Evans, 1957; Thomas et al., 1959) and might increase PEP carboxylase activity simply by virtue of PEP availability and/or its high reactivity. This relationship could account in part for the close positive association between calcium and malate as shown by DeKock (1964) and Kirkby and DeKock (1965).

The above relationships prompt further agronomic investigations wherein cation ratios are considered with respect to their influence on quality (Beeson, 1941; Grunes, 1967; Jasiorowski, 1960; Nowakowski, 1964; Reed et al., 1960; U.S. Government Printing Office, 1965; Wallace et al., 1949). But these studies must be conducted across varying inorganic anion levels. For example, Drake and Campbell (1956), Drake and White (1961), and Gray et al. (1953) show that nitrate has a profound effect on cation uptake, altering the root cation-exchange capacity, and hence the Ca/K ratio in plant tissue.

Dijkshoorn (1963) cautions that it is possible to have an anion-induced cation shortage. He states that the total cation concentration minus the total inorganic anion concentration (an estimate of the organic anions) is remarkably constant. However, Kirkby and DeKock (1965) found that this is not necessarily true. Due to the relative ease with which different cations are absorbed by different species, it may be possible to create an undesirable cation ratio. The fundamental consequences of this relationship are evident in research with corn and barley roots by Elzam and Hodges (1967). During the initial phase of transport they found calcium (or magnesium) sulfate or chloride to inhibit energy dependent potassium transport in excised corn roots. But as the absorption periods were lengthened the effect of calcium gradually changed from an inhibition to a typical promotion. Identical experiments with excerised barley roots showed that $CaSO_4$ had no effect on potassium absorption whereas $CaCl_2$ had a typical stimulatory effect. Thus according to these workers, it is hazardous to apply results obtained with one or even several plant species to all species.

C. Some Agronomic Studies

From the results of field studies by Griffith et al. (1964) and Griffith and Teel (1965) which showed potassium involvement with asparagine accumulation in orchardgrass, Cummings and Teel (1965) were prompted to examine the herbage from the same plots the following harvest season for fluctuations in malate. The analytical data and soil treatments are shown in Table 1. In striking contrast to published reports (DeKock, 1964; Splittstoesser and Beevers, 1964) potassium and malate were negatively correlated, regardless of plant age. Herbage analyzed represented the third harvest.

The negative association between potassium and malate might be expected in the light of Thomas's (1959) hypothesis. Plants adequately fertilized with potassium should theoretically fix less CO_2 nonphotosynthetically due to the preferential movement of carbon through pyruvate and further into the Krebs cycle. The lack of positive correlation between potassium and malate is not clear. Apparently there are factors operating in the soil which do not operate *in virto*.

Drake and Campbell (1956) warn that activities of cations are influenced by the presence of colloids and that cation uptake by plants from nutrient solutions omit the effect of cation competition by these colloids. Results from nutrient solution studies may therefore be misleading when applied to field studies.

As a follow up to the interesting data from Cumming's study, Pattee and Teel (1967) investigated organic acid ratios in non-nodulating soybeans. Plants were grown in soil-sand media, fertilized with nitrogen and potassium as indicated in Table 2. Seedlings were harvested 5 weeks after emergence.

With the exception of the high nitrogen level, malate was again negatively associated with potassium. The interaction at the highest nitrogen rate was highly significant. Perhaps the inorganic anion level influences the potassium-malate relationship.

The concentration of both citrate and malate as well as NPN dropped with added potassium. This suggests that potassium promotes reactions which take precedence over those leading to accumulation of acid. Malonate, at the low nitrogen level, varied reciprocally with the sum of malate plus

Table 1—Influence of N and K treatments and plant age on orchardgrass composition Cummings and Teel (1965)

Treatment, lb/acre K	N	Growth period, weeks	Total N	True protein N	Total NPN	α amino N	Residual NPN	Malic acid	K content
					%				
0	30	4	2.16	1.79	0.37	0.31	0.06	0.159	2.02
		6	2.07	1.70	0.37	0.29	0.08	0.127	2.28
0	200	4	2.84	2.21	0.63	0.32	0.31	0.213	2.03
		6	2.43	1.48	0.95	0.45	0.50	0.183	1.84
0	400	4	3.08	2.25	0.83	0.44	0.39	0.251	1.97
		6	3.04	1.66	1.38	0.49	0.89	0.135	2.21
	Avg	4	2.69	2.08	0.61	0.36	0.25	0.208	2.01
		6	2.51	1.61	0.90	0.41	0.49	0.148	2.11
166	30	4	2.11	1.73	0.38	0.25	0.13	0.089	2.51
		6	2.01	1.58	0.43	0.29	0.14	0.054	2.83
166	200	4	2.81	2.24	0.57	0.35	0.22	0.161	3.10
		6	2.46	1.54	0.92	0.38	0.54	0.068	3.30
166	400	4	2.79	2.01	0.78	0.41	0.37	0.195	3.39
		6	3.05	2.00	1.05	0.43	0.62	0.141	3.31
	Avg	4	2.57	1.99	0.58	0.34	0.24	0.148	3.00
		6	2.51	1.71	0.80	0.37	0.43	0.088	3.15
332	30	4	2.22	1.88	0.34	0.19	0.15	0.063	2.52
		6	2.24	1.85	0.39	0.21	0.18	0.052	2.90
332	200	4	2.68	2.19	0.49	0.30	0.19	0.116	3.66
		6	2.60	2.08	0.52	0.30	0.22	0.099	3.39
332	400	4	2.92	2.28	0.64	0.36	0.28	0.093	2.79
		6	2.93	1.97	0.96	0.39	0.57	0.106	3.24
	Avg	4	2.61	2.12	0.49	0.28	0.21	0.091	2.99
		6	2.59	1.97	0.62	0.30	0.32	0.086	3.18
Avg 3 values	30	4	2.16	1.80	0.36	0.25	0.11	0.180	2.35
		6	2.11	1.71	0.40	0.27	0.13	0.171	2.67
	200	4	2.77	2.21	0.56	0.32	0.24	0.221	2.93
		6	2.50	1.70	0.80	0.38	0.42	0.170	2.84
	400	4	2.93	2.18	0.75	0.40	0.35	0.218	2.71
		6	3.01	1.88	1.13	0.44	0.69	0.188	2.92
LSD 5% treatments			0.56	0.45	0.200	0.09	0.21	0.064	0.61
N and K average			0.31	0.26	0.12	0.05	0.12	0.040	0.36
Time average			0.18	0.15	0.07	0.032	0.07	0.021	0.20

citrate. These three acids accounted for approximately 72% of the total acidity.

Herbage from soils with varying potassium supplying power was analyzed with the objective of further establishing the reciprocal relationship between malonate and malate plus citrate. The data are shown in Tables 3, 4, and 5. From Table 3, note that alfalfa responded dramatically to extra potassium. As before, malate and citrate levels were lower while malonate was elevated. To further substantiate these results, alfalfa samples were hand picked, according to the expression of potassium deficiency symptoms, from three soil types within a field, known to have widely different potassium supplying capabilities. From the analytical results shown in Table 4 it is evident that that the trends found by Pattee and Teel (1967) are real. The sum of the three acids accounted for a rather constant percentage of the total acidity.

Table 2—Influence of N and K on the percent N, K, nonprotein N, and organic acid composition of soybeans Pattee and Teel (1967)

Treatment, mg/2 g of soil	N	K	NPN	Malate	Citrate	Malonate	Total
	— % dry wt —		% total wt	— % of total acidity† —			
$N_{50}K_{10}$	2.2	0.9	33.4	25.4	11.2	36.0	72.6
K_{125}	2.0	1.2	31.1	23.0	9.6	40.7	73.3
K_{250}	2.4	1.7	29.4	19.7	7.9	43.0	70.6
Avg	2.2	1.3	31.1	22.7	9.6	39.9	
$N_{150}K_{10}$	2.2	0.8	31.4	24.8	10.7	39.0	74.5
K_{125}	2.0	1.1	28.2	23.3	10.2	39.8	73.3
K_{250}	2.0	1.4	27.1	22.6	10.2	39.9	72.7
Avg	2.1	2.1	28.9	23.6	10.4	39.6	
$N_{300}K_{10}$	3.0	0.9	41.4	21.1	11.5	37.1	69.6
K_{125}	2.6	1.3	36.4	23.1	9.9	39.4	72.4
K_{250}	2.4	1.5	33.9	24.4	9.3	38.3	72.0
Avg	2.7	1.2	37.2	22.8	10.2	38.2	
S_E Within	±0.2	±0.3	±2.3	±0.9	±0.8	±1.2	
Average	±0.1	±0.1	±0.4	±0.5	±0.4	±0.7	
Statistical significance of treatments							
Nitrogen	**	--	**	--	--	--	
Potassium	--	**	*	--	*	*	
NXK	--	--	--	**	--	.15	

† Means of three replications.

In a field survey, relating soil type and fertility status to the chemical composition of soybeans, Small found much variation between upper and lower leaves with respect to both cations and acids (Table 5) (Small, Howard. 1963. Cobalt and cation-organic acid relationships in soybean plants. *Ph.D. Thesis. Purdue University, Lafayette, Ind*). Leaves were sampled over several locations in the state. Lower leaves were collected either from the third node from the plant base or the lower-most healthy trifoliate. Upper leaves consisted of the youngest fully developed trifoliate. Samples were collected at the late-bloom to early-pod stage. Each sample analyzed contained 40 randomly selected trifoliate leaves with petioles removed.

Total acidity values do not conform to the expected relationships. Dijkshoorn (1963) cites cases where high nitrogen regimes tend to produce high levels of organic acids; the theory being that when nitrogen is absorbed as an anion, the associated cation must be neutralized by an organic acid, following the reduction of nitrate to ammonium. The accumulated base component of the salt increases the pH of the root sap, a condition which presumably favors organic acid production.

In Small's study, low potassium and low nitrogen were both associated with somewhat higher total acidity. The lower leaves appeared to have lost their potassium in favor of development of upper leaves. This tended to in-

Table 3—Influence of K fertility on yield and organic acid content of alfalfa Teel (1966)

Potassium treatment	Yield second harvest	Growth rate during next 30 days	K content of third harvest	Organic acid content		
				Malate	Citrate	Malonate
lb K/acre*	lb/acre	lb/acre/day	%	microequiv/gram dry wt		
none	2,846	95	1.57	230	65	150
42 + 83	2,898	97	2.27	140	65	200
83 + 166	3,643	121	2.81†	157	54	278

* K was applied in split applications on April 15 and May 17, 1962.
† Caused severe lodging.

Table 4—Relation between K content of alfalfa and organic acid acid accumulation Teel (1966)

Soil test	Expression of deficiency symptom	K	Malate	Citrate	Malonate	Sum
lb K$_2$O/acre		%	— Percent of total acidity —			%
70-145	+*	0.70	60	26	8	94
70-145	−	0.86	46	29	17	92
145-180	+	1.42	50	20	22	92
145-180	−	1.12	43	27	20	90
180-550	+	2.65	37	14	40	91
180-550	−	1.83	39	12	38	89

* + symbol denotes presence of leaves exhibiting K deficiency. Samples were collected May 25, 1962, at the late-bud stage. Data are averages of 3 samples.

crease percent calcium in these leaves. According to Kirkby and DeKock (1965) calcium moves only in an upward direction in the transpiration stream, so that once calcium has entered a leaf, it remains there despite variations of calcium in the roots. Thus, we might expect to find the lower leaves to be richer in calcium.

Somewhat higher total acidity values in the lower leaves are associated with citrate plus malonate. Attempts to separate these two acids colorimetrically failed, consequently the sum, as determined by titration is reported. From Tables 3 and 4, and from reports by Kirkby and DeKock (1965) that citrate accumulates in aged leaves, one might safely assume that much of the acidity in the lower leaves due to citrate.

While nothing can be said about malonate fluctuations in Small's study, (Table 5) the data in Tables 3 and 4 relative to malonate merit some discussion. Most references to research on malonate deal with its synthesis (Bentley, 1952; Huffaker and Wallace, 1961; Young and Shannon, 1959; Pattee and Shannon, 1964) rather than with variations in content. Pattee et al. (1964) working with bush beans found that potassium enhanced malonate synthesis in leaf tissue, but caused a depression in root tissue. Malate accumulated under potassium-deficient conditions in both tissues.

Table 5—Cation and organic acid relationships in soybean leaves (Small, Howard G. 1963. *Ph. D. Thesis, Purdue Univ.*)

| Sample number | Leaf analysis ||||| Total acidity µeq/g | Total acidity accounted for by: |||||
| | K | Ca | Mg | K / (Ca + Mg) | Total N | | Unknown | Malate | Citrate + Malonate | Succinate | Fumarate | Total |
	%	%	%		%		%	%	%	%	%	%
						Lower Leaves						
167	0.73	1.42	0.60	0.36	3.7	826	2.4	18.6	70.8	7.0	1.2	100
227	1.06	2.90	1.19	0.26	3.9	688	4.8	9.7	74.3	7.2	4.0	100
195	1.08	1.98	0.71	0.40	3.4	766	6.9	21.7	60.4	9.5	1.3	100
43	1.35	1.95	0.73	0.50	4.8	564	3.1	9.7	77.3	6.3	3.6	100
						Upper Leaves						
163	2.40	0.50	0.37	2.76	4.1	425	11.2	8.6	34.1	15.3	30.8	100
208	2.74	0.64	0.46	2.50	5.4	451	10.5	17.0	41.3	14.9	16.3	100
6	2.95	0.52	0.49	2.92	5.3	431	18.5	20.3	31.2	13.5	16.5	100
71	3.30	0.56	0.54	3.00	5.7	535	13.4	16.1	36.6	12.3	21.6	100
88	3.43	0.74	0.60	2.56	5.9	368	9.9	12.5	52.9	14.4	10.3	100
247	3.50	0.74	0.50	2.82	5.4	637	4.4	11.9	69.5	9.1	5.1	100
168	3.66	0.65	0.49	3.21	5.8	406	8.4	16.2	38.6	10.6	26.2	100
214	3.70	0.61	0.45	3.50	5.0	447	8.3	14.4	46.8	5.1	25.4	100
177	3.84	0.76	0.50	3.02	5.2	449	8.8	15.5	58.3	8.6	8.8	100
171	5.10	0.62	0.46	4.72	5.8	468	6.6	17.0	33.3	10.7	32.4	100

de Villis et al. (1963) state that malonate may represent 2 to 3% of dry weight and up to 45% of the organic acids of legumes. They cite references to reports that malonate can be metabolized to acetyl-CoA and that it is involved in fatty acid synthesis, aromatic synthesis and mevalonate synthesis. Many metabolic pathways leading to its production are suggested. Greatest attention has been given acetyl-CoA carboxylase. Hatch and Stumpf (1962) surveyed the activity of this enzyme in embryos, roots, and leaves of various species. Activity was highest in embryos while roots were essentially devoid of activity. Leaves were about one-fifth as active as embryos. deVillis et al. (1963) investigated the source of malonate in bush bean roots and concluded that oxalacetate was the immediate precursor. Whether or not its synthesis in roots could account for fluctuations in the leaves as shown in Tables 3 and 4 is not known to this author.

III. VARIATION IN NPN

A. Potassium Relations

Steinberg (1957) warns that interpretation of the data on protein and carbohydrate metabolism of green plants subjected to mineral deficiency has been complicated by the use of a wide range of climatic conditions, differences in sampling, and the nonspecific nature of the methods of chemical analysis. Our meager information on the organic nutrition of plants limits our capacity to draw inferences from circumstantial evidence exhibited with symptom formation from mineral deficiency.

According to Evans and Sorger (1966) potassium is the only univalent cation generally indispensible for all living organisms. The average minimal and maximal concentrations in the dry matter in leaves of plants are 1.66 and 2.75%, respectively, ranging as high as 8.0%. The concentration of potassium in plant tissue exceeds that of any other cation. Relatively large quantities are obviously essential for normal metabolic processes. Fairly consistent observations indicate that potassium deficiency has an effect on protein synthesis. Free amino acids and amides accumulate. These authors published an extensive list of enzymes activated by potassium with numerous citations to research which need not be repeated here.

Steward and Bidwell (1962) emphasize that the soluble nitrogen pool is constantly subject to change. Amides usually account for most of the fluctuation in NPN. For example, in banana, the amide in the developing fruit may be either glutamine or asparagine, depending on the time of the year in which the fruit develops. Later on, after the climacteric, there is a massive conversion of the amide to free histidine.

In mint, Steward and Bidwell (1962) found that the soluble nitrogen pool was influenced dramatically by inorganic elements. Sulfur deficiency causes massive accumulation of the amide glutamine and the amino acid arginine. Low potassium-high calcium levels and/or short days work together for the production of high soluble nitrogen levels, whereas high potassium-low calcium and/or long days promote higher levels of protein with less soluble nitrogen. Overriding these factors, light promotes glutamine synthesis and protein production whereas darkness promotes asparagine synthesis and protein breakdown. Similar relationships were found in tobacco. Low night temperatures for mint causes asparagine to appear under conditions which would otherwise produce glutamine.

Glutamine synthesis appears to be associated with conditions that are favorable for growth and protein synthesis. On the other hand, orchardgrass, which accumulated asparagine, was not only lower yielding, but was less tolerant to low clipping and was, in general, less persistent (Griffith et al., 1964; Griffith and Teel, 1965). A free-choice feeding trial (Teel 1962) indicated that bromegrass-alfalfa hay adequately fertilized with potassium was consumed more readily than hay not fertilized or fertilized with nitrogen. Potassium-deficient alfalfa contained abnormally high levels of asparagine. Bromegrass contained only a trace.

The mechanism which accounts for asparagine production is not clear. Steward and Bidwell (1962) report that although glutamine can be labeled in different ways, using either exogenous urea or CO_2, the carbon from asparagine is almost impossible to label intensely. White Lupine, a classical asparagine plant, was not readily labeled from either $^{14}CO_2$ or radioactive sugar applied exogenously. Apparently asparagine arises not directly from sugar but from reworking of the products of protein breakdown. These workers successfully labeled asparagine in wheat by a small amount by using succinate labeled in both carboxyl groups. The resultant labeling of the carbon in asparagine corresponded precisely to the labeling in succinate showing that there must have been a direct conversion of succinate to aspartic acid via fumaric acid and/or oxalacetic acid, and then to asparagine (Fig. 1). This pathway would conform to the system proposed and demonstrated by Webster and Varner (1955).

B. Some Agronomic Studies

The data in Table 6 adapted from Steinberg (1957) show that considerably more than half the nitrogen in potassium-deficient plants may exist in the soluble form. Thompson et al. (1960) working with turnips found the interesting results shown in Tables 7 and 8. Potassium deficiency had the greatest deleterious effect, assuming that soluble nitrogen is a valid index to the assessment of mineral imbalance. The data in Table 8 suggest that free amino acids and amides are useful in agronomic studies for evaluating the effects of fertilizer treatment.

Table 6—Effects of potassium deficiency on amino acid nitrogen and protein nitrogen of plants*

| | | Deficiency/Control Ratio | | | |
| | | Leaves | | Stems | |
Plant	Severity of expression	Amino acid nitrogen	Protein nitrogen	Amino acid nitrogen	Protein nitrogen
Soybean	Medium	1.38	0.90	1.14	0.72
Sugar cane	Medium	3.30	0.92	1.12	0.92
Cowpea	Mild	1.09	0.94	0.95	1.18
Tomato	---	2.75	1.04	2.32	0.85
Tomato	Medium	1.38	1.02	1.32	1.07
Barley	Medium	1.30	0.92	--	--
Oats	Medium	2.91	1.46	--	--
Pineapple	(NO_3-N)	1.23	1.09	1.22	0.80
Pineapple	(NH_4-N)	2.46	1.31	1.82	0.73
Tobacco	Severe	6.87			

* Data selected from Steinberg, 1957.

Table 7—Nonprotein nitrogen content of normal and deficient turnip leaves*

Deficiency	Amino nitrogen content of NPN fraction
	micrograms/gram fresh wt.
None (entire leaf)	136
N (entire leaf)	121
P (entire leaf)	490
S (entire leaf)	336
None (midribs removed)	245
K (midribs removed)	1,066
Ca (midribs removed)	694
Mg (midribs removed)	205

* Thompson et al. 1960.

Table 8—Amino acid content of NPN fraction of normal and deficient turnip leaves (fresh weight basis)*

	Deficiency							
	Anion				Cation			
Amino acid	None	N	P	S	None	K	Ca	Mg
	micrograms/gram fresh weight				micrograms/gram fresh weight			
Aspartic acid	42	16	226	151	176	180	186	67
Glutamic acid	50	52	228	237	210	170	236	86
Serine	42	31	142	191	141	804	274	145
Asparagine	19	00	55	66	43	545	251	60
Alanine	72	50	200	136	138	147	166	89
Glutamine	39	18	1,046	1,334	213	6,036	1,884	217
Arginine	50	27	439	66	100	316	362	75

* Thompson et al. 1960.

MacLeod (1965) and MacLeod and Carson (1966) found that potassium increased grass yield up to a tissue content of 2%. The protein, non-protein, and nitrate-nitrogen content of the forage increased with nitrogen fertilization but decreased with potassium fertilization. MacLeod suggests 56kg/ha (50 lb/acre) of elemental K per each 112 to 224kg/ha (100 to 200 lb/acre) of applied N to insure a balanced nitrogen metabolism in forages.

When nitrogen fertilizer is enriched with ammonia, it is essential to supply additional potassium. In the glasshouse hydroponic experiment MacLeod and Carson (1966) found that high ammonium level (75% of the total N) and low K level (50ppm) resulted not only in yield depression but also in a reduction in grass tillering, increased mortality, and significant changes in cation and phosphorus content of the herbage. Provision of 250 ppm of potassium improved ammonium utilization, increased yield, and reduced the soluble nitrogen content of the herbage.

Barker and Bradfield (1963) fertilized corn grown in pots with pea gravel, containing about 25% limestone pebbles, with nutrient solutions containing from 10 to 500 ppm of both K and N; ammonium N constituting 12, 25, 50, and 75% of the total N supply.

Free amino acids and amides varied widely with nitrogen level, nitrogen source, and potassium treatment. Light intensity and temperature complicated the experiment to some degree, but the effects of mineral treatment persisted according to the expected patterns.

The levels of amino acids which normally occur in low concentrations were not changed, however the amides glutamine and asparagine, and corresponding amino acids, aspartate and glutamate, as well as alanine fluctuated in an interesting fashion. Asparagine level varied directly with the ammonia level of the nitrogen supply. Glutamine remained relatively unchanged with respect to the total amino acid content. Aspartic and glutamic acids showed small relative decreases as asparagine increased, suggesting that this amide was synthesized preferentially to detoxify excess ammonia.

Added potassium reduced the NPN pool at all levels of nitrogen. Fluctuations within the pool are of interest. While asparagine and glutamine levels dropped with extra potassium, aspartic and glutamic acid levels were elevated. This suggested to the authors that with higher potassium levels, there was increased utilization of the amino acids in protein synthesis, thereby increasing the need for the amino acids which take part in transamination of glutaric and oxalacetic acids (Fig. 1). Higher potassium levels enabled the corn seedlings to utilize a larger portion of the nitrogen supplied as ammonia. Both free amino acids and asparagine were decreased markedly by potassium at high ammonium levels. Under certain conditions alanine was increased by potassium. This relationship was also found in orchardgrass (Teel, 1966), suggesting enhanced pyruvate production (Fig. 1).

At 500 ppm K, Barker and Bradfield (1963) noted the soluble nitrogen pool was greater, indicating that some physiological deficiency was induced by excess potassium. Magnesium levels were reduced to as low as .18% while calcium ranged from .43 to .50%. This level of magnesium could have disrupted the activity of pyruvic kinase (McCollum 1958).

IV. IMPLICATIONS

Dr. A. B. Steward (1965), Macualay Institute for Soil Research, states "the fundamental dependence of the animal kingdom on the plant kingdom and of the plant kingdom on the soil cannot be too strongly or too frequently emphasized . ." . . ."The sectionalization of agriculture into individual soil, crop, animal, and other compartments is undoubtedly convenient for research, educational, and many other purposes of a specialized nature. In this age of specialization, however, the modern farmer faced with a bewildering choice of implements, fertilizers, herbicides, pesticides, crop varieties, animal breeds, and measures for improvement of animal health, to mention but a few components of agriculture, is concerned with the problem of integrating successfully all such components into a viable and economic farming enterprise . . . and if he is to be successful, he must somehow ensure that the various limbs move together in harmony."

Agronomists are for the most part interested in yield of plant product per unit area of land. The crop may be produced to provide energy or protein for livestock or human consumption, or fiber or other product for the industrial processor. In any case, the agronomist must be interested in all the factors which govern growth and development of the plant. He cannot study the soil and ignore the plant any more successfully than he can study the plant and ignore the soil. If his research data are to find a wide acceptance and application they must meet some current or predicted technological demand. Thus, it behooves plant and soil scientists to conduct research in practical environments as well as in laboratories so that the resulting data will be relevant to an existing or potential problem.

Inasmuch as the diet of an advanced civilizations is comprised largely of animal products, there is justification in considering research of the type discussed in this chapter from an animal viewpoint. In the final analysis, perhaps 80% of our current grain and fiber production is fed to livestock, therefore it is fitting to consider some implications for the reader interested in animal production.

A. Organic Acids

Grunes (1967) has recently discussed the literature pertaining to grass tetany as it relates to plant composition and organic acids. It is reviewed briefly here.

In California and Nevada, over 10,000 cattle have died of tetany in recent years. The tetany problem has prevailed for many years over wide areas of the world.

The problem seems to be more or less directly related to the ratio of potassium to calcium plus magnesium. At ratios greater than 2.2, tetany is likely to occur. This ratio is widened by cool temperatures. Magnesium fertilization will decrease the ratio.

Heavy nitrogen fertilization has caused low concentrations of blood serum magnesium in cattle. Long chain fatty acids in grass and hay increase with increasing concentrations of crude protein. Further, high concentrations of fatty acids in the rations seem to increase magnesium excretion in the feces by formation of insoluble magnesium soaps which the animal does not utilize.

Work in England shows that feeding 1.0% sodium citrate to calves decreased the magnesium level of blood serum. This suggests that organic acids may be casually related to tetany. Stout et al. (1967) found transaconitate in a wide number of range grasses. Levels were elevated at the time the incidence of tetany was greatest.

B. Nonprotein Nitrogen

Nowakowski's (1964) review of the literature pertaining to mineral fertilization and organic composition of herbage contains many useful references. Much research points to the potentially harmful effects of soluble nitrogen consumption by ruminants. Based on our present knowledge about essential amino acids, it can be assumed that a mixture of free amino acids present in plants has a lower nutritive value than a similar mixture present in various proteins. The value of free amino acids (and NPN in general) is determined mainly by the amount of carbohydrates supplied simultaneously in a diet.

Nowakowski stresses that ammonia, the chief nitrogenous product of protein catabolism, can be reused in the synthesis of microbial proteins which the animal later digests, or it can be absorbed through the rumen wall into the portal blood stream where it is either recycled back to rumen via the saliva or is lost as urea in the urine.

In literature cited, Nowakowski reports three-fold increases in rumen ammonia when animals are changed from a hay diet (with concentrates) to fresh pasture. It has been theorized that tetany arises from inadequate absorption of magnesium and that this is complicated by the presence of ammonia or its reaction products.

Jasiorowski (1960) emphasizes that herbage with high levels of soluble nitrogen is poorly suited for digestion by ruminants. Proneness to rapid deamination in the rumen results in high concentrations of ammonia. While this author does not discuss the direct influence of ammonia on the activity of microorganisms, he points to the possibility that work is involved when it becomes necessary for the liver to remove excess ammonia from the blood.

REFERENCES

Aronoff, S. 1962. Dynamics of amino acids in plants. *In* J. T. Holden (ed.) Amino Acid Pools. Elsevier Publ. Co. Amsterdam, London, New York. p. 657-707.

Bandurski, R. S. 1955. Further studies on the enzymatic synthesis of oxalacetate from phosphoenolpyruvate and CO_2. J. Biol. Chem. 217:317-350.

Bandurski, R. S., and C. M. Greiner. 1953. The enzymatic synthesis of oxalacetate from phosphoenolpyruvate and CO_2 J. Biol. Chem. 204:781-786.

Barber, S. A., and R. P. Humbert. 1963. Advances in knowledge of potassium relationships in soil and plant. *In* M. H. McVickar et. al. (ed.) Fertilizer Technology and Usage. Soil Sci. Soc. Amer., Madison, Wis. p. 231-259.

Barker, A. V., and R. Bradfield. 1963. Effect of potassium and nitrogen on the free amino acids content of corn plants. Agron. J. 55:465-470.

Bear, F. E. 1950. Cation and anion relationships in plants and their bearing on crop quality. Agron. J. 42:176-178.

Beevers, H., M. L. Stiller, and V. S. Butt. 1966. Metabolism of organic acids. *In* F. C. Steward (ed.) Plant Physiology. IVB. Academic Press New York. p. 117-262.

Beeson, K. C. 1941. The mineral composition of crops with particular reference to the soils in which they were grown. USDA Misc. Publication no. 369. 164p.

Bennett-Clark, T. A., and D. Bexon. 1943. Water relations of plant cells. III. The respiration of plasmolyzed tissues. New Phytologist 42: 65-92.

Bentley, L. E. 1952. Occurence of malonic acid in plants. Nature 170:843-848.

Broyer, P. D., H. A. Lardy, and P. H. Phillips. 1942. The role of potassium in muscle phosphorylation. J. Biol. Chem. 146:673-682.

Buch, M. L. 1960. A bibliography of organic acids in higher plants. Agr. Handbook no. 64. Agr. Res. Service. USDA.

Cooil, B. J., and M. C. Slattery. 1948. Effects of potassium deficiency and excess upon certain carbohydrate and nitrogen compounds in guayule. Plant Physiol. 23:425-442.

Cowie, D. B. 1962. Metabolic pools and the biosynthesis of protein. *In* J. T. Holden (ed.) Amino Acid Pools. Elsevier Publ. Co. Amsterdam, London, New York. p. 633-646.

Crooke, W. M., and A. H. Knight. 1962. An evaluation of published data on the mineral composition of plants in the light of the cation exchange capacities of their roots. Soil Sci. 93:365-373.

Cummings, G. A., and M. R. Teel. 1965. Effect of nitrogen and potassium and plant age on certain nitrogenous constituents and malate content of orchardgrass (*Dactylis glomerata* L.) Agron. J. 57:123-125.

Danner, Jean, and I. P. Ting. 1967. CO_2 metabolism in corn roots. II. Intracellular distribution of enzymes. Plant Physiol. 42:719-724.

DeKock, P. C. 1964. The physiological significance of potassium-calcium relationship in plant growth. Outlook of Agr. IV:93-96.

DeWit, C. T., W. Dijkshoorn, and J. C. Noggle. 1963. Ionic balance and growth of plants. Versl. Landbouk. Onderz. NR 69.15, Wageningen 68p.

Dijkshoorn, W. 1963. The balance of uptake, utilization and accumulation of the major elements. *In* International Potash Inst. Bern, Switzerland (ed.) Potassium in relation to grassland production. p. 43-62.

Drake, Mack, and J. D. Campbell. 1956. Cation exchange capacity of plant roots as related to plant nutrition. Amer. Soc. Hort. Sci., Proc. 67:563-569.

Drake, Mack, and J. M. White. 1961. Influence of nitrogen on uptake of calcium. Soil Sci. 91:66-69.

Evans, Harold J., and George J. Sorger. 1966. Role of mineral elements with emphasis on the univalent cations. Ann. Rev. Plant Physiol. 17:47-76.

Elzam, O. E., and T. K. Hodges. 1967. Calcium inhibition of potassium absorption in corn roots. Plant Physiol. 42:1483-1488.

Fowler, H. D. 1963. The role of potassium in plant nutrition. *In* International Potash inst., Bern, Switzerland. (ed.) Potassium in relation to grassland production p. 25-33.

Gray, B., Mack Drake, and W. G. Colby. 1953. Potassium competition in grass-legume associations as a function of root cation exchange capacity. Soil Sci. Soc. Amer. Proc. 17:235-239.

Griffith, W. K., and M. R. Teel. 1965. Effect of nitrogen and potassium fertilization, stubble height and clipping frequency on yield and persistence of orchardgrass. Agron. J. 57: 147-150.

Griffith, W. K., M. R. Teel, and H. E. Parker. 1964. Influence of nitrogen and potassium on the yield and chemical composition of orchardgrass. Agron. J. 56: 473-475.

Grunes, D. L. 1967. Grass tetany of cattle as effected by plant composition and organic acids. *In* Proc. 1967. Cornell Nutrition Conf. for Feed Manufacturers. p. 105-110.

Hatch, M. D., and P. K. Stumpf. 1962. Fat metabolism in higher plants. XVII. Metabolism of malonic acid and its *a*-substituted derivatives in higher plants. Plant Physiol. 37: 121-126.

Hendricks, S. B. 1966. Salt entry into plants. Soil Sci. Soc. Amer. Proc. 30:1-7.

Hewit, E. J. 1963. Essential nutrient elements for plants. *In* F. C. Steward (ed.) Plant Physiol. Vol. III. Academic Press, N. Y. p. 157-360.

Hiatt, A. J. 1962. Condensing enzymes from higher plants. Plant Physiol. 37: 85-98.

Hiatt, A. J. 1967. Relationship of cell sap pH to organic acid changes during ion uptake. Plant Physiol. 42: 294-298.

Huffaker, R. C., and A. Wallace. 1959. Variation in root cation exchange capacities with plant species. Agron. J. 51: 120 (note).

Huffaker, R. C., and A. Wallace. 1961. Malonate synthesis via dark CO_2 fixation in bush bean roots. Biochem. et. Biophys. Acta 46: 403-405.

Jackson, W. A., and N. T. Coleman. 1958. Ion absorption by bean roots and organic acid changes brought about through CO_2 fixation. Soil Sci. 87: 311-319.

Jackson, W. A., and N. T. Coleman. 1959. Fixation of CO_2 by plant roots through phosphoenolpyruvate carboxylase. Plant and Soil XI: 1-16.

Jacobson, L., and L. Ordin. 1954. Organic acid metabolism and ion absorption in roots. Plant Physiol. 29: 70-75.

Jasiorowski, H. 1960. The value of nitrogen compounds in meadow hay, alfalfa hay and dried alfalfa hay in feeding ruminants. Proc. 8th Intern. Grassland Congr. p. 538-543.

Kirkby, E. A., and K. Mengel. 1967. Ionic balance in different tissues of the tomato plant in relation to nitrate, urea or ammonium nutrition. Plant Physiol. 42: 6-14.

Kirkby, E. A., and P. C. DeKock. 1965. The influence of age on the cation-anion balance in the leaves of brussel sprouts (Brassica oleracea var gemmifera). Z. Pflanzenernahr., Dung., Bodenkunde 111: 197-203.

Knight, A. H., W. M. Crooke, and R.H.E. Inkson. 1961: Cation-exchange capacities of tissues of higher plants and their related uronic acid contents. Nature 192, no. 4798. p. 142-143.

Krebs, H. A., and H. L. Kornberg. 1957. Energy transformation in living systems. A survey. Springer-Verlag, Berlin. 215 p.

Latzko, E. 1954. The function of potassium in the metabolism of energy rich phosphate in plant and animal organisms. Agrochimica 3: 148-164.

MacLeod, L. B. 1965. Effect of nitrogen and potassium on the yield and chemical composition of alfalfa, bromegrass, orchardgrass and timothy grown as pure species. Agron. J. 57: 261-266.

MacLeod, L. B., and R. B. Carson. 1966. Influence of K on the yield and chemical composition of grasses grown in hydroponic cultures with 12, 50, and 75 percent of the N supplied as NH_4^+. Agron. J. 58: 52-57.

Mazelis, M., and B. Vennesland. 1957. Carbon dioxide fixation into oxalacetate in higher plants. Plant Physiol. 32: 591-600.

McCollum, R. E., R. H. Hageman, and E. H. Tyner. 1958. Influence of potassium on pyruvic kinase from plant tissue. Soil Sci. 86: 324-331.

McCollum, R. E., R. H. Hageman, and E. H. Tyner. 1959. Occurences of pyruvic kinase and phosphenolpyruvate phosphatases in seeds of higher plants. Soil Sci. 89: 49-52.

McLean, E. O., D. Adams, and R. E. Franklin. 1956. Cation exchange capacities of plant roots as related to their nitrogen contents. Soil Sci. Amer. Proc. 20: 345-347.

Mehlich, A. 1953. Factors affecting adsorption of cations by plant roots. Soil Sci. Soc. Amer. Proc. 17: 231-234.

Miller, Gene, and Harold J. Evans. 1957. The influence of salts on pyruvate kinase from tissues of higher plants. Plant Physiol. 32:346-354.

Nason, A., and W. D. McElroy. 1963. Modes of action of the essential mineral elements. *In* F. C. Steward (ed.) Plant Physiol. Vol. III. Academic Press, New York. p. 451-539.

Nowakowski, T. Z. 1964. Mineral fertilization and organic composition of herbage. *In* Le potassium et la qualite des produits agricoles. p. 63-73. International Potash Inst., Bern, Switzerland.

Pattee, H. E., and M. R. Teel. 1967. Influence of nitrogen and potassium on variations in content of malate, citrate and malonate in non-nodulating soybeans (Glycine max). Agron. J. 59: 187-189.

Pattee, H. E., L. M. Shannon, and J. Y. Lew. 1964. In vivo peroxidase inhibitor in bush beans (*Phaseolus vulgaris*) leaves. Nature 201: 1328.

Pattee, H. E., and L. M. Shannon. 1964. Malonate biosynthesis via oxalacetate in plant tissue. Bot. Gaz. 126: 179-181.

Pierce, E. C., and C. O. Applemen. 1943. Role of other soluble organic acids in the cation-anion balance of plants. Plant Physiol. 18: 224-238.

Pridham, J. B., and T. Swain. 1965. Biosynthetic pathways in higher plants. Academic Press. London, New York. 212p.

Ranson, S. L. 1965. Plant acids. *In* J. B. Pridham and T. Swain (ed.) Biosynthetic pathways in higher plants. Academic Press. London, New York. p. 179-192.

Reed, L. W., V. L. Sheldon, and W. A. Albrecht. 1960. Soil fertility and the organic composition of plants: lysine, arginine and aspartic acid variations. Agron. J. 52: 523-526.

Shear, C. B., H. L. Crane, and A. T. Meyers. 1946. Nutrient element balance: a fundamental concept in plant nutrition. Amer. Soc. Hort. Sci. Proc. 47: 239-248.

Splittstoesser, W. E., and H. Beevers. 1964. Acids in storage tissues. Effects of salts and aging. Plant Physiol. 39: 163-169.

Steer, B. T., and H. Beevers. 1967. Compartmentation of organic acids in corn roots. III. Utilization of exogenously supplied acids. Plant Physiol. 42: 1197-1201.

Steinberg, Robert A. 1957. Correlations between protein-carbohydrate metabolism and mineral deficiencies in plants. *In* Emil Truog (ed.) Mineral nutrition of plants. Univ. of Wisconsin Press, Madison, Wis. p. 359-386.

Steward, F. C., and R.G.S. Bidwell. 1962. The free nitrogen compounds in plants considered in relation to metabolism. *In* J. T. Holden (ed.) Amino acid pools. Elsevier Publ. Co. Amsterdam, London, New York. p. 667-893.

Steward, F. C., and D. J. Durzan. 1965. Metabolism of nitrogenous compounds. *In* F. C. Steward (ed.) Plant physiology. Academic Press. New York. p. 379-687.

Steward, F. C., and J. K. Pollard. 1961. Soluble nitrogenous constituents of plants. *In* J. T. Holden (ed.) Amino acid pools. Elsevier Publ. Co. Amsterdam, London, New York. p. 25-43.

Stewart, A. B. 1965. Aspects of soil, plant and animal relationships. Advance Sci. 22: 429-438.

Stout, P. R., J. Brownell, and R. G. Burau. 1967. Occurrences of trans-aconitate in range forage species. Agron. J. 59: 21-24.

Teel, M. R. 1962. Nitrogen-potassium relationships and biochemical intermediates in grass herbage. Soil Sci. 93: 50-55.

Teel, M. R. 1966. Nitrogen-potassium relationships and their influence on some biochemical intermediates and quality of crude protein in forages. Proceedings 1966, International Potash Congress, Brussels, Belgium (in print).

Thomas, G. W., N. T. Coleman, and W. A. Jackson. 1959. Influence of magnesium, potassium and nitrogen nutrition on phosphenolpyruvate-stimulated carbon dioxide fixation. Agron. J. 51: 591-594.

Thompson, J. F., C. J. Morris, and Rose K. Gering. 1960. The effect of mineral supply on the amino acid composition of plants. Qualitas Plantarum ot Materiae Vegetabiles VI, 3-1 p. 261-275.

Ting, I. P., and W. M. Dugger. 1967. CO_2 metabolism in corn roots. I. Kinetics of carboxylation and decarboxylation. Plant Physiol. 42: 712-718.

Ulrich, A. 1941. Metabolism of nonvolatile organic acids in excised barley roots as related to cation-anion balance during salt accumulation. Amer. J. Bot. 28: 526-537.

U. S. Government Printing Office. 1965. The effect of soils and fertilizers on the nutritional quality of plants. Agr. Info. Bull. no. 299. 24p.

Van Itallie, Th. 1938. Cation equilibria in relation to the soil. Soil Sci. 46: 175-186.

Vickery, H. B., and G. W. Pucher. 1940. Organic acids of plants. Ann. Rev. Biochem. 9: 529-544.

Villis, de, J., L. M. Shannon, and J. Y. Lew. 1963. Malonic acid biosynthesis in bush bean roots. I. Evidence for oxalacetate as an immediate precursor. Plant Physiol. 38: 686-690.

Walker, F. W. 1960. Uptake of ions by plants growing in the soil. Soil Sci. 89: 320-332.

Wallace, A., S. J. Toth, and F. E. Bear. 1949. Cation and anion relationships in plants with special reference to seasonal variation in mineral content of alfalfa. Agron. J. 41: 66-71.

Webster, G. C., and J. E. Varner. 1955. Aspartate and asparagine synthesis in plant systems. J. Biol. Chem. 215: 21-99.

Webster, G. C. 1956. Effect of monovalent cations on the incorporation of amino acids into protein. Biochem. et Biophys. Acta. 20: 565-566.

Young, R. H., and L. M. Shannon. 1959. Malonate as a participant in organic acid metabolism in bush bean leaves. Plant Physiol. 34: 149-152.

Zelitch, I. 1964. Organic acids and respiration in photosynthetic tissues. Ann. Rev. Plant Physiol. 15: 121-142.

The Effect of Potassium and Other Univalent Cations on the Conformation of Enzymes[1]

RICHARD H. WILSON

University of Illinois
Urbana, Illinois

HAROLD J. EVANS

Oregon State University
Corvallis, Oregon

I. INTRODUCTION

The requirement for potassium for normal growth and development in all living organisms is well established. In higher plants potassium is the most abundant mineral and the fifth most abundant element being exceeded in concentration only by carbon, oxygen, nitrogen, and hydrogen. In normal plants the content of K in leaves is 1.66-2.75% of the dry matter, whereas in leaves of deficient plants the level of K has been observed to range from 0.75-1.22% of the dry matter (Evans and Sorger, 1966). Deficiency of potassium in plants results in several characteristic metabolic lesions including chlorosis of leaves, decreased starch content of tissues, accumulation of reducing sugars and other soluble carbohydrates, and accumulation of amino acids and amides in tissues.

Although potassium is known to be a major inorganic constituent of plants and its deficiency has been observed to affect several metabolic processes, the role of potassium in cellular metabolism has remained unclear. Recently it has been proposed that the key metabolic role of potassium may involve its capacity to activate a whole series of enzyme systems (Evans and Sorger, 1966). Support for this argument is based upon the fact that the activity of

[1] This research was supported by research grant AM08123 and a Predoctoral Fellowship grant 1-E GM 34-633-01 from the U S Public Health Service and by the Oregon Agricultural Experiment Station, Corvallis, Ore.

a large group of important enzyme systems has been reported to be stimulated by or to be completely dependent upon potassium or other univalent cations (Evans and Sorger, 1966). Moreover, most metabolic symptoms of potassium deficiency can be associated with a block in one or more enzyme systems that require potassium for activity. As an example, the accumulation of amino acids in potassium deficient organisms very likely reflects the requirement for potassium in enzyme reactions associated with protein synthesis (Webster, 1956; Lubin and Ennis, 1963; and Schweet et al., 1964). It has been pointed out by Evans and Sorger (1966) that the range in concentration of potassium in normal plant material is similar to the range in concentration which results in optimal enzyme activation by potassium. Thus, there is considerable evidence to support the contention that the major role of potassium in cellular metabolism is its function as an activator of certain enzyme systems.

The first enzyme reported to exhibit a univalent cation requirement was pyruvate kinase (Boyer, Lardy, and Phillips, 1942, 1943). Since then a stimulation by or an absolute dependence upon univalent cations has been demonstrated for this enzyme from plants (Miller and Evans, 1957), animals (Boyer, 1953), and microorganisms (Seitz, 1949). Figure 1 shows the effect of increasing concentration of a series of univalent cations on the activity of pyruvate kinase from cotton seed. Potassium is the best activator but rubidium and ammonium also are effective. Sodium is a weak activator and lithium is completely ineffective. The concentration of potassium necessary for optimal stimulation is about $0.03M$. Optimal activity of pyruvate kinase from most other sources is obtained with a potassium con-

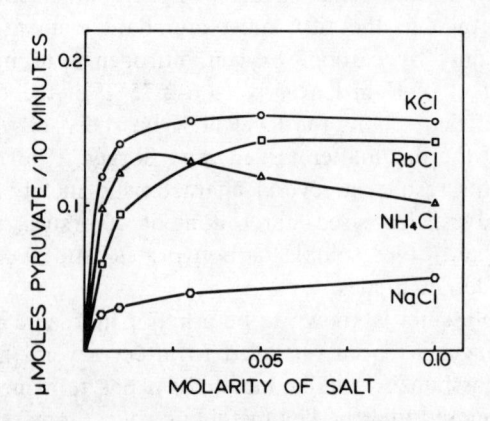

Fig. 1—The effect of increasing concentrations of univalent cation chlorides on the activity of pyruvate kinase from cotton seed. Cotton seed extract (0.8 mg of protein per reaction of 1 ml) was dialyzed for 20 hr against Tris buffer to remove endogenous salts. In each experiment the activity of the enzyme in reactions containing different concentrations of salts and $0.05M$ Tris buffer at pH 7.4 was compared with the activity of a control lacking the substrate ADP. Reactions lacking alkali salts contained $0.05M$ Tris at pH 7.4. See Miller and Evans (1957) for futher details of the procedure.

centration of about 0.05M. Investigations with several other univalent-cation-requiring-enzymes from plants and animals have revealed patterns of effectiveness of different species of univalent cations for activation, and concentrations necessary for optimal enzyme activity that are similar to those that have been established for pyruvate kinase (Evans and Sorger, 1966).

II. MECHANISMS OF ACTION OF UNIVALENT CATIONS IN ENZYME ACTIVATION

In considering the mechanism of action of univalent cations in enzyme activation, Evans and Sorger (1966) called attention to the following points: (i) the diversity of types of chemical reactions that are catalyzed by univalent-cation-requiring-enzymes; (ii) the specificity of univalent cations as enzyme activators; and (iii) the high concentration of univalent cations that are generally required for activation. Studies on the mechanism of action of univalent cations in enzyme activation have been conducted with a very few enzyme systems. The results of these investigations support the conclusion that the primary effect of univalent cations is associated with the apoenzyme or protein molecule. The evidence suggests that the univalent cations interact with the protein molecule by two basically different, but not necessarily exclusive mechanisms. It is proposed that they affect (i) the quarternary organization or subunit structure, or (ii) affect the ternary organization or enzyme conformation.

A. The Effect of Univalent Cations on Protein Subunit Structure

Glycerol dehydrase catalyzes the conversion of glycerol to β-hydroxypropionaldehyde in the presence of coenzyme B_{12}. The enzyme from *Aerobacter aerogenes* is active in the presence of potassium, rubidium, or ammonium, but is not active in a medium containing sodium. Recently, Schneider, Peck, and Pawelkiewicz (1966) have shown that glycerol dehydrase dissociates into two subunits when the enzyme is passed through a Sephadex G 100 column equilibrated with the nonactivator univalent cations, sodium or cyclohexylammonium. Neither subunit alone exhibits enzyme activity. In contrast when glycerol dehydrase is passed through a Sephadex G 100 column equilibrated with the activator univalent cations, potassium or ammonium, the enzyme molecule is eluted as an associated unit which is catalytically active. Furthermore, in the presence of the substrate, glycerol, the enzyme remains associated as one monomeric unit even when sodium, a nonactivator univalent cation, is present. In this case, however, the enzyme is catalytically inactive, which indicates that the effect of activator univalent cations on this enzyme is not limited to the subunit structure of the protein, but must have other functions. Additional studies are needed to determine the occurrence of this type of mechanism with other univalent-cation-requiring-enzymes.

B. The Effect of Univalent Cations on the Protein Conformation

The results of a number of studies with different enzyme systems requiring univalent cations for activity have supported the hypothesis that univalent cations affect the physical structure or conformation of enzyme proteins. This conclusion is based on a number of different approaches and techniques which are discussed below.

1. Kinetic Observations

Investigations of the effects of univalent cations on the kinetics of enzyme reactions have been used to provide evidence of the mechanism of activation by univalent cations (see Evans and Sorger, 1966). In 1953, Kachmar and Boyer (1953) made a detailed study of this type on pyruvate kinase from rabbit muscle. They observed that the concentration of potassium in the reaction mixture affected the V_{max} but not the K_m for substrates. Similar observations have been made with tryptophanase (Happold and Beechey, 1958), acetic thiokinase (Evans, Clark, and Russell, 1964), and fructokinase (Parks, Ben-Gershom, and Lardy, 1957). From the results of investigations with pyruvate kinase, Kachmar and Boyer (1953) considered a series of possible mechanisms from which reaction products were formed. They concluded that the results supported a reaction mechanism probably involving an independent combination of the substrate, phosphoenolpyruvate, and potassium under equilibrium reaction conditions to form an active ternary complex. The authors explained the difference in the effectiveness of univalent cations as activators by considering the size of the hydrated ionic radii as reported by Jenny (1932) and Nachod and Wood (1945). According to these workers, potassium, rubidium, and ammonium, all activators, have hydrated ionic radii of 5.32A, 5.09A, and 5.37A, respectively. Sodium, which showed a much lower capacity to activate, has a hydrated ionic radius of 7.09A, and lithium, a nonactivator, has a hydrated ionic radius of 10.03A. In their opinion, the effectiveness of these cations as activators depends on the capacity of the cations to bind at a negative site (sites) and to displace adjoining structures of the enzyme by some critical amount.

More recently, Melchior (1965) has examined a wide range of concentrations of adenosine diphosphate (ADP) and activator metals on the reaction kinetics of pyruvate kinase. She concluded that the complex MgADP^{-1} and potassium were the major species which interacted with the enzyme and that the concentration of these species was important in determining the reaction velocity. She suggested that potassium activated the enzyme and that this activation influenced the binding of substrates.

Happold and Beechey (1958) have studied the effect of potassium and other univalent cations on the kinetics of tryptophanase, a univalent-cation-requiring-enzyme from *Eschericia coli*. Consistent with the results that were obtained with pyruvate kinase, they showed that potassium affected the organization of the enzyme protein. Furthermore they agreed with the suggestion of Kachmar and Boyer (1953) that the difference in the capacities of univalent cations to function as activators was associated with different hydrated ionic radii of the univalent cations. The values of the hydrated ionic radii used by Happold and Beechey (Latimer, 1955) were correlated with the capacities of cations to activate, even though they differ from the values used by Kachmar and Boyer (1953) in their interpretation of the role of univalent cations in the pyruvate kinase reaction.

Hiatt (1964) has studied the univalent cation requirements for the acetic thiokinase reaction which has been shown to proceed through two steps (Berg, 1956):

$$\text{ATP} + \text{acetate} \rightleftharpoons \text{adenyl acetate} + \text{PP} \qquad [1]$$

$$\text{adenyl acetate} + \text{CoA} \rightleftharpoons \text{acetyl CoA} + \text{AMP} \qquad [2]$$

Potassium or rubidium was shown to activate reaction (2), but sodium and lithium failed to function in this reaction. Evans et al. (1964) reported similar results with acetic thiokinase from yeast and in addition observed that the concentration of potassium affected the V_{max} but not the K_m for substrates. They interpreted the results as indicating that univalent cations must influence the protein conformation. Their reasoning was based on the following considerations:

$$V_{max} = K_3 E_t. \qquad [3]$$

In equation (3), V_{max} is the maximum velocity of the reaction under steady-state conditions, K_3 is the rate constant for the formation of product from enzyme substrate intermediate, and E_t is the total amount of active sites on the enzyme. Since $K_m = (K_2 + K_3)/K_1$, it would be probable that a change in K_3 would also affect the K_m. They showed, however, that K_m was not affected by the univalent cation concentration. The alternative possibility that V_{max} changes due to changes in E_t, the total amount of enzyme, was considered feasible. They concluded, therefore, that the most probable explanation is that univalent cations induce a conformational change in the protein in such a way to expose additional active sites. This is equivalent to an increase in E_t, the total amount of enzyme.

2. Effect of Cations on Enzyme Stability

Stoppani and Milstein (1957) observed that the activity of the univalent-cation-requiring-enzyme acetaldehyde dehydrogenase was affected by the presence of a number of thiol reagents, but the sensitivity to these compounds varied according to the univalent cation present. The activator cations, potassium and rubidium, diminished the stimulatory effect of thiol reagents on enzyme activity, whereas, nonactivator cations, sodium and lithium, increased the stimulatory capacity of thiol compounds. Sorger and Evans (1966) observed a correlation between the capacities of univalent cations to activate acetaldehyde dehydrogenase and capacities of these cations to protect the enzyme from inactivation by heat. Moreover, when acetaldehyde dehydrogenase was dialysed against Tris buffer, a treatment which leads to loss of enzyme activity, the capacities of univalent cations to restore activity were directly correlated with capacities to activate the enzyme. In these experiments potassium and rubidium were much more effective in restoring activity than sodium or lithium.

A similar study with pyruvate kinase has shown a correlation between the capacities of univalent cations to activate and capacities to protect against loss of activity (Fig. 2). The investigations with both acetaldehyde dehydrogenase and pyruvate kinase were interpreted as indicating that univalent cations affect the conformation of the enzyme molecule. Activator univalent cations appear to stabilize the enzyme and prevent loss of activity, whereas nonactivator univalent cations contribute toward a conformation of the protein that leads to denaturation.

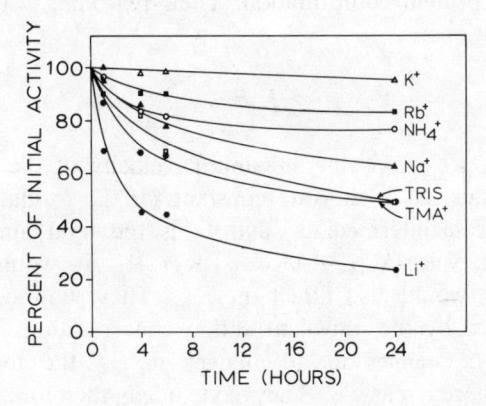

Fig. 2—The effect of a series of univalent cations on the stability of pyruvate kinase. Dialyzed pyruvate kinase was incubated at 0C in 0.05M phosphate salts (pH 7.4) of the univalent cations indicated. Aliquots of the enzyme were withdrawn and assayed for activity as indicated. See Wilson, Evans, and Becker (Fig. 3) for details of the experiment.

3. Immunoelectrophoretic Patterns of Pyruvate Kinase

Sorger, Ford, and Evans (1965) reported that the univalent cation environment markedly affected the immunoelectrophoretic patterns of pyruvate kinase. In the presence of activator univalent cations, the interaction of the enzyme with antibodies produced a relatively simple immunoelectrophoretic pattern, whereas in presence of nonactivator cations a more complex pattern was observed (Fig. 3). By suitable manipulation of the univalent cation environment in the electrophoretic buffer and agar gel the patterns could be altered from one form to the other. In similar experiments the immunoelectrophoretic patterns of catalase, which exhibits no univalent cation requirement, showed no effect of the univalent cation environment on the immunoelectrophoretic behavior. The results were interpreted to indicate that univalent cations affect the conformation of pyruvate kinase. In the presence of potassium, rubidium, or ammonium, a protein conformation presumed to be catalytically active was produced, whereas in the presence of Tris or lithium a conformation of the protein was formed that was assumed to be inactive.

4. Spectrophotometric Studies

The technique of difference spectrophotometry has been used by Suelter and co-workers to investigate the effect of certain environmental conditions

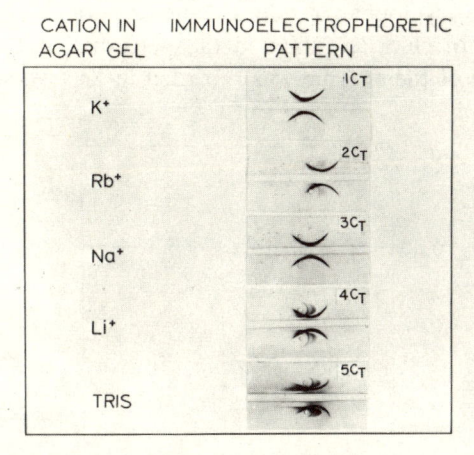

Fig. 3—The effect of univalent cations on the immunoelectrophoretic patterns of pyruvate kinase. About 1 ug of dialized pyruvate kinase was placed on slides covered with agar containing 0.05M Tris phosphate and 0.05M solutions of the univalent cations salts as indicated. After 120 min. the electrophoreses was terminated and pyruvate kinase antibodies were added to a trough in the center of the agar slide and allowed to diffuse into the agar and interact with the enzyme. The patterns of interaction were revealed after an appropriate staining procedure. For further details of these experiments see Sorger, Ford, and Evans (1965).

on the ultraviolet absorption properties of crystalline pyruvate kinase (Suelter and Melander, 1963; Kayne and Suelter, 1965; Suelter et al., 1966). A comparison of pyruvate kinase in TrisCl buffer containing either an univalent or divalent cation versus enzyme in TrisCl buffer and TMACl resulted in an ultraviolet difference spectrum characteristic of the perturbation of tryptophan residues. The magnitude of the absorption difference at 295 mμ was used as a basis for determining dissociation constants for both univalent and divalent cations. Suelter et al. (1966) did not attempt, however, to relate the perturbation of the tryptophan residues to the capacity of univalent cations to activate pyruvate kinase.

Recently the capacities of univalent cations to perturb tryptophan residues of pyruvate kinase have been investigated in relation to known capacities of univalent cations to activate the enzyme (Wilson, Evans, and Becker, 1967). A spectrum of pyruvate kinase in 0.1M KCl versus enzyme in 0.1M TrisCl showed absorption difference peaks characteristic of the perturbation of tryptophan residues (Fig. 4). The absorption differences at 295 mμ obtained from pyruvate kinase in different salt environments are presented in Table 1. As the results show, a comparison of pyruvate kinase in any one of a series 0.1M solutions of activator or nonactivator univalent cation chlorides versus enzyme in 0.1M TrisCl revealed absorption differences at 295 mμ (Table 1, exp. 1-4). The difference spectrum of pyruvate kinase in 0.1M KCl versus enzyme in 0.1M LiCl resulted in no absorption difference at 295 mμ (Table 1, exp. 5). These results suggest that the presence of TrisCl in the reference cuvette was responsible for the perturbation of the tryptophan residues. Furthermore, it was observed that phosphate instead of chloride inhibited the perturbation that was observed when the spectrum of the enzyme was recorded in KCl versus TrisCl (Table

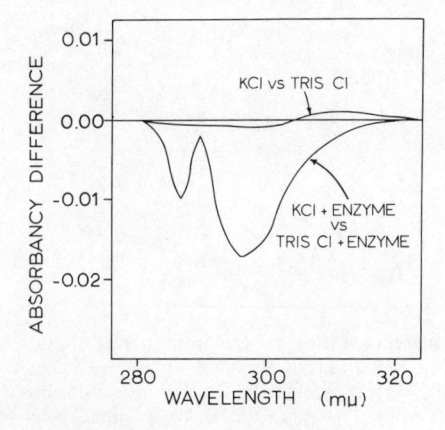

Fig. 4—Difference spectra of pyruvate kinase in 0.1M KCl in the sample cuvette versus enzyme in 0.1M TrisCl in the reference cuvette. The procedure for the experiment was essentially the same as that described in Table 1. For further details see Wilson, Evans, and Becker (Fig. 1a) (1967).

Table 1—The effect of different univalent cation salts on the absorbancy difference at 295 mμ of pyruvate kinase (after Wilson et al., 1967) *

Experiment no.	Ion comparison Sample cuvette	Ion comparison Reference cuvette	Absorbancy difference at 295 mμ ($-\Delta$ O.D. $\times 10^3$)
1	0.1M KCl	0.1M TrisCl	12
2	0.1M NH₄Cl	0.1M TrisCl	12
3	0.1M NaCl	0.1M TrisCl	11
4	0.1M LiCl	0.1M TrisCl	11
5	0.1M KCl	0.1M LiCl	0
6	0.1M K-PO₄	0.1M Tris-PO₄	0
7	0.1M K-PO₄	0.1M TrisCl	15
8	0.2M K-PO₄	0.1M TrisCl + 0.1M Tris-PO₄	6

* Each 3-ml reaction contained 0.033M Tricine buffer at pH 7.4, 6.7 × 10⁻⁴M mercaptoethanol, 6 mg of pyruvate kinase and salts (adjusted to pH 7.4) as indicated. Difference spectra were recorded with a Cary model 11 spectrophotometer using a slide-wire that gave full deflection at an absorbance of 0.1.

1, exp. 6). Moreover, a comparison of pyruvate kinase in 0.1M potassium phosphate versus enzyme in 0.1M TrisCl, resulted in an absorption difference (Table 1, exp. 7). The magnitude of the difference was diminished, however, by increasing the concentration of potassium phosphate in the sample cuvette and Tris phosphate in the reference cuvette (Table 1, exp. 8). The most logical explanation for these results is that the perturbation of the tryptophan residues of pyruvate kinase is most prominent in a TrisCl environment and that substitution of phosphate for chloride appears to inhibit the perturbation.

It was observed that pyruvate kinase in a 0.1M concentration of KCl versus enzyme in an equal concentration of LiCl produced no absorption difference at 295 mμ (Table 1). One possible explanation for this result is that the univalent cation affects the enzyme conformation only when substrate and/or magnesium are present. To examine these possibilities pyruvate kinase was compared in 0.1M KCl with respect to enzyme in 0.1M LiCl with different combinations of substrates and divalent cation activators in the cuvettes. In no case did the separate addition of the substrates, phosphoenolpyruvate, pyruvate, ADP, or ATP to both cuvettes result in an absorbance difference in the range from 280 mμ to 320 mμ (Wilson et al., 1967).

Pyruvate kinase in solutions containing different univalent cations at salt concentrations above 0.1M were compared to determine whether tryptophan residues were perturbed at these concentrations (Wilson et al., 1967). As shown in Table 2 absorption differences for all ion comparisons except potassium versus rubidium increased with increasing concentrations of salts in the two cuvettes. Furthermore, at any one concentration of salt above 0.33M it appears that the magnitude of the difference in absorbance was greatest when the enzyme in a solution of univalent cation salt known to

Table 2—The effect of different univalent cations at a series of concentrations on the absorbancy differences at 295 mu of pyruvate kinase (from Wilson et al., 1967)*

Ion comparison		Absorbance differences at 295 mμ at various concentrations of cation chloride					
Sample cuvette	Reference cuvette	0.10 M	0.33 M	0.50 M	0.66 M	0.79 M	0.86 M
		\-\-\-\-\-\-\-\-\-\-\-\-\-\-\- $-\Delta$ O.D. $\times 10^3$ \-\-\-\-\-\-\-\-\-\-\-\-\-\-\-					
K^+	Li^+	0	2	5	8	12	17
Rb^+	Li^+	0	1	7	11	13	18
K^+	TMA^+	0	2	4	7	11	16
K^+	Na^+	0	0	2	5	6	7
Na^+	Li^+	0	0	3	7	11	13
K^+	Rb^+	0	0	0	0	0	0

* The procedure was essentially the same as that described in Table 1 with the exception that the concentration of salts in cuvettes varied as indicated.

effectively activate the enzyme was compared with enzyme in a solution of univalent cation salt known to lack activating capacity. These results, which were obtained at nonphysiological concentrations of salts, show an inverse correlation of capacities of univalent cations to activate pyruvate kinase and capacities to perturb tryptophan residues of the enzyme. The authors suggested that the enzyme molecule is poised by physiological concentrations of activator univalent cations in an active conformation and that the enzyme in comparable concentrations of nonactivator cations is in an inactive conformation. When the concentrations of cations are increased in either of these environments additional effects on the physical structure of the protein may result. At least one effect of the higher concentrations of salt on the nonactive conformation of the protein is the perturbation of the tryptophan residues. With the active conformation of the protein, which is postulated to be poised differently at physiological salt concentrations, the high concentration of salts shows no evidence of causing appreciable perturbation of the tryptophan residues.

5. Ultracentrifugation Studies

It has been reported by Wilson et al. (1967) that the Schlieren pattern of crystalline pyruvate kinase in the analytical ultracentrifuge showed one homogeneous peak of protein regardless of whether the enzyme was dissolved in potassium, Tris, or lithium salt solutions. No tendency for this enzyme to dissociate into subunits was observed in environments of either activator or nonactivator univalent cations. The dissociation of glycerol dehydrase into subunits in a nonactivator univalent cation environment (Schneider et al., 1966) appears to reflect a mechanism of action of univalent cations with this enzyme that is different from the mechanism with pyruvate kinase.

III. GENERAL CONCLUSIONS

Evans and Sorger (1966) have listed over 40 enzyme systems the activities of which have been reported to be stimulated by or to exhibit an absolute requirement for a univalent cation. It seems probable that other enzyme systems with univalent cation requirements have not been recognized because potassium buffer systems are used routinely for many enzyme preparations and assays. Of those enzyme systems for which univalent cation requirements have been observed, the mechanism of action has been examined in detail in only a few cases. Pyruvate kinase is probably the most extensively studied enzyme of this type. With this enzyme the evidence from several different approaches supports the conclusion that the principal role of the univalent cations in the activation of the enzyme is involved in protein conformation. The precise mechanism of action of univalent cations with this enzyme necessitates a better knowledge of the structure and composition of the protein molecule as well as an understanding of the overall reaction mechanism.

It may be instructive to summarize our current understanding of the interaction of univalent cations and enzyme, using the knowledge obtained from studies with pyruvate kinase. The models in Fig. 5 show a flexible protein molecule, whose structure is dependent on the cationic environment. In the presence of K^+ or other activator univalent cations, the active site is shown available for substrate binding. This structure is enzymatically stable as well as relatively resistant to the effects of high salt concentrations on the perturbation of tryptophan residues. Furthermore the protein in this form produces a simple immunoelectrophoretic pattern. In contrast when the enzyme is in a nonactive univalent cation environment the active site is illustrated as being unavailable for substrate binding, thus poising the enzyme in a different and less stable conformation. One consequence of this difference in structure is the capacity of high salt concentration to further interact with the protein, resulting in the perturbation of tryptophan residues. Pyruvate kinase in this type of environment has been observed to produce a complex immunoelectrophoretic pattern, perhaps reflecting several gradations in structural alteration, all of which have in common the lack of enzymic activity.

Additional studies are necessary before it will be possible to assess the similarities and differences in the mechanism of action of univalent cations with the many univalent-cation-requiring-enzymes. With the information available it appears that the effect of univalent cations on glycerol dehydrase is at least in part on the subunit structure of the enzyme and therefore the mechanism of activation with this enzyme apparently is different from the mechanism of activation with pyruvate kinase.

Several questions remain unanswered regarding the function of potassium

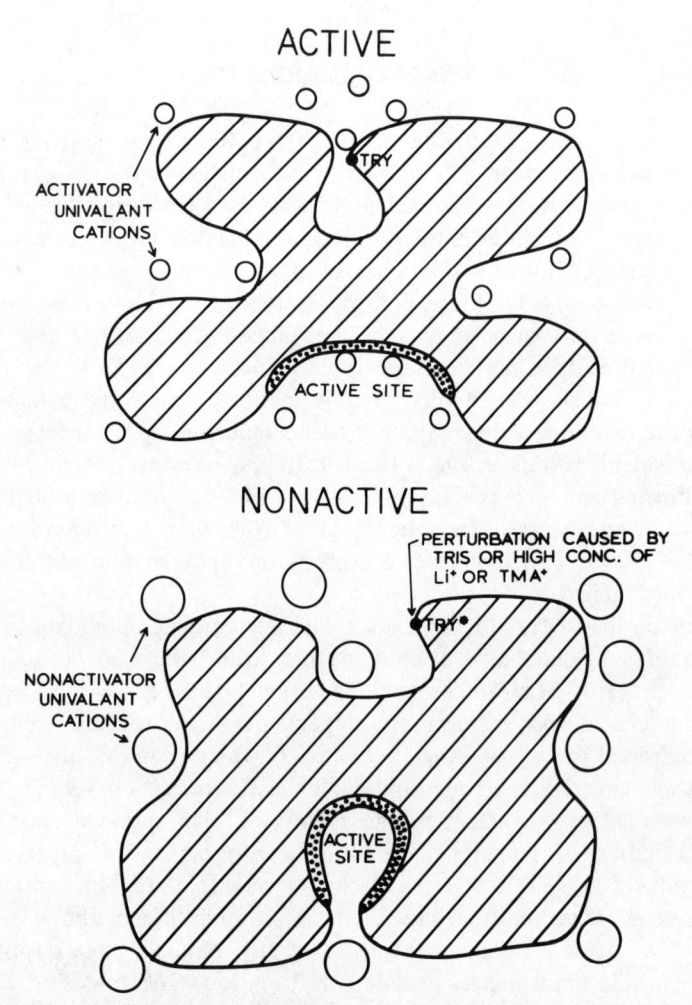

Fig. 5—Model conformational structures of an univalent-cation-requiring-enzyme in an activator and nonactivate univalent cation environment.

in enzyme activation. No studies have been reported concerning the types of chemical bonds or the forces involved in univalent cation induced conformational changes. In addition, the reason for the high concentration of activator univalent cations needed for maximum activation is not clear. Regarding this point, Evans and Sorger (1966) have suggested that univalent cations do not form highly stable chelate complexes comparable to those formed by many of the transition metals and thus a relatively high concentration of univalent cation salts would be expected to be necessary for the maintenance of an association of univalent cations and protein molecules. Finally, a basic unanswered question remains as to why certain enzyme systems exhibit requirement of univalent cations for activity. The possibility has been suggested that the requirement for potassium or other univalent cations may represent a "fine metabolic valve" controlling the rates and use of different metabolic pathways (Wyatt, 1964). The possible compartmentalization of potassium in the cell and thus the availability for activation may influence the pathways in which substrates are metabolized. In this way potassium or other univalent cations may play a role in the control of growth, differentiation, and development in living organisms.

REFERENCES

Berg, P. 1956. Acyl adenylates: an enzymatic mechanism of acetate activation. J. Biol. Chem. 222:991-1013.

Boyer, P. D. 1953. The activation by potassium and occurrence of pyruvic phosphoferase in different species. J. Cellular and Comp. Physiol. 42:71-77.

Boyer, P. D., H. A. Lardy, and P. H. Phillips. 1942. The role of potassium in muscle phosphorylations. J. Biol. Chem. 146:673-682.

Boyer, P. D., H. A. Lardy, and P. H. Phillips. 1943. Further studies on the role of potassium and other ions in the phosphorylation of the adenylic system. J. Biol. Chem. 149:529-541.

Evans, H. J., R. B. Clark, and S. A. Russell. 1964. Cation requirements for the acetic thiokinase from yeast. Biochim. Biophys. Acta 92:582-594.

Evans, H. J., and G. J. Sorger. 1966. Role of mineral elements with emphasis on the univalent cations. Ann. Rev. Plant Physiol. 17:47-76.

Happold, F. C., and R. B. Beechey. 1958. Univalent metals and other loose or nonspecific activations. *In* E. M. Crook (ed.) Metals and enzyme activity. p.52-63. Cambridge University Press, London. (Biochemical Society. Symposium no. 15).

Hiatt, A. J. 1964. Further studies on the activation of acetic thiokinase by magnesium and univalent cations. Plant Physiol. 39:475-479.

Jenny, H. 1932. Studies on the mechanism of ionic exchange in colloidal aluminum silicates. J. Phys. Chem. 36:2217-2258.

Kachmar, J. F., and P. D. Boyer. 1953. Kinetic analysis of enzyme reactions. II. The potassium activation and calcium inhibition of pyruvic phosphoferase. J. Biol. Chem. 200:669-682.

Kayne, F. J., and C. H. Suelter. 1965. Effects of temperature, substrate and activating cations on the conformation of pyruvic kinase in aqueous solution. J. Amer. Chem. Soc. 87:897-901.

Latimer, W. M. 1955. Single ion free energies and entropies of aqueous ions. J. Chem. Phys. 23:90-92.

Lubin, M., and Ennis, J. H. 1963. The role of intracellular potassium in protein synthesis. Federation Proc. 22:302.

Melchior, J. B. 1965. The role of metal ions in the pyruvic kinase reaction. Biochemistry 4:1518-1525.

Miller, G., and H. J. Evans. 1957. The influence of salts on pyruvate kinase from tissues of higher plants. Plant Physiol. 32:346-354.

Nachod, F. C., and W. Wood. 1945. The reaction velocity of ion exchange. II. J. Amer. Chem. Soc. 67:629-631.

Parks, R. E., E. Ben-Gershom, and H. A. Lardy. 1957. Liver fructokinase. J. Biol. Chem. 227:231-242.

Schneider, Z., K. Peck, and J. Pawelkiewicz. 1966. Enzymic transformation of glycerol to β-hydroxypropionic aldehyde. II. Dissociation of the enzyme into two protein fragments. Bull. Acad. Polonaise Sci. XIV:7-12.

Schweet, R. S., R. Arlinghaus, J. Shaeffer, and A. Williamson. 1964. Studies on hemoglobin synthesis. Medicine 43:731-45.

Seitz, I. F. 1949. Role of potassium and ammonium in the transfer of phosphate from phosphopyruvic acid to the adenylic system. Biokhimiya 14:134-140.

Sorger, G. J., and H. J. Evans. 1966. Effect of univalent cations on the properties of yeast NAD+ acetaldehyde dehydrogenase. Biochim. Biophys. Acta 118:1-8.

Sorger, G. J., R. E. Ford, and H. J. Evans. 1965. Effect of univalent cations on the immunoelectrophoretic behavior of pyruvate kinase. Proc. Natl. Acad. Sci. U. S. 54:1614-1621.

Stoppani, A. O. M., and C. Milstein. 1957. Essential role of thiol groups in aldehyde dehydrogenase. Biochem. J. 67:406-416.

Suelter, C. H., and W. Melander. 1963. Use of protein difference spectrophotometry to determine enzyme-cofactor dissociation constants. J. Biol. Chem. 238:PC4108-PC4109.

Suelter, C. H., Rivers Singleton, Jr., F. J. Kayne, Sandra Arrington, J. Glass, and A. S. Mildvan. 1966. Studies on the interaction of substrate and monovalent and divalent cations with pyruvate kinase. Biochemistry 5:131-139.

Webster, G. C. 1956. Effects on monovalent cations on the incorporation of amino acids into protein. Biochim. Biophys. Acta 20:565-566.

Wilson, R. H., H. J. Evans, and R. R. Becker. 1967. The effect of univalent cation salts on the stability and on certain physical properties of pyruvate kinase. J. Biol. Chem. 242:3825-3832.

Wyatt, H. V. 1964. Cations, enzymes and control of cell metabolism. J. Theoret. Biol. 6:441-470.

Role of Potassium in Human and Animal Nutrition

WALTER S. WILDE

University of Michigan
Ann Arbor, Michigan

I. INTRODUCTION

When a fertilizer specialist is chagrined that the lay world associates his product with lowly barnyard manure, he should bolster his ego by enumerating the very dramatic roles potassium plays in life processes. Let this be the chief role of my essay. To my knowledge, the facts I shall enumerate provide no practical suggestions concerning the form of potassium fertilizer you should prepare. If you can induce the plant to grow, it will always contain enough potassium for animal nutrition. Man and animals suffer potassium deficiencies only when ill for reasons other than improper food composition.

You may rightfully claim that potassium is life itself. A living cell is to the physical chemist a chemical cell or storage battery charged with the electrolyte potassium. Currents or migrations of potassium and sodium as charged ions moving across animal cell membranes generate or indeed are the action potentials or impulses that make protoplasm active and alive. These events generate the nerve impulse, the brain wave, the heart beat, the electrocardiogram.

Another role of potassium relates to protection of cells against osmotic damage such as would result if water were imbibed from watery extracellular fluid. Suppose that protoplasm first evolved as a gel made up of genetic nucleoprotein, enzyme, or other protein. Such chemicals act as weakly ionizing acids. Their hydrogens would be exchanged off as in an ion exchange resin by sodium and potassium, in the sea if they were unicellular animals, or in extracellular fluid if land animals. The ensuing ionization, of metallic cations associated with polyvalent anions, produces a high osmotic activity with swelling and damage to the cell. Actually chlorides of sodium and potassium would diffuse to equal concentration and osmotic strength between water in the interstices of the gel and in the external fluid, but the ionized gel itself would exert the excess osmotic activity that induces swelling. The protection that has evolved is a special partition or separation of potassium from sodium promulgated by the active cell membrane. A specific ion "pump" or transporting system in cell membranes moves sodium ions continually outward from the protoplasm and potassium inward across the external surface. As a result of this selective pumping, the polyvalent proteins, etc., inside the cell remain in ionic association only with potassium while the sodium is relegated to the outside medium as an inorganic salt mainly of chloride. Osmotically, polyanion-potassium inside now just balances osmotically sodium salt outside.

It is this sequestration of potassium, rich inside, and of sodium, rich outside, which because of a slow retrograde leak of each, sets up the ions currents mentioned above. A so-called "resting membrane potential" results from the potassium current flowing outward. An "action current" is a sodium current or leak inward. The two leaks are the basis of the electrical properties of cell membranes which underlie irritability or being alive. The reverse and active ion transports keep the cell charged with potassium and relatively free of sodium.

Thirdly, potassium, like many ions, is a secondary "activator" of enzymes. Thus in this third sense, the element is the "spice of life". Undoubtedly enzymes involved in protein synthesis are activated by potassium. Young rats starved of potassium do not grow well. Likely this is more than the mere obligatory association of potassium with polyanion protein as a building block (Kernan, 1965; Ussing, et al. 1960).

Physiologists are now studying a "transport enzyme" — an ATPase which splits the third or terminal phosphate group off the high energy storage form adenosinetriphosphate. The splitting may provide energy for the potassium and sodium transport mentioned above. Strangely, sodium and potassium both activate the ATPase while at the same time being bound to it in such a way that they are transported across the cell membrane in the appropriate direction.

II. CONCENTRATION OF POTASSIUM IN CELLS AND BODY FLUIDS

Since determination of the proportion of total osmotic pressure contributed by ions in cells is of central importance, ion contents are expressed in milliequivalents per unit volume of water content in cells (meq/liter). It is the ions present in unit volume of water that act osmotically, not the number of ions per gram of tissue in which dry solids occupy space which is inert osmotically. Determinations of ion and water contents in complicated tissues is very indirect, but rather simple in erythrocytes recovered from blood in the centrifuge. Equivalents (equal to moles when the ion is monovalent) are directly proportional to osmotic pressure by the factor of Avogadro's number.

In Table 1, the sum of K + Na in red cell water nearly equals that in blood plasma, but potassium is high in cells and sodium is high in plasma. This illustrates the work load of ion pumps which must develop these differences. In each system, cell and plasma, cation sum must equal anion sum. In cells, the excess of 155 cations over 97 anions (Cl plus HCO_3) is matched by anionic radicals of hemoglobin; in plasma, mostly by HCO_3 (25 to 30 meq/liter) and by plasma proteins. The sums on each side approach 150 cations plus 150 anions or 300 millimoles/liter grand sum. This is the classical value for mammalian tissues and corresponds to 0.3 mole × 22.4 atm/mole or 6.72 atm of osmotic pressure acting but balanced across a cell membrane. Similar values have been estimated for all types of cells and tissues but to cite them all would be tedious.

III. MECHANISM OF ACCUMULATION OF POTASSIUM INSIDE CELLS

It has been said that "potassium is of the soil and not the sea; it is of the cell but not the sap" (Fenn, 1940). This is an apt comparison, that potassium is rich in soil and rich in protoplasm. But there the simile ends. The vast weight of authority and of evidence does not admit that potassium is confined inside cells by firm binding or complexing as it might well be in soils. It is generally conceded that half of the osmotic pressure of cell protoplasm, that half attributed to positively charged cations, is due to potassium. Potassium could not be both bound, nonionized, and yet exert an osmotic pressure. The resting membrane potential, as fit into the ionic hypothesis of nerve conduction and recently given a Noble Award (Hodgkin, 1964), presupposes ionized potassium which generates potential by diffusion of free ions across cell membranes.

Thus the confinement or the limiting barrier is set at the cell surface within the membrane. The "carrier" hypothesis states that potassium is lifted across the cell membrane inwards by complexing to a special site or molecule on the outside surface of the membrane. By migration or rotation of the binding site, potassium is released into the cell from the inner side

Table 1—Ionic concentrations in the erythrocytes of man *

Ion concentration		Ion concentration	
——— meq/liter cell water ———		——— meq/liter plasm ———	
K = 136	Cl = 78	Na = 155	Cl = 112
Na = 19	HCO_3 = 19	K = 5	
Sum: 155	97	160	112

* Values taken from Harris (1956; cf Table 14).

of the membrane. A completely analogous carrier molecule is presumed for sodium, but carrying sodium in the opposite direction from inside to outside the cell. The end result is a potassium-rich, sodium-poor protoplasm. Whether or not their action is coupled the transfer systems for sodium and potassium are referred to as "pumps". Since a concentration difference is established for each ion energy must be expended to push an ion uphill. We speak of active ion transport. Evidence will be cited that sodium will move outwards only when potassium is available outside to move inwards, that sodium-potassium movements are in an obligatory couple, perhaps involving a common carrier molecule, that this is actually a protein or enzyme molecule, ATPase, that catalyzes removal of the terminal or high energy phosphate of adenosinetriphosphate, whose energy moves the ions.

Once potassium has accumulated inside the cell, it now possesses a concentration drive and tends to diffuse or leak outwards. This flow is referred to as passive diffusion or passive ion transport. Biochemical energy is not being expended at the time, but was spent earlier for the inward pumping. Later we will see how the outward diffusion develops resting membrane potential. The round trip ion cycle is referred to as turnover or exchange. Potassium experiences active influx and passive efflux. At moments these opposing movements are unequal but over time they average out equal or else the content of cell potassium would diminish and death would result.

By inference the influx of sodium is passive or downhill and the outflux is active.

IV. MEMBRANE ATPase

For more than a decade after high energy phosphate, adenosinetriphosphate ATP, became known as a last stage energy packet for several biological end functions, transport specialists continued to search for a particular chemical step in metabolic pathways, the molecules of which might act as carriers. The influence of inhibitors of particular steps in the Krebs and glycolytic cycles were tested to no avail. As Glynn (1966) recently commented, most cells depend upon Krebs' aerobic cycle for energy for ion transport, but red blood cells, which contain no mitochondria and no cytochrome, possess a slow transport dependent on glycolysis only. It is unlikely then that a particular metabolic step supplies carrier molecules. More likely, ATP is the energy source common to all tissue types. Caldwell et al. (1960) made a conclusive experiment with giant axons of squid. Cyanide poisoned the extrusion of sodium from the nerve fibers, but microinjection of ATP along the axoplasm core of the fiber restored sodium extrusion in proportion to the amount of ATP injected.

Strangely, the carrier associated with the ATP energy source is not a metabolic chemical but the protein enzyme ATPase which splits the terminal or high energy phosphate from ATP. In a favorite scheme, sodium ion activates binding of ATP to the enzyme. Perhaps sodium forms a link or bonding between ATP and enzyme. At this stage sodium is fixed, nonionized to enzyme. The enzyme is in the membrane. For instance, when erythrocytes, placed in hypotonic fluids, swell by imbibing water, the outer hull or membrane stretches, which allows leakage of internal hemoglobin. The hemolyzed cells when washed repeatedly still contain ATPase in the outer envelope or ghost. In normal cells, during the transfer act, internal cell sodium is presumed to affix to ATPase at the inner cell membrane. A rotation of the enzyme molecules places sodium on the outside surface of the membrane.

At the outside surface, Na is released to constitute active transport and extrusion from the cell of sodium only if potassium is present to complex with the enzyme. As outside potassium is bound, sodium is released outward, the terminal phosphate is released along with the ATP residue adenosinediphosphate ADP. The ultimate fate of the high energy bond, of ADP and of the terminal inorganic phosphate are still to be traced. The same or an opposite molecular rotation of the ATPase places the bound potassium inside the cell ready for release, where internal sodium probably augments release. Coupled sodium-potassium transfer has occurred. A model for ATPase as a rotating carrier was described by Opit and Charnock (1965).

Suffice it to say in this elemental discussion that this story was initiated by Skou (1965) who began in Aarhus, Denmark, with crab nerves and worked with a preparation of microsomes or endoplasmic reticulum (broken cytoplasmic membranes). Other early and continuing workers are Glynn in Cambridge, Post in Nashville, and Whittam in Oxford. These work mainly with red cell ghosts.

Most of the evidence derives from a measure of rate of formation of inorganic phosphate from ATP and enzyme when incubated together with appropriate concentrations of the activating potassium and sodium. The fascinating spatial requirements of sodium and potassium mentioned above for each side of the membrane, were worked out separately by Glynn (1962) and Whittam (1962). "Reconstituted" ghosts of red cells were prepared containing an elected composition of ATP and of ions. Cells are made to hemolyze or swell in a hypotonic solution. It other chosen constituents are placed in the hemolyzing fluid, they will penetrate the opened ghosts by diffusion to a desired concentration. When the cells are suddenly shrunken and sealed in a new solution, the constituents are trapped at the elected concentration. The resealed ghosts will even transport sodium and potassium back toward normal concentration values. In particular, a sealed ghost with a high internal sodium placed in a high external potassium will hydrolyze ATP, produce inorganic phosphate, at maximum rate. A ghost with high internal potassium, with or without, sodium is slow. External sodium does not accelerate.

A first type of circumstantial evidence for a transport role for ATPase compared the influence of concentration of sodium and potassium on the rate of ATP splitting with the corresponding rate of ion transport. The red cell was a convenient object because flux rates of sodium and potassium had been studied in detail using isotopes. The rate of active influx (pumping) of potassium into red cells, as measured by ^{42}K, rises as external K concentration is elevated between zero and 6 meq/liter of medium. Pumping rate rises hyperbolically with potassium in the medium, the curve first rises rapidly and then flattens to a horizontal asymptote at which 2.1 meq K are entering 1 liter of packed cells/hour (Glynn, 1956). At this maximum rate external potassium keeps filling all binding sites on the ATPase enzyme. The external K concentration at which ion entry rate is half maximum is K_M = 2.2 meq/liter. By analogy to the Michaelis-Menten kinetic analysis of combination of substrate to enzyme, this is combination of potassium to carrier or ATPase and the K_M value is a characterization of the combining tendency.

The fascinating finding is that potassium augments or activates splitting of inorganic phosphate from ATP in the presence of red cell ghosts containing ATPase, the augmentation reaches a maximum rate and that the magic K_M for half maximum rate of splitting is 3.0 meq K/liter medium (Dunham and Glynn, 1961). The binding of potassium to the outer pole of ATPase in the membrane must be the required event for ion transport as for ATP splitting and release of PO_4.

By analogous reasoning and experimentation, it is shown that the active outflux of sodium from red cells has a half maximum rate at K_M = 20 meq Na/liter of inside cellular water (Post and Jolly, 1957), that the half maximum rate of ATP splitting by suspended ghosts occurs at K_M = 24 meq Na/liter of medium (Dunham and Glynn, 1961). Again the implication is that binding of sodium to the inner pole of ATPase, ready for transport outward, is linked to hydrolysis of ATP.

Another intriguing relation is that sodium will not leave a cell via the transport system unless potassium can enter, that sodium exit rate is half maximal when external K is at 1.8 meq/liter (Glynn, 1956). This compares to an external K of 2.2 when its own entrance rate is half maximal. External potassium augments extrusion of cell sodium by the same molecular kinetics as it augments its own entry into cells. Movements of two ions are linked in a cycle.

The cardiac glycosides or digitalis preparations, are very specific inhibitors of ATPase. These are the drugs used to alleviate heart failure. They strengthen the beat. In the kidney, because of influence on ion transport, they induce losses from the patient of masses of salt and water, which in a heart patient interferes with breathing. Ouabain, as dilute as 10^{-7} moles/liter, inhibits the ATPase of red cell ghosts (Dunham and Glynn, 1961); whereas about 5×10^{-8} moles/liter blocks Na and K fluxes (Glynn, 1957). Elevation of external potassium concentration partially blocks action of ouabain on ATP splitting. The action suggests that ouabain competes with potassium in seating on the ATPase and that it interferes with the unseating of terminal phosphate. In these ways ouabain is so specific an inhibitor of these special transport steps that it interferes in no way with any step in energy metabolism or with production of ATP.

As so far prepared, only part of the ATPase system (that splits ATP) is inhibited by ouabain. Uniquely only this part is the moiety activated by sodium and potassium. Only this part is fittingly called "transport ATPase." Calcium ions inhibit this same part, but only this part. Magnesium is needed. Investigators must be fastidiously clean with respect to calcium in distilled water and on glassware.

ATPase occurs in cell membranes and endoplasmic reticulum of all tissue cells. It is especially rich in the electric organ of the electric eel, where rapid ion transport is required to support ion currents for production of electric potential.

V. POTASSIUM FLUXES DURING ACTIVITY

A. Resting Membrane Potential: Potassium Outflux

Once potassium concentration is built up by the transport to about 150 meq/liter inside muscle and nerve fibers, it has a continual tendency to leak out by passive diffusion. The slow escape of cations forms a fringe of plus charge as a blanket over the surface of the fiber. This leaves the potassium pool in the middle of the cell relatively negative. The membrane is polarized, a double layer (negative inside, positive outside) of charge surrounds the cell. Some consider the membrane to contain pores through which the potassium cations pass. The more readily potassium ions pass, the higher the permeability, the greater the potential, proportional to $\log_e$ K inside/K outside as concentrations. The leak must not be so fast as to lower the concentration ratio. In fact, the leak is not so much an escape of potassium as an intensified clustering around the surface.

Tiny glass microelectrodes can impale such fibers in the heart. They register the voltage as 90 mv, minus inside compared to an electrode outside near the membrane in the medium (Fig. 1). When the concentration of K in extracellular fluid is lowered from the normal 5 meq/liter down toward 2, the membrane potential increases; when concentration is elevated toward 10, as in over zealous intravenous infusion of K during fluid therapy, the potential falls. At the same time, dynamic changes occur in the action potential and electrocardiogram, which in part can be used to diagnose potassium de-

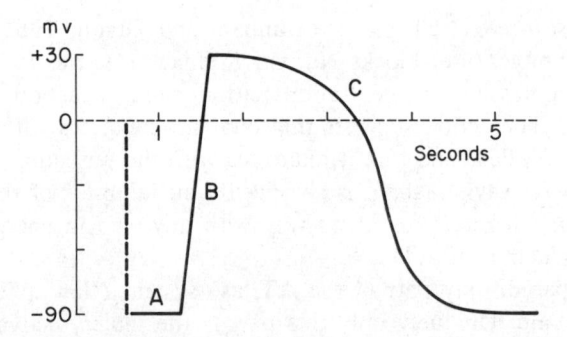

Fig. 1—Action potential recorded from turtle heart muscle fiber. Vertical scale, mv; + and − indicate sign of potential inner side of membrane relative to outside. Horizontal scale, time in seconds. As micropipette electrode pierces membrane and enters protoplasm, inside records −90 mv resting at A. As fiber 'fires', depolarization wave B rises quickly from −90 to +30 mv; late rise above zero is 'overshoot'. C is the slow repolarization wave.

ficiency in illness or to detect excess plasma potassium developing during therapy (Darrow, 1964).

B. Membrane Action Potential: Sodium Influx; Potassium Outflux

Since the active pump lowers internal muscle or nerve Na to a low level compared to outside, 12 compared to 145 meq/liter, the passive diffusion of Na is inward. The sodium charge is really outside the fiber, whereas the potassium charge is inside. The migrating tendency forms a fringe or layer of sodium atoms on the inside surface of the membrane. The sodium potential is plus inside and minus outside the membrane. During rest, the permeability or passage of ions is slow and the fringe formed by sodium very weak (low potential). But at the moment a fiber is excited the permeability and inward escape of sodium increases and the fringe becomes more intense. This new double layer grows rapidly. Its opposite charge cancels out the potassium potential to zero algebraically. The layer intensifies further until the combined potential is reversed to become 30 mv plus inside. This point on the recording of potential is called the "overshoot", resulting from the rapid upward sweep of the oscilloscope beam (Fig. 1). The membrane is said to be "depolarized".

The return of the potential toward the resting level is much slower, the "repolarization". Most feel that a slow intensification of the potassium fringe begins the event, an increased permeability and outward flux of potassium. Thus further growth of the potassium potential cancels the sodium overshoot. Later as this potential diminishes toward resting level, the Na permeability and fringe diminishes back to resting level and the recording has returned to − 90 mv.

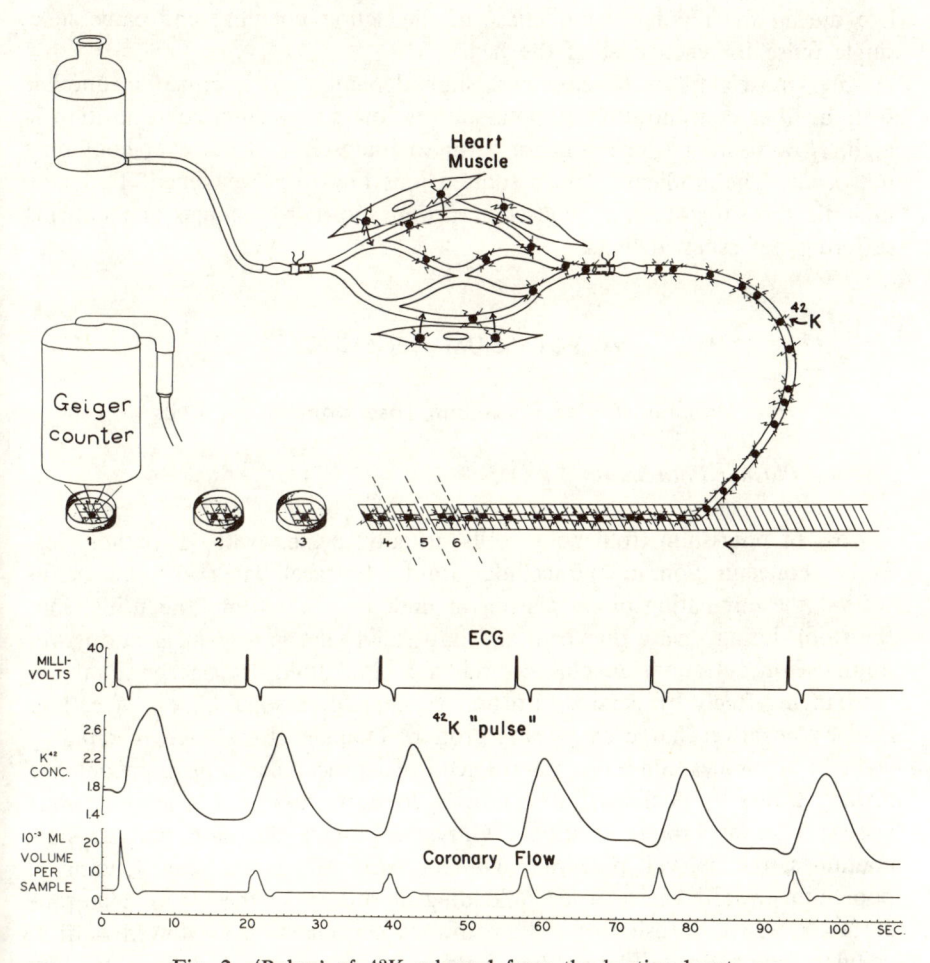

Fig. 2—'Pulses' of ^{42}K released from the beating heart.

The best evidence for these proposed ion migrations is provided by the voltage clamp (Hodgkin, 1964), but this approach is too involved to describe here. A very graphic demonstration of ^{42}K release from a beating heart (Wilde, 1957) is shown in Fig. 2. A slow beating turtle heart was labeled very intensely with radioactive ^{42}K, by intraperitoneal injection before the heart was excised. When fully labeled the ventricle was quickly excised, cannulated, and mounted on a saline pipeline system which irrigated the coronary artery and muscle (fibers diagrammed in upper figure). The effluent fluid carried, via a cannula, the escaping ^{42}K to a traveling strip of filter paper. Samples, 200 segments for one beat, could be cut from the timed strip for radioactivity counting. The recordings below show, above, the electrocardiogram; middle, the pulses of ^{42}K released in relation to beats (phases of the ECG); below, flow rate in the coronary circulation. The ^{42}K escaped late during the repolarization phase of the action potential and came in a single pulse for each beat of the heart.

After muscle fibers develop potassium depletion as described in another section, fiber concentration of potassium is low and sodium concentration is high. Now sodium fringe will not form so intensely and the overshoot may not occur. The gradient to push sodium ions inward is weakened. This may underlie the sluggish, weak muscular action shown by animals and patients suffering potassium deficits.

VI. POTASSIUM DEFICIENCY

A. Mechanisms of Potassium Loss from Tissue Cells

1. *Low Plasma Potassium*

Loss of potassium from body cells usually begins with a pathological fall in concentration in extracellular fluids. Extracellular potassium drains off via the circulating blood plasma in undue amounts into the urine, into the stool during severe diarrhea, and into saliva and gastric juice during vomiting. Concentrations in plasma and interstitial fluid are nearly equal, interchanging freely by passive diffusion (except for a small excess of cations held by negative charge on plasma protein; Donnan effect) across capillary walls in all tissues. Since the rate of feed of extracellular potassium onto its inward transport declines as external concentration falls, active influx lessens (the hyperbolic relation). Outward passive diffusion of potassium continues, the net cell potassium content falls. Meanwhile slow loading of potassium inward has retarded unloading of sodium at the outer membrane surface. Passive diffusion of sodium inward continues. With slowed outflux, cellular sodium rises. Flame photometer analyses of muscle from rats show these changes. Fast growing young rats were fed (Cooke et al., 1952) potassium-deficient diets and given the hormone desoxycorticosterone, which wastes potassium via the urine. Muscle composition is expressed in meq K/ 100 g fat-free solids (dry weight). Muscle K fell from 48.9 to 34.2 or lost 14.7 meq K. Muscle Na rose from 10.0 to 18.4, or gained 8.4 meq K. Loss of 3 K atoms accompanied gain of 2.4 Na atoms, probably 3 to 2. Some hydrogen but possibly basic (cationic) amino acids must have entered. Hydrogen may have replaced external potassium on the ATPase inward path.

Naturally these changes in cellular and extracellular potassium concentrations change resting and action membrane potentials. Body musculature is weak, intestinal motility is sluggish. The heart may develop necrotic areas. Growth is retarded. Evidently protein synthesis and deposition is retarded (Kernan, 1965).

2. Severe Exercise or Activity

Since every nerve and muscle cell action potential releases a pulse of potassium, if the frequency of these pulses is increased enormously in strenuous activity the passive outfluxes of potassium will exceed the inward pumping. Cell potassium will diminish. An elevated plasma potassium occurs in venous blood draining from an exercising muscle. Probably cell sodium rises.

3. Anoxia and Ischemia

When the energy supply to muscle is reduced by poor oxygen supply, as at high altitude mountain climbing or flying, the ATPase pumping will diminish. The passive fluxes will drain out potassium and add sodium to muscle. If the blood supply locally is rendered sluggish (ischemia) the same events occur, as with a tourniquet or any failing arterial supply.

4. Breakdown in Tissues of Fixed Counter Ions for Potassium

During untreated diabetes mellitus and during mild to severe starvation, carbohydrate stores are exhausted and tissue protein is metabolized. In muscle fibers particularly, protein acts as a fixed counter ion to hold ionized potassium. This moiety of potassium or sodium cannot easily diffuse out of muscle because of the charge on protein. If this protein charge disappears, granting that the ion pump continues, only the very diffusible chloride can act as counter ion for (Na + K). Passive outflux of potassium will increase along with the highly diffusible chloride. Cell potassium diminishes. Blood plasma potassium may be elevated and potassium enters the urine while this is happening.

5. Potassium Associated with Liver Glycogen

During any stress that excites secretion of epinephrine, potassium is released from the liver in lock step with glucose. Glycogen, phosphate, and potassium must associate with one another in the liver. When insulin augments liver glycogen, potassium too is deposited. Plasma potassium falls. In diabetes mellitus, liver potassium diminishes along with glycogen. If the diabetic is suddenly given excess insulin, so much potassium is bound with glycogen in the liver that plasma potassium may fall to harmful levels.

B. Avenues of Potassium Loss from Body

1. Saliva

Saliva contains up to 20 meq K/liter, 3 to 4 times that in plasma. During vomiting, its flow increases entailing additional loss of body potassium.

2. Gastric Juice

Investigators seek a molecular interaction between potassium and hydrogen movements during the secretion of HCl in the stomach. Space prevents detailed discussion (see Davenport, 1966). Potassium reaches about 16 meq /liter in the gastric juice of unstimulated (basal) stomachs. When the volume of juice secreted per minute is increased by intravenous histamine the concentration and mass of potassium entering the juice increases. This potassium arises mainly from gastric secreting cells and not directly from plasma. Normally the potassium in gastric juice is absorbed by the gut, but in severe vomiting enough potassium is lost to develop body deficiency. The increased alkalinity of blood produced by vomiting triggers further loss of potassium to the urine.

3. Intestine and Colon

Active absorption of potassium from the intestine has not been identified (Davenport, 1966). Movement seems to follow the drag of membrane charge and concentration gradient of potassium, that is, to follow the electrochemical gradient. In the duodenum while salt and water are being absorbed, potassium seems to follow water. Luminal concentration equals plasma during the removal of water. In the ileum, potassium seems to lag behind salt and water removal. Potassium may reach 35 meq K/liter with the plasma at 5.

The colon absorbs nearly all the salt and water entering it. Glands in the colon secrete potassium, creating a higher concentration on the luminal side, but since the residual water moved into the stool is so small only 6-20 meq K/day enter the stool. However, because of potassium secretion, a local enema of the bowel, repeated many times, can remove critical amounts of potassium.

In severe diarrhea, most of the potassium lost in the watery stool comes from the small intestine. In cholera, the victim may lose 60 liters of salt and water a day containing 1,200 meq K. Replenishment of the salt and potassium lost requires dozens of infusion bottles but is completely life saving. The normal dietary intake of K is 77 meq of which 7 meq enters the stool.

4. Kidney

A kidney handles potassium by *filtering* blood plasma potassium into a million urinary tubules, by *reabsorbing* most of this from the proximal tubules back into the blood vessels surrounding them, by at times *secreting* potassium, moving it from blood to tubule lumen in distal tubules and collecting ducts. Of 684 meq K filtered each 24 hours, 633 meq are reabsorbed, leaving 51 meq to be urinated. These are normal figures. During body deficits of potassium, when potassium has been lost by routes other than kidney, reabsorption removes nearly all potassium from the urinary tubules, hardly any goes to urine. When blood plasma is overloaded with potassium, above 5 meq/liter, the kidney secretes K into the urine. Only the kidney regulates plasma potassium and hence body supply.

Both reabsorbing and secreting transfers are active. Much effort has attempted to explain these on a molecular basis using a model similar to the ATPase system. Exceptions are so numerous that it is futile to discuss transport models in this elementary treatment (for details, see text of Pitts, 1963).

Most of the excessive urinary losses of potassium occur through increased secretion such as may lead to body deficit. Increased plasma potassium induces increased secretion and urination of K as though a carrier were further saturated. Any excessive entry into plasma of cell potassium such as described in Section VI, A, will initiate this. During convalescence, after the tissue transport defect has been repaired, the potassium will now be unavailable. As the tissue cells reclaim potassium, plasma levels will fall to harmful levels and should be restored by intravenous potassium.

The hormone aldosterone, liberated from the cortex of the adrenal gland, particularly during stress as after injury or after extensive surgery, excites excessive secretion and urinary loss of potassium. At the same time, appetite is poor and tissue cell protein is being metabolized with release of potassium, both of these because of simultaneous release of cortisol from the adrenal cortex (Moore and Ball, 1952). Again as the patient convalesces, he will have a potassium deficit. Intravenous potassium and increased dietary potassium (broths) must be given.

Plasma hydrogen ions seem to compete with potassium in a carrier for renal secretion. Plasma acidity induces potassium retention; alkalinity increases K secretion. I tell my students "when H is low, K must go". During excessive vomiting, the alkalosis set up by the loss of gastic HCl induces more loss of potassium via urine than in the vomitus. Alkalosis during excessive breathing (high altitude mountain climbing) similarly drains off potassium. Finally intravenous infusion of sodium bicarbonate initates potassium loss. Conversely, a low plasma potassium forces excessive hydrogen secretion, increases reabsorption of $NaHCO_3$. The kidney is unable to excrete alkali; excessive alkalosis develops (see Pitts, 1963).

Large, impermanent anions, such as SO_4, which cannot be rapidly reabsorbed compared to chloride, retain excessive potassium in the urine, evidently because as sodium is reabsorbed to very low levels the negative charge remaining behind in tubule lumen attracts potassium and hydrogen by charge. Urinary K may exceed 150 meq/liter, the highest under any known condition (Sullivan, Wilde, and Malvin, 1960). Such anions as PO_4 and SO_4 accumulate during protein splitting in starvation and diabetes. This mechanism may partially explain the low plasma potassium ultimately developing in these conditions.

C. Measuring Mass of Body Potassium

Since the measured level of plasma K is not a reliable reflection of the total body (cellular) K content, or degree of deficiency, clinicians have often turned to the isotope ^{42}K dilution method for estimating mass of body K. The computation is based on the assumption: (^{42}K counts/ml plasma) / (meq K/ml plasma) = (total ^{42}K counts injected intravenously) / (meq K in patient's body). The prime uncertainty is the degree of mixing of ^{42}K with all body K. I reviewed elsewhere (Wilde, 1962; see table of contents) the extensive data from rats and rabbits on ^{42}K exchange which bear on this. Seven hours after intravenous injection the isotope was uniformly mixed in 10 different organs and tissues of rats (Table II cited there from Ginsburg and Wilde, 1954). Ljunggren, Ikkos, and Luft (1957) estimated 48.5 ± 1.87 meq K/kg body weight in 15 normal men and 37.3 ± 1.38 meq K/kg in 16 normal women. Ideally K content should be related to muscle or lean body mass. Variations in body build as reflected in fat and in massive accumulation of edema fluid vitiate effort to appraise potassium mass which must be anchored to body weight.

The problem of isotope mixing is completely eliminated by measuring the naturally ocurring isotope ^{40}K, a gamma emitter. Anderson, et al., (1956) constructed a counter large enough to contain a man and measuring with 4 II geometry. An interesting commerical use for such a counter is to measure the ^{40}K content of slaughter house hams (Kupwich, et al., 1958). Nearly all of the potassium (^{40}K) is in the lean muscle. Fat content can be estimated by difference.

The total potassium in a 70 kilo man weighs 300 g as chloride and would fill 3 graduate cylinders (100 ml each). Saltpeter weighs a third more. Potassium is a macronutrient.

D. Gauging Therapeutic Dosage of Potassium

The body has no storage depot for potassium. Most body potassium is dynamic in protoplasmic muscle and nerve. Some in bone connective tissue is inert but still in dynamic equilibrium with plasma potassium. Potassium dosage must be carefully gauged otherwise plasma K rises above 10 meq/ liter and is toxic to heart and respiratory center. Lethal tetany occurs. Dosage must be metered to what the great mass of body muscle will take up.

Two modes of distribution in muscle occur, following KCl injection, (i) the behavior when plasma potassium and muscle potassium content are normal; ((ii) the behavior when muscle potassium is low and sodium high as in potassium deficient states. In the normal case, the plasma K being near 6 meq/liter, the active influx is nearly saturated or maximum in rate. If plasma K is elevated by intravenous infusion, say from 5 to 8 meq/liter, active influx of K will not increase appreciably, but diffusion inwards increases and stops when the *added* concentration inside the muscle just equals the concentration that remains *added* outside at this time.

This is a description of what I have called the "free-swimming" KCl (Wilde, 1962). Chloride enters with this moiety. The dosage potassium merely acts as though it were freely dissolved in all the water of the muscle, cellular, and extracellular. Since muscle is a huge proportion of body mass, the behavior is about the same in the entire body. If total body water is 0.7 liter in each kilo body weight, to raise plasma K (and cellular water) 3 meq/ liter, inject slowly 0.7 × 3 = 2.1 meq K/kilo body weight.

Normal muscle contains a small amount of sodium, some of which must be associated with protein, etc., with ions other than chloride. To the extent these sodium atoms are exchanged off protein by the increased potassium pressure, the dosage potassium replacing them would not enter the system above. Plasma potassium rise would fall below the predicted value by that much.

In case (2), with potassium deficiency, the muscle sodium is even much higher, much more dosage potassium would be removed from the free-swimming system by exchange for sodium. The dosage system is over cautious when used for a patient in potassium-deficient state. It is safe in normals and has been tested experimentally in cats (Wilde, 1962; p. 100). Clinicians have worked up dosage schemes by trial and error (Darrow, 1964). Their dosages limits can be predicted from the above analysis.

VII. OTHER ALKALI METALS

Of other alkali metals, rubidium and cesium compete ahead of potassium in muscle uptake. Relman, et al. (1957) fed mixtures of potassium-rubidium to one group and potassium-cesium to another group of potassium-depleted rats. For given plasma ratios, say Rb/K = Cs/K = 1.0, the ratios in muscle were Rb/K, 3.0, and for Cs/K, 4.5. Wilde (1962, p. 74, 87) discusses possible mechanisms for this. Massive amounts of these chemicals are toxic, probably relative to action potentials. Cesium-137 is a dangerous fallout radiation hazard of atomic bombs. Obviously it would accumulate more in plants growing in low potassium soils. If a plant accumulates cesium as selectively as does muscle, the food chain: soil to plant, plant to steer, steer to man, could concentrate Cs relative to K 4.5 × 4.5 × 4.5 or 91 times. Other nutrient sources could break the chain by dilution with potassium at any stage. Heavy application of potassium fertilizer would reduce the first factor.

Dr. Frank F. Hooper, Professor of Fisheries here at Michigan, has called my attention to the accumulation of fallout ^{137}Cs in fish in Finnish lakes. When lake waters have a low potassium content, ^{137}Cs uptake is greatly intensified, compared to lakes with greater mineral content. The multiplying food-chain is alga to daphnia to small fish to pike (Kolehmainen, Hasanen, and Miettinen, 1966).

VIII. OTHER REVIEWS

For a very extensive coverage of potassium in any animal or organ system, see the monumental coverage by Ussing, et al. (1960). See also the table of contents in the 4 volumes on *Mineral Metabolism* edited by Comar and Bronner, 1960-1962). *Annual Review of Physiology* carries a chapter on "Transport through Biological Membrane" every second year.

IX. CONCLUDING REMARKS

The living cell captured by active transport the potassium of the sea, as it does now the potassium of extracellular fluid. It extrudes sodium from its protoplasm. It evolved a membrane enzyme system, ATPase, which by molecular rotation simultaneously moves a K into and a Na out of its protoplasm. Sodium and potassium, when present inside and outside the membrane respectively, catalyze release of high energy PO$_4$ from ATP by ATPase as the ions themselves move their respective ways across the membrane. When so partitioned, sodium and potassium control the osmotic water content of the cell and thus guard osmotically its integrity and composition. While leaking outward, yet captured as a surface fringe, potassium sets up the resting membrane potential, as a double layer of + charge outside. During excitation and activity, while leaking inward, Na sets up an action potential "overshoot", a depolarization, positive inside the membrane. The membrane action potential returns to resting level by repolarizing. In this, sodium and potassium fringes interact until both are at resting level.

If, in the animal, extracellular potassium is very low in supply (K deficient states), the membrane ATPase will not load potassium inward or sodium outward so quickly; potassium leaks out of the cell, sodium inward;

membrane action potential and muscular activity are imparied. Gastrointestinal tract and kidney control such losses. Administration of intravenous potassium can relieve the situation. Potassium in the soil makes the "game" possible for farm animals and man.

REFERENCES

Anderson, E. C., R. Schuch, J. Perrings, and W. Langham. 1956. The Los Alamos human counter. Nucleonics 14:26-29.

Caldwell, P. C., A. L. Hodgkin, R. D. Keynes, and T. I. Shaw. 1960. The effects of injecting 'energy rich' phosphate compounds on the active transport of ions in the giant axons of Loligo. J. Physiol. 152:561-590.

Comar, C. L., and F. Bronner [Editors] 1960-1962. Mineral metabolism. 4 vols. Academic Press, New York.

Cooke, R. E., W. E. Segar, D. B. Cheek, F. E. Coville, and D. C. Darrow. 1952. The extrarenal correction of alkalosis associated with potassium deficiency. J. Clin. Invest. 31:798-805.

Darrow, D. C. 1964. A guide to learning fluid therapy. C. C. Thomas, Springfield, Ill. 280 p.

Davenport, H. W. 1966. Physiology of the digestive tract. 2nd ed. Year Book Med. Publishers, Inc., Chicago. 230 p.

Dunham, E. T., and I. M. Glynn. 1961. ATP-ase activity and the active movement of alkali metal ions. J. Physiol. 156:274-293.

Fenn, W. O. 1940. The role of potassium in physiological processes. Physiol. Rev. 20:377-415.

Ginsburg, J. M., and W. S. Wilde. 1954. Distribution kinetics of intravenous radiopotassium. Amer. J. Physiol. 179:63-75.

Glynn, I. M. 1956. Sodium and potassium movements in human red cells. J. Physiol. 134:278-310.

Glynn, I. M. 1957. The action of cardiac glycosides on Na and K movements in human red cells. J. Physiol. 136:148-173.

Glynn, I. M. 1962. Activation of adenosinetriphosphatase activity in a cell membrane by external potassium and internal sodium. J. Physiol. 160:18-19.

Glynn, I. M. 1966. The transport of sodium and potassium across cell membranes. Scientific Basis of Med. Ann. Rev. (1966). p. 217-237.

Harris, E. J. 1956. Transport and accumulation in biological systems. Academic Press, New York. 291 p.

Hodgkin, A. L. 1964. The ionic basis of nervous conduction. Science 145:1148-1153.

Kernan, R. P. 1965. Cell K. Butterworth, Washington. 152 p.

Kolehmainen, S., E. Häsänen, and J. K. Miettinen. 1966. ^{137}Cs in fish, plankton and plants in Finnish lakes during 1964-1965. Univ. of Helsinki Dept. of Radiochemistry, Ann. Report 1965. Paper no. 8. 12 p.

Kupwich, R., L. Feinstein, and E. C. Anderson. 1958. Correlation of potassium ^{40}K concentration and fat-free lean content of hams. Science 127:338-339.

Ljunggren, H., D. Ikkos, and R. Luft. 1957. Studies on body composition. I. Body fluid compartments and exchangeable potassium in normal males and females. Acta Endocrinol. 25:187-198.

Moore, F. D., and M. R. Ball. 1952. The metabolic response to surgery. C. C. Thomas, Springfield, Ill. 156 p.

Opit, L. T., and J. S. Charnock. 1965. A molecular model for a Na pump. Nature 208:471-474.

Pitts, R. F. 1963. Physiology of the kidney and body fluids. Year Book Med. Publishers, Inc., Chicago. 243 p.

Post, R. L., and P. C. Jolly. 1957. The linkage of Na, K, and ammonium active transport across the human erythrocyte membrane. Biochem. Biophys. Acta 25:118-128.

Relman, A. S., A. T. Lambie, B. A. Burrows, and A. M. Roy. 1957. Cation accumulation by muscle tissue: the displacement of potassium by rubidium and cesium in the living animal. J. Clin. Invest. 36:1249-1256.

Skou, J. C. 1965. The enzymatic basis of active Na-K transport. Physiol. Rev. 45: 596-617.

Sullivan, L. P., W.. S. Wilde, and R. L. Malvin. 1960. Renal transport sites for K, H, and NH_3. Effect of impermeant anions on their transport. Amer. J. Physiol. 198: 244-254.

Ussing, H. H., P. Kruhoffer, J. H. Thaysen, and N. A. Thorn. 1960. The alkali metal ions in biology. Handbuch Exp. Pharmakol. 13:1-598.

Whittam, R. 1962. The asymmetrical stimulation of membrane ATP-ase in relation to active cation transport. Biochem. J. 84:110-118.

Wilde, W. S. 1957 The pulsatile nature of the release of potassium from heart muscle during the systole. Ann. New York Acad. Sci. 65:693-699.

Wilde, W. S. 1962. Potassium. p. 73-107. In C. L. Comar and F. Bronner (ed.) Mineral metabolism. Vol. 2, Part B. Acad Press, New York.

11

The Effects of Potassium on Disease Resistance

ROY L. GOSS

Western Washington Research & Extension Center
Washington State University
Puyallup, Washington

I. INTRODUCTION

Plant diseases alone cost the American farmer over 3 billion dollars annually. The cost to world crops, if known, would be staggering, especially when considering that in Japan 3 to 8% of the rice crop is lost annually to disease and without the effective control measures now in use, these losses would run much higher according to von Uexkuell, 1966. In other countries, where disease control is not yet as widely practiced and where fertilizer application is less balanced, such losses may be even higher. This becomes increasingly important when we consider that rice is the staple food for over one-half of the world's population and is the major source of income for more than 60% of the world's farm families.

There are two major groups of plant diseases—parasitic and nonparasitic. The fungi, bacteria, and nematodes cause the greatest number of parasitic diseases while unfavorable environmental factors such as soil, air, moisture, nutrients, mechanical damage, etc. cause nonparasitic diseases according to Griffith and Wagner, 1966. Poor inherent qualities of the plant can predispose it to both parasitic and nonparasitic diseases.

Fungicides can be used very effectively in controlling many plant diseases and, likewise, the improvement of certain environmental factors may be totally or partially effective as well. Hence, it would not seem reasonable if we did not take the best advantage that each type has to offer.

Potassium plays an extremely important role in the control or reduction of severity of plant diseases. The nature of the action of potassium and mechanisms involved are not well known. McNew (1953) states that, "Unlike other essential nutrients potassium does not become a structural part of the plant cell. It is the immobile regulator of cell activity and promotes the reduction of nitrates and the synthesis of amino acids from carbohydrate and inorganic nitrogen. Potassium promotes the development of thicker outer walls in the epidermal cells and firms tissues, which are less subject to collapse".

McNew further states that, "A deficiency of potassium enforces the accumulation of carbohydrates and inorganic nitrogen in the plant. Eventually it retards photosynthesis and production of new tissues. More plant diseases have been retarded by the use of potassium fertilizers than any other substance, perhaps because potassium is so essential for catalyzing cell activity".

Soil fertility can affect the populations of soil-borne fungi. Kaufman and Williams (1964) found in both field and laboratory investigations that total numbers of fungi were greatly increased by nitrogen, and to a lesser extent by phosphorus and potassium.

A number of the major crops and their disease-potassium interrelationships will be reviewed in the following discussion.

II. THE EFFECT OF SOLUBLE FORMS OF POTASSIUM ON PLANT DISEASES

Potassium has a very broad amplitude in its effect upon plant diseases. As a matter of convenience in grouping the various crops and the disease-potassium interrelationships affecting them, they are divided into four categories; namely, field crops, horticultural crops, tropical crops, and turfgrasses.

A. Field Crops

1. Tobacco

Three of the major diseases of tobacco are: wildfire, caused by *Pseudomonas tobaci* (Wolf and Foster) Stevens; angular leaf spot, caused by *Pseudomonas angulata* (Fromme and Murray) Holland; and blackfire, caused by either *P. tobaci* or *P. angulata*. Plant susceptibility, as shown by Johnson and Valleau (1940), is related to soil fertility, especially to phosphorus and potassium levels. It was found by these workers that blackfire rarely occurs on soils of high fertility and especially those with high levels of potassium. They further concluded that, although this disease was partially controlled usually by treating beds with Bordeaux, the disease often occurred in epidemic form on soils of low fertility, such as eroded knolls and hillsides. They also indicated that blackfire was induced without difficulty on plants in infertile soil by inoculations with pure cultures of *P. tobaci* and *P. angulata*.

2. Corn

Hooker (1966), at the University of Illinois, in his work with the effect of potassium on corn diseases has reported that, when plants are killed by disease before full maturity, the crop does not receive full value from fertilizer. Each diseased plant loses about 16% of its yield but the full loss depends on the stage of plant development when infection occurs. *Gibberella spp.* causes root rotting and *Diplodia zeae* (Schw.) Lev. causes stalk rotting, both causing lodging or "blow down". Down stalks are difficult to harvest and ears contacting moist soil are soon rotted by fungi.

Illinois tests demonstrated that stalk rot and lodging is usually more severe when soil potassium content is low to medium in relation to nitrogen and phosphorus. The same hybrid grown with 71 kg each of N, P, and K/ha suffered only 17% stalk rot while plots with the same amount of N and P but no K showed 49% rot. Similar trends have been found in a number of other states. Adequate nitrogen or nitrogen plus phosphorus increases the potassium requirement of the growing plant. When the soil cannot supply the plant with adequate potassium, this nutrient imbalance disturbs carbohydrate and protein metabolism and speeds up tissue aging and breakdown in the lower stalk and roots. This causes greater susceptibility to disease and lodging.

It has been pointed out by Hoffer (1949) that, when potassium deficiency occurs in corn, iron accumulates at the nodes of the stalk which interferes with translocation of nutrients to the roots. In this situation, they are weakened and become susceptible to soil fungi, often resulting in root rot and consequent lodging of plants.

3. Cotton

Cotton wilt, caused by the fungus *Fusarium oxysporum* Schlecht. f. *vasinfectum* (Atk.) Snyd. and Hans. has occurred in epidemic amounts when soil potassium levels were low. Three varieties of cotton (resistant, tolerant, and susceptible) were tested by Cralley and Tharp (1938) in Arkansas to determine the effect of rate and timing of potassium on this disease. They found that significant differences in relative disease severity between potassium-fertilized and nonfertilized plants occurred only late in the growing season.

Dick and Tisdale (1938) found that applications of nitrogen and potassium alone and, especially in certain combinations, effectively reduced wilt and increased yields. They found a differential response of varieties to fertilizers under conditions of severe wilt. Their results showed that the use of phosphate materially increased wilt, particularly so in the absence of adequate amounts of nitrogen and potassium.

A physiological or nutritional disease of cotton has been reported by several researchers. According to Cooper (1939) potassium is the most common mineral deficiency in cotton and can result in leaf discoloration and dropping. This condition is commonly referred to as rust because the leaves develop a rusty brown color. When potassium is deficient, there is a reduction in growth with the tips and margins of the leaves turning yellow, brown, and finally curling. Cooper states that the leaves turn reddish-brown and shed prematurely and the bolls are small and fail to open. Soils adequately supplied with potassium do not develop this condition.

Young, Ware, and Pope (1935) reported that, "The control of potash hunger or 'rust' through the use of potash-containing fertilizers results in a marked decrease in the severity of cotton wilt attacks".

4. Forages

Leaf spot disease of timothy, caused by *Heterosporium phlei* Greg. has been reported by Laughlin (1965) to be greatly reduced by potassium applications. His results showed that the number of spots significantly decreased as potassium levels were increased.

Laughlin found that tissue N decreased from 3.09% to 2.48% as K was increased from 0 to 237 kg/ha. The higher tissue N levels at zero level of K appears to increase the susceptibility of the plants to leaf spot.

Chi and Hanson (1961) working with red clover in pots with nutrient solutions found the least wilt and root rot disease, caused by *Fusarium oxysporum* f. *solani* (Mart.) Appel and Wr. and *F. roseum* Lk. emend. Snyd. and Hans., at nutrient concentrations optimum for the host. Their results showed that a nutrient solution of 0.5 Hoagland's (H) was best for the host. In unbalanced solutions where either phosphorus or potassium had been deleted, they found that the addition of potassium caused the greatest decrease in disease severity.

5. Cereal Crops

A. WHEAT. Glynne (1959) at the Rothamsted Experimental Station found that powdery mildew, caused by the fungus *Erysiphe graminis* DC. ex Merat, was very severe in June 1958, on certain plots deficient in minerals and relatively slight in others. Observations were made on plots receiving the same rate of N (181 kg N/ha), zero and 159 kg P, and 0 and 90.6 kg K/ha, and one plot to which dung was applied. Random sampling of infected leaves showed that older second leaves were the more severly infected. The disease was consistently most severe on three plots which received nitrogen and phosphorus, but no potassium. Disease was moderately severe where nitrogen alone was applied and relatively slight on two plots which received potassium as well as nitrogen and phosphorus. Plots receiving 12,727 kg/ha of dung annually were least affected. The dung supplied about twice as much N, the same amount of P, and 2.5 times as much K as the mineral fertilizers in other plots. Glynne concluded that the lower incidence of powdery mildew can thus reasonably be attributed, at least in part, to the extra potash application.

B. BARLEY. Yellow dwarf virus disease of malting barley in the Wilamette Valley of Oregon has produced serious yield losses with some varieties such as Hannchen, according to Jackson et al. (1962). Where heavy aphid infestations and virus infections developed during the early seedling stages of growth, yields of only 267 to 445 kg barley/ha were obtained. Chemical control of aphid vectors did not reduce the incidence of the virus disease, hence, they investigated early seeding dates and fertilizer applications. As a result of their experiments, they found in 1961 that increases in yield and test weight and decreases in percent of thin kernels were evident from application of both potassium and phosphorus when yellow dwarf virus disease was severe. These responses were noted on soils that would normally be considered adequately supplied with these two elements. They also found that planting dates and nitrogen applications increased yields.

B. Horticultural Crops

Among the horticultural crops there is considerable evidence that potassium plays an important role by increasing the resistance of the plants to diseases. In some cases potassium may play a role also in suppressing a build-up of plant pathogens in the soil. Kaufman (1964) found that total numbers of fungi were increased both in the field and in greenhouse tests by nitrogen fertilization. Potassium had the greatest suppressing effect on soil fungus populations.

1. *Tomatoes*

Bacterial wilt of tomatoes is caused by the wilt organism *Pseudomonas solanacearum* E. F. Sm. and accounts for heavy losses in this crop. Gallegly and Walker (1949) conducted experiments with nutrient solutions, temperature, day length, and light intensity to determine their effect upon this disease. Their nutrient solutions were 0.1, 0.5, 1, 2, and 3 times that of Hoagland's basal salt solution. In some cultures, nitrogen, phosphorus, and potassium were adjusted to low and high levels. During the summer, disease development in the unbalanced solutions was increased at the low potassium level and decreased at the high nitrogen level as compared with that in the basal solution. These experiments point out the importance of medium to high levels of nitrogen and potassium, but also indicate that nitrogen, phosphorus, and potassium must be in proper balance.

Walker and Foster (1946), in working with nutrient solutions, found that low potassium levels brought about an increase in plants affected with wilt. They used unbalanced Hoagland's solutions with concentrations of 0.1 H, 0.5 H, 2 H, and 3 H. They also conducted experiments with unbalanced nutrient solutions, using 1 H as the basal solution but varying nitrogen and potassium from levels used in 3 H (high nitrogen or potassium) to levels used in 0.1 H (low nitrogen or potassium). When the tomato plants were 2 weeks old, they were dug and the roots were washed and dipped in a heavy suspension of mycelium and spores of a virulent strain of the fungus *Fusarium oxysporum* Schlecht. f. *lycopersici* Sacc. In the unbalanced nutrient solution experiment, they found that development of wilt was greatest in plants receiving nutrient solutions low in potassium. Nutrient solutions high in nitrogen also favored progression of the disease over that obtained in plants given the normal solution.

Horticultural crops are affected by many virus diseases. Certain tobacco mosaic virus (TMV) isolates can cause blotchy ripening of tomatoes according to Jones and Alexander (1962). They point out that there is conflicting evidence concerning the etiology and factors associated with the occurrence of the tomato disease or diseases designated as blotchy ripening, cloud, internal browning or vascular browning, and graywall. In their greenhouse experiment with pot cultures in nutrient solutions, they found significant inverse correlation between potassium levels and blotchy ripening. The total percentage of blotched fruit was highest at the lowest potassium concentration and progressively decreased as the potassium concentration was increased.

Tomato internal browning, a chlorotic and necrotic ripening disorder of tomato fruits, is caused by a strain of the tobacco mosaic virus; however, the severity is influenced by certain environmental factors according to Rich (1958). In his experiments he was able to influence the degree of infection by adjusting soil potassium levels. Soil fertility treatments included uniform nitrogen treatments as ammonium nitrate at 76 kg N/ha, 0 and 124 kg K as potassium chloride, and 0 and 66 kg P/ha as superphosphate. Tomato plants were virus inoculated when the fruits were one-half to three-fourths grown. The results of the 1957 experiment are shown in Table 1.

When plants were inoculated with virus, potassium produced significantly less internal browning than other inoculated treatments. The percentage infection was not significant between fertility treatments and check without inoculation. Rich concluded that, according to 1956 experiments, virus infection alone will not result in severe internal browning.

Table 1—Percentage of total fruits that showed virus-induced browning (1957) *

Treatment	Average of 6 replicates
Virus + potassium	9.2
Virus + phosphorus	18.3
Virus + phosphorus + potassium	12.5
Virus alone	15.9
Potassium alone	4.5
Phosphorus alone	5.4
Check	2.6

* Data were taken from Rich 1958.

2. Table Beets

Neither soil nor seed fungicidal treatments have given economic control of damping-off of table beets in Oregon according to Yale and Vaughan (1962). This disease, commonly caused by *Pythium ultimum* Trow, causes poor stands of beets and losses result from the cost of replanting, reduced yields, and delayed harvesting. When beet seedlings were grown in pots in naturally infested soils, these researchers found that, when applications of sodium nitrate, 178 kg N/ha, and 74 kg K/ha as potassium chloride, either alone or in combination, were used, the survival rate of seedlings was increased. Survival rate was highest (Table 2) with the sodium nitrate-potassium chloride combined treatment.

3. Potatoes

Struchtemeyer (1966) in field experiments at Aroostook Farms, Presque Isle, Maine, reported that fertility balance could affect the occurrence of late blight of potatoes caused by *Phytophthora infestans* (Mont.) d By. Nitrogen treatments in his experiments ranged from 67 kg N/ha to 235 kg N in 34 -kg increments. Six phosphorus treatments ranged from 0 to 392 kg P/ha in 78 -kg increments. Potassium supplied as potassium chloride varied from 0 to 280 kg K/ha in 56 -kg increments. He concluded that an increase in nitrogen fertilization increased the incidence of late blight and that increasing levels of phosphorus and potassium caused a decrease. He further stated that, while nutrition does affect the incidence of disease and other disorders in potatoes, increasing amounts of potassium generally suppress the effect.

4. Sweet Corn

The influence of nitrogen, phosphorus, and potassium nutrition on the reaction of sweet corn seedlings to bacterial wilt of maize caused by *Xanthomonas stewartii* (E. F. Smith) Dows. was studied by Spencer and McNew

Table 2—Effects of N and K in the soil on the reduction of damping-off of table beets in the greenhouse *

Fertilizer†	Average seedlings from 35 seed balls, 21 days after planting
NaNO$_3$	8.3
KCl	5.1
NaNO$_3$-KCl	18.3
NPK	14.6
No fertilizer	1.5

* Data were taken from Yale and Vaughan 1962.
† Fertilizers applied to furnish: 178 kg/ha N as NaNO$_3$, 89 kg/ha of K$_2$O as KCl, and 107 kg/ha of P$_2$O$_5$ as Ca(H$_2$PO$_4$)$_2$.

(1938). These studies were conducted in pots with nutrient solutions and the severity of infection was noted 10 days after inoculation. Inoculated seedlings from potassium-deficient pots developed a pronounced chlorosis, and the disease severity was much more severe than those supplied with low concentrations of potassium when used in the form of potassium sulfate. Seedlings deficient in nitrogen or phosphorus were only slightly infected as compared to severe infection when potassium was deficient.

5. Ornamental Bulbs

Diseases of bulb crops, such as gladiolus and narcissus, have been reported in the literature to be aggravated by various levels of nitrogen, phosphorus, and potassium.

The most serious disease of narcissus on the east coast of the USA is basal rot of the bulbs caused by *Fusarium oxysporum* Schlecht. f. *narcissi* Snyd. Hans. According to McClellan and Stuart (1947), bulbs may not show appreciable rot at time of digging but, after 8 weeks of storage, many infected bulbs become rotted, particularly if high temperature and moisture conditions prevail and the bulbs cure slowly after digging. Experiments were conducted with various growth regulators and nitrogen bases, as well as with various levels of nitrogen, phosphorus, and potassium. The results of nutritional experiments with narcissus in general confirmed the results reported previously by Walker and Hooker (1945), Tharp and Wadleigh (1939), Young and Tharp (1941), and Stoddard (1942), in their work with *Fusarium* diseases of cabbage, tomato, cotton, cantaloupe, and other crops. High nitrogen and phosphorus with low potassium produced the most disease and high potassium reduced severity. However, in the case of *Fusarium* corm rot of gladiolus, high levels of nitrogen and low levels of phosphorus greatly increased the rot. This, again, points out the need for maintaining balanced nutritional programs.

Gould and Russell (1961), in their work in western Washington, report that fertilizers applied to soils infested with *Stromatinia gladioli* (Drayton)

Whet. had only a slight effect on the disease development during the same season. They reasoned that, since gladiolus corms contained high food reserves, this suggested that their previous nutritional history might be more important than current nutrition in influencing infection by the dry rot fungus. Their field experiments were from corms that were produced from cormels grown in different nutrient solutions. Their experiments included high and low levels of nitrogen, phosphorus, potassium, and magnesium from sources of calcium nitrate, potassium sulfate, potassium chloride, ammonium sulfate, KH_2PO_4, and $MgSO_4 \cdot 7H_2O$. Necessary minor elements were added to all solutions. They found that increasing the levels of either magnesium or nitrogen increased susceptibility of corms to infection and increased potassium levels decreased susceptibility slightly. The effect of phosphorus was variable in these experiments.

C. Tropical Crops

Tropical crops suffer from a wide variety of diseases due primarily to high temperatures, humidity, and in many cases, to depleted soil fertility. A vast amount of literature is available on subjects or coconut palm, sugarcane, bananas, and other tropical crops. The literature presented in the following discussion will be concerned principally with the effect of various sources of potassium on bananas, coconuts, and rice.

1. Bananas

Soils of the banana growing regions vary from light sandy to heavy clay. Reports of various workers indicate that bananas develop more disease problems when grown on the lighter sandy soils that are more subject to leaching than on the heavier soil. Various diseases can affect bananas, also, on the heavier soils where internal drainage is poor during the wet season.

Nearly all varieties of bananas are susceptible to *Fusarium* wilt caused by *Fusarium oxysporum* Schlecht. var. *cubense* (E. F. Sm.) Wr. The susceptibility of bananas to this disease varies both with variety and soil conditions. The 'Cavendish' group is probably more resistant than others, but some infection is common among the varieties in this group. 'Gros Michel', a variety with more desirable characteristics than those of the Cavendish group, is highly susceptible to *Fusarium* wilt. Rishbeth's (1957) investigations at the West Indies Banana Research Scheme indicate positive relationships between low soil potassium and increased incidence of disease, particularly in the Gros Michel variety and, generally, in the more resistant Cavendish varieties. Some Gros Michel plantations had healthy productive plants after 20 years, while some plantations had a total loss after 2 to 5 years. Soil analyses showed that healthy plantations had much higher potassium levels than the affected ones. Varying degrees of disease injury were found even within the same field and, again, soil tests revealed lower potassium levels in areas more severly damaged. Increased nitrogen rates invariably caused increased disease incidence. This has been reported many times with other diseases and other crops, as well.

Poor drainage, leading to waterlogged soils, tends to accentuate wilt damage. Most varieties with good wilt resistance showed some infection and wilting under these conditions, but usually recovered when the drainage was improved. The relationship here is undoubtedly associated with the uptake of potassium which is reduced when oxygen is deficient.

Nutritional effects on bananas can vary from one geographic location to another as shown by the work of Pelegrin (1954). When nitrogen, phosphorus, and potassium in various levels and combinations were tested in Guadaloupe and Guinea, the rate of application was more important than the composition of the mixture in the former location, but the best and most economic mixture in Guinea had a ratio of 1 part N, 1 part P, and 2 parts K.

Croucher and Mitchell (1940) reported in their work in Jamaica with the Gros Michel variety that nutritional experiments produced variable results. On a soil derived from "Richmond Beds," favorable effects were secured only from nitrogen. On a soil derived from calcareous deposits of the "Rossa type", phosphorus proved the limiting factor but, when this element was applied adequately, potassium also became beneficial. On another soil of the "Terra Rossa type", no yields were secured in the absence of applied potassium and doubling the potassium fraction further increased yields. They also found that an excess of phosphorus without an increase in potassium resulted in short-fingered fruits and distortion of the fingers.

2. Coconuts

A number of references appear in the literature from various parts of the world on the effect of soil conditions in relation to root and leaf diseases in coconut palm, vascular wilt diseases of the oil palm, and various other parasitic and non-parasitic diseases.

The coconut tracts of India are generally found to be deficient in organic matter, nitrogen, available phosphorus, and potassium which are required in sufficient quantities for the proper and healthy growth of the palms and for economic yields. The average annual loss of plant nutrients per hectare of coconut plantation of normal stand and yielding about 3,000 nuts/year is estimated to be about 59 kg of N, 13 kg of P, and 91 kg of K. Potassium is needed in large quantities as the coconut palm is a gross feeder of this nutrient (Anon., 1957). A concentration of potassium occurs generally in the growing parts of the plant where life and growth processes are intensive. Potassium plays a very vital role in photosynthesis, in water economy, in plant food metabolism, and in the synthesis of oil in the kernel. It increases, in particular, the copra content of the nuts and the disease resistance of the trees. In general, a low level of potassium in the soil is believed responsible for making the coconut palms susceptible to shoot rot, leaf spot, and root and leaf diseases. Potassium deficiency is evidenced by the poor development of the trees and the chlorosis of the leaves.

3. Rice

von Uexkuell (1966) reported from Japan that presently 3 to 8% of the rice crop is lost to disease. He reports that potassium deficiency is considered to be a predisposing factor for brown spot disease of rice caused by *Ophiobolus miyabeanus* Ito and Kuribay. When potassium was in short supply in Japan during World War II, brown spot disease and stem rot were the most difficult diseases to control. Spores from potassium-deficient plants appeared to be more virulent than those well supplied with potassium. von Uexkuell found that the germination rate of conidiospores on rice plants well supplied with potassium was significantly reduced. He further stated that the disease spots on well nourished plants were usually small, indicating that the disease remained localized.

Stem rot caused by *Helminthosporium sigmoideum* Cav. and *H. sigmoideum* var. *irregulare* Cralley and Tullis occurs most frequently on poorly-drained or degraded paddy fields with outbreaks closely related to potassium deficiency. This relates closely to the effect of water-logged soils on the occurrence of *Fusarium* wilt of bananas, as reported by Rishbeth (1957). von Uexkuell also reports that there is some data to support the belief that sclerotial disease *(Leptosphaeria salvinii* Catt. or *Sclerotium oryzae sativae* Catt.), Cercospora disease *(Cercospora oryzae* I. Miyake), sheath blight *(Corticium sasaki* (Shirai T. Matsu.), and blast disease *(Piricularia oryzae* Cav.) occur more abundantly when potassium is deficient. von Uexkuell found that high nitrogen is usually an important or major factor in increasing the disease and ample potassium is needed to keep the plants in a healthy status of nutritional balance.

D. Nonparasitic Diseases

Many of our plant diseases are incited through nonparasitic means. Frequently the cause of the disease is not clear. That is, an organism capable of producing disease may be present as well as certain environmental factors that can also cause the same condition. Some of the nonparasitic diseases, however, are rather clear cut and a few of them are discussed as follows:

1. Potatoes

Internal black spot potatoes has been associated with mineral nutrition, particularly potassium deficiency. Kunkel (1965) in his experiments, states that "STIB" (susceptibility to internal bruising) is the new name given to the black discoloration which often occurs near the basal end of potatoes when they are bruised hard enough to crush cells. Just how potassium works is not known, but evidence indicates that it increases the size of the root system, enables the roots to take water from the soil, and decreases the tendency of the leaves to dry out in hot weather.

2. Cabbage

Tip burn of cabbage, an apparent physiological disorder, has been reported by Chupp (1930) to be associated with fertility balance and is especially affected by potassium. He describes the problem in one field as "most plants having dead, brown, or black leaf margins with the lower leaves well spotted by *Alternaria*. In the most pronounced cases, whole plants are somewhat dwarfed and loose of head". He found that superphosphate, when used without potassium, always increased the amount of tip burn. Potassium in combination with superphosphate and sodium nitrate always lessened tip burn, but always much less than any treatment when used alone.

In this experiment no attempt was made to determine the reason why the lack of potassium or an increase of phosphorus causes tip burn. Potassium chloride was the source of potassium used in these investigations.

Physiologists at that time, however, speculated that potassium is in some way connected with the respiration process and is found most abundantly in the meristematic tissue. In a plant depleted of this mineral, the potassium may be withdrawn from the older tissues and localized in the meristematic areas. On the other hand, an increase in phosphorus, an element necessary for cell division, might tend to increase the growing area and thus hasten the potassium withdrawal from the leaves. Because of the consequent incomplete respiration, this would be followed by the presence of toxic substances that might cause the dying of leaf tissue.

3. Rice

Red withering, suffocating, bronzing, Mentek, autumn decline, blue withering, and straight head diseases of rice are closely related to potassium uptake, according to von Uexkuell (1966). Most of them result from disturbances in the root zone caused by unfavorable environmental conditions such as rapid decomposition of organic matter in flooded soil under high temperatures which exhausts the oxygen in the soil and lowers the redox potential. Organic and inorganic reduction products, such as butyric acid, acetic acid, and hydrogen sulfide are toxic to the plant and inhibit the aerobic respiration of the roots. Without aerobic respiration, potassium adsorption is seriously reduced. When potassium uptake cannot keep pace with nitrogen absorption, the K/N ratio becomes too wide. Diseased plants exhibit a much wider K/N ratio than normal plants. This disturbs protein and carbohydrate metabolism, increases respiration, nonprotein nitrogen fractions in the head, weakens healthy development of roots, and prevents a selective balanced uptake of nutrients.

Preventative measures shown by von Uexkuell for nonparasitic diseases associated with potassium predominantly called for the elimination of sulfur-bearing fertilizers and high levels of potassium.

4. Cotton

Potassium is the most common mineral deficiency in cotton according to Cooper (1939) and can result in leaf discoloration and dropping of the leaves. This condition is commonly referred to as "rust" because the leaves develop a rusty brown color. When potassium is deficient, there is a reduction in growth, with the tips and margins of the leaves turning yellow, brown, and finally curling. According to Cooper the leaves turn reddish brown and shed prematurely and the bolls are small and fail to open. Soils adequately supplied with potassium do not develop this condition.

5. Tomatoes

Hayslip and Iley (1965) found in one experiment that tomato graywall could be reduced in severity by lowering the nitrogen level and increasing potassium from sources of either potassium chloride or potassium sulfate. Ratios of N to K from 1:2 to 1:8 produced tomatoes with the least percentage of graywall when phosphorus remained constant.

Several workers, including Bewley and White (1926), show that potassium helps to prevent the physiological disorder "Blotchy ripening of tomatoes." There is a correlation between blotchy ripening and carbohydrate content of the fruit. The level of available potassium, then, may influence the incidence of internal browning by being one of the factors controlling the amount of sugar or carbohydrates entering the developing fruit.

E. The Effect of Potassium on Disease Resistance of Turfgrasses

As a matter of introduction, it should be pointed out that turfgrass is a crop valued at more than 4 billion dollars in the USA (Nutter, 1965). According to a recent survey (data in press) also conducted in the state of

Washington, we find over 116,000 acres in all turfgrasses. The investment in machinery alone for maintaining turf comes to more than $210,000,000. Weed and disease control each cost the turfgrass manager about $1,226,000/year. It was estimated in this survey that turfgrass fertilizers cost the people in the state of Washington nearly $5,000,000/year. This illustrates the magnitude of this large and expanding industry in Washington and, most probably, other states as well.

Fertilizer naturally plays a most important role in the maintenance of good turf. Well-balanced nutritional programs can aid materially in helping to suppress weeds and diseases. Potassium, one of the three major plant food elements, plays an important role in turfgrass vigor which, in turn, influences disease development. Some of the roles of potassium in plant metabolism have been mentioned earlier, and these roles are as important in turfgrass nutrition and disease resistance as in other crops. MacLean (1964) found in his investigations that certain fungi could be influenced in their growth rate by adding various amino acids in different concentrations to a basal growth medium. In the case of *Fusarium nivale* (Fr.) Ces., the inoculated fungus grew to 6.0 cm in diameter on a basal medium containing DL-lysine HCl and to 10 cm, in 5 days, on one containing L-proline. The amino acids were all added at the concentration of 2 g/liter.

In the case of *Corticium fuciforme* (Berk.) Wakef., the pathogen causing red thread in turf, MacLean also found the inocluated area grew from 0.8 cm in a control with sodium nitrate to 3.23 cm with L-histidine HCl. The smallest growth was again produced with DL-lysine with 1.36 cm increase in 25 days. All amino acids in this test were added at a concentration of 1.07g/liter.

Klein (1957) reported that increased application of ammonia fertilizer in nutrient solution caused large increases in the glutamin and glutamic acid contents of tobacco leaves, explaining in part the high susceptibility to *Alternaria, Cercospora,* and *Sclerotinia* of plants receiving ammonia. He further reported that high potassium decreased the monosaccharide and free amino acid content and susceptibility to disease. Again, it is obvious from the work of MacLean and Klein that amino acids play an important role in disease development and potassium is important in amino acid metabolism.

Turfgrasses have been found to place extremely high demands on soil potassium reserves according to Goss (1965). Table 3 shows the amounts of available phosphorus, potassium, and calcium remaining in the soil after 4 years of clipping removal at the Western Washington Research and Extension Center at Puyallup, Washington. These data do not account for that which was leached, fixed by the soil, or of any amounts that were mineralized by the soil.

Several of the more common and important turfgrass diseases will be discussed and interrelationships with potassium will be shown where these exist.

Table 3—Available nutrient levels in soil at various N, P, and K levels after 4 years of removing turfgrass clippings

Nutrients applied in kg/ha/year			Initial soil nutrient levels before fertilization			Soil nutrient levels after 4 years		
N	P	K	P	K	Ca	P	K	Ca
				kg/ha/year				
775	0	0	13	445	2,359	12	191	1,718
775	69	0	14	437	2,314	21	129	1,593
775	69	129	13	428	2,225	21	178	1,593
775	69	258	13	434	2,114	21	312	1,593
775	0	129	14	445	1,958	14	249	712
775	0	258	13	443	2,136	12	387	1,126
465	0	0	12	445	2,181	14	178	1,842
465	0	129	14	429	2,047	14	240	1,842
465	0	258	14	445	2,145	14	423	1,718
464	69	0	14	445	2,123	21	151	1,967
465	69	129	15	445	2,092	21	223	1,967
465	69	258	15	445	2,092	21	343	1,967
232	0	0	15	445	1,940	16	174	1,718
232	0	129	16	445	1,954	18	294	1,718
232	0	258	16	445	1,887	19	445	1,593
232	69	0	17	445	1,896	21	169	1,842
232	69	129	16	445	1,940	21	338	1,842
232	69	258	16	445	1,869	21	392	1,718

Table 4—The effects of various combinations of N, P, K, and lime on leaf-spot disease of Bermudagrass *

Fertilizer treatment†	Spots per leaf, average
NPKL	13.5
N½PKL	19.9
NP½KL	23.1
NKL	16.0
NPK	19.6
NPL	147.5

* Data were taken from Evans et al. 1964.
† N = 178 kg/ha, P = 39 kg/ha, and K = 74 kg/ha, L = Lime to bring pH to 6.5

1. Bermudagrass Leaf Spot Disease

A leaf spot disease on Coastal bermudagrass was reported by Evans and others (1964) at Auburn, Alabama, and was found to be caused by two undisclosed fungi. This disease is much more severe when potassium levels are kept low and is one of the few reports which definitely links potassium with a turfgrass pathologic problem. Severe disease attacks were incited with zero levels of K and high N treatments. Severity was directly related to potassium deficiency as seen in Table 4.

2. Dollar Spot Disease

This disease is caused by the fungus, *Sclerotinia homeocarpa* F.T. Bennett, and inflicts severe damage to turfgrass. Pritchett and Horn (1966) at Gainesville, Flordia, reported that, when seven sources of potassium were applied to Tifway bermudagrass, they found significant differences in grass yield and dollar spot disease among the potassium sources. Table 5 gives the potassium source data.

Smith (1965) stated that applications of potassium assisted slightly in recovery from infection from *Sclerotinia* dollar spot, but they have yet to prove experimentally that applications of this material will directly assist in the control of turf diseases. The findings of Couch and Bloom (1960) at Pennsylvania State University did not agree with Pritchett and Horn and Smith. He found that neither high nor low levels of phosphorus and potassium in combination with nitrogen had any effect upon disease development in his pot culture nutrient solution studies. He reported that low moisture stress caused more disease than nutritional levels.

3. Brown Patch Disease

The effect of nutrition, pH, and soil moisture on *Rhizoctonia* brown patch disease, caused by *Rhizoctonia solani* Kuhn, was reported by Bloom and Couch (1960). They reported that disease severity in plants was much greater at high nitrogen levels than normal or low levels. They further reported that high or low levels of phosphorus or potassium did not influence disease reaction. Couch (1966) reported, however, that Seaside bentgrass, when grown under high nitrogen with concurrent increases in phosphorus and potassium to produce a high balanced nutrition, did not alter disease susceptibility. When grown under low nitrogen imbalanced nutrition, plants were more resistant to the fungus.

Table 5—Influence of sources of K on yield and Dollarspot in Bermudagrass *

Source of K	Average yield/clipping	Dollarspot rating†
Applied at 0.91 kg/ 93 sq m/season	g/93 sq m	
KCl	1,690	7.3
K_2CO_3	2,350	8.5
K_2SO_4	2,440	8.0
KNO_3	1,650	5.8
$CaK_2P_2O_7$ (frit)	1,550	6.7
FN 519 (frit)	1,630	7.4
Green Sand	1,810	4.8
Check (no K)	1,680	6.8

* Data were taken from Pritchett and Horn (1966).
† Rating of 1 = very heavy infestation, 9 = no dollarspot fungus.

4. Red Thread Disease

This disease, caused by the fungus *Corticium fuciforme* (Berk.) Wakef., attacks the foliage of turf and gives it the appearance of being scorched. Red colored, thread-like stromatia are produced near the tips of the grass blades, hence, the name red thread. Goss and Gould (Roy L. Goss and Charles J. Gould, *unpublished data*) have found that nitrogen produced statistically significant differences in the percent of diseased area and in the number of stromatia produced by *Corticium fuciforme*. Potassium was significant in bringing about a decrease in infection during one year of these investigations. These investigations were conducted with two levels of nitrogen (1.8 and 3.6 kg N/93 sq m per season) from urea, three levels of potassium from potassium chloride (0, 1.5, and 3.0 kg K/93 sq m per season) and two levels of phosphorus from treble superphosphate (0 and 0.8 kg P/93 sq m per season). As K was increased from 0 to 3 kg concurrently with N from 1.8 to 3.6 kg/93 sq m, the percent of area infected, likewise, decreased (Fig. 1). The greatest infection from *Corticium* occurs in the Pacific Northwest in late summer and during the fall period. Tissue were found by Goss (1965) to be lower in potassium during this period of time than when infection was practically non-existent. Balanced high levels of nutrition tend to help the grass escape red thread attacks. Most agronomists and pathologists agree that high nutritional levels increase the growth rate of the leaves, and the infected tissue may be removed before becoming objectionable in color.

Muse and Couch (1965) reported in moisture stress investigations that Pennlawn red fescue grown at available water range (field capacity-permanent wilting percentage) has less *Corticium* red thread than plants grown

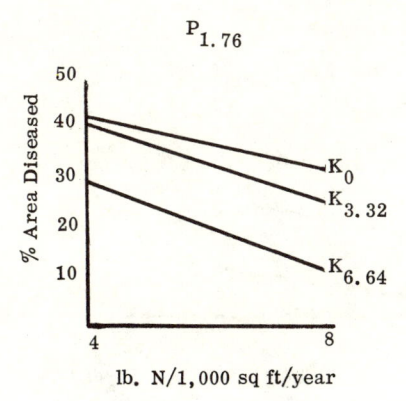

Fig. 1—Decrease in Red thread infection (*Corticium fuciforme*) with increasing rates of both N and K. (Note: both P and K are expressed in elemental form and not oxide.)

at field capacity stress. These plants, likewise, had a higher level of tissue potassium.

5. Fusarium Patch Disease

This disease, caused by *Fusarium nivale*, was found by Goss and Gould (Roy L. Goss and Charles J. Gould, *unpublished data*) to decrease with increasing levels of potassium from 0 to 3 kg K/93 sq m per season. Increasing levels of potassium tend to keep the disease incidence reduced somewhat in the 2.72- to 5.44-kg N/93 sq m range on putting green turf, but when nitrogen was increased to 9.1 kg N/93 sq m, potassium had little effect on disease incidence as seen in Fig. 2. Sources of nutrients in these studies were nitrogen from urea, potassium from potassium chloride, and phosphorus from treble superphosphate.

6. Ophiobolus Patch Disease

This disease is caused by the fungus *Ophiobolus graminis* Sacc. var. *avenae* E. M. Turner and has responded with practical significance to both phosphorus and potassium nutrition, according to Goss and Gould (1967). They reported that potassium had a suppressing effect on the amount of

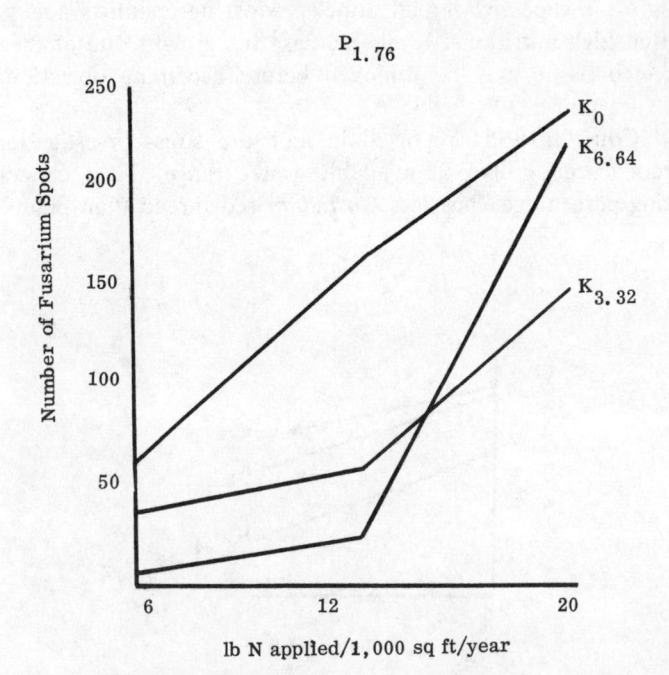

Fig. 2—Number of Fusarium patch spots (*Fusarium nivale*) are lowest with increasing rates of K but increase with increasing rates of N. (Note: both P and K are expressed in elemental form, not oxide.)

disease in 2 years of investigations, regardless of nitrogen and phosphorus levels. These studies were conducted on putting green turf in a 3 by 3 by 2 factorial experiment. Three levels of nitrogen from urea (2.7, 5.5, and 9 kg N), three levels of potassium from potassium chloride (0, 1.5, and 3.0 kg K), and two levels of phosphorus from treble superphosphate (0 and 0.8 kg P), all per 93 sq m per season, were the treatments applied. In 1964 there was five times as much disease with zero K and zero P at all N levels as there was at 1.6 kg of K and zero P at all levels of N. There was a decrease in the amount of disease in 1964 as potassium was increased at both phosphorus levels. In this case, the phosphorus probably influenced root growth somewhat while nitrogen and potassium stimulated general vigor and growth. Since this fungus is a root and crown rotter, any increases in general vigor and root development would materially help the plant to recover from infection or escape severe attack.

REFERENCES

Anon. 1957. Improving coconut yields by NPK manuring. World Crops. 9(9):379-381.

Bewley, W. F. and H. L. White. 1926. Some nutritional disorders of the tomato. Ann. Appl. Biol. 13:323-338.

Bloom, J. R., and H. B. Couch. 1960. Influence of environment on diseases of turfgrasses. I. Effect of nutrition, pH, and soil moisture on *Rhizoctonia* brown patch. Phytopathology 50(7): 532-535.

Chi, C. C., and E. W. Hanson. 1961. Nutrition in relation to the development of wilts and root rots incited by *Fusarium* in red clover. Phytopathology 51: 704-711.

Chupp, Charles. 1930. The effects of potash and phosphorus on tip burn and mildew of cabbage. Phytopathology 20(4): 307-318.

Cooper, H. P. 1939. Nutritional deficiency symptoms in cotton. Soil Sci. Soc. Amer. Proc. 4:322-324.

Couch, H. B., and James R. Bloom. 1960. Influence of environment on diseases of turfgrasses. II. Effect of nutrition, pH, and soil moisture on *Sclerotinia* dollar spot. Phytopathology 50(10): 761-63.

Couch, H. B. Relationship between soil moisture, nutrition and severity of turfgrass diseases. J. Sports Turf Res. Inst. 42: 54-64.

Cralley, E. M., and W. H. Tharp. 1938. Field studies on *Fusarium* wilt of cotton in Arkansas. The relation of "wilt" and "total infection" as influenced by potash fertilization. Phytopathology 28: 667.

Croucher, H. H., and W. L. K. Mitchell. 1940. Fertilizer investigations with the Gros Michel banana. Jamaica Dept. Sci. and Agr. Bull. 19 (N.S.). 30 p.

Dick, James B., and H. B. Tisdale. 1938. Fertilizer in relation to incidence of wilt as affecting a resistant and susceptible variety. Phytopathology 28(9):666.

Evans, E. M., R. D. Rouse, and R. T. Gudauskas. 1964. Low soil potassium sets up Coastal for leaf spot disease. Highlights of Agr. Res. 11(2). Agr. Exp. Sta. of Auburn Univ., Auburn, Ala.

Gallegly, M. E., Jr., and J. C. Walker. 1949. Plant nutrition in relation to disease development. V. Bacterial wilt of tomato. Amer. J. Bot. 36(8): 613-623.

Glynne, Mary D. 1959. Effect of potash on powdery mildew in wheat. Plant Path. 8(1): 15-16.

Goss, Roy L. 1965. Nitrogen potassium team on turfgrasses. Better Crops with Plant Food. 49(2): 34-9.

Goss, Roy L., and Charles J. Gould. 1967. Some interrelationships between fertility levels and *Ophiobolus* patch disease in turfgrasses. Agron. J. 59:149-151.

Gould, Charles J., and Thomas S. Russell. 1961. Relation of previous nutrition of gladiolus to infection by *Stromatinia gladioli*. Phytopathology 51(7):482-86.

Griffith, W. K., and R. E. Wagner. 1966. Plant nutrients and disease resistance—any relationship? Better Crops with Plant Food. 50(3):2-5.

Hayslip, N. C., and J. R. Iley. 1965. N-K ratio may influence tomato graywall. Better Crops with Plant Food. 49(2):28-29.

Hoffer, G. N., and B. A. Krantz. 1949. *In* Hunger Signs in Crops. 2nd ed. p. 59-84. The Amer. Soc. of Agron. and the National Fertilizer Assoc., Washington, D. C.

Hooker, A. L. 1966. Plant nutrients on stalk rot and lodging. Better Crops with Plant Food. 50(3): 6-9.

Jackson, T. L., W. H. Foote, and E. A. Dickason. 1962. Effect of fertilizer treatments and planting dates on yield and quality of barley. Tech. Bull. 65. Oregon Agr. Exp. Sta., Corvallis.

Johnson, E. M., and W. D. Valleau. 1940. Control of blackfire of tobacco in western Kentucky. Bull. 399. Kentucky Exp. Sta., Lexington.

Jones, John Paul, and L. J. Alexander. 1962. Relation of certain environmental factors and tobacco mosaic virus to blotchy ripening of tomatoes. Phytopathology 52: 524-528.

Kaufman, Donald D., and Lansing E. Williams. 1964. Effect of mineral fertilization and soil reaction on soil fungi. Phytopathology 54(2): 134-139.

Klein, E. K. 1957. The effect of mineral nutrition on the leaf content of free amino acids and monosaccharides and the effect of this on the susceptibility of the plant to fungal parasitism. Abst. Soils and Ferts. 20(1): 303.

Kunkel, Robert. 1965. Internal black spot of potatoes. Circ. 451. Washington Agr. Exp. Sta., Pullman.

Laughlin, W. M. 1965. Effect of fall and spring application of 4 rates of potassium on yield and chemical composition of Timothy in Alaska. Agron. J. 57(6):555-558.

MacLean, N. A. 1964. Amino acid nutrition and turf fungi. Proc. 18th N. W. Turfgrass Conf. p. 40-46.

McClellan, W. D., and Neil W. Stuart. 1947. The influence of nutrition on *Fusarium* basal rot of narcissus and on *Fusarium* yellows of gladiolus. Amer. J. Bot. 34(2): 88-93.

McNew, George L. 1953. Plant Diseases. The Yearbook of Agriculture. p. 100-114.

Muse, Ronald R., and H. B. Couch. 1965. Influence of environment on diseases of turfgrasses. IV. Effect of nutrition and soil moisture on *Corticium* red thread of Creeping red fescue. Phytopathology 55(5): 507-10.

Nutter, Gene C. 1965. Turfgrass is a 4 billion dollar industry. Turfgrass Times. 1(1): 1-2.

Pelegrin, P. 1954. The use of fertilizers in banana growing. Abst. Soils and Ferts. 17(4): 338.

Pritchett, W. L., and G. C. Horn. 1966. Fertilization fights turf disorders. Better Crops with Plant Food. 50(3):22-25.

Rich, Samuel. 1958. Fetilizers influence the incidence of tomato internal browning in the field. Phytopathology 48(8): 448-50.

Rishbeth, J. 1957. *Fusarium* wilt of bananas in Jamaica. II. Some aspects of host parasite relationships. Ann. of Bot. 21(82): 215-45.

Smith, J. Drew. 1965. Fungal diseases of turfgrasses. 2nd Ed. Revised by N. Jackson, and J. Drew Smith. A Sports Turf Res. Inst. Publication.

Spencer, Ernest L., and George L. McNew. 1938. The influence of mineral nutrition on the reaction of sweet corn seedlings to *Phytomonas stewartii*. Phytopathology 28: 213-223.

Stoddard, D. L. 1942. *Fusarium* wilt of cantaloupe and studies on the relation of potassium and nitrogen supply to susceptibility. Peninsula Hort. Soc. Trans. 31: 91-93.

Struchtemeyer, R. A. 1966. Potassium—suppressor of diseases and other disorders in potatoes. Better Crops with Plant Food. 50(3): 14-16.

Tharp, W. H., and C. H. Wadleigh. 1939. The effects of nitrogen source, nitrogen level, and relative acidity on *Fusarium* wilt of cotton. Proc. Assoc. Southern Agr. Workers. 40: 190-191.

von Uexkuell, H. R. 1966. Rice diseases and potassium deficiency. Better Crops with Plant Food. 50(3): 28-35.

Walker, J. C., and R. E. Foster. 1946. Plant nutrition in relation to disease and development. III. *Fusarium* wilt of tomato. Amer. J. Bot. 33(4): 259-264.

Walker, J. C., and W. J. Hooker. 1945. Plant nutrition in relation to disease development. I. Cabbage yellows (*Fusarium oxysporum* f. *conglutinans*). Amer. J. Bot. 32: 314-320.

Yale, John W., and Edward K. Vaughan. 1962. Effects of mineral fertilizers on damping-off of table beets. Phytopathology 52: 1285-1287.

Young, V. H., and W. H. Tharp. 1941. Relation of fertilizer balance to potash hunger and the *Fusarium* wilt of cotton. Bull. 410. Arkansas Agr. Exp. Sta. Fayetteville. p. 1-24.

Young, V. H., J. O. Ware, and O. A. Pope. 1935. Control of potash hunger and *Fusarium* wilt in cotton. Phytopathology 25(10): 969.

12

Effect of Potassium on Quality Factors— Fruits and Vegetables[1]

GEORGE A. CUMMINGS

North Carolina State University
Raleigh, North Carolina

GERALD E. WILCOX

Purdue University
Lafayette, Indiana

I. INTRODUCTION

Quality of vegetables is any property or characteristic that might be ascribed to a vegetable for the purpose of describing it relative to a standard of excellence such as, color, shape, dimensions, texture, weight, composition, etc. Vegetable refers to the fresh edible portion of a herbaceous plant which includes: roots—rutabaga, beet, turnip, radish, carrot, parsnip, sweet potato, horseradish; tuber—white potato; stem—asparagus, kohlrabi; bud—brussel sprouts; bulb—onion, garlic, shallot; petiole—celery, rhubarb; leaf—cabbage, lettuce, kale, mustard, endive, turnip leaves, chard, collards, parsley, spinach, chive; immature flower—cauliflower, sprouting broccoli, artichoke; seed—pea, lima bean; immature fruit—snap bean, okra, eggplant, summer squash, cucumber, sweet corn; fruit—squash, pumpkin, tomato, pepper. Thus, the effect of potassium on quality factors of vegetables relates to the influence that variations in potassium concentrations have on the nor-

---

[1] Paper no. 2542 of the Journal Series of the North Carolina State University Agr. Exp. Sta., Raleigh, N. C., jointly with the Purdue University Agr. Exp. Sta., Lafayette, Ind.

[2] Part II prepared by G. A. Cummings. Part III prepared by G. E. Wilcox.

mal development of the edible portion of the plant as it might cause deviation from an acceptable grade of the product.

A deficiency of potassium reduces the quality and market value of a crop as well as yields. Much of the research reported on the response of vegetables to potassium has been to determine the effect on yield first and some quality factor second. In a potassium-deficient situation, increasing the rate of potassium results in a rapid increase to maximum yield. Most of the quality characteristics of the plant are improved to optimum in the same range of potassium concentration which in turn directly or indirectly contributes to the steep yield response obtained.

The factors evaluated in determining the quality of fruit are often quite different than those utilized in the case of vegetables. Most of the work related to potassium nutrition and fruit quality has been concerned with factors that are the basis for commercial grading, such as color and size, and with factors that have been found to influence consumer acceptance. In addition to the above, there have also been attempts to relate potassium nutrition to such factors as keeping quality, constituents that influence flavor or appearance, and juice content. Another measurement of fruit quality is the biological value as an edible product. However, except in the case of Vitamin C with citrus, information on the influence of potassium nutrition upon this factor is rather meager. Within a species an evaluation of a quality factor, such as skin color of apples, may be of great importance in one variety but not applicable in another. In ascertaining the influence of potassium supply, or in correlating a quality factor with foliar or fruit content of potassium, other environmental factors cannot be overlooked. Recent reviews have discussed the influence of nutrition upon individual fruit crops and upon certain quality factors. The influence of potassium, as well as other nutrients, upon fruit quality has been reported for the following crops: peaches—Ballinger, Bell, and Childers (1966); strawberries—Boyce and Matlock (1966); citrus—Smith (1966); cherries—Westwood (1966); bush fruits—Jones (1949); and apples—Boynton and Overly (1966). The factors being evaluated were often quite different for the various crops, however, and it was impossible to ascertain the influences of potassium for a given factor over a wide range of crops.

II. FRUIT

A. Influence of Potassium Upon Fruit Quality

1. *Color*

One of the most important quality factors in regard to consumer preference is fruit color. The relationship of potassium supply and fruit skin color is not without controversy but normally the relationship is a positive one.

The enhancement of fruit color by added potassium has been observed when potassium was already sufficient for maximal production.

Reports of increased color in apples resulting from the application of potassium have been reported by Burrell and Cain (1940), but the foliage initially was extremely low in potassium. Beattie (1959) observed that soil application of KCl was more effective than foliar application of K_2SO_4 in increasing the red skin color of Rome Beauty apples. In the first of two reports, Weeks et al. (1952) noted that both nitrogen and potassium influenced fruit color with nitrogen depressing percent red skin color while potassium increased it. In a later report, Weeks et al. (1958) again demonstrated the enhancement of red color by potassium fertilization and noted that trees fertilized with high nitrogen and low potassium rates produced poorly colored fruit. They emphasized that red color was associated with both nitrogen and potassium content of the foliage and that in some instances potassium fertilization appeared to offset the depressing effect upon fruit color of moderate nitrogen levels. Fisher and Kwong (1961) reported an increase in the red color of apples resulting from potassium application but noted that this increase occurred only when K content of the foliage was below 0.5%.

There are also reports which indicate that potassium nutrition of peaches and red coloration are positively related. Stembridge et al. (1962) observed an increase in percent red coloration as potassium application was increased in 2 consecutive years. They also reported that fruit from the low nitrogen, high potassium plots received superior color rankings each year. In a potassium and magnesium factorial experiment, carried out over a 5-year period, Cummings (1965a) observed that with an increased potassium supply there was an increase in red color of the fruit. The data indicate that most of this increase in red coloration was obtained with the first increment of potassium and was diminished by magnesium applications. Bould and Catlow (1954) reported that the 'Climax' strawberry produced fruits with poor color when potassium fertilizer was omitted.

There is little doubt that potassium nutrition is related to fruit color when fruit from potassium-deficient trees are compared with fruit from trees with adequate potassium levels. However, the probability of obtaining an increase in color by supplying potassium above that needed for maximal production is small. The enhancement of red color by potassium nutrition, though evident, may be small in comparison to the effect of climatic or other nutritional factors.

2. *Fruit Size*

Many factors, but especially fruit load, are related to the ultimate size of fruit produced. Much of the work indicating an association between potassium fertilization and fruit size has been carried out with trees at or near a potassium deficient level. Increases in size resulting from increasing potassium application for several fruits are quite similar to those obtained with apples by Fisher and Kwong (1961). They noted an increase in size of 'McIntosh' apples when potassium was applied in an orchard where foliar potassium level was 0.5%. Forshey (1963) also reported increases in size related to potassium nutrition where leaves from trees in control plots contained approximately 1% K. Lilleland et al. (1962) reported a marked increase in fruit size in peaches as leaf K was increased from 0.2 to 1.0%.

The results with citrus fruit concerning increases in size has received much attention with some controversy regarding the value of potassium application above that needed for maximal yield. Smith and Rasmussen (1961) reported that fruit size was reduced consistently in 'Marsh' grapefruit by low potassium, but they found no increase in size of fruit after foliar K reached the level of 2%. Haas (1949) noted that potassium-deficient orange trees were characterized by a poor fruit set as well as by small fruit. Chapman and Brown (1943) reported that fruit from potassium-deficient orange trees grown out-of-doors in solution cultures were small but that the fruit was of good quality. They noted a marked increase in size and a decrease in eating quality when trees received excessive rates of potassium. They reported that the calcium content of the fruit from plots treated with high potassium rates was extremely low and the poor quality may have been associated with low calcium induced by a high potassium supply, rather than by any direct physiological effect of high potassium. However, later work by Smith and Rasmussen (1959) indicated that when the supply of potassium, calcium, and magnesium supplied to orange trees was varied that only potassium had appreciable effect upon fruit size. Parker and Jones (1950) reported that the increase in the size of oranges with potassium fertilization was usually greater in years when fruit size was generally small. In their work, potassium did not influence the total number of fruit; but yields were normally increased because of the increase in fruit size. Reuther and Smith (1952) also noted the increase in size of oranges with increased potassium fertilization. Their data also indicates that the effect of potassium varies greatly as nitrogen supply is changed and suggest that strong nutrient interactions may be evident in some seasons while absent in others. Similar results were reported by Embleton et al. (1956) in 'Valencia' oranges where potassium applications increased yield and reduced fruit size in 1954 but increased fruit size in 1955. Goldwebber, Boss, and Lynch (1956) noted that potassium increased the yield of 'Persian' limes and that the yield increase was accounted for by a larger number of fruit per tree and by increases in the size of individual fruit.

Kirkpatrick and Fisher (1958) reported an increase in the size of sour cherries associated with potassium levels. They attributed the increase in size to a higher juice content rather than to larger pits or a greater amount of pulp. Kwong (1965) reported that sweet cherries increased in size when the K leaf content approached 1.25%. His data indicate that the increase in size was not evident until 3 years after differential potassium treatments were initiated. Fisher et al. (1959) noted a response in size of pears to potassium fertilization as the K content of mid-summer leaf samples was increased from 0.7 to 1.0%. No increases in fruit size resulting from K application were noted in orchards where leaf samples were above 1% K.

Increase in fruit size resulting from the application of potassium to deficient trees is well documented. The level of potassium supply at which increases in fruit size would not be attained upon adding additional potassium probably extends into the mid-portion of most published sufficiency ranges. With the exception of citrus fruit, where heavy potassium applications may be associated with fruit of poor eating quality, the potassium foliar content should probably be maintained in the upper portion of the sufficiency range to assure adequate fruit size.

3. *Fruit Acidity*

Several factors contribute to the taste of fruits. Foremost among these factors are the sugar and acid content. Smith (1966) in a review of citrus nutrition emphasizes that total acid content of the juice is consistently and strongly increased by potassium fertilization. In determining the effects of the supply of nitrogen, phosphorus, and potassium upon composition of oranges, Anderssen (1937) noted that increased potassium supply was associated with higher total acids. In a study comparing the effect of different potassium sources upon composition of orange juice, Roy (1945) reported that both citric and ascorbic acid were increased by potassium fertilization. Similar results of increased total acids with potassium fertilization were reported by Jones and Parker (1949). It should be noted in their study that both total acid and pH were positively related to potassium nutrition. Reuther and Smith (1952) noted that total acid content of Valencia oranges was positively correlated with potassium treatment at low levels of potassium. This effect was similar to that observed on fruit size. The percent acid in juice increased as K supplied to the tree was increased from 0.25 to 1.5 lb, but when K was increased from 1.5 to 3 lb/tree the effect was not consistent and usually percent acid was depressed.

Work with apples indicates that the rise in total acids of fruit as potassium supply is increased is not as pronounced as with citrus crops, but is still often evident. Wilkinson (1957, 1958) reported that potassium fertilization was closely associated with high acid content of apples. Eaves and Leefe (1955) noted that fruit from trees receiving high rates of potassium contained a significantly higher acid content than fruit from control trees or trees fertilized with nitrogen. In their work dolomitic lime reduced the acid content of fruit, thus indicating that the supply of calcium and magnesium as well as potassium may be important in the acidity relationships. Fisher and Kwong (1961) reported a significant increase in acid content of 'McIntosh' apples related to increases in potassium supply, but the foliar K content initially was extremely low (0.5%). Barden and Thompson (1962) in an experiment with variable rates of potassium supplied to 'Red Delicious' apple trees over a 2-year period, reported no difference in yield or fruit size. However, they detected an increase in total acid content associated with potassium fertilization one year. This lack of a consistent relationship between total acids and potassium supply indicates that other factors possibly climatic, must be considered.

Reports relating potassium level and total acid content to fruit crops other than citrus or apples are not so numerous but the same relationship is still obtained. Knight and Wallace (1932) reported that the acid content of strawberries was low from plots receiving low potassium. Kirkpatrick and Fisher (1958) reported an increase in malic acid content of sour cherries with increases in potassium supply. In work with sweet cherries, Kwong (1965) observed a slight but not significant increase in acid content resulting from potassium fertilization.

Kwong and Fisher (1962) reported that the acid content in 'Jerseyland Peach' was increased by potassium fertilization. However, the check trees were severely potassium deficient and the significant increase in fruit acidity resulting from potassium fertilization was primarily associated with the first increment of potassium. In a factorial magnesium and potassium experiment with 'Elberta' peaches, Reeves[3] obtained increases in fruit acidity from potassium application at three levels of magnesium supply. In his work the pH of the fruit was also positively related to potassium supply. This lack of a negative relationship between pH and total acids is also evident in some of the data reported by Jones and Parker (1949) with oranges. These data indicate that a greater potassium supply not only increases total acids but may actually decrease the amount of acid existing as free acid in the fruit. This lack of correlation between measurements of acidity suggests that the proportion of the acid existing as a salt may increase with an increase in the potassium supply.

The increase in acidity associated with increased potassium supply is more pronounced with citrus than with other fruit crops. However, even with citrus there is some controversy relative to the influence of additional potassium fertilizer upon total acids when potassium is already sufficient for optimal yield. Though a positive relationship exists between potassium supply and total acids with other fruit crops, environmental factors may exert a greater influence than does potassium supply.

---

[3] John H. Reeves. 1967. The effects of three mineral elements and two management practices upon selected chemical and physical factors in Redhaven and Elberta peaches. M. S. Thesis, North Carolina State University, Raleigh, N. C.

4. Storage Quality and Shelf Life

The time elapsed between when the fruit is picked and ultimately consumed varies widely with different fruit. The industry is well aware that the most important consideration in the maintenance of quality is the stage of maturity of the fruit when harvested. However, when fruit is harvested at the same maturity stage there is evidence, though not unanimous, that potassium supply may influence storage quality or shelf life. This effect may not be the same on firmness at harvest as it is on storage quality or shelf life. Overly and Overholser (1931) reported that the use of potassium alone or in conjunction with nitrogen resulted in less storage breakdown of 'Jonathan' apples. They found no relationship between the supply of nitrogen, phosphorus, or potassium and fruit firmness and stated that losses from breakdown, regardless of nutrient supply, varied with the season. Weinberger (1929) reported that the use of potassium fertilizer, alone or in combination with nitrogen and phosphorus, did not affect the firmness or keeping quality of apples, peaches, or strawberries. Beaumont and Chandler (1933), in an extension of Weinberger's work, reported that the use of potassium decreased the firmness of peaches at picking time but the fruit low in potassium softened more rapidly in storage. Magness and Overly (1929) reported a positive, but not significant, relationship between apple storage quality and potassium fertilization. More recent information indicates that the effects of potassium fertilization on storage quality may be limited except when working at a rather low level of potassium nutrition. Hill et al. (1950) observed a positive trend between foliar potassium concentration and keeping quality with 'McIntosh' and 'Spy' apples up to a foliar content of 1.7% K. They also noted that as the nitrogen application was increased, the potassium content of the foliage decreased with a concomitant decrease in fruit quality. Bunemann, Dewey, and Kenworthy (1959) in a 2-year study of fruit from 16 Michigan apple orchards noted that differences in nutrient content of the leaves and fruit within the optimum range for good growth and yield normally did not influence the storage quality of the fruit. The only exception found was that low potassium levels were associated with an increased amount of internal breakdown in the 1 year it was evident.

Another factor affecting the storage quality of apples is the occurrence of bitter pit. This is the development of small brown spots or streaks just beneath the skin during storage. Oberly and Kenworthy (1961) reported that heavy potassium fertilization resulted in a marked increase of bitter pit. They suggest that the incidence of bitter pit may be reduced by discontinuing the indiscriminate use of potassium or by the use of dolomitic limestone to decrease potassium absorption.

The favorable response in storage quality of apples, with the exception of bitter pit development in storage, is not so evident with other fruit. Normally, but not always, firmness at harvest and storage quality are closely related. Curwen, McArdle, and Ritter (1966) noted a decrease in the firmness of 'Montmorency' cherries associated with potassium fertilization. They attributed the decrease to an induced water insoluble pectic content. Fisher et al. (1959) reported a positive relationship between the firmness of 'Bartlett' pear and potassium nutrition but the authors doubted if the small increase in firmness was of practical importance. Darrow (1932) reported that the firmness or storage quality of strawberries was not influenced by different levels of nitrogen, phosphorus or potassium. He noted a rather poor flavor of the berries 1 year, but not in another, from the plots which had received potassium.

In seven experiments over a 4-year period Wade (1956) obtained a significant reduction in incidence of brown rot (*Scherotinia fructicola*) as the potassium supply to apricot trees was increased. He presented a regression equation which indicated that if the K concentration in leaves would be raised to 4.6%, brown rot could be eliminated. Since a practical application of this was found to be impossible he suggested that the application of 2 lb of KCl/tree would improve the K content of the foliage and reduce the severity of the disease.

5. Other Quality Factors of Fruit

A. PEAL THICKNESS OF CITRUS. There have been numerous reports that support the observation of Chapman, Brown, and Rayner (1947) that greater peel thickness of oranges is associated with increases in potassium fertilization. A more recent report by Smith (1963) indicated that a lower percent juice content and greater peel thickness in 'Marsh' grapefruit was associated with increases in potassium fertilization. One exception to the increase in peel thickness in citrus resulting from potassium fertilization is a report by Embleton and Jones (1966) of a significant decrease in the peel thickness of lemons. In their work potassium fertilization was reflected in a marked increase in foliar potassium content with a concomitant decrease in peel thickness. However, even though peel thickness decreased at each location when potassium was applied, trees from the location with the lowest potassium foliar contents normally produced fruit with thinner peels than fruit from the other two locations.

B. MATURITY DATE. Change in the date of harvest of a given variety of fruit may or may not offer advantages to the producers. Roy (1945) reported that fruit from potassium-deficient orange trees matured much earlier than fruit from trees adequately supplied with potassium. Havis and Gilkesin (1951) reported a delay in maturity of peaches resulting from the application of potassium fertilizer. Cummings (1965a) obtained similar results and noted that average harvest date was delayed approximately 2 days when K fertilization was increased from 0.09 to 0.60 kg K/tree.

C. JUICINESS. The negative relationship between potassium fertilization and percent juice of citrus is well documented. In the work previously discussed where increases in potassium supply increased the size and acid content of citrus fruit either the percent juice remained about the same or was decreased. Kirkpatrick and Fisher (1958) reported a positive relationship between potassium soil levels and the size of sour cherries and attributed the increased size to a higher juice content rather than to larger pits or more pulp. Curwen et al. (1966) reported that with a higher potassium supply to Montmorency cherries there was a greater juice loss upon pitting. They stated this was probably an indirect effect associated with lower levels of fruit calcium rather than a direct influence of high potassium.

D. SOLUBLE SOLIDS. Except in the case of citrus fruits, where increasing potassium appears to have a depressing influence upon soluble solids, a specific influence of potassium upon soluble solids of other fruits has not been clearly demonstrated. Barden and Thompson (1962) in work with apples, Kirkpatrick and Fisher (1958) with sour cherries, Curwen et al. (1966) with 'Montmorency' cherries, and Kwong (1965) with sweet cherries and Reeves [2] with peaches all reported that potassium levels did not influence soluble solids. In most of the above work, though not statistically significant, there was a slight but consistent trend for potassium application to depress the soluble solids of the fruit.

The effect of potassium application on the concentration of soluble solids in strawberries may not be the same as noted with tree crops. Knight and Wallace (1932) and Kirch (1959) reported that potassium application increased the soluble solids of strawberries. However, Knight and Wallace's results were obtained by comparing adequately fertilized berries with potassium deficient berries.

6. *Relationship of Foliage and Fruit Levels of Potassium to Quality*

The influence of the soil supply of potassium may be reflected quite differently in the potassium content of the foliage than it is in the potassium content of the fruit. Work by Batjer and Westwood (1958) showed that there was a continual seasonal movement of potassium into the developing fruits and a simultaneous decline in the potassium content of the foliage. In fruit crops the foliar concentration usually follows more closely the potassium regime of the soil than does the potassium concentration of the fruit.

Different investigators have related potassium fertilization, potassium soil test values, concentration of potassium in the foliage, potassium concentration in the fruit, or a combination of two or more of these determinations with a given quality factor. In order to assess the relationship of these various measurements of potassium nutrition, Reeves [4] carried out extensive

---

[4] Ibid

work relating several quality factors of peaches produced in some long-term potassium and magnesium plots in North Carolina. The data in Table 1 clearly indicates that the results relative to firmness, shelf life, pH, titratable acidity, and sugar contents would support most of the work published relative to the effect of potassium nutrition upon the quality factors involved. It is easy to visualize the pitfalls in utilizing any measurement of potassium nutrition because of the lack of good agreement of K application with soil tests, foliar potassium concentrations, or the concentration of potassium in the fruit and in many cases the lack of any two of these measurements resulting from potassium application. The data presented in Table 2 indicate quite clearly, however, that the various measurements utilized in determining potassium nutrient status responded similarly to potassium application. Yet the results from tests of significance attempting to relate a quality factor with potassium status often varied depending on which potassium measure-

Table 1—The influence of K application to Elberta peach trees upon certain quality factors of peaches (Reeves, 1967)[3]

Quality factor	K applied, kg/tree			LSD 5%	Correlation of quality factor with K content of fruit
	0.09	0.30	0.60		
Firmness[†]	8.73	10.09	11.28	1.45	+.195
Shelf life, days	3.03	3.64	3.70	0.60	+.589*
% dry weight of fruit	14.65	14.11	13.81	0.55	+.139
Specific gravity	.961	.958	.957	.005	−.113
pH of fruit	3.71	3.80	3.85	.07	−.006
Titratable acidity meq/100g	6.11	6.67	6.90	.47	+.367*
Sugar content, %	13.39	13.23	12.97	.54	+.215
Flesh browning[‡]	12.85	12.97	13.70	.53	+.107
Soluble amino acids, %	0.304	0.236	0.225	0.041	−.023

* Significance at 5% level 0.344.
† Resistance to penetration as measured by taking direct readings from a Ballauf pressure tester with a 3/8-inch diameter blunt plunger.
‡ Yellow color measured on a Hunterlab Model 25 Hunter Color and Color Difference meter after shaking a macerated sample for 1 hour. Higher reading indicates greater browning resistance.

Table 2—The relationship of K soil application to Elberta peach trees and certain measurements of K nutrient status (Reeves)[3] and Cummings (1965b)

Measurement	K applied, kg/tree		
	0.09	0.30	0.60
Soil K* (0-15 cm, meq/100 g)	.030	.062	.084
Soil K* (15-46 cm, meq/100 g)	.021	.037	.040
Leaf K (meq/g/dry wt)	.27	.46	.55
Fruit K (meq/g/dry wt)	.191	.245	.291

* Determined by the Soil Testing Division of the North Carolina Department of Agriculture using an ammonium acetate extract.

ment was used. In some cases, especially fruit pH, there was a significant positive relationship to potassium application, but a correlation between pH of the fruit and potassium concentration in the fruit failed to indicate a relationship.

In Reeves work the effect of potassium nutrition on four factors not previously discussed (specific gravity, percent dry weight, browning resistance, and soluble amino acids) was ascertained. There has been a concerted effort in recent years in North Carolina to develop varieties of peaches that would resist browning after the fruit was sliced. This oxidation process is apparently related to polyphenolic compounds in the fruit. It is interesting to note that potassium application was associated with an increase in resistance to browning but that, though positive, the correlation of leaf potassium and browning resistance was essentially lacking. At this time, we are not certain whether this increased resistance to browning is related directly to potassium concentration in the tree or to an indirect effect of potassium such as lower magnesium and calcium content of the tree.

The decrease in protein synthesis with a concomitant increase in soluble amino acids often observed in foliage when potassium is limiting may be even more pronounced in the fruit. Reeves did not determine the soluble amino acids in the foliage but the decrease of this constituent in the fruit as potassium was increased is of such magnitude that one could speculate that soluble amino acids were translocated from the leaves and stored in the fruit to a greater extent than was potassium. Since the fruit serves as a sink for compounds metabolized in the leaves the content of certain intermediate metabolites may give a better indication of the metabolic health of the plant than would an analysis of the foliage. To further emphasize this point note that in Table 1 the decrease of soluble amino acids was highly significant when related to potassium fertilization but was not significant at the 5% level when related to potassium concentration in the fruit.

The reported values for specific gravity, sugar content, and percent dry weight were all lowered by increasing potassium application but appeared to not always follow the same relationship when correlated to potassium concentration in the leaves.

A general interpretation of this and other data could be that the degree of relationship of a given quality factor to potassium nutrition depends upon the amount of potassium supplied, the current level of potassium in the soil, and the concentration of potassium in the foliage or fruit. However, a given quality factor may be related to one or more of these potassium measurements, but not related to another. The literature is replete with references of instances which indicate variations between years, locations, varieties, and numerous cultural practices which far overshadow any effect of variation in nutrition. If the additional variables of soils, competing ions, fertilizer forms, and time of application are added to the above it becomes apparent that the concentration of potassium in the soil, foliage, or fruit is relative and must be interpreted with a knowledgeable recognition of other factors. Even in the midst of this confusion there is little doubt at the present time that the best indicator available in determining the sufficiency of minerals in the production of fruit crops is by assaying the mineral content of the foliage. Possibly at some time in the future we will be able to measure some intermediate or terminal biochemical constituent in order to establish optimal nutrient element levels.

B. Conclusion

The level of potassium supply to fruit crops is associated with several fruit quality factors. The degree to which the factors are influenced by potassium nutrition depends primarily upon the status of the potassium supply when additional potassium is supplied. The major portion of the response to K fertilization for such factors as increased fruit size, enhanced fruit color, improved keeping quality, and higher acidity has been attained in the range of what is normally considered to be a deficiency or in the lower portion of the published optimal ranges. Fruit crops other than citrus, in which excess potassium is associated adversely with several quality factors, should be supplied with potassium to maintain foliar contents in the mid or upper portions of the published sufficiency ranges. If potassium is supplied to trees that already have sufficient potassium for optimal yields there is a possibility of lowering fruit quality such as noted with the increase in bitter pit of apples with high potassium levels.

The response of a marked increase in plant content of potassium is often not evident until after potassium is supplied for more than 1 year. There is still some controversy regarding the best assay to determine the potassium status of fruit trees, but the most generally accepted assay is foliar analysis. Many factors, other than the amount of potassium applied, often influence quality factors to a greater extent than does the nutrient supply. Until we understand thoroughly, the potassium entry into the plant, its role in biochemical reactions, and its relationship to other nutrients in the leaf, fruit, and woody portions of the tree we will continue to report widely varied relationships between potassium and fruit quality.

III. VEGETABLES

A. Role of Potassium

The edible plant part can consist of any one of the plant organs for a particular vegetable species. Thus, a review of the role of potassium to determine the relation of its effect on the overall plant structure would be in order.

The subject of the role of potassium in plant growth has been well reviewed by Evans and Sorger (1966), Freeman (1965), and Lawton and Cook (1954). The metabolic processes affected by a deficiency of the potassium ion are: (i) synthesis of simple sugars and starch, (ii) translocation of carbohydrates, (iii) reduction of nitrates and synthesis of proteins, particularly in meristem tissues, and (iv) normal cell division. Freeman (1965) states that, qualitatively, the chromatograms of the various organic acids are not influenced by the potassium status. It was concluded that potassium did not completely block normal metabolic pathways but that the level of potassium nutrition in plants profoundly influenced the quantities and proportions of organic acids.

If the potassium ion concentration affects many metabolic processes in plant cells without having been found to enter directly in the reactions, an important role of the potassium ion might be as a spectator ion. Since the rate of a reaction is related to the concentration of the reacting substances, the K^+ ion might fulfill the important role of maintaining equilibrium conditions that allow optimum concentrations of organic components in the metabolic reactions.

The deficiency of potassium can cause weakened cell structure in stems and tissue, interruption of protein synthesis with the concurrent increase in amino acid content, reduction of CO_2 assimilation, carbohydrate production and leaf surface area, and translocation of the K^+ ion from mature to meristematic tissue causing older leaves to exhibit deficiency symptoms. The relation of the role of potassium to plant development can affect quality in a number of ways, depending on the plant part that is harvested as the edible portion.

B. Potassium Deficiency Symptoms

Vegetables have a high requirement for potassium and deficiency symptoms are easy to develop. For the vegetables that have leaves as the edible portion the potassium deficiency renders the quality inferior. The leaves develop marginal chlorosis with interveinal brown necrotic spots. The K^+ ion is mobile and the symptom is more severe on the older than the younger leaves. Laughlin (1961) found that spraying 'Romaine' lettuce with a 0.63% K solution eliminated leaf scorch.

E. J. Hewitt (1963) described the marked changes in growth habit produced by potassium-deficient plants, such as beets, celery, carrots, and parsnips. These plants normally grow from a crown without an extended stem, and develop an acute rosette habit when they are deficient in potassium. Marked shortening of the internodes indicates a potassium deficiency in such plants as broad bean, pea, mustard, potato, and tomato.

Tomatoes or potatoes growing on soil that contains 100 lb or less of exchangeable K per acre usually develop deficiency symptoms. One of the first characteristics observed is the bluish-green color of the leaves, as though the plant was over supplied with nitrogen. The plants have more slender stems, shorter internodes, and develop less width of plant development. Nightingale (1943) reported that, when potassium is deficient, actively dividing cells are limited to the apical tissue, the region where potassium is found when external supply is limiting. The stem cambium becomes dormant so that growth is only in length and not in width. As tubers begin to develop on the potatoes or fruit is present on the tomato plant under deficient conditions, a large part of the potassium in the plant is transferred to them. The demand for potassium exceeds the rate of supply so that interveinal bronzing and marginal necrosis of the older leaves of potatoes and ashen marginal necrosis and loss of older leaves of tomatoes occurs as the potassium in the leaves is translocated to the developing tubers and fruit.

In addition to the discoloration of the leaves and the formation of necrotic areas, potassium deficiency also results in modification of the leaf. The younger developing leaves of tomatoes display a puckered surface and are warped or turned under.

Snap beans develop an interveinal chlorosis of the leaves that progresses from the older to younger leaves as the deficiency becomes more severe.

Restricted root development is reported for plants suffering from potassium deficiency.

C. Potassium Uptake and Requirement

The uptake of potassium by vegetables is usually higher than for any other nutrient required. Harvest of the fresh edible portion also removes a large amount of potassium with the product. The uptake and removal soon depletes the soil potassium so that subsequent crops will require a potassium fertilization program to prevent potassium deficiency.

The potassium uptake curve followed the dry matter accumulation curve of tomatoes, carrots, peas, and lima beans in work reported by Hester et al. (1951). Emmert (1949) found that there was a rapid accumulation during tomato fruit development and potatoes have a steep curve of accumulation during the period of tuber sizing.

1. Tomatoes

Total potassium accumulation by tomatoes was found to vary according to soil test and fertilizer application in a check of two experimental sites by Wilcox (1964), Table 3. On potassium-deficient soil the potassium was almost quantitatively translocated from the vine to the fruit so that the vine contained only 0.28% K or a total of 9 lb/acre. With the application of 400

Table 3—Composition and K uptake of tomatoes

Tomato variety	Dry weight yield	Exchangeable soil	Fertilizer applied	K uptake	
	lb/acre	——— lb/acre ———		%	lb/acre
KC 146	Vine 3133	110	0	0.28	9
	Fruit 3864 (64,408)*			2.43	94
	Vine 3982	110	400	2.35	94
	Fruit 6307 (105,120)*			4.48	302
Roma	Vine 3920	360	100	3.70	144
	Fruit 7080 (118,000)*			4.56	326

* Fresh fruit weight.

lb/acre K the K content of the vine was 2.35% and 4.48% in the fruit. The total K uptake was 94 lb./acre in the vine and 302 lbs/acre in the fruit. The fruit grown on the soil that con ained 360 lb/acre exchangeable K had a composition of 4.56% K with an uptake of 326 lb/acre. The potassium content of the fruit grown on the two soils was about the same when fertilizer was applied although the potassium composition of the vine grown on the high potassium soil was higher. Sixty-five to 75% of the total fruit on the vine was harvested so the actual removal of potassium from the fertilized treatments was from 200 to 240 lb/acre.

Hester (1951) reported that the K uptake per acre in a 12-ton/acre crop of fruit and the supporting vine was 145 lb.

The K composition of tomato leaves increased as fertilization rate increased to 400 lb K/acre, for the 7/10 and 7/24 sampling, Table 5, the period before rapid fruit expansion and heavy fruit load occurred. The K content in the leaves of the 8/12 sampling increased as the K fertilizer rate increased to 800 lb/acre and reflected the effect of the high requirement of the fruit on transfer of K from the leaves.

2. Potatoes

The distribution of potassium-deficiency symptoms reflects the retranslocation of the element from old to young tissues, where concentrations tend to be higher, Penston (1931), in stem apices, cambia, and abscission layers of the potato plant. The author found that the apical tissue contained less potassium than the basal tissue of potato plants grown on soil that contained adequate potassium, Table 4. The composition of the plant tip increased from 4.3 to 4.9% as the K rate increased from 0 to 100 lb/acre while composition of the basal portion increased from 6.4 to 10.4%. The composition of the tip remained fairly constant, an increase of 14%, while the content of the basal tissue was increased over 60% by the application of 100 lb/acre.

Total K uptake was reported by Tyler (1959) to be 210 lb/acre with

Table 4—Composition of 4-week-old potato seedlings that received various K levels in the fertilizer bands

Rate of K in bands	K composition	
	Plant base	Plant tip
lb/acre	%	
0	6.4	4.3
20	7.7	4.7
50	8.9	4.9
80	9.0	4.8
100	10.4	4.9

170 lb in the tubers and 40 lb in the vine. Carpenter (1957) reported a total K uptake by a 400-bushel crop to be 110 lb in the vine and 90 lb in the tubers. The author found that a 600-bushel potato crop contained 161 lb. in the vine and 86 lb in the tubers or a total K uptake of 247 lb/acre.

3. Other Vegetables

Hester (1951) found that the K uptake per acre by carrots was 220 lb, peas 80 lb, and lima beans 100 lb.

Lorenz (1942) measured the uptake of K in a crop of lettuce in Salinas Valley, California. A lettuce yield of 31,662 lb/acre contained 90 lb K.

D. Plant Response to Potassium Fertilization

1. Potatoes

A. YIELD. Potatoes did not respond to potassium application on alluvial sandy loam soil that contained 161 lb exchangeable K/acre as reported by Wilcox (1961). The application of up to 190 lb K/acre did not affect tuber yield or specific gravity. One years' potato production on this soil reduced the level of exchangeable K to about 100 lb/acre. At this level of exchangeable K an application of 125 lb K/acre increased total yields over 55 cwt/acre. No significant increase was obtained to a higher rate of potassium application. Yield response to potassium fertilization has been reported, Davis and Lucas (1959), Lucas (1954), Rowberry et al. (1963), by many researchers. Murphy and Goven (1959), Sparks and McLean (1946), Terman (1949) and Tyler et al. (1959) reported that the increased tuber yield was largely to the first increment of K which was in the 40 to 100 lb K/acre range. The yield increases obtained were from 24 to 33%. A yield increase of nearly 400 bu/acre on organic soil by the application of 250 lb K/acre was reported by Davis and Lucas (1959). Van Der Paauw (1958) found that the yield response to potassium was increased on dry years. The influence of weather conditions on the response to potassium was greater during emergence and the early phases of growth than during the latter periods of growth.

B. CHIP COLOR. Chip color was quite dark for the potatoes that were suffering from potassium deficiency. The application of K did not affect chip color in the experiment that no yield response was obtained. Eastwood and Watts (1956a) found that the use of higher K rates improved chip color slightly, but was of doubtful practical value beyond the range of economical yield response. They found that KCl produced lighter chips than K_2SO_4 as did other researchers, Berger (1961), Murphy and Goven (1961) and Wilcox (1961).

C. TUBER CHARACTERISTICS. Mulder (1956) found that tubers from potassium-deficient plants were rich in tyrosine and showed stem end blackening after rough handling. Less severe potassium deficiency resulted in the blackening of tubers after boiling. A wide variation in varietal susceptibility to black spot was reported by Scudder et al. (1950). He reported that K rates from 100-400 lb/acre reduced percent black spot index on a highly susceptible variety. Van Middelen et al. (1953) found that K deficiency on Long Island soil aggravated internal black spot of potato tubers and application of 400-600 lb K/acre decreased black spot. The tyrosine, total soluble nitrogen and total amino acid in the tubers decreased as the K rate increased to 600 lb/acre. Dickens et al. (1962) found that higher potassium rates reduced blackening. Eastwood and Watts (1956) and Rowberry et al. (1963) did not find any effect of potassium on reducing sugar, chip flavor, or losses in storage.

D. SPECIFIC GRAVITY. Specific gravity of the potato tuber is one of the most widely accepted measurements of internal tuber quality. Specific gravity is closely associated with starch content, total solids, and mealiness of potato tubers.

Many researchers, Berger et al.(1961), Eastwood and Watts (1956), Lucas et al. (1954), Murphy and Goven (1959, 1961), Rowberry et al. (1963), Terman (1949, 1950), and Wilcox (1961), have reported that, at equal rates of potassium in the fertilizer band, K_2SO_4 produced tubers with higher starch content and higher specific gravity than did KCl. Also, as the rate of K in the band was increased beyond the rate that produced the major portion of increased yield the specific gravity was reduced. Murphy and Goven (1959) found that as the rate of K was increased from 0 to 250 lb/acre the specific gravity of the tubers was decreased linearly from 1.084 to 1.070 while a major part of the yield response was to 50 lb K/acre. The application of very high rates of potassium in the row was found to reduce the specific gravity of potato tubers, Wilcox (1964). The effect of increasing rates of potassium in the band was more detrimental on mineral than organic soil.

2. Tomatoes

A. YIELD. Tomatoes grown on soil with 100 lb exchangeable K/acre responded to applications of K, Wilcox (1964), in yield and quality. The total yield of tomato fruit was increased 30,000 lb/acre by an application of 200 lb K/acre, Table 5, a 60% yield increase. The yield of US no. 1 grade tomatoes was also increased 30,000 lb, an increase of 156%. At harvest it was observed that many of the ripe fruit on the potassium-deficient plants had dropped off the vine. At the third and fourth pickings the fruit from the deficient plots were 36% culls compared to 8.6% culls from fertilized plots. Total fruit yields were increased 25% by the application of 150 lb K to soil that contained 150 lb exchangeable K per acre. Thomas (1943) reported tomato fruit yield increased with the application of 40 lb K_2O/acre.

B. RIPENING. Lanham (1926) did not find any change in the rate of ripening with the addition of potassium although Winsor et al. (1961) reported more uniform ripening of fruit on vines fertilized with potassium. Wilcox (1964) found that the early fruit yields were the same on the fertilized and potassium-deficient plots, but the fruit production was higher at the later harvests for the potassium-fertilized plots. The fruit on the potassium-deficient plots were sun scalded due to exposure to the sun because of leaf loss from the deficient vines.

The first fruit to ripen on the potassium-deficient vines were as large as the fruit on the fertilized vines. However, as the season progressed the potassium-deficient fruit was smaller and of lower quality due to desiccation and because of the separation of the abscission layer in the pedicel.

Studies have been carried out on the effect of potassium on blotchy ripening. Collin and Cline (1966) found that the effect of potassium was incon-

Table 5—Tomato leaf composition, fruit yield, and quality from soil treated with various rates of K

K broadcast	Leaf K composition			Ripe fruit yield		
	7/10	7/24	8/12	Total	US No. 1	US No. 1
lb/acre	%			lb/acre		%
0	1.94	2.20	1.05	52,428	19,441	37
200	3.68	3.83	2.42	83,047	49,828	60
400	4.06	4.34	3.70	83,853	50,312	60
600	4.14	4.54	4.11	89,026	53,415	60
800	3.95	4.75	4.59	83,446	50,068	60
1,000	3.99	4.62	4.58	79,007	47,404	60
LSD (0.05)	.63	.30	.25	14,300	9,068	

sistent and their results did not support the premise that potassium deficiency causes blotchy ripening.

C. TITRATABLE, TOTAL ACIDS, AND REFRACTIVE INDEX. Titratable and total acids and the refractive index of expressed sap was increased by increased levels of potassium in the soil, Winsor (1963). Tomato fruit from potassium deficient soil did not have a different pH or refractive index compared with fruit from potassium fertilized plots in research carried out by the author.

D. TOMATO JUICE. The potassium content of the tomato juice was increased by potassium fertilizer applications to the soil in experiments by Winsor (1963). Wilcox (1964) found in 1962 that the K content was 2.9% K on unfertilized soil, 3.8% with the addition of 150 lb K/acre and 4.7% with the addition of 450 lb K/acre. In 1963 the K content of the juice was increased from 2.43% to 4.48% by the application of 400 lb K/acre.

3. Other Vegetables

A. PEAS. The potassium content of peas was increased and a slightly higher proportion of protein to total N was obtained with potassium fertilization by Jodidi and Boswell (1932). Bowers and Mahoney (1939) reported no yield response by peas to potassium fertilization on soils low in available potassium.

B. SNAP BEANS. Yield response to potassium fertilization by snap beans was reported by Forsee and Hoffman (1950) on soil that contained less than 80 lb K/acre as determined by a 0.5N acetic acid extract. Maximum yields were obtained with about 50 lb K/acre. Peck et al. (1963) reported a 1380 lb/acre increase in yield to 33 lb K/acre in bands on soil that contained less that 90 lb exchangeable K per acre. No response was obtained to K fertilizer application to soil in the 95-155 lb exchangeable K range and a yield depression was obtained to K fertilizer application on soils that contained over 170 lb exchangeable K/acre. Jenkins (1936) also reported yield depressions of snap beans fertilized with 70 and 100 lb K_2O in bands.

A study of the effect of potassium in the starter band on snap beans in the greenhouse is presented in Table 6. The beans dry weight and growth rate decreased with increased potassium rate in the band.

C. CARROTS. The effect of potassium in the starter fertilizer on carrot yields was studied in a test on organic soil, Table 7. The yield of carrots decreased as potassium rate increased in the band 2-inches under the seed. An observation shortly after the carrot seedlings emerged indicated that the salts might have been affecting stand as well as growth rate. Haworth and Cleaver (1963) found that carrot seedlings grew more rapidly on farm yard manure treated plots than on plots that received all the nutrients as inorganic salts. Davis and Lucas (1959) obtained optimum carrot yields with an

Table 6—Growth of snap bean seedlings with various rates of K in the starter fertilizer band

K rate*	Dry wt†
lb/acre	grams
0	1.711
33	.946
83	.818
133	.526
167	.462

* Band 2 inches below and 2 inches to side of seed in combination with 20-80-0.
† Tender-crop variety snapbeans harvested 4 weeks after planting.

Table 7—Carrot response to row fertilizer

K in row, lb acre	Yield, lb acre
0	76,480
50	71,768
100	65,960
120	66,310
LSD (0.05)	4,000

annual application of 160 lb K/acre. The high potassium requirement of carrots should be satisfied with a method of potassium application that prevents salt injury.

D. SWEET POTATO. Sweet potatoes were found to respond to potassium application on potassium deficient soil by Jackson and Thomas (1960). No enlarged roots were obtained from potassium deficient soil. The application of 117 lb K/acre increased the root yield to 12,580 lb/acre. The yield was increased another 34% with the application of 350 lb K/acre.

E. MUSKMELON. Total yield response of muskmelon was reported by Carolus and Lorenz (1938) to the application of up to 200 lb K/acre. The early yield was increased with K rates up to 100 lb/acre. A negative effect of potassium on flavor and total solids was found by Jacob and White-Stevens. They reported that increased potassium reduced total soluble carbohydrates and reduced sucrose more than hexose.

REFERENCES

Anderssen, F. G. 1937. Citrus manuring - its effect on cropping and on the composition and keeping quality of oranges. J. Pomol. Hort. Sci. 15:117-159.

Ballinger, W. E., H. K. Bell, and N. F. Childers. 1966. Peach nutrition, p. 276-390. In N. F. Childers (ed) Temperate to tropical fruit nutrition. Horticultural Publications, New Brunswick, N. J.

Barden, John A., and Arthur H. Thompson. 1962. Effects of heavy annual application of potassium on Red Delicious apple trees. Amer. Soc. Hort. Sci. Proc. 81:18-25.

Batjer, L. P., and M. N. Westwood. 1958. Seasonal trend of several nutrient elements in leaves and fruits of Elberta Peach. Amer. Soc. Hort. Sci., Proc. 71:116-126.

Beattie, James. 1959. Potassium aides in developing better apple color. Ohio Farm Home Res. 44:72-73.

Beaumont, J. H., and R. F. Chandler. 1933. A statistical study of the effect of potassium fertilizers upon the firmness and keeping quality of fruits. Amer. Soc. Hort. Sci., Proc. 30:37-44.

Berger, K. C., P. E. Potterton, and E. L. Hobson. 1961. Yield, quality and phosphorus uptake of potatoes as influenced by placement and composition of potassium fertilizers. Amer. Potato J. 38:272-285.

Bould, C., and E. Catlow. 1954. Manurial experiments with fruit. I. The effect of long-term manurial treatments on soil fertility and on the growth, yield and leaf nutrient status of strawberry, var. Climax. J. Hort. Sci. 29:203-219.

Bowers, J. L., and C. H. Mahoney. 1939. The influence of nitrogen, phosphorus and potassium on the yield of Alaska peas grown on soils of known fertility level. Amer. Soc. Hort. Sci., Proc. 37:707-712.

Boyce, B. R. and David Matlock. 1966. Strawberry nutrition, p. 518-548. In N. F. Childers (ed) Temperate to tropical fruit nutrition. Horticultural Publications, New Brunswick, New Jersey.

Boynton, Damon, and G. H. Oberly. 1966. Apple nutrition, p. 1-50. In N. F. Childers (ed) Temperate to tropical fruit nutrition. Horticultural Publications, New Brunswick, N. J.

Bunemann, Gerhard, D. H. Dewey, and A. L. Kenworthy. 1959. The storage quality of Jonathan apples in relation to the nutrient levels of the leaves and fruits. Michigan Agr. Exp. Sta. Quart. Bull. 41:820-833.

Burrell, A. B., and J. Carlton Cain. 1940. A response of apple trees to potash in the Champlain Valley of New York. Amer. Soc. Hort. Sci., Proc. 38:1-7.

Carolus, R. L. 1949. Calcium and potassium relationships in tomatoes and spinach. Amer. Soc. Hort. Sci., Proc. 54:281-285.

Carolus, R. L., and O. A. Lorenz. 1938. The interrelation of manure, lime, and potash on the growth and maturity of the muskmelon. Amer. Soc. Hort. Sci., Proc. 36:518-522.

Carpenter, P. N. 1957. Mineral accumulation in potato plants. Maine Agr. Exp. Sta. Bull. 562:3-5.

Chapman, H. D., and S. M. Brown. 1943. Potash in relation to citrus nutrition. Soil Sci. 55:87-100.

Chapman, H. D., S. M. Brown, and D. S. Rayner. 1947. Effects of potash deficiency and excess on orange trees. Hilgardia 17:619-650.

Collin, G. H., and R. A. Cline. 1966. The interaction effect of potassium and environment on tomato ripening disorders. Can. J. Plant Sci. 46:379-387.

Cummings, G. A. 1965a. Effect of potassium and magnesium fertilization on the yield, size, maturity and color of Elberta peaches. Amer. Soc. Hort. Sci., Proc. 86:133-140.

Cummings, G. A. 1965b. Plant and soil effects of potassium and magnesium fertilization of Elberta peach trees. Amer. Soc. Hort. Sci., Proc. 86:141-147.

Curwen, David, F. J. McArdle, and C. M. Ritter. 1966. Fruit firmness and pectic composition of Montmorency cherries as influenced by differential nitrogen, phosphorus, and potassium applications. Amer. Soc. Hort. Sci., Proc. 89:72-79.

Darrow, George M. 1932. Effect of fertilizers on firmness and flavor of strawberries in North Carolina. Proc. Amer. Soc. Hort. Sci. 28:231-235.

Davis, J. F., and R. E. Lucas. 1959. Organic soils, their formation, distribution, utilization and management. Michigan State Spec. Bull. 425.

Dickins, J. C., F. E. G. Harrap, and M. R. J. Holmes. 1962. Field experiments comparing the effects of muriate and sulfate of potash on potato yield and quality. J. Agr. Sci. 59:319-326.

Eastwood, Tom, and James Watts. 1956a. The effect of potash fertilization upon potato chipping quality III. Chip Color. Amer. Potato J. 33:255-257.

Eastwood, Tom, and James Watts. 1956b. The effect of potash fertilization upon potato chipping quality IV. Specific Gravity. Amer. Potato J. 33:265-268.

Eaves, C. A., and J. S. Leefe. 1955. The influence of orchard nutrition upon the acidity relationships in Cortland apples. J. Hort. Sci. 30:86-96.

Embleton, T. W., and W. W. Jones. 1966. Effects of potassium on peel thickness and juciness of lemon fruits. Hort. Sci. 1:25-26.

Embleton, T. W., J. D. Kirkpatrick, Winston W. Jones, and Clarence B. Cree. 1956. Influence of applications of dolomite, potash, and phosphate on yield and size of fruit and on composition of leaves of Valencia orange trees. Amer. Soc. Hort. Sci., Proc. 67:183-190.

Emmert, E. M. 1949. Tissue analysis in diagnosis of nutritional troubles. Amer. Soc. Hort. Sci., Proc. 54:291-298.

Evans, Harold J., and George J. Sorger. 1966. Role of mineral elements with emphasis on the univalent cations. Ann. Rev. Plant Phys. 17:47-76.

Fisher, E. G., K. G. Parker, N. S. Luepschen, and S. S. Kwong. 1959. The influence of phosphorus, potassium, mulch, and soil drainage on fruit size, yield, and firmness of the Bartlett pear and on development of the fire blight disease. Amer. Soc. Hort. Sci., Proc. 73:78-90.

Fisher, E. G., and S. S. Kwong. 1961. Effects of potassium fertilization on fruit quality of the McIntosh apple. Amer. Soc. Hort. Sci., Proc. 78:16-23.

Forsee, W. T. Jr., and J. C. Hoffman. 1950. The phosphate and potash requirements of snap beans on the organic soils of the Florida Everglades. Amer. Soc. Hort. Sci., Proc. 56:261-265.

Forshey, C. G. 1963. Potassium - magnesium deficiency of McIntosh apple trees. Amer. Soc. Hort. Sci., Proc. 83:12-20.

Freeman, G. G. 1965. The role of potassium in vegetable nutrition. Nat. Veg. Res. Sta. Ann. Report. 16:35-38.

Goldwebber, Seymour, Manley Boss, and S. John Lynch. 1956. Some effects of nitrogen, phosphorus, and potassium fertilization on the growth and yield of Persian limes. Florida State. Hort. Soc., Proc. 69:328-332.

Hass, A. R. C. 1949. Potassium in citrus trees. Plant Physiol. 24:395-415.

Havis, Leon, and Anna L. Gilkeson. 1951. Interrelationships of nitrogen and potassium fertilization and pruning practice in mature peach trees. Amer. Soc. Hort. Sci., Proc. 57:24-30.

Haworth, F., and T. J. Cleaver. 1963. The effects of the uptake of different amounts of potassium on the rate of growth of carrot seedlings. J. Hort. Sci. 38:40-45.

Hepler, J. R., and Kraybill, H. R. 1925. Effect of phosphorus upon the yield and time of maturity of the tomato. New Hampshire Agr. Exp. Sta. Tech. Bull. 23.

Hester, J. B., F. A. Shelton, and R. L. Isaacs, Jr. 1951. The rate and amount of plant nutrients absorbed by various vegetables. Amer. Soc. Hort. Sci., Proc. 57:249-251.

Hewitt, E. J. 1963. The essential nutrient elements: Requirements and interactions in plants. F. C. Steward (ed.) Plant physiology, a treatise. Inorg. Nut. of Plants III: 176-192.

Hill, H., F. B. Johnston, J. B. Henney, and R. W. Buckmaster. 1950. The relation of foliage analysis to keeping quality of McIntosh and Spy varieties of apples. Sci. Agr. 30:518-534.

Jackson, William A., and Grant W. Thomas. 1960. Effects of KCl and dolomitic limestone on growth and ion uptake of the sweet potato. Soil Sci. 89:347-352.

Jacob, W. C., and R. H. White-Stevens. 1941. Studies in the minor element nutrition of vegetable crop plants: II. The interrelation of potash, boron, and magnesium upon the flavor and sugar content of melons. Amer. Soc. Hort. Sci., Proc. 39:369-374.

Jenkins, J. Mitchell, Jr. 1936. Some effects of potassium on yields of snap beans. Amer. Soc. Hort. Sci., Proc. 34:471-473.

Jodidi, S. L., and Victor R. Boswell. 1932. Effect of nitrogen, phosphorus and potash on composition of Alaska peas. Amer. Soc. Hort. Sci., Proc. 29:454.

Jones, Winston W., and E. R. Parker. 1949. Effects of nitrogen, phosphorus and potassium fertilizers and of organic materials on the composition of Washington Naval orange juice. Amer. Soc. Hort. Sci., Proc. 53:91-102.

Kirch, Richard K. 1959. The importance of interaction effects in fertilizer and lime studies with strawberries. Amer. Soc. Hort. Sci., Proc. 73:181-188.

Kirkpatrick, J. D. and E. G. Fisher. 1958. Sour cherry affected by soil potassium level. New York Agr. Exp. Sta. Fam Res. 24(3):11.

Knight, Lucy O. M., and T. Wallace. 1932. The effects of various manurial treatments on the chemical composition of strawberries. J. Pomol. Hort. Sci. 10:147-180.

Kunkel, Robert. 1947. The effect of various levels of nitrogen and potash on the yield and keeping quality of onions. Amer. Soc. Hort. Sci., Proc. 50:361-367.

Kwong, S. S. 1965. Potassium fertilization in relation to titratable acids of sweet cherries. Amer. Soc. Hort. Sci., Proc. 86:115-119.

Kwong, S. S. and E. G. Fisher. 1962. Potassium effects on titratable acidity and the soluble nitrogenous compounds of Jerseyland peach. Amer. Soc. Hort. Sci., Proc. 81:168-171.

Lanham, W. B. 1926. Effect of potash fertilizer on carrying quality of tomatoes. Amer. Soc. Hort. Sci., Proc. 23:351-359.

Laughlin, Winston, M. 1961. Influence of potassium sprays on foliar necrosis and yields of Romaine lettuce, radishes, and potatoes. Can. J. Plant Sci. 41:272-276.

Lawton, Kirk, and R. L. Cook. 1954. Potassium in plant nutrition. Advance. Agron. VI: 253-303.

Lilleland, Omund, K. Uriu, T. Muraeka, and J. Pearson. 1962. The relationship of potassium in the peach leaf to fruit growth and size at harvest. Amer. Soc. Hort. Sci., Proc. 81:162-167.

Ljones, Bjarne. 1966. Bush Fruit Nutrition. In N. F. Childers (ed) Temperate to tropical fruit nutrition. Horticultural Publication. New Brunswick, N. J. p. 130-157.

Lorenz, O. A., and P. A. Minges. 1942. Nutrient absorption by a summer crop of lettuce in Salinas Valley, California. Amer. Soc. Hort Sci., Proc. 40:523-527.

Lucas, R. E., E. J. Wheeler, and J. F. Davis. 1954. Effect of potassium carriers and phosphate-potash ratios on the yield and quality of potatoes grown in organic soil. Amer. Potato J. 31:349-352.

Magness, J. R., and F. L. Overley. 1929. Effect of fertilizers on storage quality of apples. Amer. Soc. Hort. Sci., Proc. 26:180-181.

Mulder, E. G. 1956. Effect of the mineral nutrition of potato plants on the biochemistry and the physiology of the tubers. Netherland J. Agr. Sci 4:333-356. 1956 Abs. Soils and Fert. 20(1):334. Feb. 1957.

Murphy, Hugh J., and Michael Goven. 1958. Sulfate of potash will increase russeting. Maine Agr. Exp. Sta. Farm Res. 58:27-28.

Murphy, H. J., and M. J. Goven. 1959. Potash and processing potatoes. Maine Agr. Exp. Sta. Farm Res. 7(1): 3-8.

Murphy, H. J., and M. J. Goven. 1961. Varietal response of potatoes to source of potash. Maine Agr. Exp. Sta. Farm Res. April 1961. p. 19-21.

Nightingale, G. T. 1937. Potassium and calcium in relation to nitrogen metabolism. Bot. Gaz. 98:725-734.

Nightingale, G. T. 1943. Physiological-chemical functions of potassium in crop growth. Soil Sci. 55:73-78.

Oberly, G. H., and A. L. Kenworthy. 1961. Effect of mineral nutrition on the occurrence of bitter pit in northern spy apples. Amer. Soc. Hort. Sci., Proc. 77:29-34.

Overley, F. L., ad E. L. Overholser. 1931. Some effects of fertilizer upon storage response of Jonathan apples. Amer. Soc. Hort. Sci., Proc. 28:572-577.
Parker, E. R., and W. W. Jones. 1950. Orange fruit sizes in relation to potassium fertilization in a long-term experiment in California. Amer. Soc. Hort. Sci., Proc. 55:101-113.
Peck, N. H., R. F. Sansted, D. J. Lathwell, and T. E. Shultz. 1963. Potassium fertilization of snap beans. New York State Agr. Exp. Farm Res. 29:8-9.
Penston, N. L. 1931. Studies of the physiological importance of the mineral elements in plants III. A study of microchemical methods of the distribution of potassium in the potato plant. Ann. Bot. 45:673-692.
Reuther, Walter, and Paul F. Smith. 1952. Relation of nitrogen, potassium, and magnesium fertilization to some fruit qualities of Valencia orange. Amer. Soc. Hort. Sci., Proc. 59:1-12.
Rowberry, R. G., C. G. Sherrell, and G. R. Johnson. 1963. Influence of rates of fertilizer and sources of potassium on the yield specific gravity and cooking quality of Katahdin potatoes. Amer. Pot. J. 40:177-181.
Roy, W. R., 1945. Effect of potassium deficiency and of potassium derived from different sources on the composition of the juice of Valencia oranges. J. Agr. Res. 70:143-169.
Scudder, Walter T., W. C. Jacob, and H. C. Thompson. 1950. Varietal susceptibility and the effect of potash on the incidence of black spot in potatoes. Amer. Soc. Hort. Sci., Proc. 56:343-348.
Smith, Paul F. 1963. Quality measurements on selected sizes of marsh grapefruit from trees differentially fertilized with nitrogen and potash. Amer. Soc. Hort. Sci., Proc. 83:316-321.
Smith, Paul F. 1966. Citrus nutrition, p. 174-207. In N. F. Childers (ed.) Temperate to tropical fruit nutrition. Horticultural Publications. New Brunswick, N. J.
Smith, Paul F., and Gordon K. Rasmussen. 1959. Relation of potassium nutrition to size and quality of Valencia oranges. Amer. Soc. Hort. Sci., Proc. 74:261-265.
Smith, Paul F., and Gordon K. Rasmussen. 1961. Effect of potash rate on growth and production of Marsh grapefruit in Florida. Amer. Soc. Hort. Sci., Proc. 77:180-187.
Sparks, Walter C., and John G. McLean. 1946. The effect of nitrogen, phosphate and potassium on the yield of Red McClure potatoes as determined by soil analysis and fertilizer application. Amer. Soc. Hort. Sci., Proc. 48:449-457.
Stembridge, G. E., C. E. Gambrell, H. J. Sefick, and L. O. Blaricom. 1962. The effect of high rates of nitrogen and potassium on the yield, quality, and foliar composition of Dixiegem peaches in the South Carolina Sandhills. Amer. Soc. Hort. Sci., Proc. 81:153-161.
Terman, G. L. 1949. Effect of source of potash in the fertilizer on yield and starch content of potatoes. Amer. Potato J. 26:291-299.
Terman, G. L. 1950. Effect of rate and source of potash on yield and starch content of potatoes. Maine Agr. Exp. Sta. Bull. 481:1-24.
Thomas, Walter, Warren B. Mack, and Robert H. Cotton. 1943. Nitrogen, phosphorus, and potassium nutrition of tomatoes at different levels of fertilizer application and irrigation. Amer. Soc. Hort. Sci., Proc. 42:535-544.
Tiedjens, V. A., and M. E. Wall. 1938. The importance of potassium in the growth of vegetable plants. Amer. Soc. Hort. Sci., Proc. 36:740-743.
Tyler, K. B., O. A. Lorenz, and F. S. Fullmer. 1959. Soil and plant potassium studies with potatoes in Kern district, California. Amer. Potato J. 36:358-366.
Tyler, K. B., O. A. Lorenz, P. M. Nelson, and J. C. Bishop. 1959. Soil potassium for potatoes. California Agr. Exp. Sta. California Agr. 13(6):8.
Van Der Paauw, F. 1958. Relations between the potash requirements of the crops and meterological conditions. Plant and Soil IX:254-268.

Van Middelem, C. H., W. C. Jacob, and H. C. Thompson. 1953. Potassium fertilization effect on some soluble nitrogen constituents of the potato tuber in relation to the black spot problem. Amer. Soc. Hort. Sci., Proc. 61:353-359.

Wade, G. C. 1956. Investigations on brown rot of Apricots caused by Sclerotinia fructicola (wint.) Rehm. II. The relationship of potassium status of apricot trees to brown rot susceptibility. Aust. J. Agr. Res. 7:516-526.

Weeks, W. D., F. W. Southwick, Mack Drake, and J. E. Steckel. 1952. The effect of rates and sources of nitrogen, phosphorus and potassium on the mineral composition of McIntosh foliage and fruit color. Amer. Soc. Hort. Sci., Proc. 60:11-21.

Weeks, W. D., F. W. Southwick, Mack Drake, and J. E. Steckel. 1958. The effect of varying rates of nitrogen and potassium on the mineral composition of McIntosh foliage and fruit color. Amer. Soc. Hort. Sci., Proc. 71:11-19.

Weinberger, J. H. 1929. The effect of various potash fertilizers on the firmness and keeping quality of fruits. Amer. Soc. Hort. Sci., Proc. 26:174-179.

Westwood, M. N., and F. B. Wann. 1966. Cherry nutrition, p. 158-173. *In* N. F. Childers (ed.) Temperate to tropical fruit nutrition. Horticultural Publications, New Brunswick, N. J.

Wilcox, G. E. 1961. Effect of sulfate and chloride sources and rates of potassium on potato growth and tuber quality. Amer. Potato J. 38:215-220.

Wilcox, Gerald E. 1964. Row fertilization effects on potato growth on sandy organic soils. Indiana Acad. Sci. 72:227-231.

Wilcox, Gerald E. 1964. Effect of potassium on tomato growth and production. Amer. Soc. Hort. Sci., Proc. 85:484-489.

Wilkinson, B. G., 1957. The effect of orchard factors on the chemical composition of apples. I. Some effects of manurial treatment and of grass. J. Hort. Sci. 32:74-84.

Wilkinson, B. G. 1958. The effect of orchard factors on the chemical composition of apples. II. The relationship between potassium and titratable acidity, and between potassium and magnesium, in the fruit. J. Hort. Sci. 33:49-57.

Winsor, G. W., J. N. Davies and M. I. E. Long. 1961. Liquid feeding of glasshouse tomatoes; The effects of potassium concentration on fruit quality and yield. J. Hort. Sci. 36:254-267.

Winsor, G. W. 1963. The composition of tomato fruit. Glasshouse Crops. Res. Inst. Ann. Rep. pp. 57-61.

Soil Factors Affecting Potassium Availability

GRANT W. THOMAS

Texas A & M University
College Station, Texas

BILLY W. HIPP

Lower Rio Grande Valley Research &
Extension Center
Weslaco, Texas

I. INTRODUCTION

The behavior of potassium in soils and its availability to plants have been studied in some detail in nearly every part of the world. The amount of published material on these subjects seemed overwhelming when we attempted to summarize the results and define general rules of potassium behavior. As a result, we have tried to concentrate our review on material published since 1950.

We have written this chapter with the intention of presenting a reasonable view of the behavior of potassium in important soil groups. In doing so, we realize that there will be a number of published results which conflict with our conclusions. We can defend ourselves only by suggesting that the conclusions we have reached seem reasonable in the light of the best evidence we can find.

Because we have had to speculate on many points for lack of direct evidence, it is entirely possible that some of the conclusions we have drawn are incorrect. Even if this is so, we hope that this chapter will serve a purpose by challenging more intensive research on the problems of potassium behavior in soils and plants.

II. CHEMICAL FACTORS AFFECTING POTASSIUM AVAILABILITY

A. Nature of Cation-Exchangers in Soil

1. Kinds of Cation-Exchangers in Soil

The materials which have the property of adsorbing and exchanging cations in soils include the clay minerals: mica, montmorillonite, and vermiculite which are composed of an octahedral hydroxide layer sandwiched between two tetrahedral oxide layers, and kaolinite and halloysite, which are composed of one octahedral and one tetrahedral layer. In addition, in many soils, organic matter is an important source of cation-exchange capacity.

The source of cation-exchange capacity in all of the clay minerals is a negative charge caused by substitution of one structural cation for another when the mineral is formed. Examples of this substitution are Al^{3+} for Si^{4+}, Mg^{2+} for Al^{3+} and Li^+ for Mg^{2+}. For a given species of clay mineral, the amount of substitution is fairly constant and, therefore, the total negative charge is rather predictable. In micas and vermiculites formed from micas, a large part of the negative charge is balanced by potassium ions which are a part of the mineral structure. Because of their position and/or because of the high affinity these potassium ions have for the clay surfaces, other cations do not exchange places with them very readily. Only rigorous, long-time exchange (Cook and Rich, 1963a) or precipitation of the replaced potassium (DeMumbrum, 1963) can completely remove it. Under natural conditions, a small amount of this "difficultly exchangeable" potassium is continually replaced by other cations. This source of potassium for plants is found only in the micaceous clay minerals, including muscovite, biotite, illite, hydrobiotite, and vermiculute. The clay minerals montmorillonite, kaolinite and halloysite do not contain difficulty-exchangeable potassium. The result is that any soil which contains mica or mica products has a reserve supply of potassium which is not normally extracted by a salt solution. The importance of this reserve potassium is a function of the amount, stability, and weathering state of the mica or micaceous minerals.

About 60% of the cation-exchange capacity of organic matter arises from the carboxyl and the remainder arises from phenolic groups on the matrix (Broadbent and Bradford, 1952; Schnitzer and Skinner, 1963). The carboyxl groups have an acid strength of acetic acid (pK 4-5) (Martin and Reeve, 1958), whereas the phenolic groups are considerably weaker (pK 8). The pK is most conveniently regarded as the pH at which half of the sites are able to attract metallic cations. Because of the weak acid nature of charge on organic matter, the effective cation-exchange capacity is extremely sensitive to pH.

Table 1—Cation-exchange capacities of materials commonly occurring in soils

Material	Cation exchange capacity	Lattice charge
	meq/100g	meq/100g
Montmorillonite	100	100
Muscovite	20	250
Biotite	40	250
Vermiculite (di- or trioctahedral)	150	200
Kaolinite	5	5
Allophane	0-100	0
Organic Matter	50-250	0

Weak acid sites which give rise to cation exchange capacity also exist in crystalline clays, but they are not very important. Some clays, especially when weathered under acid conditions, become covered with iron and aluminum hydroxides $[R(OH)_x]_n^{n(3-x)}$ which are positively charged, but not exchangeable. These giant polycations can be replaced by increasing the pH so that they contain their full complement of three hydroxide ions per iron or aluminum (Thomas, 1960b). In such soils, the clays act as if they do have weak acid charges, since as the pH increases, more negative sites become available.

In poorly crystalline clay minerals (allophanes) which occur in quantity in Hawaii, the Pacific Northwest, Japan, and many tropical and subtropical regions, the charge on the clay is truly weakly acid and pH dependent. These materials can have effective cation exchange capacities ranging from 100 to zero. This apparently results from the shift of aluminum in the lattice from a six-coordinated to a four coordinated state, a kind of pH-induced isomorphous substitution (Tamele, 1950; Milliken et al., 1950).

A table of cation exchange capacities for some of the cation exchangers found in soils is given in Table 1. The total lattice charge, where it is different from the measurable cation exchange capacity is also given. The difference between these two figures is a measure of the potassium which is not exchangeable by normal methods.

2. Effect of Site of Charge and Density of Charge on Potassium Adsorption

When the clay mineral is formed, the substitution of one cation for another may occur in the silicon-oxygen terahedral layer, or it may occur in the aluminum or magnesium hydroxide octahedral layer. If it occurs in the former, the location of the excess of negative charge will be closer to the point where the exchangeable cations are gathered. For electrostatic charge, Couloumb's law states that its strength varies as the reciprocal of the squared distance between the charge and the ion. For tetrahedrally substituted clays, the distance between the charge and the cation is about half what it is for octahedrally-substituted clays, but the difference in strength should be four times as great as with octahedrally substituted clays. The observed differences in cation affinity between a clay substituted entirely in the octahedral layer (hectorite) and one substituted in both layers (Wyoming montmorillonite) are quite large (Fink and Thomas, 1964). Clays substituted entirely in the tetrahedral layer have even higher affinities for cations as has been shown by Barber and Marshall (1951) with beidellite.

In most comparisons between clay types, the location of the charge and the amount of charge is confounded too much to draw absolute conclusions about the effects of each. For example, micas are entirely substituted in the tetrahedral layer, but they also have a larger amount of substitution than other clays. When biotite weathers, the ferrous iron is oxidized to ferric iron and/or internal protonation occurs and the negative charge on the clay is reduced, usually from 250 meq/100 g to 200 meq/100 g, or less (Jackson, 1964). In the case of muscovite, weathering does not reduce the charge as much (Cook and Rich, 1963), but the negative sites tend to become neutralized with hydroxyaluminum ions. From the fast weathering rate of biotite compared to muscovite (De Mumbrum, 1963; Mortland et al., 1958), it might be concluded that density of charge strongly favors potassium retention by clays.

Evidence that this also is true in the case of montmorillonitic clays has been obtained by Tabikh et al. (1960) for mineral specimens and by Knibbe for soil clays (Knibbe, W.G.J. 1968. *Unpublished work. Texas A&M Univ.*). Knibbe found that soil montmorillonites have a much higher affinity for potassium (compared to calcium) than does Wyoming Bentonite. He also found that the charge density was 50% higher in the soil clays than in the mineral specimen.

It can be concluded that, for crystalline clays, the presence of a charge originating in the tetrahedral layer favors specific potassium adsorption. The adsorption of potassium is also favored by a high density of negative charge, presumably because the high charge favors a collapse of the clay lattice around the potassium ions, and because the high substitution results in deformation of voids in the clay surface so that potassium is likely to be trapped (Radoslovich and Norrish, 1962).

In the case of kaolinite and halloysite, most of the negative charge appears to arise from random isomorphous substitution in the lattice (Robertson et al., 1954; Sampson, H. R. 1953. The deflocculation of kalonite suspensions. *Ph.D. Thesis. University of London*). It has been observed that potassium salts can move in between layers of these clays, although this occurs much more easily in halloysite than in kaolinite (Andrew et al., 1960; Wada, 1959). As far as can be determined, the interlayer potassium salts in both these minerals are not held by negative charges and the salts can be completely replaced by water (Thomas, 1960a).

The affinity of organic matter for potassium is low compared to its affinity for calcium and magnesium. McGeorge (1931) found that the electrical conductivity of organic matter saturated with potassium was extremely high, indicating rather complete ionization. Field results with soils high in organic matter have tended to confirm the view that potassium is barely retained by the negative sites on organic matter (Mehlich, 1946; Spencer, 1954).

3. Relation Between Exchangeable and Soil Solution Potassium and its Significance

The distribution of potassium between negatively-charged sites on the soil and the soil solution is a function of the kinds and amounts of complementary cations, the anion concentration and the properties of the soil cation-exchange materials. In most cultivated soils, calcium is the major cation, both in the soil solution and on the soil itself. For this reason, calcium-potassium equilibria in soils have been studied most often.

Schofield (1947) proposed that the ratio of the activities of two cations such as potassium and calcium were defined by the relation $k = [K]/\sqrt{[Ca]}$. In soils where the amounts of cations in solution were negligible compared to those adsorbed by the soil, the ratio above was found to be reasonably independent of dilution (Taylor, 1958). Quite independently of Schofield, Woodruff (1955 a, b) suggested that the relation $[K]/\sqrt{[Ca]}$ could be used as an index of potassium availability in soils. A limited number of results by Woodruff and McIntosh (1960) and others (Ramamoorty and Paliwal, 1965; Thomas, 1958; MacLean, 1960) suggested that the ratio law reflected the differences in affinities for potassium which were found in soils. Earlier approximations by Barber and Marshall (1951), Bray (1942), Mehlich (1946), and Spencer (1954) of the relative amount of potassium available for plant uptake from the soil solution suggested that the type of cation exchanger in the soil has a determining effect on potassium availability to plants.

Quite recently, Beckett (1964 a, b) and his co-worker, Tinker (1964 a, b) have suggested that the ratio $[K]/\sqrt{[Ca + Mg]}$ be related to the change in exchangeable potassium to obtain a more complete picture of the effect of the quantity (exchangeable potassium) on the intensity ($[K]/\sqrt{[Ca + Mg]}$). The ratio Δ K exch.$/[K]/\sqrt{[Ca + Mg]}$ was found to be rather constant for a given soil regardless of potassium removal by plants provided that the level of nonexchangeable potassium was not significantly changed (Beckett and Nafady, 1967).

For a group of English soils which varied widely in clay mineralogy (Beckett, 1964 b) the quantity/intensity (Q/I) relationship increased rather regularly with increased clay content. No relation between clay mineralogy and variation in Q/I was evident (Beckett et al., 1966). A soil from Tanganyika behaved like the English soils, but soils from Natal and Nigeria had much lower values of Q/I. (If Q/I is low, small changes in exchange-

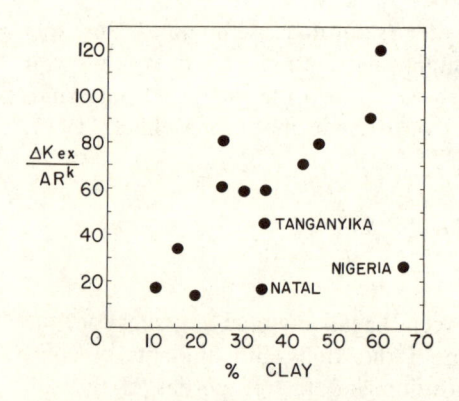

Fig. 1—The Q/I relationship for English (unlabeled) and African soils as related to clay content. (Beckett, 1964).

able potassium make large differences in potassium in the soil solution). The results are shown in Fig. 1.

In soils of the West Indies, Moss (1967) found that the Q/I relation was rather well associated with the clay mineralogy of the soils. Young, micaceous soils showed the least charge in solution potassium, montmorillonitic soils were intermediate and kaolinitic soils were the least buffered against changes in solution potassium. The last group of soils probably is similar to the soils from Nigeria and Natal cited above.

A high value of Q/I suggests that the availability of potassium will remain about the same over a long period of time. A low Q/I value suggests that frequent fertilization will be necessary. In either case, the value of the ratio $[K]/\sqrt{[Ca + Mg]}$ remains important because it is a measure of the relative activity of potassium in solution. In certain montmorillonitic soils of Texas, Nelson et al. (1960) have found that, although the supply of exchangeable potassium (Q) is nearly inexhaustible and the Q/I is very high, the value of $[K]/\sqrt{[Ca + Mg]}$ (I) is too low to support optimum plant growth. Because the value of Q/I is so large, a great deal of potassium must be added to the soil to effect any significant change.

Sandy soils, especially those in which organic matter makes up much of the cation-exchange capacity, have the opposite problem. Quantity/intensity is so low that the amount of potassium in the soil solution at a given time is virtually meaningless. Heavy rains or rapid plant growth can seriously deplete available potassium in a matter of days.

In general, it appears that the relation between exchangeable potassium and soil solution potassium is a good measure of the availability of the more labile potassium in soils to plants. It is certainly true, as pointed out by Barber (1962), that diffusion is a large factor in potassium uptake by plants. On the other hand, the total potassium that can diffuse through the soil solution is directly related to the proportion which is in the soil solution at any given time. For that reason, the "intensity" of potassium is of great importance in plant nutrition.

B. Clay Mineralogy, State of Soil Weathering and Release of Potassium from the Nonexchangeable Form

The equilibrium between exchangeable potassium and soil solution potassium is of great importance in determining the potassium nutrition of plants for a single season or, in some cases, for several years. For sustained plant growth over long periods of time, however, the equilibrium between nonexchangeable and exchangeable potassium is of greater importance. Probably more work has been done on the reactions between nonexchangeable and exchangeable potassium in soils than on any other facet of potassium chemistry. The results, taken as a whole, are utterly confusing. This is so because most of the studies have been conducted on soils in which the clay mineralogy was inadequately characterized.

In soils where a knowledge of clay mineral characteristics is combined with equilibrium data, there is a fairly distinct picture of potassium behavior. Extrapolation of these data to similar soils gives reasonable explanations for changes in soil potassium during plant growth.

Studies carried out by Barshad (1948), studies by the junior author in Texas (Hipp, B. W. *Unpublished work. Texas Agr. Exp. Sta., 1967.*), and work on biotite and vermiculite (Mortland and Ellis, 1959; Mortland et al., 1958) suggest that in virtually unweathered soils, high in trioctahedral mica and its derivatives, release of soil potassium from nonexchangeable forms occurs almost as rapidly as potassium is taken up by plants. Such soils make up many of the irrigated regions of the West and Southwest USA.

In soils where there is less trioctahedral mica and more dioctahedral mica and montmorillonite, the release is appreciable, especially under intensive cropping, but the rate of release is inadequate to maintain the levels of exchangeable potassium over a long period of time. Soils of this type are common in the Great Plains, parts of the Midwest, and in the Mississippi Delta. Examples of this group of soils (Cameron and Harlingen) are compared with two unweathered, micaceous soils (Hidalgo and Laredo) in Fig. 2. This figure shows the effect of successive cropping on the level of exchangeable potassium in the four soils. Notice the very small change in exchangeable potassium compared to potassium removal in the micaceous soils ($b = -0.286$ and -0.435) compared to b values of -0.78 and -0.815 for the montmorillonite-mica soils.

Work in other parts of the USA generally shows soils in Iowa (Pratt, 1951) and Minnesota (Weber and Caldwell, 1965) to release much more potassium than soils in Indiana (Rouse and Bertramson, 1950) Ohio, New York (Garman, 1957), and Pennsylvania (Grissinger and Jeffries, 1957) Kentucky results (Cook and Hutcheson, 1960) suggest that many soils in that state release potassium in much larger amounts than those in Alabama

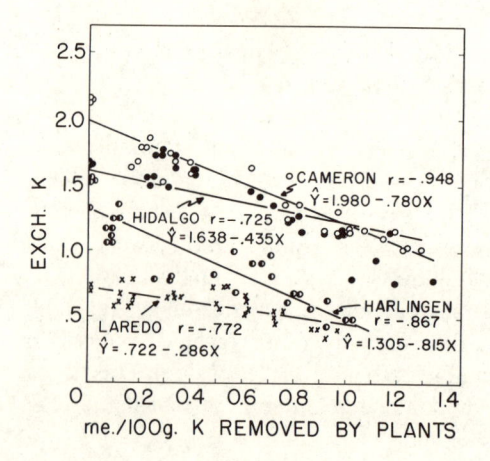

Fig. 2—The loss in exchangeable potassium compared to plant removal of potassium from four Texas soils.

(Pearson, 1952). Gholston and Hoover (1948) in Mississippi, found that the release of potassium is directly related to the type of soil parent material and its state of weathering.

In many of the soils of the eastern part of the Midwest, the Northeast and Southeast USA, and in lateritic soils in general, the release of potassium from nonexchangeable to exchangeable form is largely of academic interest. In soils weathered to this degree, the nonexchangeable potassium content is low and the complete removal of exchangeable potassium results in a soil which is so deficient in potassium that it will not allow plants to mature.

In each region, the generalized picture drawn above is markedly influenced by the amount of mica clay in the soil. As an example, Jones found that in two Virginia Piedmont soils, cropped to corn, wheat, red clover rotations, the potassium supply patterns were markedly different (Jones, G. D. *Unpublished work. Virginia Agr. Exp. Sta., Orange, 1964, 1967.*). In the Davidson soil, formed from basic rock, low in potassium, applications of 50 lb K/year would barely maintain good plant growth, whereas in a nearby Nason soil formed from muscovite schist, no potassium response occurred until 14 years had elapsed.

The results obtained in the USA suggest that a trioctahedral mica parent material and a soil with a low intensity of weathering favor release of sufficient potassium for crop needs, a dioctahedral mica parent material and/or a moderate state of weathering allow enough potassium release so that potassium fertilization requirements are significantly reduced (Milford and Jackson, 1966), whereas a low mica parent material and/or intensive weathering produce soils in which potassium requirements for plants must be met by applications of potassium fertilizers, for the most part.

C. Replaceability of Potassium by Other Cations

The exchangeable potassium obtained by replacement with ammonium only represents a value obtained under a given set of circumstances. It is useful because it has been found to correlate well with potassium taken up by plants in a wide range of soils. Ammonium-exchangeable potassium does not represent the amount of potassium exchanged under field conditions, however. From the evidence now in hand, it appears that ammonium exchanges more potassium than do most other cations from soils containing dioctahedral micaceous clays. In soils containing hydrobiotite and trioctahedral vermiculite, ammonium appears to replace potassium less efficiently than most other cations. The reasons for these observations are discussed below.

1. Dioctahedral Mica and Vermiculite

The weathering of muscovite is difficult, and it tends to be changed to an expanding clay from the edges inward as Jackson (1964) has shown. This mode of weathering produces many wedge-shaped zones in which potassium is tightly held, and into which only potassium-sized cations such as ammonium, hydronium (H_3O^+) and silver can enter. From the work of Merwin and Peech (1950), Rich (1964), and Rich and Black (1964) it appears that cations such as calcium and magnesium cannot replace any of the potassium held in such positions. It further appears that in some soils, a very large proportion of the potassium exchangeable to ammonium is held in wedge-shaped voids so that virtually no potassium is released by any naturally-occurring soil cations, except ammonium and hydronium.

Evidence that this is not merely a laboratory observation was obtained by Murdock and Rich (1965). They found that in a limed soil high in muscovite containing very little hydronium and no ammonium, a severe potassium deficiency occurred in oats. Addition of either ammonium or lower lime rates greatly improved potassium uptake by oats. This observation fits in well with the so-called lime-induced potassium fixation which has been observed in micaceous soils of the eastern USA. It is possible that one of the effects of overliming in many soils containing dioctahedral micas is a poor replacement of potassium by calcium and magnesium.

2. Trioctahedral Mica and Vermiculite

In contrast to the dioctahedral micas, potassium tends to be weathered out of biotite a whole layer at a time, in alternate layers, producing hydrobiotite (Rhoades and Coleman, 1967). When most of the potassium is gone, the mineral becomes vermiculite. Because nearly all of the potassium is stripped out of a layer and replaced by calcium and magnesium, there are fewer wedge-shaped zones and most of the exchangeable potassium exists in fully expanded layers. When ammonium in large quantity is added to such a clay, it tends to pinch the edges down (Barshad, 1948), trapping whatever cations are between the layers. This action is almost exactly the opposite of what occurs in muscovite-like clays.

Potassium exchange in trioctahedral mica derivatives is assured by cations which keep the lattice expanded. Therefore, in these clays calcium and magnesium are superior replacers of potassium.

D. The Effects of Soil Reaction on Potassium Availability

1. Effect of pH on Potassium Fixation

The effect of soil pH on the fixation of potassium against removal by ammonium was first noted by Volk (1934). In this very early work he observed a marked increase in potassium fixation in soils where the pH was elevated to about 9 or 10 with sodium carbonate. A more comprehensive study of the pH effect on potassium fixation was made by Martin et al. (1946). A graph showing their results is presented in Fig. 3. At pH values up to 2.5 there was essentially no fixation; between pH 2.5 and 5.5, the amount of potassium fixation increased very rapidly. Above pH 5.5 the amount of fixation increased more slowly. These results can be explained rather well by the work on soil acidity and aluminum done during the past 15 years.

At pH values above 5.5, aluminum cations are precipitated as hydroxy polycations which increase in hydroxide groups with increasing pH until they assume a form like gibbsite ($Al(OH)_3$) (Thomas, 1960 b). At this point, (about pH 8) the aluminum does not neutralize charge on the clay, and cannot prevent collapse of the clay by potassium. At pH's below 5.5 a combination of trivalent aluminum and hydroxy aluminum cations dominates, with the proportion of the latter dropping as the pH is lowered. Below pH 3.5, the hydronium ion begins to come into importance (Coleman and Harward, 1953). The increase in potassium fixation between pH 5.5 and 7.0 can be attributed to decreased numbers of hydroxy aluminum polymer cations, which can effectively block collapse of the clay (Rich, 1960). At very low pH values, the lack of fixation probably is due to the large numbers of hydronium ions and their ability to replace potassium effectively (Rich, 1964). The importance of interlayer hydroxyaluminum on potassium fixation was first emphasized by Rich and Obenshain (1955).

Fig. 3—The effect of pH on potassium fixation (Martin et al., 1946).

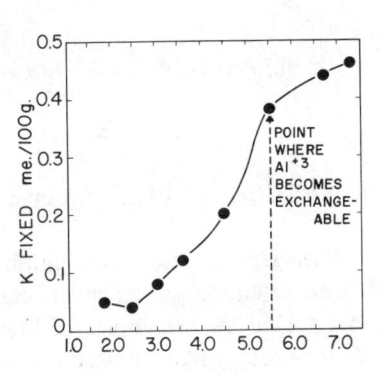

2. Effects of pH On Potassium Availability

The effect of calcium on potassium in the soil solution has been a matter of controversy for many years. It seems self evident that the addition of calcium to a soil will replace potassium, but in practice, it has been found that liming a soil reduces the amount of potassium in the soil solution (Peech and Bradfield, 1943). When a soluble calcium salt is added to a soil, all the cations in the soil are replaced to some degree, so that potassium in solution is increased. When an acid soil is limed, however, the exchangeable aluminum is precipitated by the hydroxide ions formed. In addition, the hydroxyaluminum cations are progressively "hydroxylated" by the lime until they have no charge (e.g., $Al(OH)^+_2 \rightarrow Al(OH)_3$). Thus, the addition of calcium carbonate removes the trivalent ions from competition with potassium and it frees blocked sites so that potassium can compete with calcium for them. The combination of these effects greatly increases the potassium held by the clay and decreases the amount of potassium in the soil solution (Thomas, 1958).

As one result, it has been found that potassium leaching is a much less serious problem in limed soils than in acid ones (Baver, 1943; Thomas and Coleman, 1959).

It might be inferred that liming acid soils also decreases the availability of potassium to plants, as indeed it sometimes does (*see* section II C). On the whole, however, the increased root vigor and volume in limed soils compared to soils in which aluminum dominates more than offsets the lime effect on potassium availability. In a practical way it will always be found that liming a very acid soil to a pH of 6 will increase potassium uptake by plants. Liming a soil having a pH of 6 to a pH of 7.5 generally will decrease the potassium uptake by plants.

III. PHYSICAL FACTORS AFFECTING POTASSIUM AVAILABILITY

A. Water Content of the Soil

1. *The Effect of Water Content on the Ratio of Cations in the Soil Solution*

The effect of the soil solution on plant nutrition has recently received renewed attention, primarily because modern laboratory methods have made analysis of small samples of low concentration possible. Simplified methods of diagnosing plant nutrient uptake have resulted in increased interest in the proper ratios and concentrations of cations in solution for optimum plant growth.

In the case of potassium, particular attention has been given the ratio $[K]/\sqrt{[Ca + Mg]}$ (*see* section II, A, 3). As soil becomes wet, the concentration of ions in the soil solution decreases because of a dilution effect, but a net adsorption of divalent cations and exchange of monovalent cations occurs because the ratio above tends to remain constant. This is the well-known dilution effect (Wiklander, 1964). As the soil moisture content is reduced, the concentration of calcium, magnesium, and potassium increases, but the concentration of calcium and magnesium increases faster than does the concentration of potassium. This results in a decreasing value of $K/(Ca + Mg)$ with increasing soil moisture tension. The increase of soil moisture tension and resulting decrease in $K/(Ca + Mg)$ was directly related to the $K/(Ca + Mg)$ in plants in a study by Moss (1963). These results are plotted in Fig. 4.

2. *The Effect of Water on Potassium Diffusion Coefficients*

Even though abundant potassium may be found in the soil, there is no assurance that plant roots will be able to utilize it. Barber (1962) calculated that potassium in solution was inadequate to account for the amount of potassium found in growing plants, and suggested that the rate of diffusion in the soil was the limiting factor in plant uptake of potassium. Since potassium-42 has a short life, most investigations on diffusion of potassium have been conducted using rubidium-86. Fried et al. (1959), have indicated that rubidium is a good indicator of potassium behavior in certain instances.

Evans and Barber (1964) have suggested critical diffusion rates of rubidium for corn. A linear increase of uptake of rubidium by corn was obtained with increased diffusion coefficients (Dp/b) until the value of Dp/b reached 1×10^{-6} cm^2 sec^{-1}.

Place and Barber (1964) showed that rate of rubidium-86 diffusion was increased by increasing soil moisture and the increased diffusion rate resulted in increased Rb uptake by plants. An r^2 value of 0.987 was obtained between

Fig. 4—The relation between pF of soil moisture, K/Ca + Mg in the soil solution and K/Ca + Mg in the plant (After Moss, 1963).

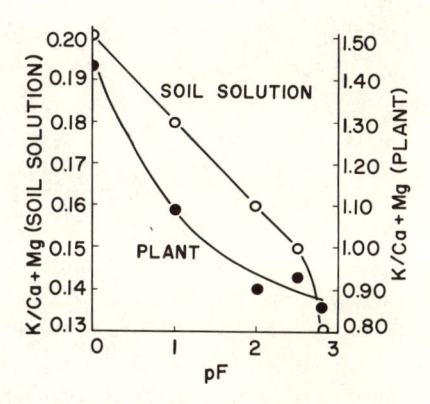

self diffusion and plant uptake. Diffusion coefficients of rubidium-86 also were found to decrease markedly as the moisture content decreased when measurements were made in a system of glass beads (Klute and Letey, 1958).

It is quite evident that increased soil moisture content results in increased diffusion as well as increased uptake of potassium. It is not entirely clear, however, whether the increased uptake is caused directly by the increased diffusion rate.

3. Release and Fixation of Exchangeable Potassium as Affected by Water

Determination of the influence of soil moisture on exchangeable soil potassium can be performed with little difficulty. Relating the change in exchangeable potassium to plant availability, however, is extremely difficult since diffusion rate, soil oxygen content, root growth, and possibly several other factors are also a function of soil moisture.

The influence of soil moisture on the exchange of potassium in several soils has been demonstrated by Brown (1953). He found that as soil moisture was increased, more potassium was exchanged from soil to hydrogen-saturated resin but that the effect on moisture varied with different soils. This study was carried out in the range between field capacity and the wilting point.

Under Iowa conditions (Scott and Smith, 1957) it has been found that the exchangeable potassium in the surface and subsoil was doubled upon drying and that potassium uptake by plants was always less on continually moist soil than from soil that had been dried. Luebs et al. (1956), however, found that under field conditions the change in exchangeable potassium due to changes in soil moisture was limited to the surface inch of soil. Studies in Wisconsin (Attoe, 1946) have indicated that the increase in exchangeable potassium upon drying varied from 4 to 90% over moist soil. Van der Paauw (1962) has suggested that exchangeable potassium levels fluctuate depending on rainfall distribution, and that wet periods are accompanied by a decrease in exchangeable potassium.

Luebs et al. (1956) reported that plants grown on Iowa soils that had been subjected to drying absorbed 60 to 120 lb/acre more K than those not subjected to drying. Laboratory studies from the same investigations indicate that in the vicinity of 5% moisture there is a sharp increase in exchangeable potassium with decreasing moisture content.

Investigations regarding the release of nonexchangeable potassium to exchangeable and solution potassium have been made by Scott and Hanway (1960). They found that under humid conditions less potassium was extracted from Marshall subsoil samples (Fig. 5) than at low humidity. Furthermore, they found that the reaction was partially reversible. Similarly, Attoe (1946) found an increase in exchangeable potassium in Miami silt loam as humidity was decreased, whereas more of the applied potassium was fixed as the moisture decreased. Hanway and Scott (1957) showed that the release of potassium on air drying subsoil was quite large and the effect was even larger when oven dried. More potassium release was observed on drying of subsoils than of topsoils, which suggests that Iowa subsoils are less weathered than the topsoils.

Dowdy and Hutcheson (1963) found that, for Kentucky soils that originally contained small amounts of exchangeable potassium, there was more potassium released to plants if the soils were kept continually moist than if they were previously air dried.

4. *Influence of Soil Moisture on Potassium Nutrition of Plants*

The ultimate goal of studies involving soil potassium is the precise prediction of potassium absorption by plants. Plants and soils are so variable and the factors governing potassium uptake are so interdependent that there is disagreement on the influence of soil moisture on potassium uptake. The general trend, however, is an increase in potassium uptake by plants as soil

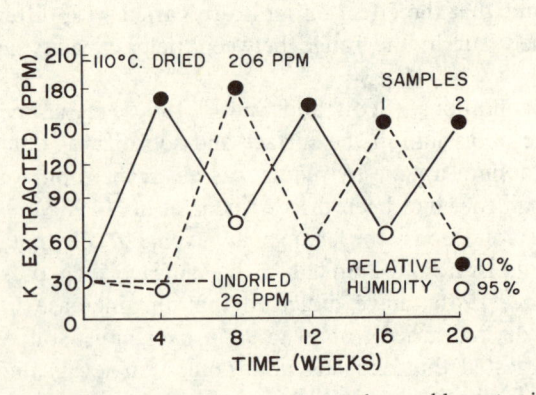

Fig. 5—The effect of relative humidity on the exchangeable potassium level in Marshall subsoil (After Scott and Hanway, 1960).

moisture increases. Jenne et al. (1958) found that corn plants with low moisture supply contained only 71% as much potassium as corn plants with adequate moisture. Branton et al. (1961) sampled apricot leaves from trees grown on Yolo silty clay loam that had been subjected to different irrigation treatments for 10 years. They found that high potassium contents were associated with low soil moisture stress. Mederski and Wilson (1960) found a correlation coefficient of 0.97 between percent of potassium in corn plants and soil moisture percentage.

These findings are not in agreement with expectations from equilibria between fixed and exchangeable potassium and soil moisture carried out in the Midwest. From those results, exchangeable soil potassium generally is more abundant with dry conditions and the expectation would be that under dry conditions the potassium uptake would be increased. This would suggest that the higher level of exchangeable potassium in dry soils cannot be relied upon for universal prediction of potassium uptake by plants. Explanation must be found in the relative increase of potassium compared to calcium and magnesium in the soil solution as moisture content increases and in the similar effect on the potassium diffusion rate.

B. Oxygen Content of the Soil

The relationship between the availability of potassium to plants and soil aeration seems to be one which involves the ability of the plant to utilize potassium under certain levels of soil oxygen. A great deal of work has been done to relate the uptake of soil potassium by plants to the rate of oxygen diffusion in soils. Methods of measuring oxygen diffusion by the platinum electrode have been well developed.

The general trend indicates that as aeration is decreased the uptake of potassium by plants is decreased. There is not complete agreement on the rate of oxygen diffusion necessary for adequate growth of plants. An increase in potassium concentration of bluegrass with increased oxygen supply has been indicated by Letey et al. (1964). Similar results have been indicated for snapdragon shoots growing on Yolo soil (Letey et al., 1961a) but the rate of increased uptake with increased potassium was not as rapid for snapdragon as for bluegrass. Letey et al. (1962a) found that as the diffusion rate of O_2 was increased to 40×10^{-8} g cm^{-2} min^{-1} there was a corresponding increase in potassium content of sunflower shoots. They suggested that this is the optimum rate for sunflower nutrition. Cline and Erickson (1959) found that high potassium levels would tend to offset reduced uptake brought about by poor aeration. They sugested that 70 to 100 $\times$ 10^{-8} g cm^{-2} min^{-1} O_2 is necessary for optimum potassium uptake by peas.

A decrease in soil oxygen supply in the field generally is associated with an increase in carbon dioxide concentration of the soil atmosphere. Much of the research has been done keeping carbon dioxide constant, but Harris and Van Bavel (1957) made studies on tobacco nutrition in sand cultures with the root zone subjected to variations in O_2 and CO_2 from about 20% O_2 and 0% CO_2 to 0% oxygen and 20% CO_2. Decreasing oxygen and increasing carbon dioxide resulted in decreased potassium uptake by tobacco plants, but there was not a serious potassium decrease until the O_2 was below 10%. Letey et al. (1961b) have suggested that potassium uptake by sunflower will be increased until the O_2 atmosphere around containers of soil reaches 21%, whereas uptake of potassium by cotton was influenced very little after the surrounding air reached 7% O_2. The critical level of soil O_2 for potassium uptake by barley grown in Yolo silt loam has been shown to be about 5% (Letey et al., 1962b). Danielson and Russell (1957) have shown with short-term experiments that the accumulation of Rb by corn seedlings was not affected by increases in O_2 concentration above 10%. Hammond et al. (1955) observed that reducing the supply of O_2 to 5% did not affect potassium absorption by corn roots grown in nutrient solution.

One of the primary hazards to crop growth as a result of poor soil structure is lack of aeration. Poor aeration can be a result of improper tillage, waterlogged conditions, dense or impervious zones, or small pore spaces from numerous causes. Larson (1954) has suggested that poor aeration was important in potassium nutrition of sugar beets grown on Montana soils that contained a small amount of large pores and with high moisture content. Plants grown on these soils showed potassium deficiency and contained 31% less petiole K than plants grown in soils with lower moisture content. Soil compaction has an obvious influence on aeration and Phillips and Kirkham (1962) have shown that tractor traffic can reduce total potassium in corn leaves at indigenous levels of potassium as well as with added fertilizer. Saturated soils or soils containing high amounts of water reduce the oxygen supply available to plant roots. Lawton (1945) increased the potassium content of corn growing on high moisture soils merely by forcing air through the soil.

The potassium nutrition of citrus trees as well as row crops and vegetable crops is influenced by the level of aeration in the root zone. Labanauskas et al. (1965) have shown that the uptake of potassium by citrus seedlings decreased as the oxygen supply to the roots was reduced. Citrus seedlings growing on Ramona sandy loam were found to have less potassium in both tops and roots as a result of decreased soil oxygen (Labanauskas et al., 1966). Shapiro et al. (1956) have suggested that the decreased uptake of potassium by plants as a result of low soil oxygen may be because of decreased translocation rather than decreased absorption by the roots. They observed a decrease in potassium content of corn tops grown on Clermont silt loam with decreased aeration, but the potassium content of the roots was increased.

There is general agreement that the potassium content of plants will be decreased as the oxygen content of the soil atmosphere is decreased, but the percentage of oxygen required or the oxygen diffusion rate for adequate absorption of potassium is apparently not the same for all plants. Little indication is given regarding the mechanism of this reduced uptake by plants. As to whether the decrease is a result of decreased absorption from the soil or is involved with transport within the plant, is not yet clear.

C. Temperature Effects

1. The Effect of Temperature on Potassium Equilibrium Shifts

An increase in temperature increases the rate of chemical reactions. Applied to soil conditions it would be expected to increase the rate of cation exchange which would in turn increase the amount of potassium absorbed by plant roots, providing the reactions of potassium in the soil are a major factor in potassium absorption by plants.

Woodruff's (1955b) equation, $\Delta F = RT \ln (K/\sqrt{[Ca]})$, infers temperature dependence of $K/\sqrt{[Ca]}$ ratio in the soil solution as long as ΔF remains constant for a given soil. The effect is in the direction of increased potassium in solution. Golden (1962) found that increasing extractant temperature increased the amount of potassium extractable from several alluvial and terrace soils of Louisiana. Slopes of curves obtained between plots of temperature of extracting solution (0.1N HCl) and ppm K extracted with a 1:20 soil-extractant ratio ranged from 0.45 to 1.66 ppm K/degree C. Based on this value of temperature dependence, an extracting solution temperature variation of 10C would result in an exchangeable K variation of 4.5 to 16.6 ppm. The implications regarding standardized extraction temperatures in routine soil testing are obvious.

The rate of release of nonexchangeable potassium has been found to be temperature dependent by Haagsma and Miller (1963). They found that release of nonexchangeable potassium to a cation-exchange resin increased with increasing temperature. The increase was not linear, however, since more potassium was released when the temperature was increased from 50 to 80C than from 5 to 50C. Laboratory studies by Barber (1960) also have indicated an increase in the rate of release of nonexchangeable potassium with increasing temperature.

2. The Effect of Temperature on Plant Uptake of Potassium

In considerations regarding potassium availability and temperature, a factor that must not be overlooked is the increase in metabolic activity of plant roots as a result of increased temperature, at least up to a point that heat damage to roots occurs. It is difficult to separate potassium availability in soils and solution and increased plant activity due to temperature.

Worley et al. (1963) found that the relation between metabolism rate and potassium uptake from nutrient solution of four species was very different. All four species increased in metabolism rate with increasing temperature, but potassium uptake by the cool season species was reduced at higher temperatures.

Proebsting (1957) has shown the effects of root temperatures of 45 to 90F on the absorption of potassium by strawberries. Potassium in the roots was reduced at the higher temperatures. The maximum concentration of potassium in leaves and petioles was found at 85F but the maximum total potassium absorbed by the plants was at 75F.

Martin and Wilcox (1963) studied the influence of temperature on two varieties of tomatoes and found that percent potassium in the plants increased markedly when the soil temperature was increased from 56F to 70F with low rates of phosphorus but the temperature effects on potassium content were not as great when high rates of phosphorus were used. Cannell et al. (1963) suggested that the influence of temperature on potassium uptake by tomatoes was curvilinear. Nielsen et al. (1960) have found that potassium uptake by oats in temperature-controlled soils was increased with increases in the soil temperature from 41 to 67F. Wallace (1957) working with barley and soybeans at various temperatures, found that the potassium content of barley increased from 12 to 22C but decreased at 32C. Soybeans, however, showed an increase in potassium content at the expense of divalent cations with increasing temperature. Nielsen et al. (1961) usually found an increase in uptake of potassium by corn, bromegrass, and potatoes as the soil temperature increased to 67F.

Growth of crops at different seasons of the year would lead one to speculate that the absorption of nutrients by plants is influenced by other climatic factors besides temperature. Army and Miller (1956) found that soil temperature variations from 60 to 80F did not influence the potassium content of turnip leaves in fall or spring but that the potassium of the leaves was higher in the fall than in the spring.

In summary, it appears that potassium uptake by plants (when separated from soil effects) increases up to some optimum temperature, specific for a given plant. At higher temperatures, some internal damage to the plant absorption mechanism occurs.

REFERENCES

Andrew, R. W., M. L. Jackson, and K. Wada. 1960. Intersalation as a technique for differentiation of kaolinite from chloritic minerals by X-ray diffraction. Soil Sci. Soc. Amer. Proc. 24:422-424.

Army, T. J., and E. V. Miller. 1956. The interaction of kind of soil colloid, fertility status and seasonal weather variation on the cation of turnip leaves. Soil Sci. Soc. Amer. Proc. 20:57-59.

Barber, S. A. 1960. The influence of moisture and temperature on phosphorus drying conditions. Soil Sci. Soc. Amer. Proc. 11:145-149.

Attoe, O. J. 1946. Potassium fixation and release in soils occurring under moist and and potassium availability. Intern. Congr. Soil Sci. Trans. 7th Congr. (Madison, Wis.) 3:435-442.

Barber, S. A. 1962. A diffusion and mass-flow concept of soil nutrient availability. Soil Sci. 93:39-49.

Barber, S. A., and C. E. Marshall. 1951. Ionization of soils and soil colloids: II. Potassium-calcium relationships in montmorillonite group clays and in attapulgite. Soil Sci. 72:373-385.

Barshad, I. 1948. Vermiculite and its relation to biotite as revealed by base exchange reactions, X-ray analyses, differential thermal curves, and water content. Amer. Mineral. 33:655-678.

Barshad, I. 1954. Cation-exchange in micaceous minerals: I. Replaceability of the interlayer cations of vermiculite with ammonium and potassium ions. Soil Sci. 77:463-472.

Baver, L. D. 1943. Practical applications of potassium interrelationships in soils and plants. Soil Sci. 55:121-126.

Beckett, P.H.T. 1964a. Studies on soil potassium I. Confirmation of the ratio law: Measurement of potassium potential. J. Soil Sci. 15:1-8.

Beckett, P.H.T. 1964b. Studies on soil potassium. II. The immediate Q/I relations of labile potassium in the soil. J. Soil Sci. 15:9-23.

Beckett, P.H.T., J. B. Craig, M.H.M. Nafady, and J. P. Watson. 1966. Plant and Soil 25:435-455.

Beckett, P.H.T., and M.H.M. Nafady. 1967. Effect of K release and fixation on the ion-exchange properties of illite. Soil Sci. 103:410-416.

Branton, D., K. Lilleland, Uries, and L. Werenfels. 1961. The effect of soil moisture on apricot leaf composition. Amer. Soc. Hort. Sci., Proc. 77:90-96.

Bray, R. H. 1942. Ionic competition in base-exchange reactions. J. Amer. Chem. Soc. 64:954-963.

Broadbent, F. E., and G. R. Bradford. 1952. Cation-exchange groupings in the soil organic fraction. Soil Sci. 74:447-457.

Brown, D. A. 1953. Cation-exchange in soils through the moisture range, saturation to the wilting percentage. Soil Sci. Soc. Amer. Proc. 17:92-96.

Cannell, G. H., F. T. Bingham, J. C. Lingle, and M. J. Garber. 1963. Yield and nutrient composition of tomatoes in relation to soil temperature, moisture and phosphorus levels. Soil Sci. Soc. Amer. Proc. 27:560-565.

Cline, R. A., and A. E. Erickson. 1959. The effect of oxygen diffusion rate and applied fertilizer on the growth, yield and chemical composition of peas. Soil Sci. Soc. Amer. Proc. 23:333-335.

Coleman, N. T., and M. E. Harward. 1953. The heats of neutralization of acid clays and cation-exchange resins. J. Amer. Chem. Soc. 75:6045-6046.

Cook, M. G., and T. B. Hutcheson. 1960. Soil potassium reactions as related to clay mineralogy of selected Kentucky soils. Soil Sci. Soc. Amer. Proc. 24:252-256.

Cook, M. G., and C. I. Rich. 1963a. Formation of dioctahedral vermiculite in Virginia soils. *Clays and Clay Minerals*. 10:96-106. (Pergamon Press, New York)

Cook, M. G., and C. I. Rich. 1963b. Negative charge of diocatahedral micas as related to weathering. *Clays and Clay Minerals* 11:47-64. (Pergamon Press, New York)

Danielson, R. E., and M. B. Russell. 1957. Ion absorption by corn roots as influenced by moisture and aeration. Soil Sci. Soc. Amer. Proc. 21:3-6.

De Mumbrum, L. E. 1963. Conversion of mica to vermiculite by potassium removal. Soil Sci. 96:275-276.

Dowdy, R. H., and T. B. Hutcheson, Jr. 1963. Effect of exchangeable potassium level and drying upon availability of potassium to plants. Soil Sci. Soc. Amer. Proc 27:521-523.

Evans, S.D., and S. A. Barber. 1964. The effect of cation-exchange capacity, clay content and fixation on rubidium-86 diffusion in soil and kaolinite systems. Soil Sci. Soc. Amer. Proc. 28:53.56.

Fink, D. H., and G. W. Thomas. 1964. X-ray studies of crystalline swelling in montmorillonites. Soil Sci. Soc. Amer. Proc. 28:747-750.

Fried, M., G. Hawkes, and W. F. Mackie. 1959. Rubidium-potassium relations in the soil plant system. Soil Sci. Soc. Amer. Proc. 23: 360-362.

Garman, W. L. 1957. Potassium release characteristics of several soils from Ohio and New York. Soil Sci. Soc. Amer. Proc. 21:-52-58.

Gholston, L. E., and C. D. Hoover. 1948. The release of exchangeable and nonexchangeable potassium from several Mississippi and Alabama soils upon continuous cropping. Soil Sci. Soc. Amer. Proc. 13:116-121.

Golden, L. E. 1962. Effect of temperature of extractant on P, K, Ca and Mg removed from different soil types. Soil Sci. 93:154-160.

Grissinger, E., and C. D. Jeffries. 1957. Influence of continuous cropping on the fixation and release of potassium in three Pennsylvania soils. Soil Sci. Soc. Amer. Proc. 21:409-412.

Haagsma, T., and M. H. Miller. 1963. The release of non-exchangeable soil potassium to cation-exchange resins as influenced by temperature moisture and exchanging ion. Soil Sci. Soc. Amer. Proc. 27:153-156.

Hammond, L. C., W. H. Allaway, and W. E. Loomis. 1955. Effect of oxygen and carbon dioxide levels upon absorption of potassium by plants. Plant Physiol. 30:155-161.

Hanway, J. J., and A. D. Scott. 1957. Soil potassium moisture relations: II. Profile distribution of exchangeable K in Iowa soils as influenced by drying and rewetting. Soil Sci. Soc. Amer. Proc. 21:501-504.

Harris, D. G., and C. H. M. van Bavel. 1957. Nutrient uptake and chemical composition of tobacco plants as affected by the composition of the root atmosphere. Agron. J. 49:176-181.

Jackson, M. L. 1964. Chemical composition of soils. *In* F. E. Bear (ed.) Chemistry of Soil. 2nd Ed. ACS Monograph No. 160. Reinhold Publishing Corp., N. Y. p. 71-141.

Jenne, E. A., H. F. Rhoades, C. H. Yien, and O. W. Howe. 1958. Change in nutrient element accumulation by corn with depletion of soil moisture. Agron. J. 50:71-74.

Klute, A., and J. Letey. 1958. The dependence of ionic diffusion on the moisture of nonadsorbing porous media. Soil Sci. Soc. Amer. Proc. 22:213-215.

Labanauskas, C. K., J. Letey, L. H. Stolzy, and N. Valoras. 1966. Effects of soil-oxygen and irrigation on the accumulation of macro and micronutrients in citrus seedlings (*Citrus sinensis* var. Osbeck). Soil Sci. 101:378-384.

Labanauskas, C. K., L. H. Stolzy, L. J. Klotz, and T. A. De Wolfe. 1965. Effects of soil temperature and oxygen on the amounts of macronutrients and micronutrients in citrus seedlings. (*Citrus sinensis* var. Bessie) Soil Sci. Soc. Amer. Proc. 29:60-64.

Larson, W. E. 1954. Response of sugar beets to potassium fertilization in relation to soil physical and moisture conditions. Soil Sci. Soc. Amer. Proc. 18:313-317.

Lawton, K. 1945. The influence of soil aeration on the growth and absorption of nutrients by corn plants. Soil Sci. Soc. Amer. Proc. 10:263-268.

Letey, J., O. R. Lunt, L. H. Stolzy, and T. E. Szuszkiewiez. 1961a. Plant growth, water use and nutritional response to rhizosphere differentials of oxygen concentration. Soil Sci. Soc. Amer. Proc. 25:183-186.

Letey, J., L. H. Stolzy, G. B. Blank, and O. R. Lunt. 1961b. Effect of temperature on oxygen-diffusion rates and subsequent shoot growth, root growth and mineral content of two plant species. Soil Sci. 92:314-321.

Letey, J., L. H. Stolzy, O. R. Lunt, and V. B. Youngner. 1964. Growth and nutrient uptake of Newport Bluegrass as affected by soil oxygen. Plant and Soil 20:143-148.

Letey, J., L. H. Stolzy, N. Valoras, and T. E. Szuszkiewiez. 1962a. Influence of oxygen diffusion rate on sunflower growth at various soil and air temperatures. Agron. J. 54:316-319.

Letey, J., L. H. Stolzey, N. Valoras, and T. E. Szuszkiewiez. 1962b. Influence of soil oxygen and growth and mineral concentration of barley. Agron. J. 54:538-540.

Luebs, R. E., G. Stanford, and A. D. Scott. 1956. Relation of available potassium to soil moisture. Soil Sci. Soc. Amer. Proc. 20:45-50.

MacLean, A. J. 1960. Water-soluble K, percent K-saturation, and pK-'p(Ca + Mg) as indices of management effect on K status of soils. Int. Congr. Soil Sci. Trans. 7th (Madison, Wis.) 3:86-91.

Martin, A. E., and R. Reeve. 1958. Chemical studies of podzolic illuvial horizons: III. Titration curves of organic matter suspensions. J. Soil Sci. 9:89-100.

Martin, George, C., and G. E. Wilcox. 1963. Critical soil temperature for tomato plant growth. Soil Sci. Soc. Amer. Proc. 27:565-567.

Martin, J. C., R. Overstreet, and D. R. Hoagland. 1946. Potassium fixation in soils in replaceable and nonreplaceable forms in relation to chemical reactions in the soil. Soil Sci. Soc. Amer. Proc. (1945) 10:94-101.

McGeorge, W. T. 1931. Organic compounds associated with base exchange reactions in soils. Arizona Agr. Exp. Sta. Tech. Bull. 31.

Mederski, H. J., and J. H. Wilson. 1950. Relation of soil moisture to ion absorption by corn plants. Soil Sci. Soc. Amer. Proc. 24:149-152.

Mehlich, A. 1946. Soil properties affecting the proportionate amounts of calcium, magnesium, and potassium in plants and in HCl extracts. Soil Sci. 62:393-409.

Merwin, H. D., and M. Peech. 1950. Exchangeability of soils potassium in the sand, silt, and clay fractions as influenced by the nature of the complementary exchangeable cation. Soil Sci. Soc. Amer. Proc. (1951) 15:125-128.

Milford, M. H., and M. L. Jackson. 1966. Exchangeable potassium as affected by mica specific surface in some soils of North Central United States. Soil Sci. Soc. Amer. Proc. 30:735-739

Milliken, T. H., G. A. Mills, and A. G. Oblad. 1950. Chemical characteristics and structure of cracking catalysts. Discuss. Faraday Soc. 8:279-290.

Mortland, M. M., and B. Ellis. 1959. Release of fixed potassium as a diffusion controlled process. Soil Sci. Soc. Amer. Proc. 23:363-364.

Mortland, M. M., K. Lawton, and G. Uehave. 1958. Alteration of biotite to vermiculite by plant growth. Soil Sci. 82:477-481.

Moss, P. 1963. Some aspects of the cation status of soil moisture. Part I: The ratio law and soil moisture content. Plant and Soil 18:99-113.

Moss, P. 1967. Independence of soil quantity-intensity relationships to changes in exchangeable potassium. Similar potassium exchange constants for soils within a soil type. Soil Sci. 103:196-201.

Murdock, L. W., and C. I. Rich. 1965. Potassium availability in Nason soil as influenced by ammonium and lime. Soil Sci. Soc. Amer. Proc. 29:707-711.

Nelson, L. A., G. W. Kunze, and C. L. Godfrey. 1960. Chemical and mineralogical properties of San Saba Clay, a Grumusol. Soil Sci. 89:122-131.

Nielsen, K. F., R. L. Halstead, A. J. MacLean, R. M. Holmes, and S. J. Bourget. 1960. The influence of soil temperature on the growth and mineral composition of oats. Can. J. Soil Sci. 40:255-263.

Nielsen, K. F., R. L. Halstead, A. J. MacLean, S. J. Bourget, and R. M. Holmes. 1961. The influence of soil temperature on the growth and mineral composition of corn, bromegrass and potatoes. Soil Sci Soc. Amer. Proc. 25:369-372.

Pearson, R. W. 1952. Potassium-supplying power of eight Alabama soils. Soil Sci. 74:301-309.

Peach, M. and R. Bradfield. 1943. The effect of lime and magnesia on the soil potassium and on the absorption of potassium by plants. Soil Sci. 55:37-48.

Phillips, R. E., and D. Kirkham. 1962. Soil compaction in the field and corn growth. Agron. J. 54:29-34.

Place, G. A., and S. A. Barber. 1964. The effect of soil moisture and rubidium concentration on diffusion and uptake of rubidium-86. Soil Sci. Soc. Amer. Proc. 28:239-243.

Pratt, P. F. 1951. Potassium removal from Iowa soils by greenhouse and laboratory procedures. Soil Soc. 72:107-117.

Proebsting, E. L. 1957. The effect of soil temperature on the mineral nutrition of the strawberry. Amer. Soc. Hort. Soc. Proc. 69:278-281.

Radoslovich, E. W., and K. Norrish. 1962. The cell dimensions and symmetry of layer-lattice silicates. I. Some structural considerations. Amer. Mineral 47:599-616.

Ramamoorthy, B. and K. V. Paliwal. 1965. Potassium adsorption ratio of some soils in relation to their potassium availability to paddy. Soil Sci. 99:236-242.

Rhoades, J. D., and N. T. Coleman. 1967. Interstratification in vermiculite and biotite produced by potassium sorption. I. Evaluation by simple X-ray diffraction pattern inspection. Soil Sci. Soc. Amer. Proc. 31:366-372.

Rich, C. I. 1960. Aluminum in interlayers of vermiculite. Soil Sci. Soc. Amer. Proc. 24:26-32.

Rich, C. I. 1964. Effect of cation size and pH on potassium exchange in Nason soil. Soil Sci. 98:100-106.

Rich, C. I., and W. R. Black. 1964. Potassium exchange as affected by cation size, pH and mineral structure. Soil Sci. 97:384-390.

Rich, C. I., and S. S. Obenshain. 1955. Chemical and clay mineral properties of a Red-yellow Podzolic soil derived from mica schist. Soil Sci. Soc. Amer. Proc. 19:334-339.

Robertson, R. H. S., G. W. Brindley, and R. C. Mackenzie. 1954. Mineralogy of kaolin clays from Pugu, Tanganyika. Amer. Mineral. 39:118-138.

Rouse, R. D., and B. R. Bertramson. 1950. Potassium availability in several Indiana soils: Its nature and methods of evaluation. Soil Sci. Soc. Amer. Proc. (1951) 14:113-123.

Schnitzer, M., and S. I. M. Skinner. 1963. Organo-metallic interactions in soils: I. Reactions between a number of metal ions and the organic matter of a Podzol Bh horizon. Soil Sci. 96:86-93.

Schofield, R. K. 1947. A ratio law governing the equilibrium of cations in the soil solution. Proc. 11th Int. Congr. Pure Appl. Chem. 3:257-261.

Scott, A. D., and J. J. Hanway. 1960. Factors influencing the change in exchangeable soil K observed on drying. Int. Congr. Soil Sci. Trans. 7th Madison, Wis. 4:72-79.

Scott, T. W., and F. W. Smith. 1957. Effect of drying upon availability of potassium to soil moisture. Soil Sci. Soc. Amer. Proc. 20:45-50.

Shapiro, R. E., S. Taylor, G. W. Volk. 1956. Soil oxygen contents and ion uptake by corn. Soil Sci. Soc. Amer. Proc. 20:193-197.

Spencer, W. F. 1954. Influence of cation-exchange reactions on retention and availability of cations in sandy soils. Soil Sci. 77:129-136.

Tabikh, A. A., I. Barshad, and R. Overstreet. 1960. Cation-exchange hysteresis in clay minerals. Soil Sci. 90:219-226.

Tamele, M. W. 1950. Chemistry of the surface and the activity of alumina-silica cracking catalysts. Discuss. Faraday Soc. 8:270-279.

Taylor, A. W. 1958. Some equilibrium solution studies on Rothamsted soils. Soil Sci. Soc. Amer. Proc. 22:511-513.

Thomas, G. W. 1958. Plant and soil factors influencing the mineral nutrition of plants. Ph.D. Dissertation. North Carolina State College. 186 pp.

Thomas, G. W. 1960a. Factors affecting the removal of salts from halloysite. Soil Sci. 90:344-347.

Thomas, G. W. 1960b. Forms of aluminum in cation exchangers. Int. Congr. Soil Sci. Trans. 7th (Madison, Wis.) II:364-369.

Thomas, G. W., and N. T. Coleman. 1959. A chromatographic approach to the leaching of fertilizer salts in soils. Soil Sci. Soc. Amer. Proc. 23:113-116.

Tinker, P. B. 1964a. Studies on soil potassium. III. Cation activity ratios in acid Nigerian soils. J. Soil Sci: 15:24-34.

Tinker, P. B. 1964b. Studies on soil potassium. IV. Equilibrium cation activity ratios and responses to potassium fertilizer of Nigerian oil palms. J. Soil Sci. 15:35-41.

Van Der Paauw, F. 1962. Periodic fluctuations of soil fertility, crop yields and of responses to fertilization as effected by alternating periods of low or high rainfall. Plant and Soil 17:155-182.

Volk, N. J. 1934. The fixation of potash in difficultly available form in soils. Soil Sci. 37:267-287.

Wada, K. 1959. Oriented penetration of ionic compounds between the silicate layers of halloysite. Amer. Mineral. 44:153-165.

Wallace, Arthur. 1957. Influence of soil temperature on cation uptake in barley and soybeans. Soil Sci. 83:407-411.

Weber, J. B., and A. C. Caldwell. 1965. Potassium-supplying power of several Minnesota surface soils and subsoils. Soil Sci. 100:34-43.

Wiklander, L. 1964. Cation and anion exchange phenomena, p. 188-189. *In F. E. Bear* (ed.). Chemistry of the soil. 2nd ed. Reinhold Publishing Corp., New York.

Woodruff, C. M. 1955a. Ionic equilibria between clay and dilute salt solutions. Soil Sci. Soc. Amer. Proc. 19:36-40.

Woodruff, C. M. 1955b. The energies of replacement of calcium by potassium in soils. Soil Sci. Am. Proc. 19:167-171.

Woodruff, C. M., and J. L. McIntosh. 1960. Testing soil for potassium. Proc. 7th Int. Congr. Soil Sci. 3:80-85.

Worley, R. E., R. E. Blaser, and G. W. Thomas. 1963. Temperature effect on potassium uptake and respiration by warm and cool season grasses and legumes. Crop Sci. 3:13-16.

14

Mechanism of Potassium Absorption by Plants

STANLEY A. BARBER

Purdue University
Lafayette, Indiana

I. INTRODUCTION

Potassium is a monovalent cation that is absorbed in larger quantities by plant roots than any other cation. In spite of this it is usually present in much smaller quantities in the root medium than calcium or magnesium, two other cations also absorbed in large amounts. Because of the large quantities of potassium absorbed by the plant, many soils are rapidly depleted by cropping so that the supply of potassium to the plant root becomes a very important process.

When crops are grown in the field, potassium may go through many steps in moving from its association with the soil particle until it reaches the leaves of the growing plant. Some of these processes are, release by the soil particle, movement to the root surface, adsorption on the root surface, transfer across the semipermeable cell membrane to the inside of the plant root, movement through the cells of the root, and translocation to the top of the plant. In this review, we will consider the movement of the ion from the soil up to the root and into the cytoplasm or vacuole of the root cell. In following these steps we will discuss absorption by both excised roots and intact plants from stirred solutions, from stirred clay suspensions and from soil. In addition to the effect of the concentration of potassium in the substrate on potassium uptake rate and plant growth, the influence of the concentration of companion ions on these processes will also be discussed.

II. MOVEMENT OF POTASSIUM INTO THE PLANT ROOT

Plants usually contain most of the ions present in soil solution; however, there is little relation between the relative quantities of available ions in the soil and the relative quantities absorbed by the plant root. Potassium is a prime example. Plants commonly contain 10 times as much potassium as calcium, yet the level of available calcium in the soil is frequently of the order of 10 times that of potassium. Plant roots are highly selective in the absorption of nutrients.

Potassium has frequently been used as the indicator ion in research investigations of the mechanisms of the uptake process. In this review, the main theories of ion uptake will be discussed with specific reference to their influence on potassium absorption by the root.

Ion absorption has been divided into two parts; passive adsorption and active absorption. Passive adsorption refers to the equilibration of ions with the root without expenditure of metabolic energy. Active absorption refers to the movement of ions across cell barriers into the cytoplasm and vacuole of the cell and requires metabolic energy. These metabolically accumulated ions are positionally available for translocation to the other parts of the plant.

The absorption processes are related to the structure of the root. About 10% of the root consists of spaces in the cell walls where the solution has the same composition as the surrounding solution. This space is called water-free space since it is freely penetrated nonmetabolically by the solution that bathes the roots. There are additional spaces within the root that have cation-exchange sites on their bonding surfaces. These exchange sites hold cations and influence the concentration of ions in the solution within the space due to Donnan effects.

Passive adsorption may precede or occur simultaneously with active absorption. Ions may need to diffuse into the root interstices before active absorption can occur. There is not, however, evidence for the dependence of active absorption on prior passive adsorption.

Absorption of potassium into the cytoplasm or vacuole of the root cells is termed active absorption and requires an expenditure of metabolic energy. The barrier membrane through which active absorption occurs is believed to be the plasmalemma membrane which surrounds the cytoplasm.

Two theories that have been proposed to explain the metabolic-dependent selective absorption of potassium into the root are the carrier theory and a theory based on the interdiffusion of ions across the membrane as the result of a potential gradient created by metabolically controlled anion absorption. The carrier theory, which is widely used, proposes that particular molecular species exist which are instrumental in transporting the ions through the cell membrane. These particular molecular combinations, called carriers, combine with an ion such as potassium outside of the barrier membrane and, as a potassium-carrier combination, can pass through the membrane. Some process on the inside releases the potassium from the carrier, presumeably by destroying it. Metabolic energy is required for the formation and destruction of the carrier. In addition, it is theorized that there are a multitude of different carriers each with a particular ion specificity. There may be more than one type of carrier for a single ion. Also one carrier may be able to move several different ions into the root.

Evidence for the existence of carriers comes from kinetic studies of ion absorption. The rate of uptake of an ion increases as the concentration of the ion in the substrate is increased until a maximum uptake rate is reached. This is interpreted to mean that there are a definite number of carrier sites, and more of the sites are saturated as the concentration of the ion in the absorption solution is increased until at high concentrations they are all saturated, and the uptake rate reaches a maximum. The kinetics of the relation between ion uptake rate and ion concentration in the substrate agrees with that used for the kinetic treatment of enzyme action. The Michaelis-Menten Equation fitting the relation used for enzyme action is

$$v = V(S)/[Ks + (S)]$$

where v is the reaction velocity, S is the substrate concentration, V is the maximum velocity when all sites are saturated, and Ks is a constant. When related to ion uptake Ks is believed to be the apparent dissociation constant of the ion-carrier complex.

When the reciprocal of the uptake rate is plotted versus the reciprocal of the substrate concentration we have a linear relation known as the Linweaver-Burke plot. The Linweaver-Burke plot is used to evaluate Ks and V. In this relation, the slope of the line is Ks/V and the intercept is $1/V$.

Further evidence for the carrier theory for potassium absorption comes from the kinetics of ion competition. Using the Michaeles-Menten equation, uptake data can be evaluated to determine whether two ions compete for the same carrier site. Uptake studies have shown that rubidium, cesium, hydrogen, and ammonium compete with potassium for the same carrier sites used in potassium uptake. Sodium and lithium do not compete. These competitive effects are evaluated by assuming that their competitive uptake follows the same kinetics as competition in enzyme action.

Epstein et al. (1963) present evidence for two separate potassium carriers. One carrier (that referred to in the previous paragraph) is very selective for K and operates at substrate concentrations below 0.2mM, while a second carrier with a lower affinity for potassium operates only at concentrations above 0.2mM. Sodium competes with potassium on the latter carrier site. Uptake by the first carrier is independent of the accompanying anion; uptake by the second is influenced by the accompanying anion.

A second theory for active potassium absorption (Hendricks, 1966; Robertson, 1960; Schaidle and Jacobson, 1967) by the plant root cells is based on active absorption of anions by respiration energy which create a potential gradient along which potassium can diffuse into the cell even though it is against a potassium concentration gradient. When the anion concentration is greater within the cell than without, it creates the potential gradient along which potassium may interdiffuse for hydrogen produced within the plant root. A proposed scheme for this type of salt transport and the reactions which could be associated with the respiration energy involved in anion transport are shown in Fig. 1. The uptake of potassium is by diffusion across the negatively charged positions on the cellular membrane in exchange for hydrogen which is produced metabolically in the cell as a result of the metabolically controlled anion absorption.

Evidence supporting this theory is that potassium moves both into and out of the root cell. The net flux is the difference between influx and eflux. This movement has been demonstrated with the use of radioactively labeled potassium. The relation between potassium concentration in the root medium and potassium uptake follows a Langmuir absorption relationship. This would indicate potassium adsorbed on the root surface determined the rate of potassium diffusion into the root. Once potassium has reached the cytoplasm or vacuole it then may move to the other parts of the plant by translocation within the plant.

The rate of salt uptake by cells is directly associated with the amount of respiration of the root. As the solution concentration is increased and salt uptake increases, the rate of root respiration also increases.

In order to satisfy the requirement for selective absorption of potassium versus ions such as sodium, selective potassium-sodium exchange is theor-

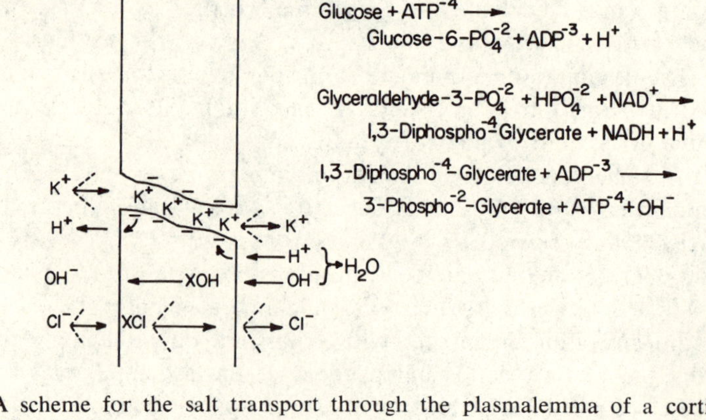

Fig. 1.—A scheme for the salt transport through the plasmalemma of a cortical cell. (Hendricks, 1966)

ized. By a repetition of the selective exchange absorption along a series of exchange sites, a process analagous to chromotography, a high degree of selectivity in the rate of movement of cations into the plant root can be obtained.

Both theories have their strong and weak points. Further developments in this field will undoubtedly occur in the future. These theories are mainly based on evidence from uptake by excised roots. Uptake of potassium by intact plants is also influenced by the rate of translocation of potassium from the root to the shoot as well as the processes occuring at the root surface.

III. DEPENDENCE OF PLANT GROWTH RATE ON POTASSIUM CONCENTRATION OF THE SUBSTRATE

There is a relation between the potassium concentration of the root medium and the absorption of sufficient potassium to support the maximum growth rate of the plant. Solutions expressed from soil contain concentrations of K varying from 20 to 1,000 μM. A concentration in the range of 100 μM is common on soils of humid regions (Barber et al., 1963). This concentration is about 0.01 the concentration of approximately 10 μM frequently used in nutrient culture.

In order for a solution of low concentration to supply the plant root and not be reduced appreciably in potassium concentration by plant uptake, the solution must be replenished rapidly. In comprehensive nutrient culture experiments, Asher and Ozanne (1967) maintained low concentrations about plant roots by using large volumes of rapidly flowing solution. They investigated the relation between plant growth and the potassium content of the nutrient solution used for the root medium and found that 8 of 14 species investigated reached maximum yield with a potassium concentration of 24 μM. The remaining species reached maximum yield with a concentration of 95μM. If soils can maintain these concentrations in the solution at the root surface, presumeably we would not have to invoke mechanisms such as contact exchange or exchange diffusion to explain the supply mechanism for potassium to plant roots. It is difficult, however, to visualize how this solution concentration could be maintained during plant growth in many soils since the soil solution is not flowing rapidly past the root. The soil acts as a buffer to maintain the potassium level in the solution; however, the soil's buffer capacity is limited in localized regions and the soil particles next to the root surface usually do not have enough available potassium to maintain the solution concentration at that point. The potassium uptake is so large that these soil particles immediately adjacent to the root may be entirely depleted of potassium in order to supply the amount removed by the plant root. On many soils, potassium must diffuse through the soil toward the root in order to supply the potassium requirement.

IV. INFLUENCE OF CALCIUM ON POTASSIUM ABSORPTION

Some (Beckett, 1964; Woodruff, 1955) have proposed to measure potassium availability to the plant root in terms of the relationship between potassium and calcium or potassium and the sum of calcium plus magnesium. This relationship would indicate that potassium uptake would be reduced as calcium and magnesium are increased and conversely calcium and magnesium uptake would be reduced as the available supply of potassium is increased. According to this concept the availability of potassium would be determined not so much by the absolute amount present as by the amount relative to that of calcium and magnesium.

Not all of the experimental evidence supports this contention. Some experiments have shown that an increase in calcium in the solution causes an increase in potassium uptake. Viets (1942) used three levels of calcium in solution with a constant level of potassium. Within 10 hours the potassium uptake of excised barley roots increased as calcium concentration in the solution increased. In nutrient culture experiments Collander (1941) found little effect on K uptake rate when Ca was increased from 0.1 to 1.0 mM. Conversely he also found little influence of K on Ca uptake when K concentration was varied from 0.2 to 2.0 mM. The plants were grown for 2 months in these experiments.

In experiments in soil, Van Itallie (1938, 1948) observed that calcium uptake was dependent upon the ratio of calcium to magnesium and potassium and not on the absolute quantity of calcium present. Potassium uptake, on the other hand, was affected only by potassium additions to the soil. Calcium had little influence on potassium uptake. More than 20 years ago several reviews (Albrecht, 1943; Peech and Bradfield, 1943; Pierre and Bower, 1943) discussed the variable effects of calcium on potassium uptake. There were also proposals (Bear, 1950; Wallace et al., 1948) of a constant value for the sum of the equivalents of the cations, potassium, sodium, calcium and magnesium within the crop. An increase in the concentration of one would result in an equivalent reduction in one or more of the other three. There was apparently more influence of potassium on calcium, magnesium, and sodium, than of the other ions on potassium in the plant. Even so, the reciprocal relationships that did occur led to the approximate constancy of the total number of equivalents of these cations within the plant. The total number of equivalents varied for crops and experimental conditions but remained somewhat constant as the levels of calcium, magnesium, and potassium in the soil medium were varied.

The idea of a constant number of equivalents appears to be only partial ly satisfied. Increasing potassium does cause a reduction in the calcium and magnesium uptake as potassium uptake increases. However, when calcium or magnesium are increased, their content in the plant may increase without a corresponding decrease in the potassium content. The availability of potassium appears to be more closely related to the total quantity of potassium present than to its relation to the concentrations of calcium and magnesium.

V. EXCHANGE CAPACITY OF ROOTS AND POTASSIUM UPTAKE

Plant roots have cation-exchange properties and many attempts have been made to relate the magnitude of this property to the ability of plants to absorb K (Drake, 1964). Plant roots vary in cation-exchange capacity per unit weight of root by a factor of at least 10. Actual values depend somewhat on methods of analysis; however, values ranging from 94 meq/100 g dry weight for Larkspur to 9 for winter wheat have been obtained when measured by a common procedure. In general, the cation-exchange values are much higher for dicotyledons than for moncotyledons.

The theory of selectivity in uptake due to root cation-exchange capacity is based on the Donnan principle as it relates to the distribution of monovalent versus divalent ions absorbed on the exchange sites of the root. According to this principle the ratio of potassium to calcium in systems in equilibrium with each other is as the ratio of the activities of $K/(Ca)^{1/2}$ Hence in two systems in equilibrium, the system at the higher concentration, or in this case with the higher exchange capacity roots, would have a ratio of potassium to calcium that would be much lower than in a system at lower concentration or where the roots have a lower exchange capacity. In comparing plant species such as alfalfa and bromegrass, the alfalfa has the higher exchange capacity root and is the more susceptible to potassium deficiency which argues in favor of this principle. The argument for the high cation-exchange capacity roots of alfalfa being less efficient in potassium uptake than the low exchange capacity roots of bromegrass is that due to Donnan effects on the surfaces and in the Donnan free-space there is less potassium and more calcium present at the alfalfa root surfaces than at the bromegrass root surfaces. Uptake, whether due to the absolute amount of potassium present or the ratio of the potassium to calcium, would be less.

The maximum Donnan effects occur in very dilute solutions which are much lower in concentration than those found in most fertile soils. The differences between species in the rates at which they absorb potassium from solution may be due to the differences in the cation-exchange capacities of the roots. These differences may also be due to differences in other root properties such as surface area, and these may in turn be related to the exchange capacities of the root.

Evidence has been obtained both for and against a role for cation-exchange capacity of plant roots in the absorption of potassium (Cunningham and Nielsen, 1963; Smith and Wallace, 1956). It is not known whether these exchange sites serve as the initial points of entry of potassium into the root or whether they compete with ion carriers for potassium which will be moved into the plant.

VI. QUANTITY-INTENSITY MEASUREMENTS AND POTASSIUM ABSORPTION

It is popular to consider the rate of uptake of ions by a plant as being influenced both by the concentration present in the medium or the intensity and by the ability of the system to maintain this concentration or the capacity of the system. The intensity value for potassium in soil has been proposed by Schofield (1947) as being the ratio of the activity of potassium to the square root of the sum of the activities of calcium plus magnesium. This is the so called Ratio Law. For a particular soil, this ratio remains constant for solutions in equilibrium with the soil over a wide range of dilutions so long as the amounts of potassium, calcium, and magnesium in the soil remain the same. The use of this value as an intensity measurement for determining the rate of uptake of potassium by a plant root presupposes that the uptake rate is governed by the ratio of the activity of potassium to the square root of the activity of calcium plus magnesium rather than to the absolute level of potassium activity in the root medium. For this system the difference between the chemical potentials of potassium and calcium is $RT \ln a_K/(a_{Ca})^{1/2}$. If in measuring the ratio we add sufficient calcium to the system so that the activity of calcium is independent of the activity of calcium initially in the soil, then we have a relative value for the chemical potential of potassium in the solution phase of the soil at a common level of calcium for all soils. In measuring the activity ratio, which is defined as $a_K/(a_{Ca})^{1/2}$, Schofield suggested, adding $0.002M$ $CaCl_2$ to the soil to give a relatively uniform quantity of Ca in the system.

Beckett (1964) has developed a procedure for obtaining a measure of the buffering capacity of the soil for K or the ability of the soil to maintain its original activity ratio. He added uniform amounts of soil to Ca-K chloride solutions that varied in activity ratio from 0.01 to 0.07. He measured the activity ratio after the systems achieved equilibrium and, by the difference between the original and final potassium levels in solution, he determined the release or absorption of potassium (ΔK) by the soil. He then plotted ΔK versus the final activity ratio, and the slope of the line was a measure of the capacity of the soil to supply potassium.

For plant roots absorbing potassium from a stirred soil suspension, the chemical potential of potassium may be the dominant external factor in determining the rate of potassium absorption. The ability of the soil to maintain this potassium potential would be a measure of the ability of the soil to maintain the supply. There is little evidence to indicate that the activity ratio of potassium to the square root of calcium plus magnesium is the governing factor in determining the rate of potassium absorption by the root.

This same general activity ratio relationship has been used by Woodruff (1955) to determine the ability of the soil to supply potassium to the plant root. He determined the free energy change that accompanies the exchange of calcium for potassium on the exchange sites of the soil. The equation he uses for this relation is

$$\Delta F = RT \ln a_K/(a_{Ca})^{1/2}$$

where R is the gas constant and T the absolute temperature. He proposes that the difference in free energy can be used as a measure of potassium availability. As in the previous discussion on activity ratio, (this is no more than a function of the activity ratio) this presupposes that the rate of K absorption is dependent upon the activity ratio rather than the absolute quantity, activity, or chemical potential of potassium in the root medium. In experiments that provide evidence for this belief, both the activity ratio and potassium level in the soil are confounded so that it is difficult to determine a cause and effect relationship.

As noted previously, research by Viets, Collander and Van Itallie has shown that increasing the calcium level in the soil or the solution has increased or had no effect on the rate of potassium uptake rather than decrease it as would be the case if the activity ratio were the determining factor.

VII. EFFECT OF CLAY SUSPENSIONS ON K UPTAKE

When plants are grown in rapidly flowing nutrient solutions, the rate of potassium uptake by the plant roots is related to the K concentration in solution. When the plants are grown in clay suspensions, the K in solution is a small part of the K in the system since a large part of the K present may be held as exchangeable K by the clay. Jenny (1951, 1966), Jenny and Overstreet (1939), and Scheuring and Overstreet (1961) have proposed and given evidence for ion movement directly from the exchange positions to the plant root in addition to that absorbed from the solution. This exchange of ions directly from the exchange spot on the clay to the root has been termed "contact exchange" since presumeably the root and the clay particle would have to approach very close to allow the direct interchange to occur. Evidence for this effect comes from the measurement of a greater rate of uptake from the suspension than from its equilibrium solution (Scheuring and Overstreet, 1961). Barlett and MacLean (1959),

Helmy and Oliver (1961), Lagerwerff (1958, 1965) and Olsen and Peech (1960) have obtained evidence both for and against the concept of contact exchange. The differences between experiments were sometimes due to the experimental parameters used. Contact exchange effects are greatest when the solution concentration is kept low and the possibility for root-soil contact is optimized. In addition, the ion flux reaching the root must be limiting the uptake rate. The greater uptake of potassium from a suspension over that from solution indicates a greater flux. This greater flux is believed due to ions taken up from both the solution and the clay by the root.

Helmy and Oliver (1961) have shown a greater rate of potassium uptake by excised barley roots from an illite suspension than from its equilibrium solution. Additional evidence to support the direct movement of potassium between the clay and the root is also shown (Jenny, 1951) by the effect of clay suspensions in increasing the rate of potassium movement out of the root when the potassium-containing roots are placed in hydrogen-clay over the rate obtained by placing the roots in an HCl solution of similar pH.

Plant roots are also able to remove nonexchangeable potassium from the soil that is not removed by repetitive leaching with a displacing salt. The amounts of nonexchangeable K absorbed from the soil by plant roots has been correlated with the K released from the soil by boiling with $1N$ HNO_3. Hence it takes a rather drastic acidic treatment of the soil to remove as much potassium from it as can be removed by plant roots when the soils are exhaustively cropped.

VIII. EXCHANGE DIFFUSION AND POTASSIUM UPTAKE

Measurement of the mechanisms responsible for the supply of potassium from the soil to the plant root surface indicates that, for many soils, diffusion is the dominant mechanism. This diffusion may result from the movement of ion pairs such as K and NO_3 to the root or it may result from the interdiffusion of potassium with another cation such as hydrogen moving away from the root. Both the potassium ions balanced with anions in solution and the potassium ions held as exchangeable ions by the soil may reach the root by interdiffusion. The interdiffusion of exchangeable ions on clay surfaces will be called exchange diffusion. The supply of potassium to the root by contact exchange depends upon exchange diffusion to replenish the soil exchange sites at the root surface that are supplying potassium directly to the root.

The rate of diffusion of K to the root surface depends upon the phases that it diffuses through. The amount of potassium diffusing through any phase per unit time depends upon the concentration gradient of potassium, upon the diffusion coefficient for potassium in the phase and the area for diffusion. When 95% of the K is held by the soil and 5% is in solution, the rate of diffusion in the solution would need to be 20 times faster than that of the exchangeable K in order to supply K at the same rate per unit area since the concentration gradient is 0.05 of the exchangeable K.

Jenny (1966) has discussed the several possibilities for diffusion from the soil to the root. Diffusion may occur within the root, through the surface water film, and by exchange diffusion along the particle surface. When concentrations in solution are low the significance of exchange diffusion increases.

When the soil particle is in close proximity to the plant root the distance for diffusion through the water film is less and the shorter distance gives a larger concentration gradient that leads to a greater amount of diffusion than when the soil particle and the root are farther apart.

IX. POTASSIUM UPTAKE FROM SOILS

Plant root absorption of potassium from stirred solutions assumes uptake from a medium in which the concentrations of ions at the root surface is the same as that throughout the medium. In contrast potassium uptake from soils represents uptake from a dynamic system that is not at equilibrium, hence rate processes involved with ion movement may become the limiting factor in determining absorption rates by plant roots.

While the soil may have a more or less uniform distribution of ions, initially the root alters this uniformity. The influence of the plant root on the ionic conditions of the soil begins when the root starts to force its way through the soil. Since the diameter of the root is frequently larger than the diameter of the majority of soil pores, the root moves soil particles aside and in so doing increases the density of soil in the immediate vicinity of the root so that it has greater than average density. This will also increase the concentration of exchangeable potassium per unit volume of soil. In addition to pushing the soil aside, the root will intercept ions in its path and absorb them as it occupies a volume element of the soil. The magnitude of the influence of the root, termed root interception, is in direct relation to the volume of displaced soil and this is equal to the volume of the root.

The root absorbs water and causes a flow of water through the soil toward the root. Since this water contains potassium ions, these ions are transported to the root by mass-flow. The amount reaching the root will be the product of the water used multiplied by the potassium concentration of the water. Since the supply by mass-flow and root interception is often less than the amount required, the plant absorbs potassium from the soil at the root interface, reduces its concentration and creates a gradient along which potassium will diffuse toward the root.

The contribution of each of the three mechanisms of root interception, mass-flow, and diffusion to the supply of potassium to the plant root depends upon the plant, the soil, and the environment. A general evaluation of the relative significance of each mechanism can be obtained by measuring the appropriate physical and chemical parameters of a number of crops and soils and of the effect of environment on the transpiration rate and rooting charactertistics of the crop.

The contribution of root interception is related to root volume because it is the volume of the root that determines the amount of potassium displaced. This displaced potassium may be absorbed at the time the root contacts it or it may be pushed aside with the soil particles to which it is attached. This increases the potassium concentration in the soil near the plant root, and this potassium moves back either by diffusion or mass-flow. While much of it may move by mass-flow or diffusion the evaluation of the amount depends on root volume. Root volumes are of the order of 1% of the volume of the surface 7 inches of soil. They may vary from 0.4 to 2.0%. Hence root interception contributions amount to about 1% of the amount of available potassium in the soil. Since soils commonly have 100 to 300 kg of exchangeable K/ha, root interception contributions are of the order of 1 to 3 kg/ha. For many crops this is < 5% of the requirement.

Mass-flow contributions depend upon the concentration in soil solutions and the amount of water used. Soil solutions on humid region soils vary from a concentration of 1 to 40 ppm with 4 ppm being a common value. Crops use about 2 to 3 million kg water/ha, hence the supply of K by mass-flow may commonly be the order of 10 kg/ha. This is usually less than 10% of the K required by the plant. Soils of arid regions may contain higher amounts of both potassium and soluble salts so that the amounts of potassium in solution are much higher and mass-flow may supply very significant quantities of potassium to the plant root.

Since the combined amount of potassium supplied by the mechanisms of root interception plus mass-flow may be 15% or less of the K requirement on crops grown on humid soils, most of the K reaches the root by diffusion. The factors affecting the amounts of potassium reaching the root by diffusion are very important in determining potassium availability.

Evidence of the significance of diffusion for the supply of potassium to the plant is shown in plant growth experiments. Oliver and Barber (1966) grew soybeans on five soil-sand mixtures of one soil under three environments which gave a 3-fold variation in the transpiration rate. This variation in transpiration gave a 3-fold variation in the supply of K by mass-flow. The appropriate parameters of the soil and the plant and its root system were measured so that the supply of potassium to the plant by root interception and mass-flow could be evaluated. The data averaged across soils and given in Table 1 show that only a small portion of the K uptake was supplied by either root interception or mass-flow, and the major part of the potassium moved to the root by diffusion. Since the rate of transpiration had little influence on the total amount of potassium absorbed by

Table 1—The supply of potassium to soybean roots by root interception, mass-flow, and diffusion

	Transpiration rate		
	Low	Medium	High
		mg K/pot	
K uptake	61.6	65.1	61.7
K supply by root interception	1.1	1.7	2.4
K supply by mass-flow	1.4	3.3	5.7
K supply by diffusion	59.1	60.1	53.6

the plant, this also indicated that mass-flow had little influence on the supply of potassium to the plant root.

The significance of potassium diffusion has also been verified by measuring the distribution of potassium about the roots of the growing plant by the use of autoradiographs (Barber, 1962; Kurihara et al., 1964; Walker and Barber, 1962). Rubidium was used in place of potassium because its radioactive isotope, ^{86}Rb, has a more convenient half life than ^{42}K. The soil was treated uniformly with ^{86}Rb and placed in a box with one side supported by a 1 mil thick mylar film. Some of the roots grew through the soil next to the mylar film. Autoradiographs were made by placing X-ray film against the mylar film and exposing it for an appropriate period of time (of the order of 30 min in this case). These autoradiographs showed the influence of the plant root on the distribution of ^{86}Rb in the soil. An example is shown in Fig. 2. In this autoradiograph the dark areas are areas where ^{86}Rb was present in the soil or in the roots, and the light areas were depleted areas. The area about the root was apparently very low in Rb and the concentration of Rb increased in a diffusion pattern out from the root. Although root hairs influence this picture, the depleted area extended much further than the length of the root hairs. Similar short-term experiments were conducted with ^{42}K, and the autoradiographs obtained were similar indicating that the autoradiographs obtained with Rb could be used to interpret the reactions of K with the soil.

Since diffusion is important in potassium uptake, the supply of potassium to the root will depend upon the factors in equation (1)

$$J = -ADdc/dx \qquad (1)$$

where J is the flow or flux of K to the root, A is the root area, D is the rate of diffusion and dc/dx is the potassium concentration gradient out from the root. In this discussion, D is the effective rate of diffusion of K in the soil system as measured by the rate of movement of potassium or rubidium from a block of treated into a block of untreated soil as indicated by the potassium or rubidium distribution from the boundary. Of course, with the root, cylindrical rather than the simpler linear co-ordinates expressed here will apply. The supply to the root will depend upon the type of plant root

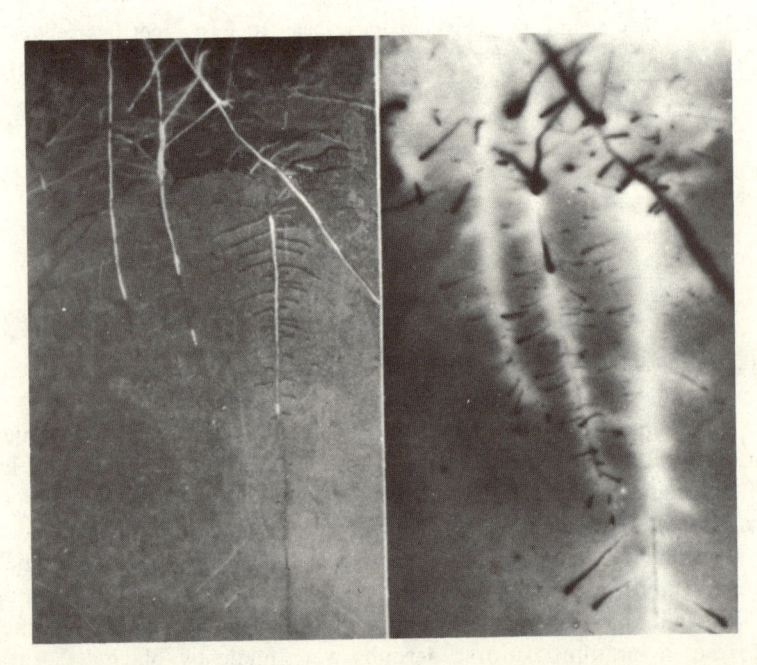

Fig. 2—A photograph (*left*) and an autoradiograph (*right*) showing the effect of corn roots on the distribution of ^{86}Rb in the soil. Light areas show ^{86}Rb depletion about corn roots.

system since the larger the area of the root the more area for diffusion of potassium to the root. This may be why some crops such as small grains and grasses are more effective feeders for potassium in the soil than a crop such as alfalfa. The availability of potassium also depends upon the diffusion coefficient of potassium in the soil and on the concentration gradient which will be related to the level of available potassium in the soil and the ability of the root to reduce the level of potassium at the soil-root interface. The difference between the potassium concentration at the root surface and that in the unaffected soil will determine the concentration gradient.

Evans and Barber (1964) obtained a high correlation ($r^2 = 0.94$) between the diffusion coefficient, D, and the rate of absorption of ^{86}Rb, in a soil-plant root system where the root area and the initial concentration in the soil were constant. Soils and soil-agar systems varying in their rates of diffusion were used in this experiment. The relationship is shown in Fig. 3.

The rate of diffusion of potassium or rubidium in soil is lower than that in water principally because of the reaction of the potassium ions with the soil. Diffusion coefficients are of the order of 10^{-7} to 10^{-8} in soil whereas in solution they are approximately 1.9×10^{-5} cm^2 sec^{-1}. There is some variation between soils in the rate of diffusion of potassium. Usually the higher the exchange capacity of the soil, the lower the rate of diffusion.

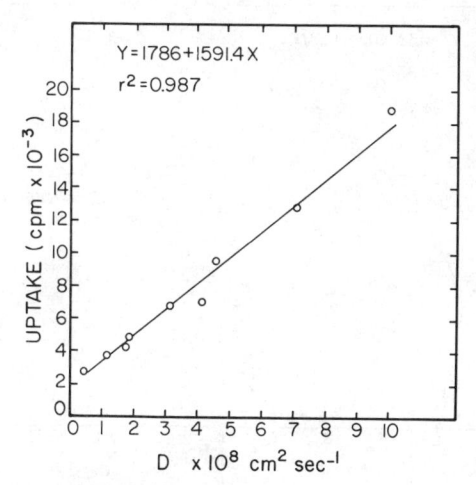

Fig. 3—The relation between ^{86}Rb diffusion coefficient and ^{86}Rb absorption.

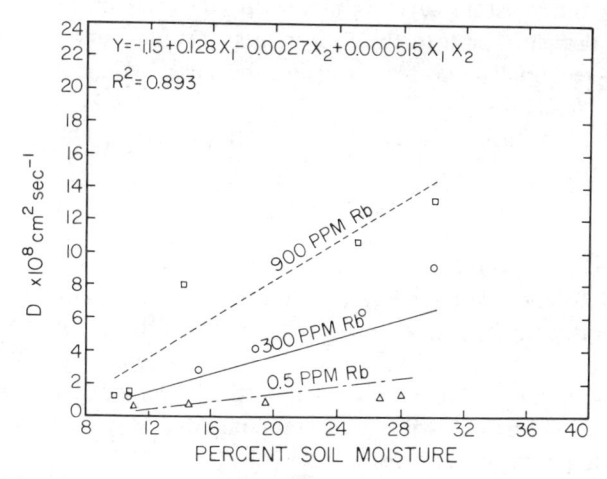

Fig. 4—The effect of soil moisture and Rb level in the soil on Rb diffusion.

The rate of diffusion is influenced by both the soil moisture content and the level of exchangeable potassium. Using ^{86}Rb, Place and Barber (1964) found a high degree of correlation between the level of rubidium in the soil and the diffusion coefficient and also between the soil moisture content and the diffusion coefficient. These relationships are shown in Fig. 4. The uptake of ^{86}Rb by the corn root was also influenced in the same manner by the level of exchangeable rubidium and the soil moisture level so that variations in rubidium uptake was highly correlated with the variation in rubidium diffusion coefficient that was caused by these variables (Fig.

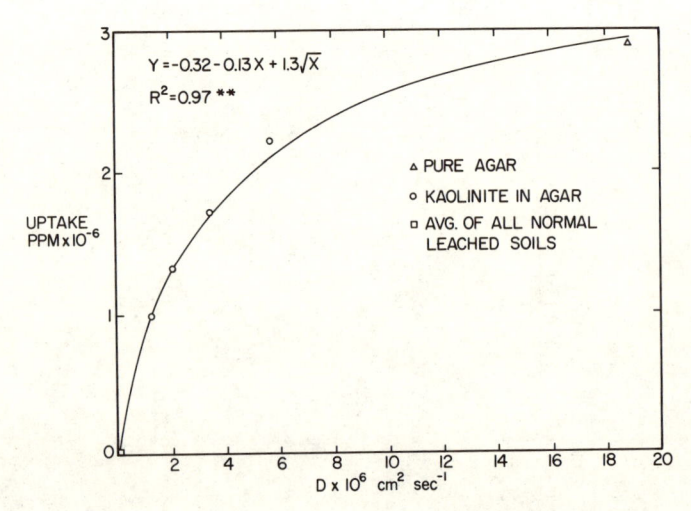

Fig. 5—The relation between the variation in diffusion coefficient resulting from variations in soil moisture and Rb level and the rate of Rb absorption by corn roots.

5). As the soil moisture level is increased, the rate of diffusion increases because the diffusion path to the root is less tortuous and there is also a larger cross-sectional area for diffusion because of the reduction in air space.

The rate of diffusion is affected by temperature. The Stokes-Einstein equation,

$$D = kT/6\pi rn \qquad (2)$$

shows that the diffusion coefficient is directly proportional to the absolute temperature T. The other symbols k, r and n indicate, respectively, the Boltzmann constant, radius, and viscosity. This is for diffusion in a medium such as water. Potassium diffusion in soil is also affected by the reaction of potassium with the soil particles so that the effect of temperature on the rate of diffusion in soil will be more complex. However, it will usually increase as temperature increases.

The relationship between potassium uptake and the supply of potassium to the root by mass-flow and diffusion was investigated by Drew et al. (1967). They measured diffusion coefficients for potassium in the soil and uptake rates of potassium by the plant. They used these values to calculate the concentration of potassium that should occur at the soil root interface by substituting them in the appropriate diffusion equation. They calculated that with the experimental conditions they used, that potassium absorption reduced the K level at the root 15% in one experiment and 42% on the average in a second experiment where the initial K level in the soil was lower. This is less depletion than was apparent in the experiments by Walker and Barber (1961) where the concentration gradient was observed with autoradiographs.

Drew et al. (1967) also raised the question as to whether the rate of diffusion to the root is rate limiting in potassium uptake or whether the rate of uptake by the root is rate limiting. Where a diffusion gradient occurs it would seem that diffusion would limit at least to some extent the rate of uptake since rate of uptake is usually related to the concentration of potassium at the root surface. Where mass-flow causes accumulation about the root, the rate of uptake being slower could certainly be termed the rate limiting step.

REFERENCES

Albrecht, W. A. 1943. Potassium in the soil colloid complex and plant nutrition. Soil Sci. 55:13-21.

Asher, C. J., and P. G. Ozanne. 1967. Growth and potassium content of plants in solution cultures maintained at constant potassium concentrations. Soil Sci. 103:155-161.

Barber, S. A. 1962. A diffusion and mass-flow concept of soil nutrient availability. Soil Sci. 93:39-49.

Barber, S. A., J. M. Walker, and E. H. Vasey. 1963. Mechanisms for the movement of plant nutrients from the soil and fertilizer to the plant root. J. Agr. Food Chem. 11:204-207.

Bartlett, R. J., and E. O. McLean. 1959. Effect of potassium and calcium activities in clay suspensions and solutions on plant uptake. Soil Sci. Soc. Amer. Proc. 23:285-289.

Bear, F. E. 1950. Cation and anion relationships in plants and their bearing on crop quality. Agron. J. 42:176-178.

Beckett, P. H. T. 1964. Studies on soil potassium. I. Confirmation of the ratio law: Measurement of potassium potential. J. Soil Sci. 15:1-8.

Collander, R. 1941. Selective absorption of cations by higher plants. Plant Physiol. 16:691-720.

Cunningham, R. K., and K. F. Nielsen. 1963. Evidence against relationships between root cation exchange capacity and cation uptake by plants. Nature 200:1344-1345.

Drake, M. 1964. Soil chemistry and plant nutrition. In F. E. Bear (ed.) Chemistry of the Soil. Reinhold Publishing Co. New York, p. 395-439.

Drew, M. C., L. V. Vaidyanathan, and P. H. Nye. 1967. Can soil diffusion limit the uptake of potassium by plants? Trans. Intl. Soil Sci. Soc. Commissions II and IV. Soil Chemistry and Fertility. p. 335-344.

Epstein, E., D. W. Rains, and E. O. Elzam. 1963. Resolution of dual mechanisms of potassium absorption by barley roots. Natl. Acad. Sci. Proc. 49:684-692.

Evans, S. D., and S. A. Barber. 1964. The effect of ^{86}Rb diffusion on the uptake of ^{86}Rb by corn. Soil Sci. Soc. Amer. Proc. 28:56-57.

Helmy, A. K., and S. Oliver. 1961. Cation absorption by excised barley roots from soil suspensions and their equilibrium true solutions at different time intervals. Soil Sci. 91:339-340.

Hendricks, S. B. 1966. Salt entry into plants. Soil Sci. Soc. Amer. Proc. 30:1-7.

Jenny, H. 1951. Contact phenomena between adsorbents and their significance in plant nutrition. In E. Truog (ed.) Mineral Nutrition of plants. The University of Wisconsin Press. p. 107-132.

Jenny, H. 1966. Pathways of ions from soil into root according to diffusion models. Plant and Soil 25:265-289.

Jenny, H., and R. Overstreet. 1939. Contact interchange between plant roots and soil colloids. Soil Sci. 47: 257-272.

Kurihara, K., M. Yokomizo, and S. Mitsui. 1964. Dynamic studies on the nutrient uptake of crop plants. XLIII Mineral nutrients in the rhizosphere. 2. Distribution of various ions surrounding corn roots as a result of uptake. Nippon Dozo-Hiryogaka Zasshi 35:363-366.

Lagerwerff, J. V. 1958. Comparable effects of adsorbed and dissolved cations on plant growth. Soil Sci. 86:63-69.

Lagerwerff, J. V. 1965. Multiple-rate effect in ion transfer. Soil Sci. 100:25-33.

Olsen, R. A., and M. Peech. 1960. The significance of the suspension effect in the uptake of cations by plants from soil-water systems. Soil Sci. Soc. Amer. Proc. 24:257-261.

Oliver, S., and S. A. Barber. 1966. An evaluation of the mechanisms governing the supply of Ca, Mg, K and Na to soybean roots (*Glycine max*). Soil Sci. Soc. Amer. Proc. 30:82-86.

Peech, M., and R. Bradfield. 1943. The effect of lime and magnesia on the soil potassium and on the absorption of K by plants. Soil Sci. 55:37-48.

Pierre, W. H., and C. A. Bower. 1943. Potassium absorption by plants as affected by cationic relationships. Soil Sci. 55:23-36.

Place, G. A. and S. A. Barber. 1964. The effect of soil moisture and Rb concentration on diffusion and uptake of ^{86}Rb. Soil Sci. Soc. Amer. Proc. 28:239-243.

Robertson, R. N. 1960. Ion transport and respiration. Biol. Rev. 35:231-264.

Schaidle, M., and L. Jacobson. 1967. Ion absorption and retention by *Chlorella pyrenvidosa* III. Selective accumulation of Rb, K and Na. Plant Physiol. 42:953-958.

Scheuring, D. G., and R. Overstreet. 1961. Sodium uptake by excised barley roots from sodium bentonite suspensions and their equilibrium filtrates. Soil Sci. 92:166-171.

Schofield, R. K. 1947. A ratio law governing the equilibrium of cations in the soil solution. Trans. Faraday Soc. 45:612.

Smith, R. L., and A. Wallace. 1956. Cation exchange capacity of roots and its relation to calcium and potassium content of plants. Soil Sci. 81:97-110.

Van Itallie, Th. B. 1938. Cation equilibria in plants, I. Soil Sci. 46:175-186.

Van Itallie, Th. B. 1948. Cation equilibria in plants in relation to the soil, II. Soil Sci. 65:393-416.

Viets, F. G., Jr. 1942. Effects of Ca and other divalent ions on the accumulation of monovalent ions by barley root cells. Science 95:486-487.

Walker, J. M., and S. A. Barber. 1961. Ion uptake by living plant roots. Science 133:881-882.

Walker, J. M., and S. A. Barber. 1962. Uptake of Rb and K from the soil by corn roots. Plant and Soil 17:243-259.

Wallace, A., S. J. Toth, and F. E. Bear. 1948. Further evidence supporting cation equivalent constancy in alfalfa. J. Amer. Soc. Agron. 40:80-87.

Woodruff, C. M. 1955. Ionic equilibrium between clay and dilute salt solutions. Soil Sci. Soc. Amer. Proc. 19:36-40.

15

Exchangeable Cations of Plant Roots and Potassium Absorption by the Plant

K. MENGEL

Landwirtschaftliche Forschungsanstalt Büntehof
Hanover, Germany

I. INTRODUCTION

The cations of the plant roots may be divided into (i) cations which can diffuse from the root system rather easily and can be replaced by other cations, and (ii) those which are held more strongly by special bonds or by cell membranes (Briggs and Robertson, 1957; Marschner and Mengel, 1962). The latter group of cations belongs to the "inner space", and the easily diffusible or exchangeable cations belong to the "outer space" or "free space", which is mainly located in the cell walls of the root and its intercellular compartments (Vakhmistrov, 1965; Levitt, 1957).

The sites of exchange of the free space are mainly represented by carboxyl groups of the cell wall and by phosphoryl groups of the plasmalemma. On their way from the soil solution to the metabolic centers of the cell, ions have to pass the free space. The question, therefore, arises whether the exchange of cations in the free space is the first step of this inwardly directed transport and whether the exchangeable fraction of the free space has an influence on the cation transport to the inner space. As K^+ and Ca^{2+} play a dominant role in plant nutrition, and as especially Ca^{2+} is present in the soil solution in rather high concentrations, the question was investigated whether the exchangeable Ca^{2+} of the root influences the fluxes of K^+.

II. MATERIALS AND METHODS

Seeds of barley and oats were allowed to germinate for 24 hours in aerated tap water at room temperature. The germinating seeds were put on a nylon screen, which was spread over a vessel containing an aerated $10^{-4}M$ $CaCl_2$ solution. The germinating seeds were covered by a wet cloth, on which filter paper was spread dipping at two sides into the solution so that the cloth remained moist. The cloth and the filter paper were removed as soon as the roots dipped into the solution. At this stage the roots were about 2-3 cm long.

After some days of growth the roots were excised just below the screen and rinsed several times with distilled water. Roots used for Donnan investigations were in addition rinsed with $0.01N$ HCl in order to remove the Ca^{2+} adsorbed in the free space. Then the roots were blotted gently on dry filter paper and 1-g samples were weighed. Each sample was put into a beaker and a 50ml of nutrient solution was added. Each treatment comprised 4 root samples (4 replications). The absorption process was terminated by decanting the nutrient solution and rinsing the root samples 4 times with 50-60 ml of distilled water. Each root sample was then put into a beaker, 25 ml of $0.02M$ $CaCl_2$ solution was added and the sample was shaken by hand for 1 min. Once more 25 ml of the solution was added and shaking was repeated for 2 min. Both "exchange solutions", containing the exchangeable cations of the roots, were joined and the amounts of the respective cations were determined. If the nutrient solution was not labelled with radioisotopes, $0.02M$ $BaCl_2$ instead of 0.02 $CaCl_2$ was used, in order to achieve a clear distinction between the Ca^{2+} of the roots and the exchanging cation. The whole procedure of determination of the exchangeable cations was checked in detail (Mengel, 1961).

These investigations also showed that the exchangeable cations are not influenced by metabolic processes, whereas the nonexchangeable fraction is reduced significantly at low temperatures. Cations which were not removed by this exchange method were regarded as nonexchangeable (nonexchangeable fraction).

In one experiment, intact oat plants were used. The plants were grown for 5 weeks in a nutrient solution. For the experiment, plants of even size were selected. Each plant was studied individually, 4 samples forming an experimental treatment. They were fixed by a collar of foam rubber in an Erlenmeyer flask containing 32 ml of a nutrient solution.

The plant material was dried and ashed. The ash was dissolved by $0.1N$ HCl. In this solution, total potassium and calcium were determined by flame photometry. The labeled potassium was determined by counting the β-radiation of the ^{42}K with a Geiger counter (liquid tube). Labelled Ca (^{45}Ca) was precipitated as oxalate, which was spread on planchets as an infinite layer and the β-radiation was counted with an end-window Geiger-Mueller tube.

III. RESULTS

A. The Donnan Distribution in the Free Space of Plant Roots and its Significance for the Uptake of Potassium

Due to the indiffusible anions, it is supposed that the distribution of the cations in the free space follows the Donnan distribution (Briggs and Robertson, 1957; Vervelde, 1953; Mengel, 1961). If this is true, the preferred adsorption of divalent cations should be higher the more diluted the nutrient solution. This means that under these special conditions, the calcium/potassium ratio of the exchangeable fraction should decrease with an increasing concentration of the nutrient solution. This assumption was supported by experimental data. Excised roots of barley and oats and also intact roots of oats showed decreasing calcium/potassium ratios of the exchangeable fraction with an increasing concentration of the nutrient solution (Fig. 1).

In further experiments in which Ca^{2+} and K^+ were labelled with ^{45}Ca and ^{42}K, the amounts of nonexchangeable K and Ca absorbed during the uptake period were determined. The results of these experiments show that the exchangeable fraction follows the Donnan distribution, i.e., decreasing calcium/potassium ratios with increasing concentration, while the nonexchangeable fraction shows an inverse tendency.

As can be seen from Table 1, 10-day-old excised roots of barley absorbed high qualities of exchangeable and low amounts of nonexchangeable Ca^{2+}.

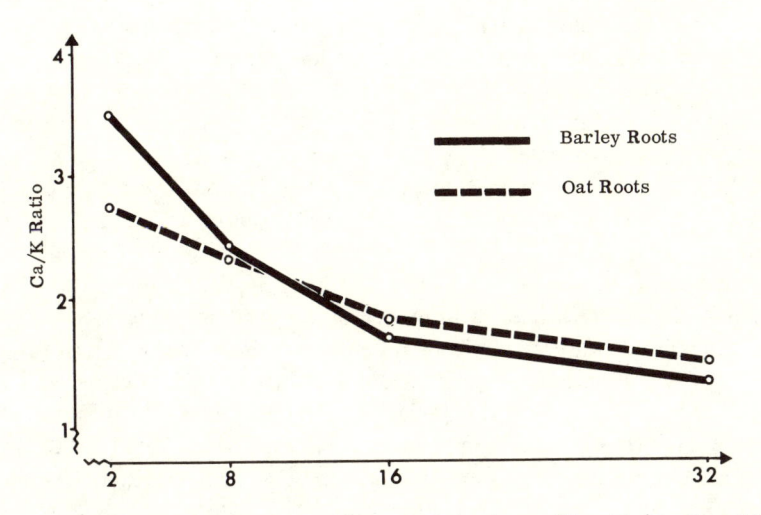

Ca and K Concentration of the Nutrient Solution in meq/liter (Ca/K ratio = 1:1)

Fig. 1—The influence of an increasing concentration of the nutrient solution on the Ca/K ratios of the exchangeable root fraction (excised barley and oat roots, 16 days old, absorption period 3 min)

Table 1—The influence of the concentration of the nutrient solution on the K and Ca uptake and the Ca/K ratios (excised barley roots, 10 days old, absorption period 3 min)

Nutrient solution		Exchangeable			Nonexchangeable		
Ca	K	Ca	K	Ca/K	Ca	K	Ca/K
— meq/liter —		— Equiv.$\times 10^{-9}$/b fresh root matter —					
1.0	0.5	880	8.1	109	42.8	773	0.055
3.0	1.5	810	34.1	23.8	55.0	670	0.082
9.0	4.5	926	49.3	18.8	43.4	760	0.057

Regarding K^+, the situation was inverse. The amount of exchangeable K^+ was low and that of nonexchangeable K^+ high. The calcium/potassium ratios of the nonexchangeable fraction did not correspond to the Donnan distribution. This was also the case in the experiment with 35-day-old intact oat plants (Table 2). Likewise, the calcium/potassium ratios of the shoots did not follow the Donnan distribution (decreasing calcium/potassium ratios with increasing concentration) of the exchangeable fraction. The absolute calcium/potassium ratios of the two experiments differ rather widely, which is due to the different plant material used and especially to the various experimental techniques. In the experiment with excised roots the samples had been treated with 0.01N HCl to remove the original Ca^{2+} from the free space (see Materials and Methods); whereas the intact oat seedlings had not been treated with HCl.

From the data of these two experiments it was concluded that the exchangeable fraction of the roots has no marked influence on the potassium migration into the "inner space". In order to prove this, another experimental series was carried out in which the potassium concentration of the nutrient solution remained constant, whereas the different treatments received additional cations in form of chlorides (Table 3). Potassium ions were present as $10^{-3}M$ KNO_3, labelled with ^{42}K. The added cations also had a concentration of $10^{-3}M$. The control solution contained only the labelled KNO_3. After an absorption period of 2 hours the exchangeable and nonexchangeable potassium was determined.

The results of this experiment (Table 3) clearly demonstrate that the addition of NaCl and KCl had no influence on the exchangeable labelled potassium, whereas it was affected significantly by the addition of $MgCl_2$ and $CaCl_2$. On the other hand, the nonexchangeable K^+ was only significantly decreased by the addition of unlabelled KCl. From these data it is concluded that the transport of K^+ into the inner space is not influenced by Ca^{2+} and Mg^{2+} adsorbed in the free space. The lower uptake of nonexchangeable potassium effected by the addition of unlabelled KCl can be explained by a competition at the sites of the active potassium uptake mechanism, e.g., a carrier competition (Epstein and Leggett 1954).

The above described experiment is only an example of a series of similar

Table 2—The influence of the solution concentration on the Ca/K ratios of the exchangeable and nonexchangeable fractions of the roots and the Ca/K ratios of the shoots (intact oat plants, 35 days old, absorption period 24 hours)

Nutrient solution		Ca/K ratio		
Ca	K	Roots		Shoots
meq/liter		exch.	nonexch.	
1.2	0.6	4.5	0.82	1.66
1.6	0.8	3.8	0.80	1.80
3.2	1.6	3.4	0.78	1.77
6.4	3.2	3.2	1.68	1.88
12.8	6.4	2.4	2.25	1.84

Table 3—The influence of different cations on the exchangeable and nonexchangeable K (μM K/g fresh matter) (excised barley roots, 12 days old, absorption period 2 hours)

Added cation	K exchangeable	rel.	K nonexchangeable	rel.
Control	0.27	100	1.77	100
Na	0.27	100	1.60	90
K	0.27	100	0.91*	51
Mg	0.07**	26	1.83	103
Ca	0.06**	22	1.76	100

** Significant difference at 1% level.   * Significant difference at 5% level.

ones which all showed that the addition of Ca^{2+} to the nutrient solution decreased the exchangeable potassium but did not decrease the potassium transport into the inner space (nonexchangeable potassium fraction) (Mengel, 1963). In some cases Ca^{2+} had a positive effect on the potassium uptake of the inner space. In order to investigate this Ca effect in detail, 4-day-old barley plants were grown without Ca^{2+} in the nutrient solution. Before the uptake of labelled K^+ began, the excised root samples were shaken for 3 min. in a $CaCl_2$ solution with different concentrations according to the experimental design. The control samples were shaken with distilled water. After this procedure, part of the samples were analyzed for exchangeable and nonexchangeable Ca^{2+}. The remaining samples were used for the absorption experiment (2 hours) with labelled potassium ($10^{-3}M$ KNO_3).

The results of this experiment show that the short-time application of $CaCl_2$ hardly influenced the nonexchangeable calcium fraction, while the exchangeable Ca^{2+} increased with the increasing calcium concentration of the applied solution. This increase of exchangeable Ca^{2+} corresponds to the absorption of nonexchangeable labelled K^+ (Fig. 2). Additional experiments showed that the release of K^+ by the plant roots was influenced considerably by the exchangeable Ca^{2+}, as with low exchangeable Ca^{2+} the release rate of potassium was rather high (Mengel and Helal, 1967). Therefore it was supposed that not the potassium influx but the potassium efflux might be influenced by the exchangeable Ca^{2+}.

In the experiment focused on this problem, the potassium influx was mea-

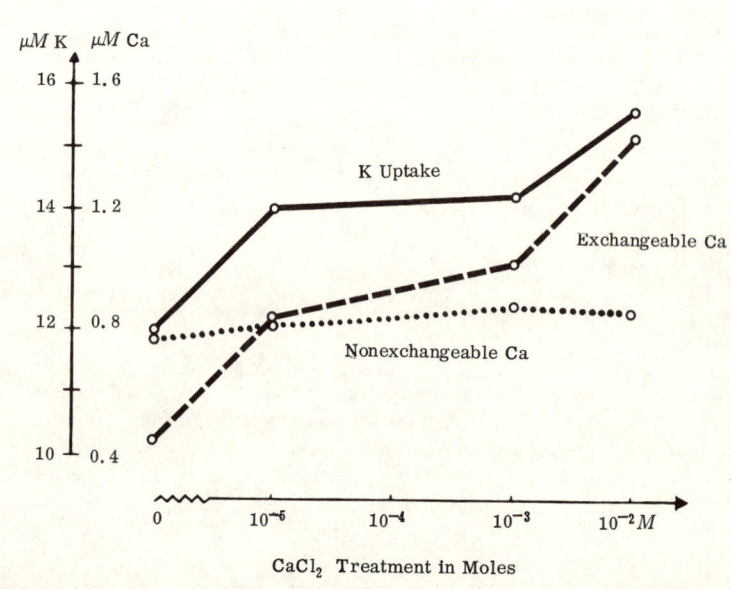

Fig. 2—Effect of a short-time pretreatment with Ca on K uptake, exchangeable, and nonexchangeable Ca (excised barley roots, 4 days old, absorption period 2 hours)

sured by labelled K^+ and the potassium efflux by unlabelled potassium. Excised barley roots of 4-day-old seedlings, grown without Ca^{2+}, absorbed K for 1 hour from an unlabelled $10^{-3}M$ KNO_3 solution. After removal of the adhering KNO_3 by rinsing, one-half of the root samples were shaken for 3 min with $10^{-3}M$ $CaCl_2$ (roots high in exchangeable Ca) and the second half was shaken in distilled water for 3 min (roots low in exchangeable Ca). Then the total potassium for one part of the samples was determined. The other root samples were used for the absorption experiment with labelled KNO_3 ($10^{-3}M$ KNO_3, 1-hour absorption period). Afterwards, these samples were analyzed for labelled potassium and total potassium. By using this experimental design, it was possible to determine the uptake of labelled potassium and simultaneously the release of potassium of roots high and low in exchangeable Ca. As it can be seen from Table 4 the uptake of labelled K (influx) was hardly influenced by the exchangeable Ca^{2+} of the roots while the K^+ release was affected considerably.

It can, therefore, be supposed that the exchangeable calcium does not affect the potassium influx, but that the potassium retention is significantly increased by the exchangeable calcium.

Roots low in exchangeable calcium had a potassium efflux which was approximately half as high as the potassium influx, while roots high in exchangeable Ca^{2+} released only about 5% of the amount absorbed. These calcium effects on the retention or release of potassium are especially marked with root material low in Ca^{2+}.

Table 4—The influence of exchangeable Ca of young barley roots on the K influx and on the K release (μM K/g fresh matter) (excised barley roots, 4 days old, absorption period 1 hour)

Ca of the roots	Unlabelled K experiment Start	End	K release	Uptake of labelled K
Low	16.6	14.7	1.9	3.52
High	16.1	15.9	0.2***	3.74

*** Significant difference at 0.1% level.

IV. DISCUSSION

The experiments with increasing concentrations and constant calcium/potassium ratios in the nutrient solution clearly show a Donnan effect (concentration effect) on the exchangeable calcium/potassium fraction of plant roots. As the K^+ and Ca^{2+} of the inner space and of the shoots behave differently the areas for which the Donnan distribution is valid should not be relevant for the K transport into the inner space. These data are supported by experiments in which the influence of different cations on the exchangeable and nonexchangeable K fraction was studied. The cation adsorption in the free space seems to be rather unspecific, and monovalent cations are replaced easily by bivalent cations (Marschner, 1963). Without being influenced by the calcium adsorption at the sites of exchange in the free space, K^+ will diffuse from the soil solution through the pores of the cell wall to the outer plasma membranes. It is assumed that the active uptake mechanism of K^+ is located at or in these membranes (Luttge and Laties, 1966). This active uptake is not directly influenced by Ca^{2+}.

On the other hand, Ca^{2+} may also be adsorbed at sites of the plasmalemma. This adsorption may influence its permeability since with a rather low Ca adsorption the permeability of the membranes is increased which results in a lower retention of diffusible ions and molecules of the inner space. The different investigations (Pitman, 1965; Rains et al., 1964; Hooymans, 1964; Elzam and Epstein, 1965; Jacobson et al., 1960; Viets, 1944) showing a positive effect of Ca^{2+} and other divalent cations on the net uptake of K^+ and phosphate (Viets effect) are probably based on a better retention caused by a lower permeability of the membranes (Marschner, 1964; Kavanau, 1965; Marschner and Mengel, 1966). This effect can only be demonstrated when the exchangeable fraction of calcium (or other divalent cations) of the plant roots is very low. Under natural conditions, this will occur only in extreme cases. Usually the Ca concentration of the soil solution is about 5-20 times higher than the K concentration (Fried and Shapiro, 1961). Furthermore, the absolute calcium concentration of the soil solution is usually high enough for a sufficient saturation of the sites of exchange of the free space.

The potassium migration into the inner space which may be regarded as an active uptake, is very specific and not influenced by the calcium adsorption in the free space. Therefore, the exchange processes in the free space cannot be considered as the first step of active potassium uptake. This conclusion is in agreement with findings of other authors (Lagerwerff and Peech, 1961; Winter, 1961; Marschner, 1961).

V. CONCLUSIONS

1) Investigations with excised roots of barley and oats showed that the exchangeable fraction of Ca^{2+} and K^+ follows the Donnan distribution. The Ca/K ratios of the exchangeable fraction decreased with increasing Ca^{2+} and K^+ concentrations of the nutrient solution. The same pattern of Ca and K distribution in the free space was found in roots of intact young oat plants.

2) The Donnan distribution of the exchangeable ions did not influence the K and the Ca of the nonexchangeable fraction of the roots. The transport of K and Ca into the shoots was not affected by the exchangeable fraction.

3) Additions of Ca^{2+} and Mg^{2+} lowered the exchangeable K of the roots significantly while the nonexchangeable K fraction was not influenced.

4) Roots low in exchangeable Ca showed an increased release of K (efflux), but the K migration (influx) into the inner space was not influenced by the rate of Ca adsorption in the free space.

5) It is concluded that the exchangeable Ca fraction of the roots does not influence the K uptake of the inner space and that the exchange adsorption in the free space cannot be regarded as the first step of active K uptake. If the exchangeable Ca fraction is very low, the permeability of the plasma membranes can increase, resulting in a lower K retention (increased efflux). This explains the phenomenon of the Viets effect.

REFERENCES

Briggs, G. E., and R. N. Robertson. 1957. Apparent free space. Ann. Rev. Plant Physiol. 8:11-29.

Elzam, O. E., and E. Epstein. 1965. Absorption of chloride by barley roots: Kinetics and selectivity. Plant Physiol. 40:620-624.

Epstein, E., and J. E. Leggett. 1954. The absorption of alkaline earth cations by barley roots: Kinetics and mechanism. Amer. J. Bot. 41:785-791.

Fried, M., and R. E. Shapiro. 1961. Soil-plant relationships in ion uptake. Ann. Rev. Plant Physiol. 12:91-112.

Hooymans, J. J. M. 1964. The role of calcium in the absorption of anions and cations by excised barley roots. Acta Bot. Neerl. 13:507-540.

Jacobson, L., D. P. Moore, and R. J. Hannapel. 1960. Role of calcium in absorption of monovalent cations. Plant Physiol. 35:352-358.

Kavanau, J. L. 1965. Structure and function in biological membranes. Holden-Day, Inc., San Francisco, London, Amsterdam. Vol. 1. 132 p.

Lagerwerff, J. V., and M. Peech. 1961. Relation between exchange absorption and accumulation of calcium and rubidium by excised barley roots. Soil Sci. 91:84-93.

Levitt, J. 1957. The significance of "Apparent Free Space" (A.F.S.) in ion absorption. Physiol. Plantarum 10:882-888.

Lüttge, U., and G. G. Laties. 1966. Dual mechanisms of ion absorption in relation to long distance transport in plants. Plant Physiol. 41:1531-1539.

Marschner, H. 1961. Untersuchungen über die Aufnahme von Cäsium durch isolierte Gerstenwurzeln. Z. Pflanzenernähr., Düng., Bodenk. 95:30-51.

Marschner, H. 1963. Untersuchungen über die Aufnahme von Cäsium durch isolierte Wurzeln und intakte Pflanzen. Agrochimica 7:75-88.

Marschner, H. 1964. Einfluss von Calcium auf die Natriumaufnahme und die Kaliumabgabe isolierter Gerstenwurzeln. Z. Pflanzenernähr., Düng., Bodenk. 107:19-32.

Marschner, H., and K. Mengel. 1962. Apparent free space. Z. Pflanzenernähr., Düng., Bodenk. 98:30-44.

Marschner, H. and K. Mengel. 1966. Der Einfluss von Ca- und H-Ionen bei unterschiedlichen Stoffwechselbedingungen auf die Membranpermeabilität junger Gerstenwurzeln. Z. Pflanzenernähr., Düng., Bodenk. 112:39-49.

Mengel, K. 1961. Die Donnan-Verteilung der Kationen im Freien Raum der Pflanzenwurzel und ihre Bedeutung für die aktive Kationenaufnahme. Z. Pflanzenernähr., Düng., Bodenk. 95:240-259.

Mengel, K. 1963. Die Bedeutung von Kationenkonkurrenzen im free space der Pflanzenwurzel für die aktive Kationenaufnahme. Agrochimica 7:236-257.

Mengel, K., and M. Helal. 1967. Der Einfluss des austauschbaren Ca junger Gerstenwurzeln auf den Flux von K und Phosphat - eine Interpretation des Viets-Effektes. Z. Pflanzenphysiol. 57:223-234.

Pitman, M. G. 1965. Sodium and potassium uptake by seedlings of Hordeum vulgare. Aust. J. Biol. Sci. 18:10-24.

Rains, D. W., W. E. Schmid, and E. Epstein. 1964. Absorption of cations by roots. Effects of hydrogen ions and essential role of calcium. Plant Physiol. 39:274-278.

Vakhmistrov, D. B. 1965. On the magnitude of the "Apparent Free Space" of plant roots. Sov. Plant Physiol. 12:705-711.

Vervelde, G. J. 1953. The Donnan-principle in the ionic relations of plant roots. Plants and Soil 4:309-322.

Viets, F. G. 1944. Calcium and other polyvalent cations as accelerators of ion accumulation by excised barley roots. Plant Physiol. 19:446-480.

Winter, H. 1961. The uptake of cations by Vallisneria leaves. Acta Bot. Neerl. 10:341-393.

16

Interaction of Potassium and other Ions

ROBERT D. MUNSON

American Potash Institute, Inc.
St. Paul, Minnesota

I. INTRODUCTION

Under true "nutrient balance" conditions exist for optimum plant growth, differentiation, and maturation. With proper climate and nutrient balance, maximum genetic potential can be achieved. However, optimum nutritional levels for one plant specie in no way implies the best levels for all crop species or even varieties of a given specie. Furthermore, because of fluctuations in atmospheric and soil conditions and fertilization, the environment of the crop is dynamic, both within and among days and seasons. Yet, working with extremely divergent soils it is our obligation to use soil and plant analyses to understand and adjust soils toward the "optimum" in the context of a balanced ionic soil environment (Heimann, 1958), which will produce high-yielding, quality crops. It is over this phase of production that man can exert control.

Hope (1960) indicated that the following external and internal factors influence the amounts of an ion or ions taken up by plants and their interaction:

External (soil or culture) Factors
1) Availability (plant-soil relation)
2) Nature and concentration of the ion
3) Temperature
4) Oxygenation
5) Water status
6) Presence of other ions
7) pH, a special case of 6

Internal (plant) Factors
1) The nature of the root surface, its area and number of secondary roots and hairs
2) The free space available to the ion
3) The number and concentration of ion-exchange centers
4) The rate of transpiration
5) Age and rate of growth
6) The rate of respiration
7) Internal ion status

Note that in the soil, several of the factors are interrelated. For example, water status influences oxygenation, temperature, and the biological oxidation and release of elements in organic matter. In addition to those listed, one might add light quality, intensity, and duration, as well as other factors.

In plants individual ions, such as potassium, attain identity only through their interaction with other ions or molecules as they maintain electrochemical neutrality, respond to chemical or thermal gradients, and influence or are influenced by metabolism. As interactions occur, changes are initiated at the subcellular level which may ultimately be manifested through changes in rates of respiration, photosynthesis, cell division and expansion, utilization, and translocation of carbohydrates and organic acids. The net influence of these interactions and processes produces the final yield of a crop.

When potassium ions or other cations are added to the soil or cultural system, they are accompanied by an anion and added as an inorganic salt. Comprehensive coverage of the effects of inorganic salts on plant growth has been presented by Steward and Sutcliffe (1959). Hendricks (1964 and 1966) has directed attention to salt uptake and transport. Hewitt (1963) has discussed the requirements and interactions of essential elements and Jackson (1967) has reviewed the physiological aspects of plant nutrition in connection with liming and soil acidity. Nason and McElroy (1963) have related the modes of action of ions at the enzyme level. The latter briefly discuss the enzymatic action of potassium, but cover the influence of magnesium and calcium well. Evans and Sorger (1966) have provided the most comprehensive review of enzymes influenced by potassium and other univalent cations written to date.

It is the primary purpose of this paper to discuss the interactions of potassium, calcium, magnesium, and sodium as related to concentrations and influence on growth, yield, and other characteristics, and point out some of the factors influencing the interrelations among them. In discussing these interactions, because of the scope of the subject and literature on various crops, attention will primarily be directed to understanding relations found in and influencing corn. Other crops will be discussed where appropriate to compare specific interactions or relationships.

II. CHARACTERISTICS AND EXPRESSION OF POTASSIUM AND OTHER CATIONS

A. Physical and Chemical Characteristics

As one studies plant nutrition it becomes apparent that the action of elements, within limits, are less specific than one might think. For example, substitution of one ion for another has often been discussed. Different cations may serve as co-factors for a specific enzyme (Dixon and Webb,

1964; Nason and McElroy, 1963; and Evans and Sorger, 1966), although in many cases enzymes are activated by only one ion. In some cases, a combination of two or more ions are required for maximum activity. However, if an ion functions in nutrition, it must do so because of physical and chemical characteristics that make it unique in action or interaction with other ions or molecules.

In order to compare several of these characteristics for many of the cations present or taken up by plants, the information in Table 1 was assembled.

A whole discussion could revolve about the characteristics of the cation shown. The importance of ionic radii, mobilities, and potentials all play an important role in the action of potassium with other ions. Many similarities are shown among the essential elements and those that may partially substitute for one another. The repeating sequence of characteristics as one moves through a valence group is of interest. These relationships are left to the reader to relate to the discussion.

B. Expression of Relations Among Ions

Because ions are taken up by plants in chemical equivalents, it is important that they be expressed in that form. Use of equivalents allows comparisons other than empirical. Lucas and Scarseth (1947) criticized the use of percentages for evaluating the concentration or composition of ions in plants. They pointed out that percentages tend to minimize variations in magnesium, while accentuating variations in potassium. This is due to the respective equivalent weights of the two elements. They emphasized that over three times as much potassium as magnesium, and nearly twice as much potassium as calcium, are required on a weight basis to obtain equal chemical equivalency. Percent of an element based on dry matter content in plant material can be readily converted in meq/100 g by multiplying the percent by the reciprocal of the meq weight of the element. Conversion factors for several elements are given:

(Percent of element in dry matter x conversion factor = meq/100g)

	Factor		Factor
K	25.57	Zn	30.59
Na	43.50	Cu	31.48
Ca	49.90	Fe	35.81
Mg	82.24	Mn	36.42
Al	111.23		

In this discussion, meq/100g of dry matter will be the primary means of expressing data, but other means of expression will be included. In most cases data from original papers were recalculated to compare and evaluate interrelations.

Table 1—Physical and chemical characteristics of elements or cations that interact to influence plant growth. (Lang, 1967; Daniels and Alberty, 1955; and Bould, 1963)

Element or cation	Atomic number	Atomic weight	Radii Covalent (Å)	Radii Atomic (Å)	Radii Ionic (Å)	Radii Ionic hydrated (Å)	Ionic potential	Ionic mobilities cm²volt sec⁻¹	Ionization potential volts	Activity coefficients	Oxidation potentials volts	Terrestrial abundance g/ton
H	1	1.00797	--	0.37	2.08	--	--	363	13.595	0.914	0.000	--
Li	3	6.939	1.34	1.50	0.60	2.30	1.3	40.1	5.390	0.907	3.045	65
Na	11	22.9898	1.54	1.86	0.95	1.79	1.0	52.0	5.138	0.902	2.714	28,300
K	19	39.102	1.96	2.27	1.33	1.22	0.71	76.2	4.339	0.899	2.925	25,900
Rb	37	85.47	2.11	2.43	1.48	1.17	0.67	77.6	4.176	0.898	2.92	310
Cs	55	132.905	2.25	2.62	1.69	1.17	0.61	78.1	3.893	0.898	2.92	7
NH₄⁺	--	--	--	--	1.42	--	--	--	--	--	--	--
H₃O⁺	--	--	--	--	1.27	--	--	--	--	--	--	--
Mg	12	24.312	1.30	1.595	0.65	--	2.6	--	7.644	0.690	2.37	20,900
Ca	20	40.08	1.74	1.97	0.99	--	1.9	61.6	6.111	0.675	2.87	36,300
Sr	38	87.63	1.92	2.14	1.13	--	1.6	--	5.70	--	2.89	300
Ba	56	137.34	1.98	2.17	1.35	--	1.4	66.0	5.210	0.67	2.90	250
Mn	25	54.938	1.17Mn²⁺ Mn⁴⁺ Mn⁷⁺	1.24 1.26 1.29	0.80 0.50 0.46	--	--	--	7.432	--	1.18	1,000
Fe	26	55.847	1.17Fe²⁺ Fe³⁺	1.238 1.26	0.75 0.67	--	--	--	7.896	0.675 0.445	0.440	50,000
Co	27	58.9332	1.16	1.75	0.82	--	--	--	7.86	--	0.277	23
Cu	29	63.54	1.35	1.275Cu⁺ Cu²⁺	0.96 0.70	--	--	--	7.723	0.675	-0.337	70
Zn	30	65.37	1.31	1.38	0.74	--	--	--	9.391	0.675	0.763	132
Al	13	26.9815	1.18	1.43	0.50	--	5.3	--	5.984	0.445	1.66	81,300
Si	14	28.09	1.11	1.32	0.41	--	10.0	--	8.149	0.67	0.86	277,200

III. FACTORS WHICH INFLUENCE THE INTERRELATIONS AMONG IONS

A. Crop Grown

1. Cation Relations Among Crops

Newton (1928) evaluated differences of ion uptake among crops grown in the same culture. He grew sunflowers, beans, wheat, barley, peas, and corn and found wide variations in concentrations of the individual ions, as well as the interrelations among and between them in the different crops (Table 2). Largest differences in concentrations among the crops were noted for Ca and K. Calcium and K showed a range of 84 meq and 78 meq, respectively. All crops except beans took up more K than Ca. The sum of the three cations studied was not constant among crops, but showed a range of 157 meq. The cation concentration of barley was double that of corn. A range of ratios was exhibited by the crops. Wheat had the highest K/Mg ratio, followed by barley, peas, and corn. The high K concentrations of corn and wheat relative to Ca and Mg produced (Ca+Mg)/K ratios of of about 0.6 or less. Nitrogen/potassium (percent) ratios among the crops were less than one and showed the least variation of the ratios evaluated.

Collander (1941) also found crops to be highly selective in cation uptake. He conducted solution culture studies in which 4 meq/liter of K,

Table 2—Uptake and interrelations between and among cations for crops grown together in the same medium. (Newton, 1928)

Crop	Cations K	Ca	Mg	sum	K/Ca	K/Mg	Ca+Mg / K	N/K(%)
	meq/100g							
Sunflowers	128	109	53	290	1.2	2.4	1.26	0.72
Beans	103	105	49	257	1.0	2.1	1.50	0.90
Wheat	172	39	33	244	4.4	5.2	0.42	0.67
Barley	177	93	44	314	1.8	4.0	0.78	0.67
Peas	134	77	41	252	1.7	3.2	0.88	0.86
Corn	99	25	33	157	3.9	3.0	0.59	0.75

Table 3—Cation content and interrelations in crops when grown in cultures with equal concentrations of each cation. (Collander, 1941)

Plant	Cations K	Na	Ca	Mg	sum	K/Ca	K/Mg	Ca+Mg / K
	meq/100g							
Oats	149	8	16	29	202	9.3	5.1	0.36
Buckwheat	126	3	107	87	323	1.2	1.4	1.56
Spinach	339	29	85	202	655	4.0	1.7	0.93
Corn	169	7	27	39	242	6.2	4.3	0.43

Na, Ca, and Mg were supplied to 13 different crops. The ranges in concentration among the crops for the ions were: K, 109-339; Na, 3-125; Ca, 16-116; and Mg, 26-202 meq/100g. Because of these variations, interrelationships were equally striking. The K/Ca and K/Mg ratios found were particularly high and the (Ca+Mg)/K values accordingly were low. Data for several crops are shown in Table 3. Thus, it is apparent that cations are absorbed preferentially by different crops.

2. Potassium Fertilization of Different Crops

Bower and Pierre (1944) applied potassium to several different crops grown on a "high-lime" calcareous soil. Stanford et al. (1941) found that these soils had a pH of 8 and were high in carbonates and nitrate-nitrogen. Soil solution extracts contained 348 ppm Ca, 83 ppm Mg, 2 ppm K, and 13 ppm Na. Cation contents of the crops were determined. Oats and flax were found to be sodium accumulators. The results for corn, soybeans, and sorghum are shown in Table 4.

As potassium was added magnesium was decreased, but yields were increased. For example, with corn Mg was decreased 52 meq and K was increased 58 meq while yields were more than doubled. The cation sum in the case of corn did not change materially. Potassium/calcium and K/Mg ratios increased, while (Ca+Mg)/K decreased. Similar relations were apparent in sorghum and soybeans, except soybeans, a legume, tended to contain higher contents of the divalent cations. This was reflected in the ratios. Calcium was materially reduced by potassium only in sorghum, but the sum of the cations was reduced in both sorghums and soybeans.

Table 4—Effect of K on yield and content and interrelations of cations in sorghum and soybeans grown on a calcareous "high-lime" soil.* (Bower and Pierre, 1944)

Potassium treatment	Dry matter yield	K	Na	Ca	Mg	sum	K/Ca	K/Mg	Ca+Mg / K
kg/ha	g/pot			meq/100g					
				Corn					
0	4.5	11	<1	58	105	174	0.19	0.10	14.9
420	9.6	69	<1	48	53	170	1.48	1.30	1.4
				Sorghum					
0	6.5	14	<1	64	72	150	0.21	0.19	10.0
420	14.7	49	<1	35	34	118	1.39	1.46	1.4
				Soybeans					
0	7.9	15	<1	92	141	248	0.16	0.11	15.1
420	11.7	45	<1	88	64	197	0.53	0.70	3.4

* $CaCO_3$- equivalent 30.9%, exch. Na, 0.3 meq, and exch. K 0.17 meq/100g.

3. Sodium and Potassium Fertilization of Different Crops

Larson and Pierre (1953) applied sodium and potassium to table beets, flax, oats, and corn grown on Carrington loam and Harpster silt loam soils. They found wide variations in the ability of these crops to take up cations from soils with and without added sodium and potassium. Table beets and oats readily took up sodium, while corn essentially excluded it, which agrees with results of Collander (1941) and Wehunt et al. (1957). Data for table beets and corn are compared in Table 5. Sodium increased beet yields 3.8 g without K and increased the Na concentration 86 meq, while decreasing Ca and K by 26 and 20 meq, respectively. With K, Ca and K concentrations remained about the same, but the Na content increased 81 meq and Mg was decreased. The K treatment decreased Ca and Mg by 26 and 20 meq, respectively, while increasing the K concentration 44 meq indicating the equivalent behavior of ions. For crops taking up sodium, it may be helpful to include it along with calcium and magnesium in evaluating ratios. Beet yields increased as $(Ca+Mg+Na)/K$ ratios increased with added sodium.

Sodium treatments on corn produced essentially no change in the calcium, magnesium, or potassium concentrations. Potassium reduced the Ca and Mg by 22 meq. Even though K increased yield by about 50%, the sum of the three ion species changed only 4 meq and higher yields were associated with $(Ca+Mg)/K$ ratios of less than 3.

Cooper et al. (1953) conducted sodium and potassium fertilizer studies on cotton. Sodium nitrate and KCl had been applied over 18 years. Cation relationships of the whole plant were determined. Some of the results are shown in Table 6. Potassium and sodium both increased the yield of seed cotton. Indications are that higher rates of both may have increased yields. Sodium with potassium decreased the calcium concentration and increased yields. At the highest yield level Na decreased Ca 42 meq, while the Na content did not change. Yields increased as the $(Ca+Mg)/K$ ratios decreased. Based on the low Na concentrations, it would appear that yields were being influenced more by the nitrogen applied with the sodium, than sodium *per se*. Analyses the final year also indicate there is little doubt that magnesium was a limiting factor. Page and Bingham (1965) found that cotton responded best with petiole K/Mg ratios of 2.0 to 3.2.

Lancaster et al. (1953) found that sodium increased cotton yields on potassium-deficient soils, but that there was little, if any, response to sodium unless potassium was applied. Improvements in seed cotton quality were obtained with added potassium. Joham (1955) found sodium reduced wilting in cotton deprived of calcium. The apparent substitution of sodium for calcium was thought to be due to the mobilization of calcium from older tissue. Roots contained higher concentrations of sodium than stem or leaf

Table 5—Effect of added Na and K on yields and contents and interrelations of cations of table beets and corn grown on a calcareous soil.* (Larson and Pierre, 1953)

Treatment		Dry matter yield	Cations					K/Na	K/Ca	K/Mg	Ca+Mg+Na	Ca+Mg
			K	Na	Ca	Mg	Sum				K	K
K	Na											
kg/ha		g/pot	meq/100g									

Table beets

0	0	13.07	64	30	157	207	458	2.1	0.4	0.3	6.2	5.7
0	98	16.88	44	116	131	212	503	0.4	0.3	0.2	10.4	7.8
168	0	17.29	108	31	131	187	457	3.5	0.8	0.6	3.2	2.9
168	98	18.34	103	112	136	160	511	0.9	0.8	0.6	4.0	2.9

Corn

0	0	15.7	16	—	50	78	144	—	0.3	0.2	—	8.0
0	98	15.7	21	—	52	75	148	—	0.4	0.3	—	6.0
168	0	23.4	38	—	40	62	140	—	1.0	0.6	—	2.7
168	98	22.7	38	—	40	62	140	—	1.0	0.6	—	2.7

* Soil pH 8.0, C.E.C. 30.3 meq/100g, CaCO$_3$-equivalent 21%, exch. Na 0.07 meq, and exch. K 0.19 meq/100g.

Table 6—Effect of K and Na on the yield of seed cotton and cations and their relations analyzed the final year of the study. (Cooper, et al. 1953)

Treatment		Seed cotton yield	Cations							Ca+Mg
K	Na*	18-yr avg	K	Na	Ca	Mg	sum	K/Ca	K/Mg	K
kg/ha		kg/ha			meq/100g					
0	0	292	20	2	125	12	159	0.16	1.6	7.1
	83	499	19	12	113	12	156	0.17	1.6	7.3
14	0	747	24	1	114	11	150	0.21	2.2	5.3
	83	922	28	6	95	9	138	0.29	3.2	4.0
42	0	1,076	41	1	108	8	158	0.38	5.0	2.9
	83	1,221	46	5	98	8	157	0.47	5.8	2.4
56	0	1,149	41	1	138	8	188	0.30	4.9	3.6
	83	1,310	50	2	96	8	156	0.52	6.5	2.1

* Sodium added as sodium nitrate.

tissue. With the high Na treatment the roots contained about 72 meq of Na and the stems and leaves 17 meq. With adequate potassium, sodium was more uniformly distributed in the tissues. Calcium concentrations were greatest in the old leaves. Joham included Na with K in studying ratios and found a Ca/(K + Na) ratio of 0.7 was associated with the best yields. Highest yields were associated with a ratio about 1 in the studies of Cooper et al. (1953).

Sodium and lithium apparently have adverse effects on some crops. Bingham et al. (1964) have discussed lithium toxicity in a number of crops. It may be that the presence of sodium and lithium increases the requirement for other cations.

4. Cation Relations Within a Crop

Many examples could be cited even at the varietal level which indicate variation in cation content. Corn exhibits wide variations among lines. Research on hybrid corn at the Tennessee Experiment Station is indicative of results. Freeman [1] used a number of inbred lines and crosses of corn and applied three annual applications of potassium. Cation relations of the ear leaf at silking and yields for several hybrids for the third year of the study are shown in Table 7. Among the hybrids there was a definite repetitive pattern of decreased leaf calcium and magnesium contents with applied potassium, with magnesium being reduced the most. Added potassium decreased the sum of the cations in the leaf as yields were increased. Among the various hybrids, Mg content of the controls ranged from 58 to 107 meq without K and from 15 to 20 meq with the highest level of K.

---

[1] Freeman, Charles E. 1965. The effect of potassium fertility levels on the uptake and utilization of potassium, calcium and magnesium by corn inbreds and hybrids. M.S. Thesis. University of Tennessee, Knoxville.

Table 7—Influence of three annual rates of K on third year grain yields and leaf cation contents and interrelations of several corn hybrids (Freeman[1])

Potassium treatment avg annual* kg/ha	Corn grain yield kg/ha	K	Ca	Mg	Sum	K/Ca	K/Mg	Ca+Mg / K
			meq/100g					
		T101 × T111						
0	4,267	19	34	58	111	0.6	0.3	4.8
40	7,907	31	28	37	96	1.1	0.8	2.1
121	8,911	51	23	15	89	2.2	3.4	0.8
		T220 × T111						
0	4,016	20	49	76	145	0.4	0.3	6.3
40	7,279	31	40	41	112	0.8	0.8	2.6
121	8,786	50	28	18	96	1.8	2.8	0.9
		Dixie 22						
0	4,895	19	44	72	135	0.4	0.3	6.1
40	7,907	39	36	37	112	1.1	1.0	1.9
121	8,158	50	37	22	109	1.4	2.3	1.2
		T105 × T115						
0	1,506	15	49	72	136	0.3	0.2	8.3
40	5,522	32	43	39	114	0.8	0.8	2.5
121	6,840	46	47	26	119	1.0	1.9	1.6

* Annual applications of N and P were 168 and 39 kg/ha, respectively. Two tons of limestone was applied once and 22 kg of Zn sulfate was applied twice during the three years.

Wide shifts occurred in the cation ratios. It would appear that acceptable ratios in this experiment were 1.4 to 2.0 for K/Ca, 2.5 to 3.4 for K/Mg and less than 1 for (Ca + Mg)/K.

Moss and Peaslee (1965) determined the critical levels of K and Mg associated with optimum stomatal openings for photosynthesis to be 1.5% K and 0.15% Mg. These critical levels were determined in solution without interferring ions or moisture stress. On a meq basis the critical K/Mg ratio would be 3.11.

Grosline et al. (1961) have studied potassium, calcium and magnesium inheritance of corn using the ear leaf. They found a high correlation between inbreds and their diallel progeny for Ca and Mg, indicating heritability through additive gene action. Heritability of K was apparent, but was believed to be more complicated and nonadditive. Baker et al. (1966) also studied the range in potassium, calcium, and magnesium among different genotypes. The range between the low and high values for the different elements was as follows: K, 29 meq; Ca, 29 meq; and Mg, 22 meq. The similarity of values clearly indicates ion substitution or interaction.

Loúe (1963) conducted K fertility research on corn at several locations in France. He used leaf analysis and found the leaf tissue to reflect interrelations among cations (Fig. 1). As the nutrient content of the leaves changed, so did the yield (Fig. 2). For example, in one experiment as grain yields were increased from 2,353 kg to 7,061 kg/ha by adding 132 kg of K, the following changes were noted in ratios: K/Ca, 0.25 to 1.53;

PERCENTAGES OF Ca AND Mg IN CORN LEAVES AS RELATED TO LEAF K

Fig. 1—Calcium and Mg contents of corn leaves change with the percent K content. (Loúe, 1963).

CORN YIELD AS RELATED TO % LEAF K AT TASSELING

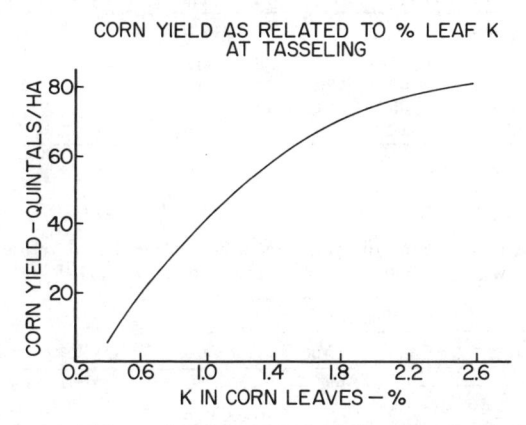

Fig. 2—Corn yield as related to corn leaf K at tassel. (Loúe, 1963).

K/Mg, 0.22 to 3.10; and (Ca+Mg)/K, 8.4 to 1.0. The values for the higher yields are not widely different than found by Freeman [2] for his highest yielding hybrid.

---

[2] Ibid.

B. Age of Crop

Several investigators have observed shifts in the nutrient content of crops with age. Sayre (1955) studied the nutrient content of corn leaves sampled through the season. Variation in cation contents and interrelations of leaves with the age of the crop are shown in Table 8. The potassium content of the leaves dropped sharply between the June 20 and July 20 samplings during which time N/K ratios increased. Magnesium was low early in the season and may have been deficient, but increased with age. The calcium content was also low. The K/Ca and K/Mg ratios decreased with age or maturity, while (Ca + Mg)/K values tended to increase. The grain yield was 7,214 kg/ha (115 bu/acre). The sum of the cations continued to decrease with maturity.

Jenne et al. (1958) obtained data on leaf cation contents of corn with and without irrigation (Table 9). On the unirrigated corn, potassium content decreased with time as did the K/Ca and K/Mg ratios, while calcium and magnesium increased. Because of these relations, (Ca + Mg)/K increased as the season progressed.

Table 8—Effect of age on leaf dry matter yield, cation content and ratios of corn - Ohio 135 (Sayre, 1955)

Sampling date	Dry weight of leaves	N/K(%)	K	Ca	Mg	Sum	K/Ca	K/Mg	Ca+Mg / K
	g/plant			meq/100g					
June 20	5	0.71	107	22	13	142	4.9	8.2	0.33
July 20	46	1.07	72	14	15	101	5.1	4.8	0.40
August 19	62	1.35	51	17	20	88	3.0	2.6	0.72
September 18	61	1.12	43	15	20	78	2.9	2.2	0.81

Table 9—Influence of sampling date on cation content and ratios of corn leaves with and without irrigation (Jenne et al., 1958).

Sampling date	K	Ca	Mg	Sum	K/Ca	K/Mg	Ca+Mg / K
	meq/100g						
			Unirrigated				
June 9	130	32	34	196	4.0	3.8	0.51
July 28	105	22	27	154	4.8	3.9	0.47
August 11	84	55	40	179	1.5	2.1	1.13
August 25	69	50	34	153	1.4	2.0	1.22
September 27	82	63	42	187	1.3	1.9	1.28
			Irrigated				
July 28	77	34	27	138	2.3	2.8	0.80
August 11	66	38	34	138	1.8	1.9	1.09
August 25	66	45	40	151	1.5	1.7	1.28
September 27	61	46	42	149	1.4	1.5	1.44

Irrigation decreased the cation contents and reduced fluctuations among sampling dates. This may have been partially due to a dilution effect because of increased dry matter production. The contents and ratios with irrigation should have been more nearly optimum for corn growth because moisture stress was reduced. With these contents and ratios, total yield of dry matter was 17,593 kg/ha and the grain yield was 9,601 kg/ha (153 bu/acre). Total accumulation for the various nutrients at harvest was as follows: N, 225 kg; P, 28 kg; K, 232 kg (maximum, 248 kg); Ca, 43 kg; and Mg 35 kg/ha.

Benne et al. (1964) analyzed different portions of the corn plant for various elements at different times during the season which reflects aging. The results of the analyses of the upper and lower leaves and stalks at the various sampling dates are shown in Table 10. The young corn contained high potassium and low magnesium at the first sampling. Calcium and magnesium concentrations of leaves increased with aging, but the calcium content of the lower stalk remained low, as the aging process progressed. The lower leaves and stalk tended to be higher in potassium and magnesium than the upper, indicating adequate levels of these nutrients. A sharp drop in potassium was noted in the upper stalk on the August 27th sampling. This occurred following pollination and at a time during which carbohydrates were rapidly being translocated to the ear.

The ratios reflect the behavior of the individual cations. In the leaves K/Ca and K/Mg ratios decrease with maturity and the upper and lower values are similar. These ratios for the stalk were higher particularly in the lower stalk because of the lower Ca contents, and the relationship between the two ratios changed with maturity. The (Ca + Mg)/K ratios for the leaves increased by a factor of 10. The ratios of the stalk tissue tended to

Table 10—Effect of sampling date on cation interrelations of lower (L) and upper (U) leaves and stalks of corn.* (Benne et al., 1964)

Sampling date	K L	K U	Ca L	Ca U	Mg L	Mg U	K/Ca L	K/Ca U	K/Mg L	K/Mg U	(Ca+Mg)/K L	(Ca+Mg)/K U
	meq/100g											
					Leaves							
June 22	--	97	--	17	--	6	---	6.0	---	17.0	---	0.23
July 24	82	57	36	23	28	20	2.3	2.5	2.9	2.9	0.78	0.75
Aug 27	59	47	43	41	35	30	1.4	1.2	1.7	1.6	1.32	1.50
Oct 6	37	32	42	37	48	34	0.9	0.9	0.8	0.9	2.42	2.20
					Stalks							
June 22	--	190	--	23	--	8	---	8.3	---	23.0	----	0.16
July 24	89	51	11	10	19	17	8.1	4.8	4.7	3.0	0.34	0.54
Aug 27	57	21	10	16	25	16	5.4	1.4	2.3	1.3	0.61	1.50
Oct 6	85	47	9	16	28	22	9.0	3.0	3.0	2.1	0.44	0.82

* Area received animal manure. Row fertilizer at planting 19.6 kg N, 34.5 kg P and 65 kg K/ha. Corn was sidedressed with 92.4 kg N.

increase, but the data for August 27 sampling for the upper portion of the plant showed an increased ratio, again perhaps reflecting changes due to transport of carbohydrates and nutrients during ear fill.

Roots were also analyzed during the season in the studies of Benne et al. (1964). Potassium content was uniformly high until the last sampling, when it decreased by a factor of two. Magnesium content increased during the season, as did the (Ca + Mg)/K ratio. The (Ca + Mg)/K ratio was 0.56 for the August 27 sampling date.

Grosline et al. (1965) prepared a comprehensive bulletin on the accumulation elements by corn. They evaluated the nutrient contents of various plant parts for several hybrids, changes in individual leaf contents from the bottom to the top of the plant and seasonal changes of the leaves at different sampling dates. Their data indicate potassium accumulation continues until 50% silk and then essentially levels off, but calcium and magnesium continue to accumulate to the later stages of maturity. The potassium accumulation pattern agrees with studies conducted by Hanway (1962) and the nutrient redistribution research of Kissel and Ragland (1967).

In other studies conducted by the same group, Grosline et al. (1963) found leaf cation contents of different lines of corn to be related to leaf blight. Potassium was negatively correlated and Zn was positively correlated to the blight. The blight was related to early maturity or senescense.

DeKock (1964) used percent-leaf-ash-weight of corn leaves from the top to the bottom of the plant, which is essentially an aging sequence. His results indicate definite decreasing K/Ca ratios, from the top to the bottom leaves, which he linked metabolically to the P/Fe ratio in the plant. DeKock relates these changes to shifts in the organic acid contents of the leaves to growth, aging, and senescence. He also points out the importance of the K/Ca ratio to the interrelationships of micronutrients such as copper, manganese, and iron.

C. Rate and Source of Nitrogen

Viets et al. (1954) applied rates and sources of N in an irrigated corn experiment and determined the percent calcium, magnesium, and potassium in the second leaf below the upper ear at silking. Nitrogen more than doubled corn yields (Table 11), but N/K(%) ratios only reached a high of 0.80. With increasing rates of nitrogen, K/Ca ratios increased with the highest yield being associated with a ratio of 3.87. The K/Mg ratios were relatively stable among nitrogen rates, but varied with the source of nitrogen; with ammonium sulfate producing a ratio of 5.9, and with calcium nitrate a ratio of 3.60. Highest corn yields were associated with the lowest (Ca + Mg)/K ratios and these varied with the source of N.

Boawn et al. (1960a) applied rates and sources of nitrogen annually

Table 11—Effect of rates and sources of N on corn yield, leaf cation and N/K ratios of Iowa 939 corn, grown under irrigation.. (Viets et al., 1954)

Nitrogen rate	Corn grain yield	N/K(%)	Cation sum	K/Ca	K/Mg	$\frac{Ca+Mg}{K}$
kg/ha	kg/ha		meq/100g			
Control	4,782	.50	88.1	2.34	3.53	0.71
			Ammonium Sulfate			
45	6,972	.42	108.4	2.87	5.95	0.52
90	9,024	.69	99.7	3.47	5.90	0.46
180	10,028	.78	98.9	3.87	5.92	0.43
			Ammonium Nitrate			
45	6,721	.53	101.7	2.51	4.47	0.62
90	8,334	.67	101.1	2.83	4.84	0.56
180	9,758	.80	96.4	3.01	4.42	0.56
			Calcium Nitrate			
45	6,376	.51	99.6	2.13	3.72	0.74
90	7,066	.62	96.2	1.98	3.86	0.76
180	8,691	.78	103.7	2.46	3.60	0.68

* Grown on a virgin Ephrata sandy loam. Soil analyses were as follows: pH 7, CEC 13 meq/100g, Ca 7.78 meq, Mg 2.80 meq, Na 0.10 meq, and K 1.14 meq/100g.

Table 12—Influence of rates and sources of N on grain sorghum yield, zinc content, N/K ratios, cation contents, and ratios of the second leaf at heading. (Boawn et al., 1960)

Nitrogen treatment	Grain sorghum yield	Zn	N/K(%)	K	Ca	Mg	Sum	K/Ca	K/Mg	$\frac{Ca+Mg}{K}$
kg/ha	kg/ha	ppm		— meq/100g —						
Control	2,330	--	1.18	29	31	22	82	0.94	1.31	1.83
				Ammonium Sulfate						
45	3,485	8.3	1.55	23	23	16	62	1.00	1.44	1.70
90	6,051	11.1	1.50	29	26	17	72	1.11	1.70	1.48
180	8,578	17.1	1.76	34	32	21	87	1.06	1.62	1.56
				Ammonium Nitrate						
45	3,849	10.6	1.46	25	23	16	64	1.09	1.56	1.56
90	4,824	11.0	1.70	26	27	17	70	0.96	1.53	1.70
180	7,351	13.6	1.78	32	30	21	83	1.07	1.52	1.59

for 6 years. Using irrigated sorghum, the second leaf from the top of the plant was sampled at heading and analyzed (Table 12). Nitrogen rates more than tripled yield. The N/K ratios were all above 1.00, and increased to nearly 1.80 with added N. The sum of the cations initially decreased and then increased with added nitrogen, following a pattern similar to the individual cations. Ammonium sulfate appeared to be a more effective source of nitrogen than ammonium nitrate. The zinc content of the leaf was greater with the sulfate source of nitrogen. These differences were shown to be due to an effect of the N carrier on soil pH, and subsequent availability of man-

Table 13—Effect of sources of N on soil, pH, sorghum yield, N recovery and leaf Zn at heading. (Boawn et al., 1960)

Nitrogen source 180 kg N/ha	1958 Grain sorghum yields kg/ha	July Soil pH	Nitrogen recovery %	Zn leaf content −Zn ppm	+Zn ppm
Ammonium sulfate	7,480	6.5	44.1	18.8	27.0
Ammonium nitrate	6,371	6.8	32.8	15.2	23.5
Calcium nitrate	5,177	7.0	23.4	14.3	18.3
Calcium nitrate + lime	4,292	7.6	12.7	--	--

Table 14—Effect of K and soil moisture on cation content, ratios, and yields of corn grown in the greenhouse.* (Lawton, 1945)

Nutrient treatment	Soil moisture %	Cations K	Ca	Mg	Sum	K/Ca	K/Mg	Ca+Mg K	Corn yield tops	roots
		— meq/100g —							— g/pot —	
NP	15	50	16	29	95	3.2	1.8	0.8	20.8	12.9
NP	25	44	22	34	100	2.0	1.3	1.3	20.9	9.9
NP	40	24	18	33	75	1.4	0.7	2.0	13.3	5.6
NPK	15	80	18	23	121	4.4	3.5	0.5	24.5	14.2
NPK	25	78	18	26	122	4.5	3.0	0.6	23.9	11.5
NPK	40	53	24	30	107	2.2	1.7	1.0	18.1	9.0

* Clarion loam soil having the following analyses: pH 5.6, moisture equivalent 25%, Exch. K 123 kg/ha, and extractable P 15 ppm.

ganese and zinc (Table 13). In these studies interactions in the soil were occurring simultaneously with those in the plant, influencing the availability, interaction, and utilization of ions by the crop. In this regard, Wolcott (1964) has indicated how the acidifying effects of nitrogen can influence micronutrients.

D. Soil Moisture, Compaction, and Aeration

Lawton (1945) studied the effects of soil moisture, packing, and aeration on the uptake of potassium, calcium, and magnesium by corn with fertilizer treatments (Table 14). Moisture caused shifts in the cation contents and ratios. Above 15% moisture, root yields were decreased more than those of tops. Excess moisture reduced potassium concentrations of the corn and K/Ca and K/Mg ratios. With added potassium, corn withstood excess moisture conditions better. Added potassium increased the cation content of the corn. Excess moisture increased (Ca + Mg)/K ratios and decreased yields. Highest yields were obtained with (Ca + Mg)/K ratios of about 0.50.

Compaction markedly reduced the potassium content and increased cal-

INTERACTION WITH OTHER IONS

Table 15—Effect of K and soil compaction on cation content, ratios and corn yield at 15% soil moisture. (Lawton, 1945)

Nutrient and compaction treatment	Composition of corn plants			Cation sum	K/Ca	K/Mg	Ca+Mg / K	Corn yield	
	K	Ca	Mg					Tops	Roots
	meq/100g							g/pot	
NP	54	16	29	99	3.4	1.9	0.82	20.8	12.9
NP-Packed	21	20	38	79	1.1	0.6	2.66	8.7	4.3
NPK	80	18	23	121	4.4	3.5	0.52	24.5	14.2
NPK-Packed	73	28	39	140	2.6	1.8	0.93	10.7	5.2

Table 16—Effect of K and soil aeration on cation content, ratios, and corn yields at 40 percent soil moisture. (Lawton, 1945)

Nutrient and aeration treatment	Cations				K/Ca	K/Mg	Ca+Mg / K	Corn yield	
	K	Ca	Mg	Sum				Tops	Roots
	meq/100g							g/pot	
NP	24	18	33	75	1.39	.74	2.05	13.3	5.6
NP-aeration	33	16	29	78	1.98	1.14	1.38	17.3	8.7
NPK	53	24	30	107	2.20	1.73	1.03	18.1	9.0
NPK-aeration	84	16	26	126	5.45	3.21	0.49	27.7	17.8

Table 17—Effect of irrigation on corn grain yield and cation content and ratios of leaves at tassel.* (Jenne et al., 1958)

	Grain yield	Cations				K/Ca	K/Mg	Ca+Mg / K
	kg/ha	K	Ca	Mg	Sum			
		meq/100g						
Nonirrigated	4,330	84.4	54.8	40.4	179.6	1.54	2.09	1.13
Irrigated	9,601	66.5	37.9	34.5	138.9	1.75	1.93	1.09

* Tripp very fine sandy loam. Soil pH 7.3; CEC 16.5 meq/100g and Exch. K 2.1 meq/100g. Fifteen tons of manure and 90 kg N/ha were applied.

cium and magnesium with a resulting shift in the ratios and corn yield (Table 15). The lowest (Ca + Mg)/K ratio was associated with the highest yield. Potassium increased the cation content and yield, but did not overcome the effects of compaction.

Aeration at 40% soil moisture with or without added K increased the K content and sum of the cations while decreasing the Ca or Mg contents (Table 16). These changes were associated with increased yield. Potassium with aeration gave the highest yield of tops and roots. Yield of the roots was more than tripled and yield of the tops was increased 60% by adding K with aeration. The optimum ratios under the conditions of the study appear to be as follows: K/Ca, 5.45; K/Mg, 3.21; and (Ca + Mg)/K, 0.49.

Jenne et al. (1958) obtained data on corn that give further insight into

Table 18—Effect of soil temperature and K on corn yield, cation contents, and ratios of whole plants at 31 days.* (Nielsen et al., 1961)

Soil temperature	Nutrient treatment	Yield	K	Ca	Mg	Sum	K/Ca	K/Mg	Ca+Mg / K
°C		g/pot		meq/100g					
5	NP	4.4	22	77	120	219	0.28	0.18	8.95
	NPK	10.3	101	44	42	187	2.30	2.40	0.85
26.6	NP	34.3	27	23	47	97	1.17	0.57	2.59
	NPK	51.5	23	21	47	91	1.10	0.49	2.96

* Castor silt loam.

Table 19—Influence of increasing osmotic pressure and salts on corn yield, cation content, and interrelationships. (Kaddah and Ghowail. 1964)

Salt added	Osmotic pressure	Corn yield	K	Ca	Mg	Na	K/Ca	K/Mg	Ca+Mg / K
	atm	g/culture		meq/100g					
None		124.2	135	34	34	4	4.0	4.0	0.50
NaCl	1	76.1	137	29	32	13	4.7	4.3	0.44
	3	38.6	106	25	29	61	4.2	3.6	0.51
$CaCl_2$	1	104.1	148	58	21	3	2.6	7.0	0.54
	3	17.1	118	113	13	6	1.0	9.1	1.07
$MgCl_2$	1	117.9	146	19	66	3	7.7	2.2	0.58
	3	12.6	97	18	143	3	5.4	0.7	1.68
Na:Mg:Ca	1	102.4	146	35	46	5	4.2	3.2	0.55
	3	46.6	142	48	51	6	3.0	2.8	0.69

the effect of soil moisture on the interrelation of cations (Table 17). The cation content of the leaves was higher in the non-irrigated treatment. Irrigation more than doubled yield. The K/Ca and K/Mg ratios appear to be lower than necessary for optimum yield. Little relation exists between yield and the ratios. Based on responses to zinc that have been obtained on the Tripp soil series (Ward et al., 1963), it may be that zinc deficiency influenced the results, although the manure application may have alleviated that possibility.

E. Soil Temperature

Nielsen et al. (1961) studied the effects of soil temperature and nutrient application on corn yields and cation content (Table 18). Yields and cation content were markedly influenced by soil temperature and nutrient treatments. Calcium and magnesium contents were drastically reduced on the nitrogen-phosphorus treatment by the higher temperature as yields were increased. The sum of the cations was reduced. The K/Ca and K/Mg ratios are far below those suggested as being optimum. Potassium was reduced in the nitrogen-phosphorus-potassium treated corn as the temperature was increased. Based on knowledge of nutrient contents of young corn, it would appear that these plants may have been under N and K levels that were too low because the highest yield, which was 51.5 g, contained only 1.1% N and 3.96% K. Nevertheless, these data present conclusive evidence of the importance of soil temperature on cation content and relations.

Epstein et al. (1962) have shown the importance of temperature in rubidium uptake by barley in kinetic carrier studies. Rubidium uptake is similar to potassium. Uptake was linear with time and the rate was markedly increased from 5 to 30C.

F. Osmotic Pressure and Associated Cations

Kaddah and Ghowail (1964) added increased salt concentrations to a base solution to obtain osmotic pressures of 1, 2, and 3 atm. The salts used were chlorides or sulfates of sodium, calcium, and magnesium, added singly or in combination. Data for the chloride salts are shown in Table 19. Yields were decreased by each level of osmotic pressure regardless of the salt. High Na, Ca, or Mg concentrations were associated with the lowest yields. Also, when significant concentrations of sodium were added, (Ca + Mg)/K ratios showed little relationship to yield, but if (Ca + Mg + Na)/K ratios were used an inverse relationship to yield was observed. The initial level of sodium seemingly produced a deleterious effect on yield far greater than could be accounted for on the basis of changes in potassium, calcium, or magnesium composition or osmotic pressure. At equal osmotic pressures, calcium and magnesium were more deleterious to yield than sodium. With isomotic mixtures of $NaCl:CaCl_2:MgCl_2$, yields were reduced less by the higher osmotic pressure than when individual salts were used alone.

Regarding the use of a combination of nutrients to improve yields under high osmotic pressure, Heimann and Ratner (1962) have discussed the interrelationships of potassium and sodium under saline conditions. They found proper applications of potassium reduced the adverse effects of sodium. This was in agreement with an earlier proposal of Heimann (1958) on the concept of a balanced ionic environment. Francios and Bernstein's (1964) studies on salt tolerance of safflower indicate that Ca/K ratios were important. In the highest yielding variety, yield was decreased 25% by increasing EC_e from 0.9×10^3 to 11.2×10^3 milliohms/cm as K/Ca ratios decreased from 0.98 to 0.45.

Soofi and Fuehring (1964) conducted corn fertility studies on a calcareous soil (32% $CaCO_3$ equivalent). The highest grain yield obtained was 9,124 kg/ha (145 bu/acre). This was attained by applying 610 kg N and 82.7 kg of P, K, Mg, and S/ha. Corn leaf analyses at silking for the treatment had the following cation contents and ratios: K, 128.6 meq; Ca, 88.8 meq; Mg, 12.6 meq; N/K(%), 0.54; K/Ca, 1.4; K/Mg 10.2; and (Ca + Mg)/K, 0.79.

G. Micronutrients

Micronutrients appear to play a special role in plant nutrition, being associated with enzyme activation anl cation transport. Copper, for example, has been shown to influence calcium transport and movement in plants (Brown and Foy, 1964; and Brown, 1965). Phosphorus has been shown to increase the severity of copper deficiency (Brown and Foy, 1964), just as it does zinc deficiency (Olson et al., 1962; Langin et al., 1962; and Ward et al., 1963). Willis and Piland (1934) studied the interrelationship of lime, copper, and iron. Micronutrients show interrelationships much the same as observed among the major cations. For example, the reciprocal relationship between iron and manganese has often been observed (Somers and Shive, 1942). Because of these relations, it is likely they will influence the interrelationship among potassium, calcium, and magnesium.

1. Influence of Potassium and Boron on Relations

Woodruff et al. (1960) worked with soybeans and studied potassium and boron using various combinations of potassium, calcium and magnesium saturations. Earlier studies indicated yields of soybeans grown to maturity increased up to 2% K saturation, but dropped to nearly zero at 20% K saturation. With high potassium saturation a dead growing tip and development of axil shoots suggested boron deficiency. Varying K saturation from 0 to 20%, and Ca from 56 to 76%, with Mg being kept constant at 24% saturation, soybeans were grown with and without boron. Results of analyses

Table 20—Effect of soil K saturation and B on cation content and interrelationships of the second trifoliate of soybeans. (Woodruff et al., 1960)

Potassium saturation	B	K	Ca	Mg	Sum	K/Ca	K/Mg	Ca+Mg / K
%	ppm	meq/100g						
Without Boron								
0.5	42	13	60	90	163	0.2	0.1	12.0
4.0	11	54	36	33	123	1.5	1.6	1.3
8.0	6	54	30	25	109	1.8	2.2	1.0
With Boron								
0.5	78	8	85	99	192	0.1	0.1	23.5
4.0	50	41	55	68	164	0.7	0.6	3.0
8.0	49	69	65	68	202	1.0	1.0	1.9

of the second trifoliate leaf after the fifth trifoliate leaf had developed are in Table 20.

Without B, increasing K saturation decreased the B content of the sampled trifoliate 36 ppm and with B the decrease was 29 ppm. Higher boron content was associated with enhanced calcium and magnesium concentrations with higher potassium saturation. With boron the increasing potassium saturation did not reduce the cation concentration nearly to the extent it did without boron. The contents of the three cations were essentially equalized by the added boron, as evidenced by the K/Ca and K/Mg ratios of about 1 for the 8% K saturation treatment. The (Ca + Mg)/K ratios decreased both with and without boron, the ratios tending to be lower without added boron.

2. Relation of Potassium and Zinc

The stunting effects of potassium or zinc deficiency on corn are very similar, although specific leaf symptoms are different. Ward et al. (1963) and Stuckenholtz et al. (1966) found that the depressive effects of phosphorus on the zinc content of corn decreased as the percent potassium saturation or soil potassium increased. Zinc and potassium appear to be related in action through the pyruvic kinase enzyme system (Miller and Evans, 1957; McCollum et al. 1958; and Weber, et al. 1965). Insulin in animals is essential for the formation of pyruvic kinase (Weber et al., 1965). Zinc is intimately associated with the insulin molecule. A similar substance undoubtedly exists in plants. Through this system both potassium and zinc are related to energy transfers involving phosphorus.

3. Relation of Potassium and Molybdenum

Relations between potassium and molybdenum have been reported. Jones (1965) found that increasing applications of row-applied KCl decreased the molybdenum content of corn leaves. In a private communication, analysis of the data showed that as row-applied K was increased from 0 to 112 kg/ha, yields increased from 2,573 kg (2,296 lb) to 6,087 kg/ha (5,431 lb/acre) and molybdenum decreased from 4.0 to 0.9 ppm (J.B. Jones, Agronomy Dept., Ohio Agricultural Research and Development Center, Wooster.). The ratios for the control and high KCl treatment were as follows: K/Ca, 0.32 to 2.8; K/Mg, 0.23 to 3.4; and (Ca + Mg)/K, 7.5 to 0.66. Certainly the effects were striking and the interrelationships apparent. The decrease in molybdenum could have been due to the associated anion. Stout et al. (1951) have shown that with potassium, the anion effect on decreasing molybdenum uptake is $SO_4^{2-} > Cl^- > NO_3^- > H_2PO_4^-$.

IV. CRITICAL LEVEL OF POTASSIUM SHIFTED BY INTERACTING IONS

Hylton et al. (1967) evaluated the interaction of K and Na in the growth and mineral content of Italian ryegrass. They found that a reciprocal relationship existed. Sodium partially substituted for potassium, when potassium in the medium was low, but potassium influenced growth more than sodium. Using the youngest fully open leaf blade, they found that to properly evaluate the critical level of potassium in ryegrass it was necessary to determine the concentration of sodium also. The critical concentration of the blade tissue was 0.8% K or 20.4 meq/100 g with 2.4% or 104 meq of Na, and 3.5% K or 89.5 meq when the tissue contained 0.3% or 13 meq/100g of Na on a dry basis.

Page and Bingham (1965), while not specifically looking at critical nutrient levels, obtained evidence with potassium and magnesium on cotton which would indicate the possibility of such changes. Using petioles they found that with K and Mg contents of 89 and 12 meq/100g, respectively, no cotton bolls were produced, but fair growth occurred. Growth and boll production improved when K and Mg concentrations were 36 and 58 meq/100g respectively, but the optimum growth and boll production occurred with 84 meq K and 41 meq Mg.

Shifts in the critical levels of potassium undoubtedly entered into the studies conducted by Stanford et al. (1941) with corn on calcareous soils. High magnesium concentrations required more potassium before yields could be increased.

Also, Walker (*private communication*) has obtained data which indicate that a higher concentration of corn leaf K may be required for optimum yield of corn as the concentration of N is increased (W. M. Walker, Department of Agronomy, University of Illinois, Urbana.). Nitrogen/potassium ratios have been related to corn stalk diseases (Parker and Burrows, 1959) and corn lodging problems (Liebhardt and Murdock, 1965). It is usually thought they should be 1 or less.

V. CATION RELATIONS AND ORGANIC ACIDS

A. Effect of Nutrient Deficiencies

Clark (1968) has evaluated the effect of mineral deficiencies on organic acid and cation concentrations and ratios of young corn (Table 21). Nutrient culture and growth chambers were used, but in the non-deficient treatments variation among the three experiments were noted. Calcium and manganese deficiencies decreased the yield the most. Boron deficiency increased yields, which may indicate contamination. Potassium, magnesium,

Table 21—Effect of nutrient deficiencies on N/K(%) ratios, cation contents and relations and organic acids in corn. (Clark, 1968)

Treatment	Dry weight g/culture	N/K(%)	Cations meq/100g K	Ca	Mg	Total*	K/Ca	K/Mg	Ca+Mg / K	Trans-aconitic acid meq/g fresh weight	Malic acid	Citric + isocitric
Nondeficient												
Exp. 1	3.43	0.81	71	36	33	141	2.0	2.2	0.97	14.0	4.3	2.1
Exp. 2	3.14	1.27	53	18	32	103	2.9	1.7	0.93	10.7	3.9	3.0
Exp. 3	3.11	0.98	100	41	40	181	2.4	2.5	0.83	15.0	6.0	1.3
Deficiency												
K Exp. 1	2.02	5.10	16	68	79	164	0.23	0.20	9.3	3.3	16.8	8.5
P Exp. 2	1.72	0.91	87	23	26	138	3.8	3.3	0.56	5.5	5.3	1.3
Ca Exp. 2	0.89	0.87	135	5	36	178	27.0	3.7	0.30	5.2	3.6	3.5
Mg Exp. 1	1.65	0.66	125	30	5	160	4.2	2.0	0.28	5.4	7.0	2.2
S Exp. 3	1.60	0.65	171	49	33	254	3.5	5.2	0.48	17.3	4.4	1.2
Zn Exp. 1	2.23	1.12	87	34	34	157	2.6	2.5	0.79	20.2	3.4	0.9
Mn Exp. 3	0.56	1.20	104	25	21	151	4.1	4.8	0.45	14.1	6.2	1.3
Fe Exp. 3	3.30	0.71	84	34	29	147	2.5	2.9	0.75	21.3	4.9	2.8
Cu Exp. 3	3.62	0.89	109	40	33	185	2.7	3.3	0.66	27.6	10.4	3.0
B Exp. 3	3.82	0.98	79	33	36	149	2.4	2.2	0.88	28.1	6.8	2.4

* Total cations including micronutrients.

and sulfur deficiencies, followed by iron, had the greatest influence on N/K(%) ratios. Calcium, magnesium, or sulfur deficiency caused the accumulation of large amounts of potassium. The highest cation concentration occurred with sulfur deficiency. Cation ratios were affected accordingly. The (Ca + Mg)/K ratios of the non-deficient plants were less than one.

The nutrient deficiencies influenced the organic acids as well as the cation concentrations. *Trans*-aconitic, malic, and citric-isocitric were the dominant acids. Potassium deficiency decreased *trans*-aconitic acid by a factor of four and increased malic and citric-isocitric fractions by four. Phosphorus, calcium and magnesium deficiencies also decreased the *trans*-aconitic acid content of corn. Sulfur, zinc, iron, copper, and boron deficiencies all caused substantial increases in the *trans*-aconitic acid content and copper and boron deficiencies also increased malic acid.

It is not clear as to why *trans*-aconitic acid is formed rather than the *cis*-aconitic. It could be a genetic influence of the particular line of corn used in the study or an influence of the growth chamber. However, it is clear that nutrients deficiencies do have a marked influence on interrelations of cations and play a definite role in the formation of *trans*-aconitic and other acids. This acid has been linked to problems concerning grass tetany in animal nutrition (Grunes, 1967). Temperature and light quality may enter into its formation, at least in other species of grass (Stout et al., 1967; and Grunes, 1967).

B. Effect of Nitrogen

1. Rate of Nitrogen

Goates[3] studied the influence of rates of nitrogen in solutions on cation content and organic acid anions. His results for corn are shown in Table 22. Nitrogen increased the total cations and concentrations of calcium and magnesium relative to potassium and the organic acids anions. Nitrogen rates appear to have been too low for optimum growth. With phosphorus addition in another experiment, organic acid anions were not increased. Goates found a correlation between cation absorption and plant protein. He indicated, based on the attraction between protein and hydrated cations, the predicted absorption with increasing protein content would be calcium, magnesium, potassium, and sodium, according to the decreasing charge density of the ions. He suggested that a plant will absorb cations that precipitate as insoluble compounds or are preferentially absorbed by protein, the pectic acids of the cell walls or other base binding compounds, such as organic acids.

---

[3] R. J. Goates, Cations of Plant tissue: Determination, organic anion relations, and constancy of equivalent total content. Ph.D. Thesis. University of Wisconsin. 1947.

Table 22—Influence of N rate on cation relations and organic acids anions in corn (Goates[3]).

Nitrogen added	N/K(%)	Cations K	Ca	Mg	Sum	K/Ca	K/Mg	Organic acid anions	% N
				meq/100g					
21	0.27	165	23	26	214	7.2	6.3	97	1.74
63	0.50	169	33	34	236	5.1	5.0	114	3.32
126	0.60	174	37	44	255	4.7	4.0	118	4.06

Accumulation of cations was concluded to be a function of the organic acids of plants, as well as other factors. If nitrogen increases the relative uptake of concentration of calcium and magnesium more than potassium, this could possibly cause a shift in the critical K requirement and may partially explain nitrogen-potassium interactions that have been observed.

McLean et al. (1956) found a high correlation between the cation-exchange capacity roots and the nitrogen content. This effect has been shown to cause preferential absorption of divalent cations over monovalent cations and to be related to the uronic acid contents of roots (Knight et al., 1961 and Crooke and Knight, 1962). Williams (1962) and Jenny and Grossenbacher (1963) have shown how increased quantities of protein with exposed amino groups

$$(R-NH_3^+) \text{ and pectic gels } (R-COO^-)$$
$$|\phantom{^+) \text{ and pectic gels } (R-}|$$
$$COO^-\phantom{+) \text{ and pectic gels } (R-}COO^-$$

can account for both anion and cation-exchange properties of roots.

2. Source of Nitrogen

DeKock (1964) reported the influence of the source of nitrogen on the potassium, calcium and oxalic acid content of fresh leaves of buckwheat. The relations are shown in Table 23. Higher amounts of oxalic acid were associated with the nitrate nitrogen, which favored calcium accumulation over potassium, decreasing the K/Ca ratio. This is in agreement with the ideas of Vladimirov, presented by Goates.[4] Vladimirov reasoned that in the reduction of nitrate to protein, oxidation of other organic compounds to organic acids should occur, giving rise to more organic acids and greater cation uptake. Ammonium nitrogen on the other hand, being fully reduced, would produce a lower organic acid and cation content. These data bear out the proposal.

Kirkby and Mengel (1967) studied the influence of nitrate-, ammonium-, and urea-nitrogen on the ionic balance and organic acids in different parts of tomato plants (Table 24). Source of nitrogen influenced leaf yield, cation

---

[4] Ibid.

Table 23—Influence of source of N on K, and Ca and oxalic acid contents of buckwheat leaves. (DeKock, 1964)

Source of nitrogen	K	Ca	Oxalic acid	K/Ca
		meq/kg		
NH_4^+	30	21	25	1.48
$NH_4^+ + NO_3^-$	34	40	90	0.82
NO_3^-	40	63	170	0.63

Table 24—Influence of source of N on yield, cation content, and relationships and organic acids in tomatoes. (Kirkby and Mengel, 1967)

Form of nitrogen	Leaves yield	Cation content				K/Ca	K/Mg	Ca+Mg K	Malic	Citric
		K	Ca	Mg	Na					
	g/10 plants		meq/100g						meq/kg fresh weight	
NO_3^-	30.2	58	161	30	19	0.36	1.9	3.6	130.2	47.4
Urea	15.3	34	134	27	15	0.25	1.3	5.2	25.5	55.6
NH_4^+	9.6	29	62	25	15	0.47	1.2	3.5	5.9	10.6

contents, and ratios and organic acid contents, all of which decreased in order going from nitrate- to urea- to ammonium-nitrogen. High malic acid content was associated with the highest leaf yield and a high divalent cation content. Cation ratios were insensitive to the results.

DeWit et al. (1963) suggest that plants can maintain their cation-anion balance in the top of plants only if a part of the nitrogen is transported upward as nitrate-nitrogen. Hylton et al. (1967) indicated that potassium played an important role in getting nitrate-nitrogen into the plant. Also, potassium is needed for protein formation (Griffith et al., 1964 and Teel, 1962). Ammonium-nitrogen as the only nitrogen source puts plants under stress apparently because of uptake competition with potassium. Also, when ammonium-nitrogen is incorporated into protein, a H^+ is released. Therefore, ammonium-nitrogen is of little value for maintaining the cation-anion balance of plants. Data of Kirkby and Mengel (1967) on the pH of macerated tissue grown with different sources of nitrogen bear this out. With roots, for example, they found the pH to be 5.6, 5.0, and 4.7, with nitrate, urea, and ammonium sources of N, respectively. The pH of other tissue followed the same pattern. The effects of these forms of N on yield were given in Table 24.

Barker et al. (1967) found that with low potassium and high amounts of ammonium-nitrogen, lesions developed on the stems and leaves of tomatoes. Potassium and rubidium were completely effective in reducing the lesions, while cesium was toxic. The authors suggested that K reduced ammonium toxicity or prevented the formation of toxic organic compounds. With ammonium-nitrogen, 60% of the free amino acids found were from protein, indicating that proteolytic action caused protein degradation.

Fortunately in the field, ammonium-nitrogen is usually rapidly converted to nitrate, unless reducing conditions persist because of excess moisture or low oxygen. High soil pH can lead to nitrite accumulation because ammonia is toxic to nitrite oxidizing bacteria (Frederick and Broadbent, 1966). Low soil pH could lead to slower conversion of ammonium to nitrate to cause a problem. Both conditions could conceivably lead to nutritional problems.

VI. CATION-ANION BALANCE

Initially, Bear and Prince (1945) observed the cation-equivalent constancy in crops. Many others have observed this phenomenon, but as pointed in this discussion many factors can change this "constancy". Wallace et al. (1949), Bear and Wallace (1950), and Bear (1950) indicated that the ratio of cations to anions for a crop was constant. They found that interrelations among anions were similar to those observed for cations. Others have studied the concept of nutrient balance between cations and anions and included the involvement of organic acids in nutrition (DeKock, 1958; deWit et al., 1963; Palmer et al., 1963; DeKock, 1964; Kirkby and DeKock, 1965; Noggle, 1966; and Kirkby and Mengel, 1967).

Rather than using ratios, deWit et al. (1963) proposed and found that the difference between the cation content, C, and the anion content, A, was constant. Because of electroneutrality, organic acid anions should account for the difference between the C-A. They found that the difference between C-A was constant until plants were stressed by nutrient deficiency. They theorized that the greater the difference between C and A, the greater would be the yield of the crop. Therefore, any treatment which increased the difference or increased the total organic anion content would increase yields. This seems reasonable since enhanced CO_2 fixation has to occur before organic anions are produced.

Kirkby and DeKock (1965) studied cation-anion balance and found that with increasing leaf age of Brussel sprouts calcium decreased and sulfur increased, while phosphorus and nitrogen decreased. These changes were primarily associated with increases in malate concentration which increased from 18 meq/100g in the youngest leaf to 131 meq in the oldest leaf. As pointed out earlier, these changes resulted in large shifts in the K/Ca and (Ca + Mg)/K ratios. Also, the cation-anion differences were only equal in the very youngest leaves and increased with leaf age primarily because of the increased calcium content.

Noggle (1966) tested the hypothesis of deWit et al. (1963) on 16 crops grown in the greenhouse and indeed found that treatments that decreased the inorganic anion content of crops and increased the organic anions did increase yields the most.

Potassium chloride and K_2SO_4 were the two salts compared. Potassium sulfate produced lower concentrations of inorganic anions and higher yields. It should be remembered that this was in the closed system of greenhouse containers, which would prevent downward movement of mobile anions as would usually be the case in an open system or the field.

The approach of cation-anion difference seems to hold promise for including all of the nutrients and at least more of the organic constituents in information. Therefore, it will or should receive serious attention in the future.

VII. SUMMARY AND CONCLUSIONS

There is little doubt of the interaction of potassium and other ions in crops grown on the same or different media. Fertilizer treatments influence ion relations within and among crops. Added potassium can increase, reduce or not change the sum of the cations in a crop, while sodium often increases the sum of the cations unless it is excluded, as tends to be the case with corn. Added potassium usually reduces calcium and magnesium content of corn, with the greatest effect being exerted on magnesium. Therefore, shifts occur in the cation ratios. As these shifts occur, yields are often increased. Other effects of ratio are now being studied. Shugarov (1966), for example, has found that reflectance of near infrared radiation decreased as K/Ca ratios decreased, causing leaf temperatures to rise. Nitrogen rates and sources interact to influence cation contents and ratios. Ammonium-nitrogen sources influence cations differently than nitrate sources in this regard and both can cause acidifying effects which may influence micronutrient availability. Micronutrients appear to influence the uptake and translocation of potassium and other cations.

The aging process has been shown to influence K and its relation with other ions. The ability of cells to hold potassium apparently decreases with the aging process and, therefore, calcium and magnesium under moderate potassium levels increase. One can look at the effect of aging during the season or sample from the youngest to the oldest leaf on a plant and find changes in the ion content. These changes are usually associated with shifts in the organic acid content of the portion sampled. Furthermore, specific organic acids are related to high or low amounts of various cations and these vary with the crop grown. Organic acids, in turn, are related to metabolic pathways and are used to form plant substances or accumulate if pathways of utilization are blocked.

Physical conditions in the soil such as moisture, temperature, compaction, aeration, and salinity exert a marked influence over the potassium and the crop content of other ions. These factors themselves are interrelated. Fertilizer treatments appear to overcome some of the adverse effects of excess moisture. Potassium is particularly sensitive to temperature and at lower temperatures uptake is reduced. Also, in the case of salinity, the proportions of the associated cations appears to be important. Yields are reduced less if a proper combination of ions is present.

The relations of cations and organic acids appear to be a phase of study that will receive increasing emphasis. However, the broader approach of looking at both the cations and anions and the organic acid anions found by difference may hold more promise. Analyses of organic acids should be conducted in this context. This approach may provide valuable and needed information for improving fertilizers and increasing growth rates, yields, and food quality. The inceased availability of emission spectrograph, atomic absorption, and chromatographic instruments make it much easier to determine the elements taken up by a crop. Linked with genetics and molecular biology, such studies promise to open new avenues of understanding for crop and soil scientists.

The observation that critical levels for elements shift as cation contents vary, as indicated by the Na and K example, should make one further aware that we are dealing with 16 or more elements in a plant that are simultaneously interacting. This is an important fact, because unless liming and fertilizing programs are carefully evaluated and monitored, nutrient excesses may cause shifts in the requirements for other elements. Such effects are already being observed on a limited scale. This makes it doubly important to consider these interrelations in soil-fertilizer-crop research being conducted. Schuffelen et al. (1967) have stated that, "Only inexpert fertilizing practices can have detrimental effects on crops", and indicate that in the pursuit of high crop production correct fertilizer use will improve both yield and quality. It will be a challenge during the next decade to study interaction of ions to see to it that such is the case.

REFERENCES

Baker, D. E., B. R. Bradford, and W. I. Thomas. 1966. Leaf analysis of corn - tool for predicting soil fertility needs. Better Crops with Plant Food. 50:36-40. Summer.

Barker, A. V., D. N. Maynard, and W. H. Lachman. 1967. Induction of tomato stem and leaf lesions, and potassium deficiency, by excessive ammonium nutrition. Soil Sci. 103:319-327.

Bear, F. E. 1950. Cation and anion relationships in plants and their bearing on crop quality. Agron. J. 42:176-178.

Bear, F. E., and A. L. Prince. 1945. Cation-equivalent constancy in alfalfa. Agron. J. 37:210-222.

Bear, F. E., and A. Wallace. 1950. Alfalfa, its mineral requirements and chemical composition. Better Crops With Plant Food. 34(7):6-12.

Benne, E. J., E. Linden, J. D. Grier, and K. Spike. 1964. Composition of corn plants at different stages of growth and per acre accumulation of essential nutrients. Michigan Quarterly Bull. 47:69-85.

Bingham, F. T., G. R. Bradford, and A. L. Page. 1964. Toxicity of lithium to plants. California Agr. p. 6-7. Sept.

Boawn, L. C., C. E. Nelson, F. G. Viets, Jr., and C. L. Crawford. 1960a. Nitrogen carrier and nitrogen rate influence on soil properties and nutrient uptake by crops. Washington Agr. Exp. Sta. Bull. 614.

Boawn, L. C., F. G. Viets, Jr., C. L. Crawford, and J. C. Nelson. 1960b. Effect of nitrogen carrier, nitrogen rate, zinc rate and soil pH on zinc uptake by sorghum, potatoes, and sugar beets. Soil Sci. 90:329-337.

Bould, C. 1963. Mineral nutrition of plants in soil. p. 16-133. In F. C. Steward (ed). Plant physiology, Vol. III: Inorganic nutrition of plants. Academic Press, New York.

Bower, C. A., and W. H. Pierre. 1944. Potassium response of various crops on a high-lime soil in relation to their contents of potassium, calcium, magnesium and sodium. Agron. J. 36:608-614.

Brown, J. C. 1965. Calcium movement in barley and wheat as affected by copper and phosphorus. Agron. J. 57:617-621.

Brown, J. C. and C. D. Foy. 1964. Effects of Cu on the distribution of P, Ca and Fe in barley plants. Soil Sci. 98:362-370.

Clark, R. B. 1968. Organic acids of maize (*Zea mays* L.) as influenced by mineral deficiencies. Crop Sci. 8:165-167.

Collander, R. 1941. Selective absorption of cations by higher plants. Plant Physiol. 16:691-720.

Cooper, H. P., W. R. Paden, and M. M. Phillipe. 1953. Effects of applications of sodium in fertilizer on yields and composition of the cotton plant. Soil Sci. 76:19-28.

Crooke, W. M., and Knight, A. H. 1962. An evaluation of published data on the mineral composition of plants in light of the cation-exchange capacities of their roots. Soil Sci. 93:365-373.

Daniels, F. and R. A. Alberty. 1955. Physical chemistry. p. 392, 423. John Wiley & Sons, New York.

DeKock, P. C. 1958. The nutrient balance in plant leaves. Agr. Progress. 23:88-95.

DeKock, P. C. 1964. The physiological significance of the potassium-calcium relationship in plant growth. Outlook on Agriculture. 4, 2:93-98. Imperial Chem. Inc., Ltd. London.

DeWit, C. T., W. Dijkshoorn, and J. C. Noggle. 1963. Ionic balance and growth of plants. Versl. Landbouwk. Onderz. NR. 69. 15. Wageningen.

Dixon, M., and E. C. Webb. 1964. Enzymes. 2nd ed. Academic Press, New York. p. 421-422.

Epstein, E., D. W. Rains, and W. E. Schmid. 1962. Course of cation absorption by by plant tissue. Science 136:1051-1052.

Evans, H. J., and G. J. Sorger. 1966. Role of mineral elements with emphasis on the univalent cations. Ann. Rev. Plant Physiol. 17:47-76.

Francois, L. E., and L. Bernstein. 1964. Salt tolerance of safflower. Agron. J. 56:38-40.

Frederick, L R., and F. E. Broadbent. 1966. Biological interactions, p. 207. In M. H. McVickar et al (ed). Agricultural anhydrous ammonia technology and use. Amer. Soc. of Agron. Madison, Wis.

Griffith, W. K., M. R. Teel, and H. E. Parker. 1964. Influence of nitrogen and potassium on the yield and chemical composition of orchardgrass. Agron. J. 56:473-475.

Grosline, G. W., D. E. Baker, and W. I. Thomas. 1965. Accumulation of eleven elements by field corn (*Zea mays* L.). Pennsylvania Agr. Exp. Sta. Bull. 725.

Grosline, G. W., J. L. Ragland, and W. I. Thomas. 1961. Evidence for inheritance of differential accumulation of calcium, magnesium and potassium by maize. Crop Sci. 1:155-156.

Grosline, G. W., W. I. Thomas, D. E. Baker, and C. C. Wernham. 1963: Corn leaf blight severity associated with genetically controlled element content. Plant Disease Reporter. 47(5):359-361.

Grunes, D. L. 1967. Grass tetany of cattle as affected by plant composition and organic acids. p. 105-110. Proc. Cornell Nutrition Conf. for Feed Manufacturers. Cornell University.

Hanway, J. J. 1962. Corn growth and composition in relation to soil fertility: II. Uptake of N, P, and K and their distribution in different plant parts during the growing season. Agron. J. 54:217-222.

Heimann, H. 1958. Irrigation with saline water and the ionic environment, p. 173-220. In Potassium symposium. International Potash Inst., Berne, Switzerland.

Heimann, H., and R. Ratner. 1962. The influence of potassium on the uptake of sodium by plants under saline conditions. Israel Institute of Technology. Bull. Res. Council of Israel, 10A, No. 2:55-62.

Hendricks, S. B. 1964. Salt transport across cell membranes. Amer. Sci. 52:306-333.

Hendricks, S. B. 1966. Salt entry into plants. Soil Sci. Soc. Amer. Proc. 30:1-7.

Hewitt, E. J. 1963. The essential nutrient elements: Requirements and interactions in plants, p. 137-360. In F. C. Stewart (ed). Plant physiology, Vol. III: Inorganic nutrition of plants. Academic Press, New York.

Hope, A. B. 1960. Radioisotopes in plants: Root entry and distribution, p. 60-72. In R. S. Caldecott and L. S. Snyder (ed). Radioisotopes in the biosphere. University of Minnesota. Minneapolis.

Hylton, L. O., A. Ulrich, and D. R. Cornelius. 1967. Potassium and sodium interrelations in growth and mineral content of Italian ryegrass. Agron. J. 59:311-315.

Jackson, W. A. 1967. Physiological effects of soil acidity. In F. Adams and R. W. Pearson (ed). Soil acidity and liming. Agronomy 12:43-124.

Jenne, E. A., H. F. Rhoades, C. H. Yien, and O. W. Howe. 1958. Change in nutrient element accumulation by corn with depletion of soil moisture. Agron. J. 50:71-74.

Jenny, H., and K. Grossenbacher. 1963. Root-soil boundary zones as seen in the electron microscope. Soil Sci. Soc. Amer. Proc. 27:273-277.

Joham, H. E. 1955. The calcium and potassium nutrition of cotton as influenced by sodium. Plant Physiol. 30:4-10.

Jones, J. B., Jr. 1965. Molybdenum content of corn plants exhibiting varying degrees of potassium deficiency. Science 148:94.

Kaddah, M. T., and S. I. Ghowail. 1964. Salinity effects on the growth of corn at different stages of development. Agron. J. 56:214-217.

Kissel, D. E., and J. L. Ragland. 1967. Redistribution of nutrient elements in corn (*Zea mays* L.) N, P, K, Ca and Mg redistribution in the absence of nutrient accumulation after silking. Soil Sci. Soc. Amer. Proc. 31:227-230.

Kirkby, E. A. and P. C. DeKock. 1965. The influence of age on the cation-anion balance in the leaves of brussel sprouts (*Brassica olerace* var. gemmifera). Z. Pflanzenähr Düng-Bodenk. 111:197-203.

Kirkby, E. A. and K. Mengel. 1967. Ionic balance in different tissues of the tomato plant in relation to nitrate, urea or ammonium nutrition. Plant Physiol. 42:6-14.

Knight, A. H., W. M. Crooke, and R. H. E. Inkson. 1961. Cation-exchange capacities of tissues of higher and lower plants and their related uronic acid contents. Nature 192:142-143.

Lancaster, J. D., W. B. Andrews, and U. S. Jones. 1953. Influence of sodium on yield and quality of cotton lint and seed. Soil Sci. 76:29-40.

Langin, E. J., R. C. Ward, R. A. Olson, and H. F. Rhoades. 1962. Factors responsible for poor response of corn and grain sorghum to phosphorus fertilization: II. Lime and P placement effect on P-Zn relations. Soil Sci. Soc. Amer. Proc. 26:574-578.

Lang, N. A. 1967. Lang's handbook of chemistry. Rev. 10th ed. McGraw-Hill. New York. p. 122-125, 1216, 1224-1225.

Larson, W. E., and W. H. Pierre. 1953. Interaction of sodium and potassium on yield and cation composition of selected crops. Soil Sci. 76:51-64.

Lawton, K. 1945. The influence of soil aeration on the growth and absorption of nutrients by corn plants. Soil Sci. Soc. Amer. Proc. 10:263-268.

Liebhardt, W. C., and J. T. Murdock. 1965. Effect of potassium on morphology and lodging of corn. Agron. J. 57:325-328.

Loúe, A. 1963. A contribution to the study of the inorganic nutrition of maize, with special attention to potassium. Fertilite 20, Nov-Dec.

Lucas, R. E., and G. D. Scarseth. 1947. Potassium, calcium and magnesium balance in plants. Agron. J. 39:887-896.

McCollum, R. E., R. H. Hageman, and E. H. Tyner. 1958. Influence of potassium on pyruvic kinase from plant tissue. Soil Sci. 86:324-336.

McLean, E. O., D. Adams, and R. E. Franklin, Jr. 1956. Cation exchange capacities of plant roots as related to their nitrogen contents. Soil Sci. Soc. Amer. Proc. 20:345-347.

Miller, G., and H. J. Evans. 1957. The influence of salts on pyruvate kinase from tissues of higher plants. Plant Physiol. 32:346-354.

Moss, D. C. and D. E. Peaslee. 1965. Photosynthesis of maize leaves as affected by age and nutrient status. Crop Sci. 5:280-281.

Nason, A., and W. D. McElroy. 1963. Modes of action of essential mineral elements, p. 451-536. In F. C. Stewart (ed). Plant physiology, Vol. III: Inorganic nutrition of plants. Academic Press, New York.

Newton, J. D. 1928. The selective absorption of inorganic elements by various crop plants. Soil Sci. 26:85-91.

Nielsen, K. F., R. L. Halstead, A. J. MacLean, S. J. Bourget, and R. M. Holmes. 1961. The influence of soil temperature on the growth and mineral composition of corn, bromegrass and potatoes. Soil Sci. Soc. Amer. Proc. 25:369-372.

Noggle, J. C. 1966. Ionic balance and growth of sixteen plant species. Soil Sci. Soc. Amer. Proc. 30:763-766.

Olson, R. A., A. F. Dreier, C. A. Hoover, and H. F. Rhoades. 1962: Factors responsible for poor response of corn and grain sorghum to phosphorus fertilization: I. Soil phosphorus level and climatic factors. Soil Sci. Soc. Amer. Proc. 26:571-574.

Page, A. L., and F. T. Bingham. 1965. Potassium-magnesium interrelationships in cotton. California Agr. 11(19):6-7. Nov.

Palmer, M. J., P. C. DeKock, and J. S. D. Bacon. 1963. Changes in the concentrations of malic acid, citric acid, calcium and potassium in the leaves during the growth of normal and iron-deficient mustard plants (*Sinapis alba*). Biochem. J. 86:484-494.

Parker, D. T., and W. C. Burrows. 1959. Root and stalk rot in corn as affected by fertilizer and tillage treatment. Agron. J. 51:414-417.

Sayre, J. D. 1955. Mineral nutrition of corn. In George F. Sprague (ed.) Corn and corn improvement. Agronomy 5:293-314. Academic Press, New York.

Schuffelen, A. C., M. R. Rosanow, and A. van Diest. 1967. Plant composition and mineral nutrition. Potash Rev. Subj. 3, 24th Suite.

Shugarov, Y. A. 1966. The influence of potassium on the absorption of near infrared radiation by plant leaves. Agrokhimiya No. 11. Moscow. In Potash Review. Subj. 3, 25th Suite. 1967.

Somers, I. I., and J. W. Shive. 1942. The iron-manganese relation in plant metabolism. Plant Physiol. 17:582-602.

Soofi, G. A., and H. D. Fuehring. 1964. Nutrition of corn on a calcareous soil: I. Interrelationships of N, P, K, Mg, and S on the growth and composition. Soil Sci. Soc. Amer. Proc. 28:76-79.

Stanford, G., J. B. Kelly, and W. H. Pierre. 1941. Cation balance in corn grown on high-lime soils in relation to potassium deficiency. Soil Sci. Soc. Amer. Proc. 6:335-341.

Steward, F. C., and J. F. Sutcliffe. 1959. Plants in relation to inorganic salts, p. 253-478. *In* F. C. Steward (ed). Plant physiology, Vol. II: Plants in relation to water and solutes. Academic Press, New York.

Stout, P. R., J. Brownell, and R. G. Burau. 1967. Occurrence of transaconitate in range forage species. Agron. J. 59:21-24.

Stout, P. R., W. R. Meagher, G. A. Pearson, and C. M. Johnson. 1951. Molybdenum nutrition of crop plants: I. The influence of phosphate and sulfate on the absorption of molybdenum from soils and solution cultures. Plant and Soil. 3:51-87.

Stukenholtz, D. D., R. J. Olsen, G. Gogan, and R. A. Olson, 1966. On the mechanism of phosphorus-zinc interaction in corn nutrition. Soil Sci. Soc. Amer. Proc. 30:759-763.

Teel, M. R. 1962. Nitrogen-potassium relationships and biochemical intermediates in grass herbage. Soil Sci. 93:50-55.

Viets, F. G. Jr., C. E. Nelson, and C. L. Crawford. 1954. The relationships among corn yields, leaf composition and fertilizer applied. Soil Sci. Soc. Amer. Proc. 18:297-301.

Wallace, A., S. J. Toth, and F. E. Bear. 1949. Cation-anion relationships in plants with particular reference to the seasonal variation in the mineral content of alfalfa. Agron. J. 41:66-71.

Ward, R. C., E. J. Langin, R. A. Olson, and D. D. Stukenholtz. 1963: Factors responsible for poor response of corn and grain sorghum to phosphorus fertilization: III. Effects of soil compaction, moisture level and other properties on P-Zn relations. Soil Sci. Soc. Amer. Proc. 27:326-330.

Weber, G., N. B. Stamm, and E. A. Fisher. 1965. Insulin: Inducer of pyruvate kinase. Science 149:65-67.

Wehunt, R. L., M. Stelly, and W. O. Collins. 1957. Effect of Na and K on corn and crimson clover grown on Norfolk sandy loam and two residual K levels. Soil Sci. 83 (3): 175-183.

Williams, D. E. 1962. Anion-exchange properties of plant root surfaces. Science 138:153-154.

Willis, L. G., and J. R. Piland. 1934. The influence of copper sulfate on iron absorption by corn plants. Soil Sci. 37:79-83.

Wolcott, A. R. 1964. The acidifying effects of nitrogen carriers. Agr. Ammonia News. July-Aug. (Reprint).

Woodruff, C. M., J. L. McIntosh, J. D. Mikulcik, and H. Sinha. 1960. How potassium caused boron deficiency in soybeans. Better Crops With Plant Food 44(4):4-11.

17

Plant Factors Affecting Potassium Availability and Uptake

WERNER L. NELSON

American Potash Institute, Inc.
West Lafayette, Indiana

I. INTRODUCTION

The plant is the final judge as to the availability of potassium in a soil and integrates the many factors in the environment into one final product. In addition to the external environmental factors, many internal characteristics of the plant itself affect potassium availability and uptake. Among these are extent of the root system, variety, yield level, species, and other elements. The purpose of this chapter is to consider some of the internal factors.

II. UPTAKE OF POTASSIUM

As has been indicated in earlier chapters, salt accumulation, including potassium, is an energy-requiring process with the energy being supplied by respiration. This means that oxygen is required and temperature is important. With rate of root respiration being a determining factor, active absorption will be most effective in moist, warm, well-aerated soils. Potassium deficiency has been observed in crops growing on compact (poorly aerated) or poorly drained soils (again poorly aerated as well as cold), even when the soil is high in potassium.

The rate of absorption of potassium by slices of corn leaf tissue in the light has been found to be about twice the rate in the dark (Fig. 1). When the light was turned off the decrease in rate of absorption was quite rapid. The source of energy for active accumulation of nutrients in the light was

ABSORPTION OF K BY CORN LEAF
TISSUE INFLUENCED BY LIGHT

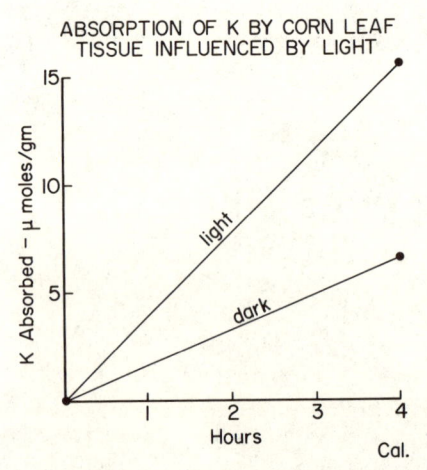

Fig. 1—Potassium absorption by slices of corn leaf tissue is twice as fast in the light as in the dark (Rains, 1967).

closely linked to photosynthetic reactions. In the dark it was linked to respiratory processes. What direct or indirect connotations might this have for potassium nutrition of plants grown under less than optimum lighting conditions?

Any factor which limits the formation and translocation of carrier compounds or respiratory substrates will eventually limit root growth and ion

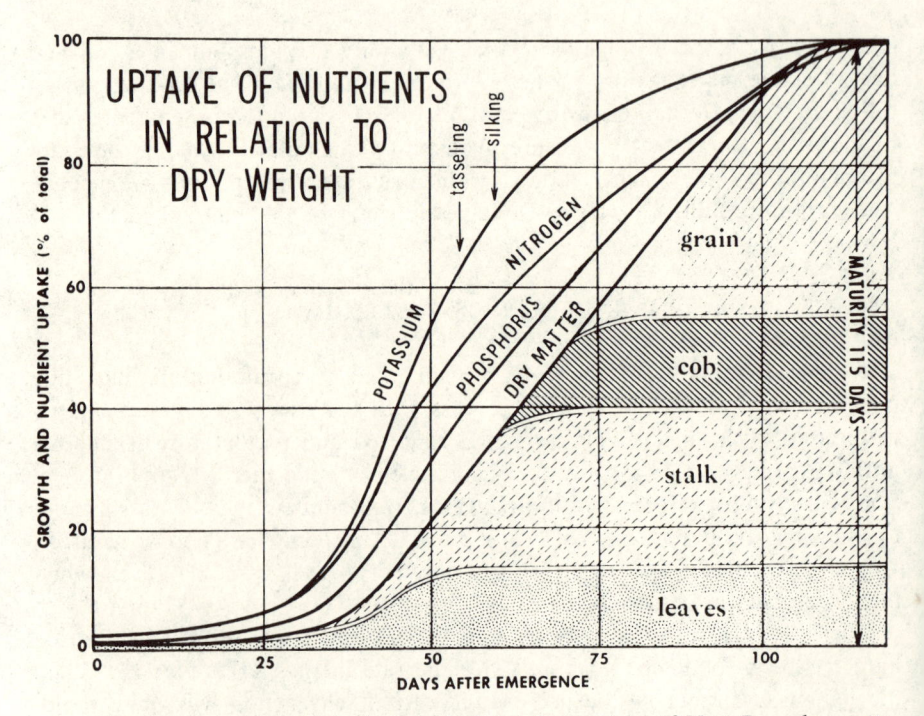

Fig. 2—The rate of uptake of K by corn is faster than uptake of N or P or dry matter accumulation (Iowa State University, 1960).

absorption. If potassium is limiting in the plant, a series of changes takes place in the plant and it is even less able to absorb the potassium which is available in the soil.

Most crop seeds contain from 0.3 to 1.0% K and while this quantity is sufficient for germination and very early development, it will not long sustain growth. Young seedlings take up K very rapidly and S. A. Barber has found from 2 to 6% K in corn plants 15 cm high depending on the amount of K supplied(*personal communication, 1967*).

Potassium is usually taken up earlier than nitrogen and phosphorus and uptake increases faster than dry matter (Fig. 2). This means that the potassium essentially accumulates early in the growing period and then is translocated to other plant parts.

III. ROLE OF ROOT SYSTEMS

The root system of a plant is often as distinctive in structure as its above-ground portions. However, roots are generally much more irregular in shape. The root development of a given crop will proceed according to a genetically determined pattern as modified by the chemical and physical environment. Since potassium enters the plant through the root system, factors affecting the nature and development of root systems affect potassium uptake. Anything which restricts root growth reduces uptake (Fig. 3).

Through the use of the electron microscope it appears that soils and roots are not separate entities, one dead and the other alive. Rather it is not clear where the root ends and the soil begins. Jenny (1967) indicates "In fact soil colloidal particles appear to be chemically bonded to root macromolecules, creating a unified system. Likewise, soil microbes show little respect for scientific fence lines. They seem to enjoy the rich boundary zone and prosper near the root's mucilaginous layer".

A. Region of Maximal Ion Uptake

There are four distinct regions in a root's anatomy (Fig 4). The root cap protects the meristematic tissue (region of cell division). The region of

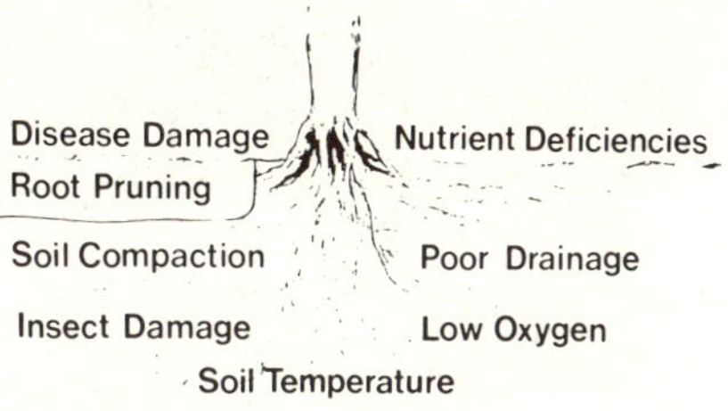

Fig. 3—Factors which restrict root growth reduce nutrient uptake.

Disease Damage   Nutrient Deficiencies
Root Pruning
Soil Compaction   Poor Drainage
Insect Damage   Low Oxygen
Soil Temperature

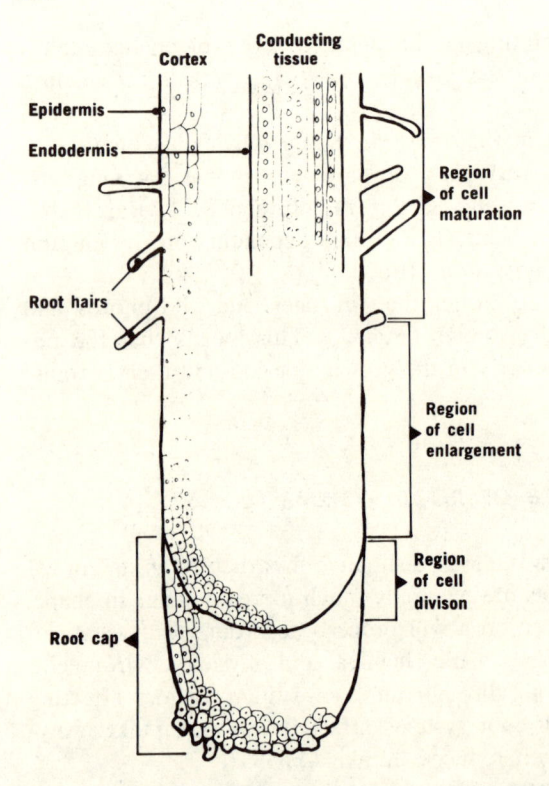

Fig. 4—There are four distinct regions in the anatomy of a root (Tanner, 1967. Reprinted from Plant Food Review).

cell enlargement plus the region of cell division bring about root extension. The root hairs develop in the young part of the maturation zone.

Root hairs grow fairly rapidly and their life span is only 1 or 2 days for many crops. As the root extends new root hairs are continually produced as the older ones become nonfunctional. Root hairs greatly increase the surface area of roots and function as the main site for water and nutrient absorption. The rapid turnover enables the root to come into contact with new soil areas containing nutrients and water. S. A. Barber indicates that less than 5% of the plant's K would be supplied by root interception. Hence, the majority of the K must come to the root by diffusion through the water films rather than by actual contact of the root with K adsorbed on the soil particles.

B. Extent of Roots

Proliferation of roots in fertilizer bands has often been observed. Wiersum (1958) found that NO_3, P, and K promoted branching in pea roots and suggested that the response occurred from growth regulatory substances responsible for regulation of root initiation. Wilkinson and Ohlrogge (1962) reported that nitrogen fertilization increased growth hormones in soybean roots.

Fig. 5—Potassium deficiency reduces extent of root systems in low K soils.

It is well recognized that a deficiency of potassium reduces the extent of the root system of crops (Fig. 5). Hence the crops are less able to forage for potassium, other nutrients, and water.

Deficiencies of other nutrients such as nitrogen likewise reduce the extent of the root system. On the other hand, addition of nitrogen to a plant deficient in nitrogen enables the plant to explore the soil more completely for potassium. On a low potassium soil addition of nitrogen will probably cause a lowering of the percentage of potassium in the plant because of dilution. On a high potassium soil addition of nitrogen could well increase the percentage of potassium in the plant. Obviously the magnitude of such effects will depend on the fertility of the soil.

Adequate fertilization, including lime, nitrogen, phosphorus, and potassium, of the claypan soils in southern Illinois, has markedly increased crop yields (Fehrenbacher et al., 1960). With low fertility, root penetration by corn was around 90 cm and with adequate fertility around 150 cm. In addition to the effect of higher fertility on root extension, a steeper gradient of potassium to the root surface should be beneficial in moving more potassium into the plant by diffusion.

The effect of deeper penetration of roots on K absorption will be related in part to the amount in the subsoil. General subsoil fertility groups have been established in some states and a map for Wisconsin is shown in Fig. 6. In the humid regions subsoil potassium is generally not high. How-

GENERAL SUBSOIL FERTILITY GROUPS

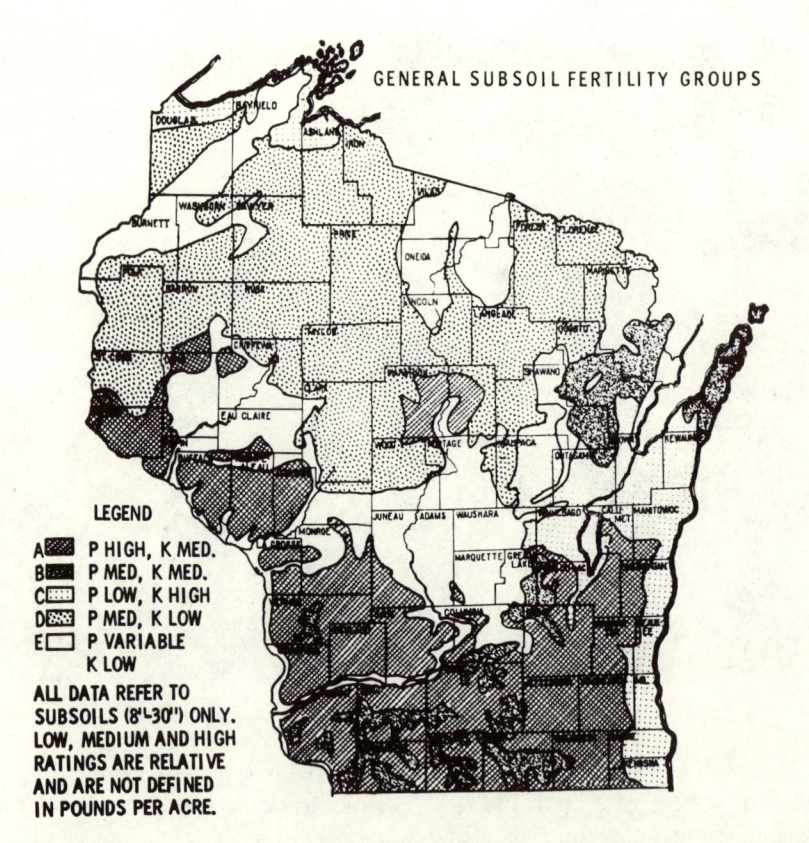

LEGEND
A ▓ P HIGH, K MED.
B ▓ P MED, K MED.
C ▒ P LOW, K HIGH
D ▓ P MED, K LOW
E ☐ P VARIABLE
    K LOW

ALL DATA REFER TO SUBSOILS (8"-30") ONLY. LOW, MEDIUM AND HIGH RATINGS ARE RELATIVE AND ARE NOT DEFINED IN POUNDS PER ACRE.

Fig. 6—General subsoil fertility groups for P and K in Wisconsin have been established (Courtesy of M. T. Beatty and R. B. Corey, University of Wisconsin).

ever, aeration of the subsoil will determine how well the plant can absorb the potassium which is present.

Another effect of deeper penetration of roots is that the more active absorbing regions of the root will more likely be in the moist zones in the soil a greater proportion of the growing season. In Iowa, Dumenil et al. (1965), emphasize that plow-under phosphorus and potassium is usually more efficient than row fertilizer in dry seasons.

Brace root development of corn is enhanced by added potassium on soils low to medium in potassium. An example for corn on a soil containing 148 kg exchangeable K/ha is shown in Table 1. About 10 days after silking, pith parenchyma disintegration started. Such disintegration with maturity is a natural phenomenon hastened by potassium deficiency.

Greenhouse studies on alfalfa shows that applied potassium aids in root extension (Table 2). The change in extractable fractions of soil potassium was directly related to the root distribution pattern of the forage crop.

The classic work by Dittmer (1937) on a single winter rye plant 20 inches tall with 80 tillers showed the plant to have 380 miles of roots at 4 months

Table 1—Influence of fertility treatment on lodging and brace root development of field corn (Liebhardt and Murdock, 1965)

N	P	K	Lodging Root	Lodging Stalk	Plants with BR above ground	Above ground BR/plant	Area above ground
kg/ha	kg/ha	kg/ha	%	%	%	no.	sq. in.
180	78	0	50	28	42	7.1	10
180	78	149	10	1	85	15.0	46

Table 2—Effects of applied K on root development of alfalfa in Hillsdale sandy loam in the greenhouse (Lawton and Tesar, 1958)

K applied	Dry weight of roots 0-8 inches	8-16 inches	16-24 inches
kg/ha		g/pot	
56	55.6	15.8	3.5
347	63.0	18.5	5.1
	Soil analysis - change from original		
		kg/ha	
Exchangeable K	−76	−47	+ 6
Acid soluble K	−45	−18	+39

of age. The average daily increase in length of roots was more than 3 miles. There were 14 billion root hairs with a total area of 4000 sq ft. Nearly 55 miles of new root hairs were formed each day on the average. Little other detailed work of this nature has been conducted because of the complexity of the task.

It is of interest to consider at what stage in plant growth active root extension ceases. Foth (1962) observed in corn that lateral growth of roots was completed 1 or 2 weeks before tassel emergence. Extensive growth of roots below 38 cm then occurred. By early roasting ear stage, root growth was completed with the cessation of brace root growth.

C. Type of Root Systems and Potassium Placement

Root systems are classified as fibrous or tap. If a tap root is produced early, part of the fertilizer might best be placed directly under the seed. Alfalfa, tomatoes, onions, and sugar beets are examples of such crops. On the other hand if many lateral roots are formed early sideband placement may be best.

Soybean fertilization has been a difficult problem to solve. It appears now, however, that soybeans are more tap rooted in nature. Much more work is needed on placement with or under the seed, combined with heavy broadcast plowdown, as compared with sideband placement.

Progressive farmers are adopting the practice of heavy broadcast application of fertilizer, including potassium, to build up the fertility level of the plow layer. In addition, they are plowing deeper and the crop may have a highly fertilized layer 30 cm deep on which to grow rather than 15 cm of a low fertility soil. The modern trend then is to minimize sizeable amounts at planting in favor of large amounts broadcast and plowed down. With a small amount of fertilizer with or directly under the seed, the type of root system as related to placement may become less important.

D. Spacing of Plants

The trend is toward more plants per acre for most crops. Several factors are operating to affect uptake of potassium. First, the higher plant populations give a higher yield (Fig. 7). With 224 kg K/ha (200 lb/acre) yield response of corn was 1,320 kg/ha (21 bu) with the low plant population and 2,450 kg/ha (39 bu) with the higher population. This of course results in greater uptake and removal of potassium.

Second, with closer spacing the plant roots are competing with each other as well as with the exchange sites in the soil. Presumably the roots might permeate the soil more thoroughly and extract potassium more completely from the soil. On the other hand we know little about the effects of roots and their products on each other.

Too, as row crops are grown in narrower rows the soil should be moist a greater portion of the time. This should render the potassium in the soil more available to the plant.

In work on an irrigated Leon fine sand in Florida several hybrids were compared at 47,000, 69,000, and 96,000 plants/ha (19,000, 28,000, and

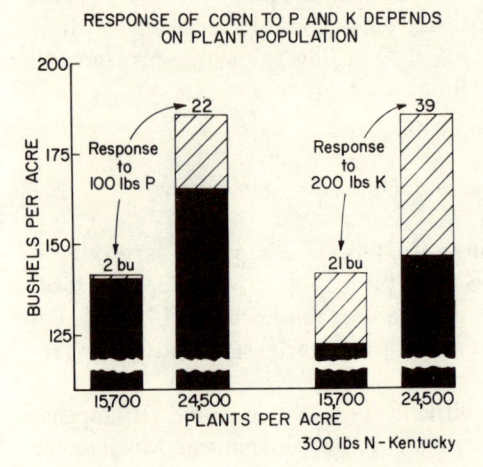

Fig. 7—Corn gives greater response to K at the higher plant population (Hatfield and Ragland, 1966).

Table 3—Effect of plant population and hybrid on percent K, K uptake, dry weight, and percent recovery of K* (Robertson et al., 1965)

Plants/acre	1963 Fla. 200	1963 DeKalb 805	1964 Fla. 200	1964 DeKalb 805	1964 Pioneer 309B	1964 Dixie 18	1964 Coker 67
			K content, %				
19,000	1.40	1.44	0.98	1.07	1.27	1.15	1.20
28,000	1.44	1.34	1.08	0.94	1.36	1.20	1.03
39,000	1.63	1.51	1.00	1.14	1.18	1.02	1.31
			K uptake, lb/acre				
19,000	150	128	125	86	120	121	131
28,000	164	135	134	96	150	159	130
39,000	228	170	142	125	156	145	192
			Dry weight, 1,000 lb/acre				
19,000	9.9	8.3	11.8	7.8	8.8	9.6	10.2
28,000	10.6	9.3	12.6	9.8	9.9	12.8	12.0
39,000	12.8	10.3	13.6	10.4	11.4	13.3	12.5
			Recovery of K, %				
19,000	58.9	55.4	40.6	33.6	45.6	39.3	51.3
28,000	60.2	58.1	53.3	32.6	59.6	58.6	48.0
39,000	85.5	65.7	51.0	53.0	56.6	48.6	65.0

* Average of 3 fertilizer rates.

39,000 plants/acre) (Table 3). In all seven comparisons one of the two higher populations was always higher in percent K and in pounds of K per acre absorbed. Percentage of applied potassium taken up was always lowest at the lowest plant population, probably because of less complete exploration of the soil by the roots.

E. Soil Aeration

It is well known that compact soil layers or poorly drained conditions limit root development. Brown (1963) states that inadequate soil drainage has long limited alfalfa production in Maine. His work indicates that on fields of marginal internal drainage, lime and/or K were effective in increasing yields (Table 4). He attributes this to a more extensive rooting habit and lateral spreading of individual plants.

The inhibitory effect of poor aeration on uptake is more pronounced with potassium than any other nutrient. Work by Lawton (1946) showed

Table 4—Yields of alfalfa-timothy hay (10% H_2O) in fifth harvest year (Brown, 1963)

Soil drainage	pH 5.8 4,500 kg/ha-lime 37 kg K/ha	pH 5.8 4,500 kg/ha-lime 150 kg K/ha	pH 6.5 20,000 kg/ha-lime 37 kg K/ha	pH 6.5 20,000 kg/ha-lime 150 kg K/ha
		kg/ha		
Somewhat poor	7,400	10,700	8,100	12,300
Moderately well	9,200	10,700	9,800	11,900

the following effect on uptake of nutrients by corn when grown in non-aerated and aerated cultures of a silt loam soil containing 50% water.

	Relative uptake nonaerated/aerated
Potassium	0.3
Nitrogen	0.7
Magnesium	0.8
Calcium	0.9
Phosphorus	1.3
Dry matter	0.6

Compaction affects soil aeration and Bower et al. (1944) reported that corn grown on plowed plots was much less deficient in potassium than plants on listed, disked, or subsurface-tilled Fayette, Clarion, and Webster soils. An average of these soils showed the corn plants to have 1.8% K in the plowed plots and 1.3% in the subsurface-tilled areas.

Some research has shown that calcium gain by roots is primarily through exchange absorption. Potassium on the other hand is highly dependent on oxygen-controlled metabolic action. Roots in low oxygen solution induced by bubbled gas lose potassium to the solution. This is reversible, however, in that when oxygen is restored to the solution the roots resume normal uptake of K (T. Ando, 1967. Cation absorption by barley roots in relation to root cation exchange capacity. *M.S. Thesis. Univ. of Massachusetts*).

Labanauskas et al. (1965) showed a marked effect of reduced oxygen in the soil on potassium uptake by citrus (Fig. 8).

F. Soil Temperature

Temperature as an environmental factor strongly affects the uptake of potassium by plants. Just how much of this is due to actual availability of

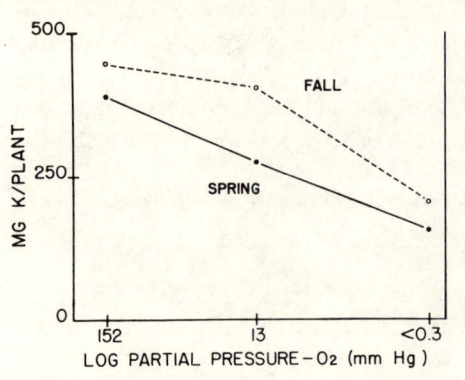

Fig. 8—Oxygen in the soil affects uptake of K by citrus (Labanauskas et al, 1965).

Fig. 9—Potassium uptake by sorghum is much greater at 32 than at 16C with zero or with added K on this Floyd soil (Weber and Caldwell, 1964).

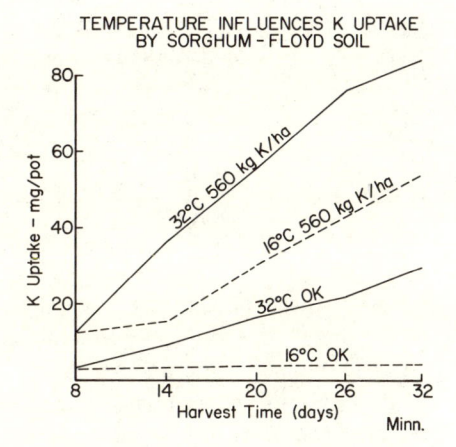

potassium in the soil and how much is due to extension of plant roots and speed of plant physiological processes is not clear. Another possibility is that higher order phase transitions take place in the structure of water and aqueous solutions as temperature increases. Miller and Davey (1967) found breaks at around 15 and 30C in roots of intact wheat plants during an 8-hour period.

Weber and Caldwell (1964) compared uptake of potassium by grain sorghum at 16 and 32C on a somewhat poorly drained soil (Floyd). The plants grown at 32C without additional potassium took up much more potassium over time than did those grown at 16C (Fig. 9). With 560 kg K/ha applied, the uptake was also much greater at 32C than at 16C.

The percentage potassium in first crop alfalfa as related to exchangeable potassium in northern and southern Wisconsin soils is of interest (Fig.

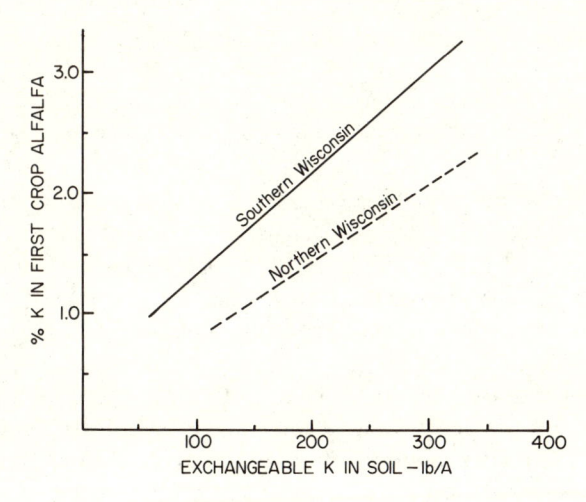

Fig. 10—With a given soil test level the percent K in the first crop alfalfa is much less in northern Wisconsin (*personal communication*, Leo Walsh, 1967).

Table 5—Effect of temperature and P on percent K in 5-week-old Fireball tomato plants (Martin and Wilcox, 1963)

P rate	13 C	14 C	16 C	20 C
ppm		% K		
5	1.55	2.30	2.93	3.46
50	2.48	3.20	3.24	3.71
100	2.70	2.83	3.45	3.30

10). As to be expected, the potassium level of the plant increases with increasing level of exchangeable potassium. However, with a given soil test level the amount of potassium in the alfalfa is much less in northern Wisconsin. Early in the growing season this difference would be accentuated. Evidently the low soil temperatures and perhaps low soil oxygen levels limited uptake of potassium in northern Wisconsin.

Martin and Wilcox (1963) observed that there was a sharp growth response of tomatoes between 13 and 14C. With a low phosphorus level, increasing the temperature from 14 to 20C markedly increased the percent K in the plants (Table 5). At a high phosphorus level there was no further increase above 16C. The percent phosphorus in tomatoes did not increase with increasing temperature above 14C.

Cannell et al. (1963) found the K content in tomato plants to be 4.15, 4.89, 5.79, and 4.50% K at 12, 20, 28, and 37C, respectively.

With 30-day-old corn grown outdoors at Wooster, Ohio, heating the soil when it dropped below 30C increased N and K level in the plant about 25%, P about 100%, while Ca and Mg were decreased by about 25 to 30% (Mederski and Jones, 1963). There were no differences in 60-day-old or mature plants.

Much work needs to be done on the effects of both temperature and oxygen supply on root extension. Top growth is affected and root growth must be too.

G. Rhizosphere and Nutrient Uptake

Potassium is essential for the growth of all microorganisms. Bacteria may contain up to about 2% K by dry weight while the mycelium of fungi may contain as little as 0.1% (Alexander, 1964). However, the total amount of K immobilized in this fashion probably does not exceed 10 to 15 kg/ha so it is unlikely that there would be serious competition for potassium between microorganisms and plants.

The area in the soil under the influence of plant roots is called the rhizosphere. The root surface and the accompanying microbial population is of most interest in this discussion. There is a vast microbial population, including bacteria, fungi, actinomycetes, and protozoa, on the surface of roots and root hairs. Increased concentrations of bacteria in rhizosphere soil of

120 times that of nonrhizosphere soil have been reported on certain root crops.

Different plant species often establish different subterranean floras. For example Azotobacter do not develop in the root zone of corn but do on wheat.

The plant contributions to the microflora are excretion products and sloughed-off tissue to help supply energy, carbon, and nitrogen and growth factors (Rovira, 1962). In a soil low in oxygen, potassium may leak out of the root and be partially used by the microflora. Too, as a plant matures potassium is lost. Large quantities of CO_2 are liberated by both the roots and microbial population and this may cause a solubilization of certain nutrients including potassium. Oxidation-reduction properties have also been suggested for plant roots as well as for the microflora.

The microflora contribution to the plant may involve a variety of substances including growth-regulating materials. Plant growth is typically more rapid in nonsterile environments providing a pathogen is not present. However, it is postulated that the microbiological buffer zone may serve to protect the plant against a pathogen. The effect of nodule bacteria on legumes is well known. Less well known is the effect of mycorrhiza on fixation of nitrogen for such crops as the *Pinus* species.

Moss (1957) inoculated apple seedlings and cuttings growing in sterilized soil with sporocarps of an *Edogone* species. The inoculated plants became mycorrhizal and grew better than the uninoculated plants. The mycorrhizal plants had a significantly higher content of potassium, calcium, iron and copper. Gerdemann (1964) grew corn with and without *Endogone*. The mycorrhizal plants produced about four times as much dry matter and contained a higher percentage of phosphorus. Potassium, magnesium, boron and manganese were present in higher concentrations in the nonmycorrhizal plants but because of the much greater weight of the mycorrhizal plants, the latter contained much larger quantities of all nutrients.

The mycorrhizal fungus infects the root and derives its food from the plant. While the fungus is growing through the roots, it also produces a network of hyphae in the soil. While the reason is not well known, the mycorrhizae increase the plant's ability to absorb nutrients from the soil. It has been postulated that the hyphae function as root hairs. Too, they grow some distance into the soil and may increase the distance from which a root can obtain nutrients.

Water relations is another aspect about which little is known in the rhizosphere. The change in amount of water in this zone must be sometimes quite marked and could affect potassium absorption.

Microbial surfaces have active exchange sites. Hence the effective exchange capacity in the vicinity of the roots is likely to be much higher than that of the inorganic soil colloids. This must surely affect the concentration and array of ions that can enter the roots.

IV. VARIETY

Varieties within a given species may vary considerably in content of potassium. While this might be expected with wide variations in yield this may occur even with approximately the same yield. The extent and activity of the root systems may be prime reasons for these differences, but as yet few of the details are known.

A. Corn

There is much evidence that concentrations of several nutrients are highly heritable in corn. A few examples will be cited.

Results of a detailed study on effect of nitrogen-phosphorus-potassium rate and population of five hybrids grown on an irrigated Leon fine sand in Florida are reported in Tables 3 and 6. The corn was harvested as sillage at the early dent stage when it contained 70% H_2O. At the intermediate fertilizer rate the five hybrids in a given year varied from 1.06 to 1.43% K, 147 to 208 kg/ha (131 to 187 lb/acre) total uptake and 42.6 to 60.3 in percent recovery of applied K. The variation between two seasons with the same hybrid was also quite marked. At the intermediate fertilizer rate one

Table 6—Effect of rate of NPK and hybrid on percent K, K uptake, dry weight, and percent recovery of K* (Robertson et al., 1965)

N+P+K lb/acre	1963 Fla. 200	1963 DeKalb 805	1964 Fla. 200	1964 DeKalb 805	1964 Pioneer 309B	1964 Dixie 18	1964 Coker 67
\multicolumn{8}{c}{K content, %}							
0-0-0	0.91	0.77	0.75	0.68	0.62	0.77	0.75
150+44+125	1.22	1.20	0.82	0.84	1.16	0.97	0.84
300+88+250	1.52	1.76	1.07	1.06	1.43	1.15	1.32
600+176+500	2.32	2.02	1.40	1.61	1.86	1.58	1.80
\multicolumn{8}{c}{K uptake, lb/acre}							
0-0-0	46	31	41	23	31	38	35
150+44+125	153	126	116	90	116	103	101
300+88+250	208	190	170	131	174	174	187
600+176+500	317	230	209	164	246	250	279
\multicolumn{8}{c}{Dry weight, 1,000 lb/acre}							
0-0-0	4.8	4.1	5.0	3.6	4.3	5.2	4.7
150+44+125	12.4	10.6	13.2	10.4	10.0	11.5	12.1
300+88+250	13.6	11.3	15.4	12.5	12.2	15.0	14.3
600+176+500	13.6	11.2	17.1	10.8	13.7	15.8	15.1
\multicolumn{8}{c}{Recovery of K %}							
150+44+125	85.4	76.1	60.3	52.3	64.3	51.0	55.3
300+88+250	64.9	63.9	51.6	42.6	55.6	53.6	60.3
600+176+500	54.2	39.3	33.0	27.4	42.0	42.0	48.6

* Average of 3 plant populations.

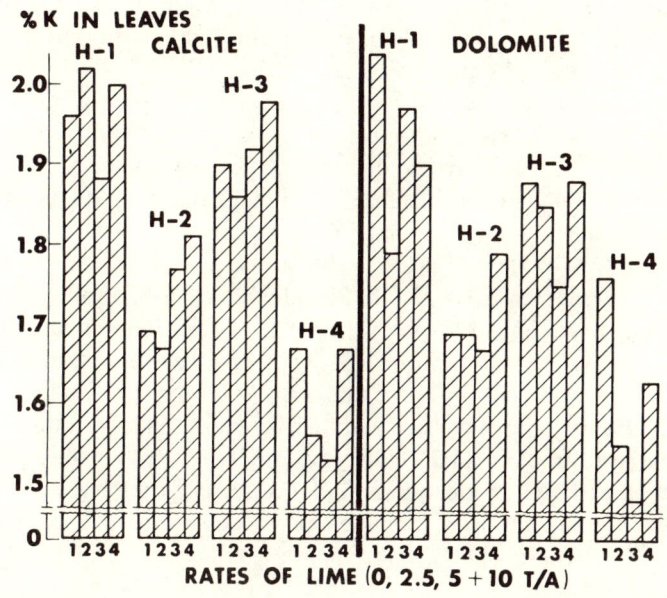

Fig. 11—Percentage K in the ear leaf varies among four hybrids (H) and is influenced by liming (Baker et al., 1966).

hybrid varied from 1.52 to 1.07% K, 233 to 190 kg/ha (208 to 170 lb/acre) total uptake, and 64.9 to 51.6% in recovery of applied K.

At the intermediate population the five hybrids in a given year varied from 0.94 to 1.36% K, 107 to 178 kg/ha (96 to 159 lb/acre) total uptake, and 32.6 to 58.6% in recovery of applied K. The variation between the two seasons with the same hybrid was quite marked. At the intermediate population, one hybrid varied from 1.44 to 1.08% K, 184 to 150 kg/ha (164 to 134 lb/acre) total uptake, and 60.2 to 53.3% in recovery of applied K.

Considerable variation in percent of potassium among four hybrids has been shown in Pennsylvania (Fig. 11). Calcitic lime increased the percent

Table 7—Effect of source of lime and hybrid on content of K and Mg (Bradford et al., 1966)

	Hybrid I		Hybrid II		Hybrid III		Hybrid IV	
	\multicolumn{8}{c}{Rate of K, ppm}
	0	400	0	400	0	400	0	400
	\multicolumn{8}{c}{% K}
Calcitic lime	3.24	5.79	3.08	5.11	3.34	5.73	3.59	5.75
Dolomitic lime	2.99	5.65	2.32	4.90	2.83	5.56	3.17	5.62
Decrease	0.25	0.14	0.76	0.21	0.51	0.17	0.42	0.13
	\multicolumn{8}{c}{% Mg}
Calcitic lime	0.25	0.17	0.22	0.14	0.25	0.16	0.26	0.15
Dolomitic lime	0.76	0.36	0.66	0.34	0.75	0.34	0.75	0.36
Increase	0.51	0.19	0.44	0.20	0.50	0.18	0.49	0.21

potassium in three of the four hybrids while dolomitic lime only increased it in one hybrid. In another study the concentration in the ear leaf in a group of hybrids ranged from 1.7 to 2.8% K.

The effect of dolomitic lime on content of potassium as related to hybrid is of interest (Table 7). Hybrid II showed the greatest depression in potassium when dolomitic lime was added and Hybrid I the least. However, Hybrid II showed no greater increase in magnesium with dolomitic limestone applied than other hybrids and this indicates genetic control.

In Tennessee, researchers indicate a difference among hybrids in content of potassium (Charles E. Freeman. 1965. The effect of K fertility levels on the uptake and utilization of K, Ca, and Mg by corn inbreds and hybrids. *M.S. Thesis, University of Tennessee*). Hybrids showing little deficiency on low potassium soils had low potential yield with applied potassium.

Generally speaking the potassium content of corn grain is little affected by potassium level in the plant. However, differences have been found in potassium content between opaque-2 and translucent corn seeds. The following potassium contents were obtained in 63 comparisons in second generation segregates where opaque inbreds were crossed (Goodsell, 1968).

	K content	
	Opaque-2	*Translucent*
	%	%
Seeds	0.52	0.37
Embryo	2.18	1.94
Endosperms	0.15	0.09

The implications are not clear but in the next decades there will be increasing emphasis on opaque-2 or high lysine corn.

B. Soybeans

One of the goals of researchers is to find varieties which will respond to fertilizer. High fertility plots were established at the Oblong Soil Experiment Field in Illinois by applying 410 kg P and 186 kg K/ha one year and 186 kg K/ha the next year (Dunphy et al., 1966). With 30 genotypes tested, responses to higher fertility ranged from 74 to 1,660 kg/ha. At another location with 24 genotypes, responses to higher fertility ranged from −74 to 610 kg/ha. In general, genotypes with capacity for high yield usually exhibited capacity to respond to higher fertility. For example, of the 8 genotypes that were significantly more responsive in the former trial, 7 were among those which gave significantly higher yields. It is of interest that varieties that exhibited the most severe potassium deficiency symptoms were usually not the ones that gave the greatest yield response to fertilizer.

New varieties with capacity for higher yields are continually coming on the scene. At the same time improved management of present varieties, including closer row spacings and weed control, will increase yields and thus increase the demand for nutrients.

C. Rice

Variations exist among rice varieties in response to potassium. Work in Taiwan shows the following ranges among varieties in response to potassium (Tsai-Fua Chiu and Lan-TiLi, 1965).

	First crop	Second crop
Percent yield response to 76 kg K/ha	4.5 to 7.9	−0.4 to 2.1
Total uptake, K/ha	60 to 78	71 to 96

V. YIELD LEVEL

The yield level within a species is a major factor influencing the amount of potassium needed. It is well recognized that higher yielding crops take up and remove larger amounts of potassium. One point which often is not recognized, however, is that under conditions of a yield response to potassium the percent potassium in the plant goes up also. Similar increases occur with additions of many other elements.

Increased content of potassium in plants has often been referred to as "luxury consumption". It is normal that potassium should increase in plants as potassium is applied. The key point is whether or not the value of the increased yield or quality of the crop goes up faster than the cost of each increment of potassium.

Unfortunately most of the published information on nutrient content of crops has been at a considerably lower yield than progressive growers obtain. However, since there is much interest in the nutrient content of high yielding crops, attempts have been made to extrapolate using known data on lower yields. One approach is to use the technique employed by R. D. Munson (*personal communication, 1967*). The total uptake of nitrogen-phosphorus-potassium per acre in the above-ground portion is divided by the actual yield in bushels per acre. To extrapolate to 12,544 kg/ha. (200 bu/acre) the values are then multiplied by 200.

An example based on the data of Benne et al. (1964) is shown below:

	N	P	K
Lb/bu (Mich. 139 bu/acre)	1.52	0.22	1.48
200 bu - total uptake-lb	304	44	296
Assumed nutrient recovery			
(Soil + Fertilizer)	60%	20%	50%
Available nutrients needed*-lb/acre	507	220	592

*Total uptake value divided by the assumed nutrient recovery i.e., $304 \div 0.60 = 507$.

The stover portion and the roots will probably be higher in K and other nutrients at the 12,544 kg/ha (200 bu) yield level than at the 8,700 kg/ha (139 bu) level. However, there will not be a proportionately higher amount of stover and roots at the higher yield so there is some compensation.

Similar calculations can be made with other grain crops. However, where the entire above-ground portion is harvested allowance must be made for the increased percentage content of nutrients at the higher yield level.

A. Soybeans

Soybean yields increased rather rapidly up to 70 kg K/ha with a relatively slow increase in percent potassium in the plant (Fig. 12). Above that point yield did not increase rapidly while percent potassium did. Soybeans

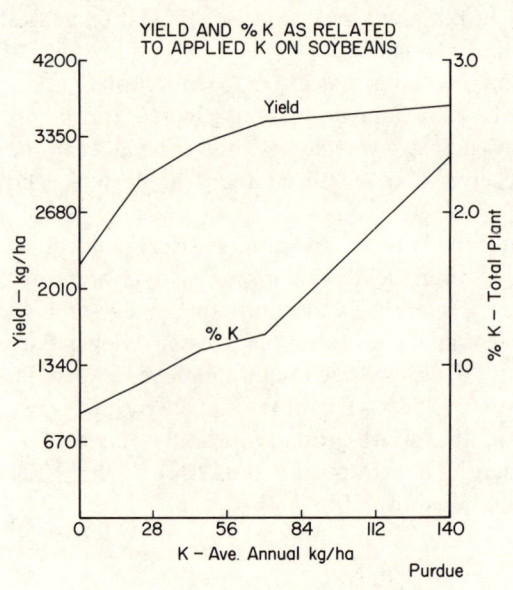

Fig. 12—Percentage K increases along with yield of soybeans (*personal communication*, S. A. Barber 1967).

AVAILABILITY AND UPTAKE BY PLANTS 373

were grown in a 4-year rotation with the potassium applied ahead of the corn. S. A. Barber calculated the optimum rate to be about 93 kg K/ha on this soil.

In an Iowa study, as the potassium content of the leaves and petioles increased due to treatment, the yields also increased to a maximum point, above which yields decreased as the potassium content continued to increase (Fig. 13). Obviously some other limiting factors caused the yields to level off at 2,620 kg/ha (39 bu/acre). Perhaps another element such as magnesium became limiting.

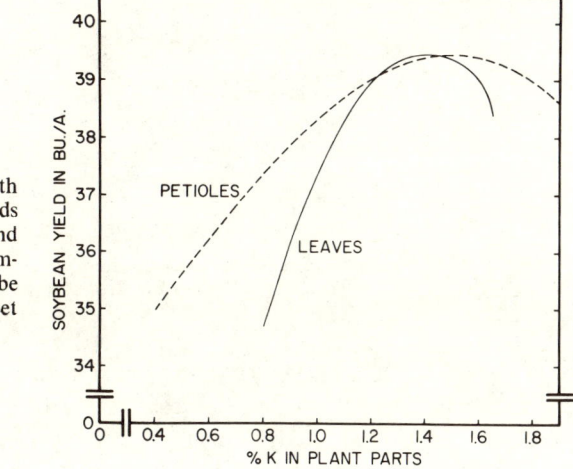

Fig. 13—Up to a point, with increasing soybeans yields percent K in petioles and leaves increase. The optimum point appeared to be about 1.5% K. (Miller et al., 1961).

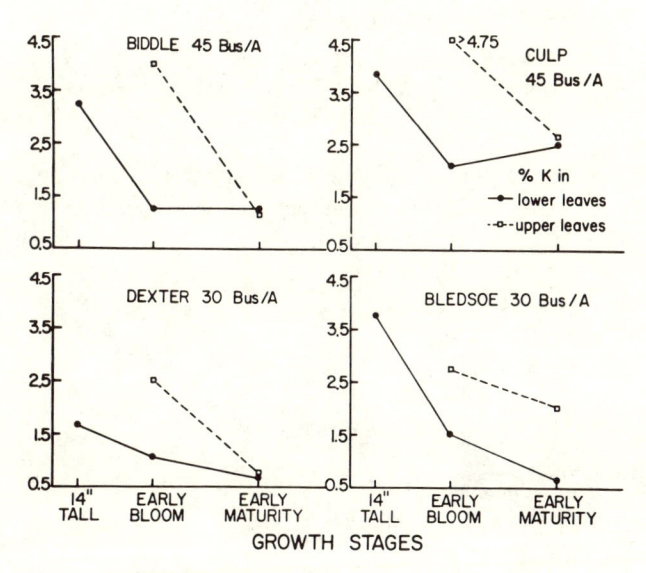

Fig. 14—A survey in Indiana showed the leaves of soybeans yielding 45 bu/acre to be higher in K than those yielding 30 bu/acre (*personal communication*, A. J. Ohlrogge, 1967).

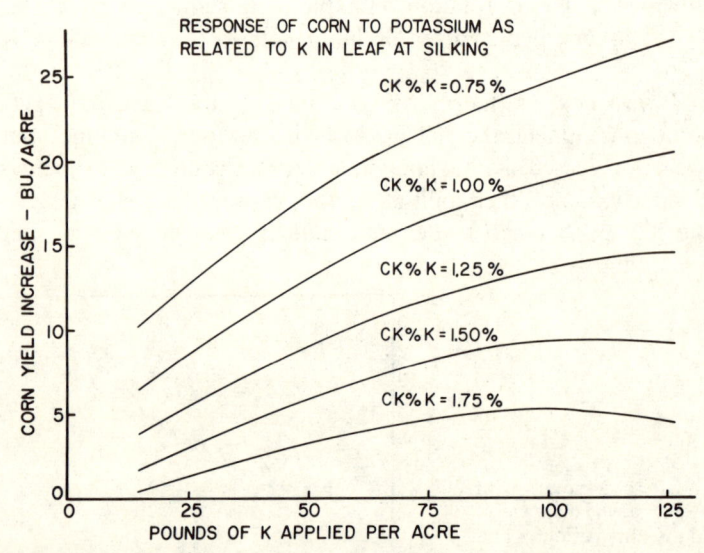

Fig. 15—Predicted yield increases of corn grain from K applications in relation to the percent K in corn leaves at silking time from plots that received no K (Hanway et al., 1962).

A survey was conducted in Indiana to identify soil and plant characteristics associated with high and low soybean yields. Among other nutrients K was found to be distinctly higher in the leaves of plants at 3,024 kg/ha (45 bu/acre) than at 2,016 kg/ha (30 bu/acre) (Fig. 14). Of course there are many other factors which might be associated with higher yields.

B. Corn

A North Central Regional study involved 41 field experiments on corn (Hanway et al., 1962). The relation between potassium content of corn leaves at silking time from plots receiving no K fertilizer and the predicted

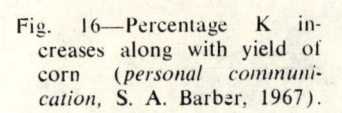

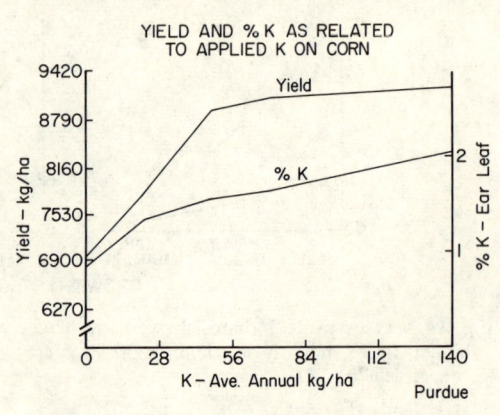

Fig. 16—Percentage K increases along with yield of corn (*personal communication*, S. A. Barber, 1967).

yield increases obtained from different rates of potassium is shown in Fig. 15. The regression equation indicated that no increase in yield would be expected when the K content of the leaves is slightly above 2.0%. A similar relationship between yield and percent potassium in the ear leaf in Indiana is shown in Fig. 16.

C. Other Crops

Under the conditions of a response of alfalfa to applied potassium in Minnesota the yield continued to increase with percent $K - r_2 = 97.6$ (Fig. 17). The highest percent K in this first crop alfalfa was 2.4%. In Indiana in a 3-year experiment yielding 19,264 kg/ha (8.6 tons/acre) one year, 40% of the cuttings over the 3-year period contained 3% K or above (George et al., 1967). The concept of adequate contents of K has continued to climb over the years—from 0.75, to 1.2, to 1.5, to 2.0%, and higher. Part of this change is related to the fact that alfalfa is being cut more frequently and with younger plants being harvested, higher potassium levels in the plant can be expected.

Cotton shows much the same behavior as other crops. Bennett et al. (1965) found that in general the yield of cotton increased with potassium content of plants. At the 90-day sampling period 2.75-3.25% K was found at the highest production.

Rates of K to 740 kg/ha increased yields of Coastal Bermuda up to 16,857 kg/ha with an accompanying increase in K content to 2.33% (Fig. 18). Yields began to level out at about 1.7% K, however.

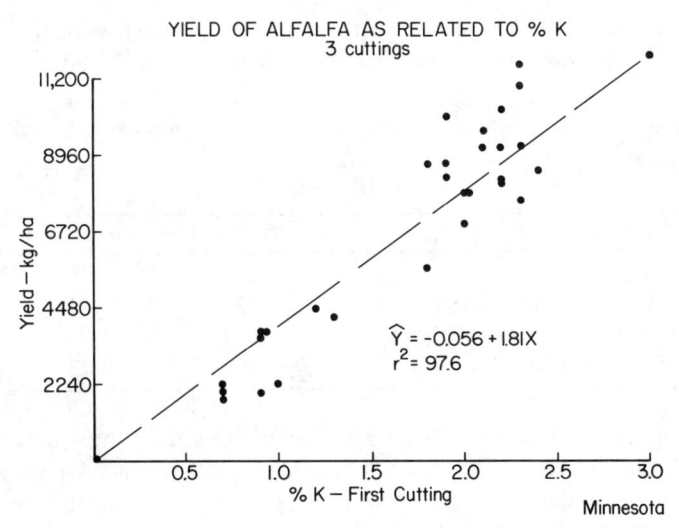

Fig. 17—Yield of alfalfa is related to percent K (*personal communication*, C. J. Overdahl, 1967).

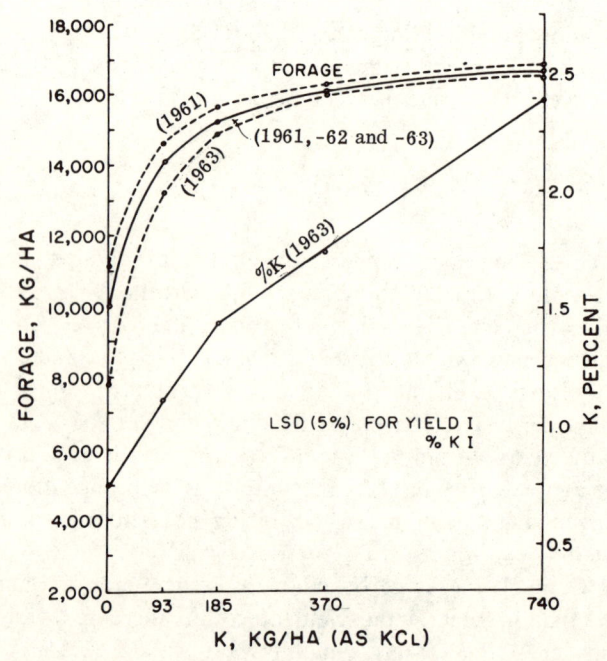

Fig. 18—Yield of Coastal Bermudagrass is related to percent K (Adams et al., 1967).

VI. SPECIES

Obviously there is much difference in the total amount of K required by various species. Data on representative crops are shown in Table 8. Yield level, soil fertility level, variety, and many other environmental factors would obviously affect the values but the values shown do give an indication.

Method of harvest is a factor. Below is an example for corn based on data of Benne et al. (1964).

Method of harvest	Percentage removal of N-P-K		
	N	P	K
Field chopper (silage)	93	95	84
Picker (grain + cobs)	60	75	15
Picker-sheller (grain)	57	73	11

In evaluating potassium needs length of growing season must be considered. Snap beans reach maturity in 55 days and probably need potassium at a faster rate than the apple for example. Tomatoes and sugar beets are examples of crops with high total uptake in a period of about 120 days. This is in contrast to sugar cane which may grow all year long. Alfalfa requires

Table 8—Potassium contained in total crop and in the harvested portion of representative crops (Amer. Potash Inst. and Nat. Plant Food Inst.)

	Yield	K in total crop	K in harvested portion
		kg/ha	
Grapes	22,400	75	45
Cotton	1,400	85	35
Wheat	4,000	100	15
Rice	5,000	100	20
Soybeans	3,350	110	65
Snapbeans	13,400	130	45
Corn	8,400	180	35
Orange	(800 boxes)	165	130
Tobacco	3,150	175	85
Alfalfa	13,400	250	250
Potato	45,000	300	200
Coastal Bermuda	22,400	370	370
Tomato	67,000	450	335
Sugar beets	67,000	500	185
Sugar cane	224,000	550	320
Celery	168,000	700	630

and absorbs potassium quite rapidly. For example a 6,720 kg/ha (3 tons/acre) second cutting of alfalfa allowed to grow 35 days could contain 170 kg K. This would be an average daily uptake of 4.8 kg.

Corn has been estimated to take up about 70% of its total K requirement in the first 60 days, with 65% coming in the 30-60 day period. Early potatoes absorb about 50% of the total K requirement in 40 days, while tomatoes absorb 50% in 80 days. Soybeans require about 20% of their total K in the first 50 days while snap beans take up about 75% of their requirement in 40 days. These comparisons help to illustrate the wide variation among species in time and rate of potassium uptake.

Extent of root development among species is a large variable, e.g., contrast snap beans and apples or sugar beets and sugar cane. The high requirement of alfalfa has been mentioned. Alfalfa notably has a deep root system and in some soils penetrates 600 to 900 cm deep. Just how deep alfalfa effectively absorbs potassium is not well known, however.

A. Legumes vs. Nonlegumes

The growth of a legume in association with a nonlegume companion crop is generally less than when a legume is grown in pure stand. Explanations include competition for light, moisture, nutrients, excretion of toxic materials and microflora effects. The greater content of K in grasses and/or weeds as compared to a legume is well illustrated in Fig. 19 and 20. Under low potassium conditions in legume-grass mixtures the legumes tend to disappear. This is often called the "legume-grass battle".

Lambert and Linck (1964) studied the efficiency of intact roots of al-

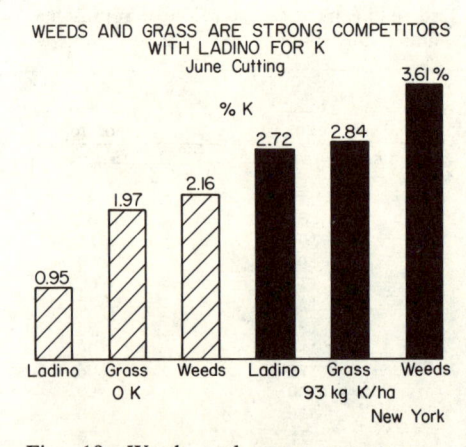

Fig. 19—Weeds and grass are strong competitors with alfalfa for K (Blaser and Brady, 1950).

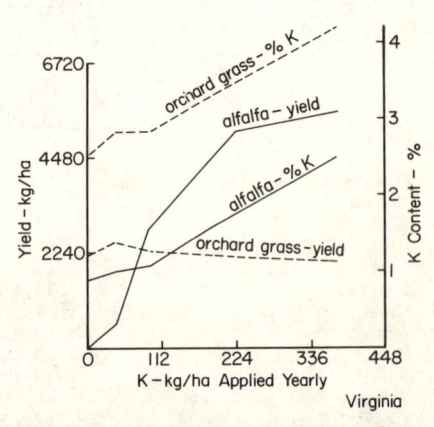

Fig. 20—Orchard grass maintains about the same advantage over alfalfa in percent K regardless of K application (Blaser, 1962).

falfa and oats in absorbing ^{42}K. With whole root systems only 9% of the ^{42}K was found in the alfalfa while 91% was in the oats (Table 9). With segments of attached roots, alfalfa root segments and shoots contained more ^{42}K than comparable portions of the oat plants of the same age (Table 10). It appears from this that the effective competition for K by oats must be due primarily to the greater volume of roots or to the greater number of root hairs.

B. Root Cation-Exchange Capacity

There has been considerable discussion as to the importance of root cation exchange capacity in absorption of divalent and monovalent cations (Crooke and Knight, 1962 and Drake, 1964). A major problem has been measuring the exchange capacity accurately. Although plant physiologists tend to regard cation-exchange properties of roots as being of little significance in development of ion uptake theories, soil chemists have been able to utilize this root property in a qualitative way as a possible explanation of differential uptake of mono-and divalent cations by different plants.

The general idea is that the greater the cation-exchange capacity of the plant root the greater will be the divalent-monovalent ratio in the plant. This relatively greater absorption of divalent cations by colloids with high exchange capacity is thought to be related in part to the greater bonding or absorption energy between the divalent cation and the high exchange colloid. On the other hand the low cation-exchange capacity colloid appears to absorb monovalent cations with relatively greater energy than divalent cations.

Table 9—Amount and percentage of the total ^{42}K in 7-day-old alfalfa and oat plants using whole root systems (Lambert and Linck, 1964)

	Radioactivity in cpm and % of the total K^{42} in plants			
	Alfalfa		Oats	
	count/min	%	count/min	%
Shoots	2,724	6	16,780	35
Roots	1,432	3	27,070	56

Table 10—Amount of ^{42}K present in alfalfa and oat plants using segments of attached roots (Lambert and Linck, 1964)

	Radioactivity (cpm $\times 10^3$)			
	Alfalfa age in days		Oats age in days	
	4	15	4	15
Root tip (1cm)	4	3	4	3
Root segment (3 mm)	9	7	7	10
Remainder of root	36	92	21	51
Shoots	305	232	140	187

Gray et al. (1953) found potassium uptake by plant species at low levels of soil potassium to be closely correlated with root cation-exchange capacity. Generally legumes have a higher cation-exchange capacity than nonlegumes.

McLean et al. (1956) showed a correlation of $r = 0.866$ between cation-exchange capacities of plant roots and their percentage of N. As more nitrogen was supplied to a given species through the growth medium, an increase in percentage of nitrogen was obtained and this was accompanied by a corresponding increase in the cation exchange capacity of plant roots (Table 11). McLean (1957) observed a progressive increase in cation-exchange capacity of four nonlegumes as nitrogen was increased in solution cultures.

Table 11—Cation-exchange capacities and percent N in roots and tops as related to N level of the nutrient solution (McLean et al., 1956)

	Age days	N level	Cation-exchange capacity of roots meq/100g	N content Roots %	N content Tops %
Corn	15	Low	21.6	1.61	3.16
		Med	24.6	1.75	3.21
		High	27.4	1.96	3.84
Oats	15	Low	17.7	1.29	3.50
		Med	21.0	1.51	4.15
		High	24.4	2.14	5.08
Cotton	25	Low	44.7	2.35	4.84
		Med	45.9	2.65	5.39
		High	50.0	3.03	6.40
Soybeans	25	Low	54.6	3.93	7.18
		Med	56.3	4.24	7.10
		High	60.9	4.21	7.21

The percentage potassium in the plants at the high nitrogen level was always lower than in the plants grown at the low or medium nitrogen level. However, Franklin (1966) found that increasing nitrogen level in the nutrient solution increased cation-exchange capacity of corn roots about 50% but reduced that of barley and soybeans.

The higher exchange capacity in the roots tends to increase the divalent cation/monovalent ratio in the plants. Hence, according to this theory research tends to support that adding nitrogen to a deficient plant would make it less capable of competing for potassium but more capable of absorbing calcium and magnesium.

Bennett et al. (1953) observed a trend to decreased leaf potassium in corn as N application was increased to 90 kg/ha. Holmes observed that N rates of 0, 112, and 224 kg/ha gave 1.9, 1.6, and 1.5%, respectively, as an average for 15 single cross corn hybrids (Holmes, John C. 1956. Effect of N, stand, and genotype on leaf composition and corn yield. *M.S. Thesis, Iowa State University*). Whether these results were due to effects on cation-exchange capacity and subsequently on K absorption, to dilution, to organic acid content, or to roots exploring a soil lower in potassium of course is not known. Obviously if the additional nitrogen caused the root system to expand into areas of higher potassium the effect on cation-exchange capacity would probably be overshadowed.

VII. IN THE FUTURE

1) The primary avenue of entrance of potassium into the plant is through the root system. There is much to be learned on the extent and behavior of roots and the subsequent effects on potassium uptake.

2) Root characteristics of high yielding genotypes should be evaluated. Through this breeders may obtain guides for even further improvement. Pest resistance by roots must be of prime importance.

3) The effects of depth of tillage and depth of placement of plant nutrients on absorption of potassium by plants may provide information for more efficient use of potassium.

4) Subsoil fertility maps will be of help in predicting responses of crops to potassium.

5) Progressive growers are continuing to increase soil fertility levels. Level of potassium fertility as related to need for banded potassium must be studied.

6) Closer spacing of plants increases root contact with the soil and results in less water loss by evaporation from the soil. How does this affect potassium needs?

7) Absorption of potassium by plants is greatly influenced by soil aeration. What is the effect of improved management, minimum tillage, and

higher amounts of crop residues on soil physical properties and the subsequent effect on K uptake?

8) Absorption of K is influenced greatly by soil temperature. The effect of the trend to earlier planting on K absorption should be evaluated in view of the cooler soils encountered by the young seedlings.

9) Indications are that increased nitrogen content in a given species affects absorption of cations both as to total uptake and percentage of potassium in the plant. With greatly increased use of nitrogen in the future there is a need to study effects on cation relationships.

10) Genotypes vary in percentage content of K but little is known as to the relation to yield and quality potential.

11) Under conditions of a yield response to K, the percent potassium in the plant increases. The relationship between yield level and potassium content in the plant under various environments is a wide open field for vigorous research.

REFERENCES

Adams, W. E., A. W. White, R. A. McCreery, and R. N. Dawson. 1967. Coastal Bermudagrass forage production and chemical composition as influenced by potassium source, rate of frequency of application. Agron. J. 59:247-250.

Alexander, Martin. 1964. Soil microbiology. John Wiley & Sons, Inc. New York.

Baker, Dale E., R. R. Bradford, and W. I. Thomas. 1966. Leaf analysis of corn. Better Crops with Plant Food 50(3):36-40.

Benne, E. J., Elizabeth Linden, J. D. Grier, and Karen Spike. 1964. Composition of corn plants at different stages of growth and per-acre accumulation of essential nutrients. Quarterly Bull., Mich. Agr. Exp. Sta. 47 (1) 69-85.

Bennett, W. F., George Stanford, and Lloyd Dumenil. 1953. Nitrogen, phosphorus and potassium content of the corn leaf and grain as related to nitrogen fertilization and yield. Soil Sci. Soc. Amer. Proc. 17:252-258.

Bennett, O. L., R. D. Rouse, D. A. Ashley, and B. D. Doss. 1965. Yield, fiber quality and potassium content of irrigated cotton plants as affected by rates of potassium. Agron. J. 57:296-299.

Blaser, R. E., and N. C. Brady. 1950. Nutrient competition in plant associations. Agron. J. 42:128-135.

Blaser, R. E. 1962. Yield and uptake in alfalfa. Better Crops with Plant Food 46(3):6-15.

Bower, C. A., G. A. Browning, and R. A. Norton. 1944. Comparative effects of plowing and other methods of seedbed preparation on nutrient element deficiencies in corn. Soil Sci. Soc. Amer. Proc. 9:142-146.

Bradford, R. R., Dale E. Baker, and W. I. Thomas. 1966. Effect of soil treatments on chemical element accumulation of four corn hybrids. Agron. J. 58:614-617.

Brown, Cecil S. 1963. Speed alfalfa adoption with fertilizer. Better Crops with Plant Food 47(2):4-13.

Cannell, G. H., F. T. Bingham, J. C. Lingle, and M. J. Garber, 1963. Yield and nutrient composition of tomatoes in relation to soil temperature, moisture and phosphorus levels. Soil Sci. Soc. Amer. Proc. 27:560-565.

Crooke, W. M., and A. H. Knight. 1962. An evaluation of published data on the mineral composition of plants in the light of the cation exchange capacities of their roots. Soil Sci. 93-365-373.

Dittmer, H. J. 1937. A quantitative study of the roots and root hairs of a winter rye plant. Amer. J. Bot. 24:417-420.

Drake, M. 1964. Soil chemistry and plant nutrition. *In* F. E. Bear (ed.) Chemistry of soil. Reinhold, New York.

Dumenil, Lloyd, John Pesek, J. R. Webb, and J. J. Hanway. 1965. P and K fertilizer for corn: how to apply. Iowa Farm Science 19(10):11-14.

Dunphy, E. J., L. T. Kurtz, and R. W. Howell. 1966. Responses of different lines of soybeans to high levels of phosphorus and potassium fertility. Soil Sci. Soc. Amer. Proc. 30:233-236.

Fehrenbacher, J. B., P. R. Johnson, R. T. Odell, and P. E. Johnson, 1960. Root penetration and development of some farm crops as related to soil physical and chemical properties. Int. Congr. Soil Sci., Trans. 7th 3:243-252.

Foth, H. D. 1962. Root and top growth of corn. Agron. J. 54:49-52.

Franklin, R. E. 1966. Exchange and absorption of cations by excised roots. Soil Sci. Soc. Amer. Proc. 30:177-181.

Gerdemann, J. W. 1964. The effect of mycorrhiza on the growth of maize. Mycologia 56:342-348.

George, J. R., C. L. Rhykerd, and C. H. Noller. 1967. Luxury K use - fact or myth. Better Crops with Plant Food 51(2):2-5.

Goodsell, S. F. 1968. Potassium levels in mature seeds of normal and opaque-2 maize. Crop Sci. 8 (In press).

Gray, Bryce, Mack Drake, and W. G. Colby. 1953. Potassium competition in grass-legume associations as a function of root cation exchange capacity. Soil Sci. Soc. Amer. Proc. 17:235-239.

Hanway, J. J., S. A. Barber, R. H. Bray, A. C. Caldwell, M. Fried, L. T. Lutz, K. Lawton, J. T. Pesek, K. Pretty, M. Reed, and F. W. Smith. 1962. North Central regional potassium studies III. Field studies with corn. North Central Regional Publication No. 135.

Hatfield, A. L., and J. L. Ragland. 1966. New concepts of plant growth. Plant Food Rev. 12(1):2, 3, 16.

Iowa State University. 1960. Growth and nutrient uptake by corn. Iowa State University Pamphlet 277.

Jenny, Hans. 1967. Roots — underground space frontiers. Plant Food Review 13(1):1.

Labanauskas, C. K., L. H. Stalzy, L. J. Klotz, and T. A. DeWolfe. 1965. Effects of soil temperature and oxygen on the amounts of macronutrients and micronutrients in citrus seedlings (*Citrus sinenis*). Soil Sci. Soc. Amer. Proc. 29:60-64.

Lambert, R. C., and A. J. Linck. 1964. Comparison of uptake of ^{32}P and ^{42}K by intact alfalfa and oat roots. Plant Physiol. 39:920-924.

Lawton, Kirk. 1946. The influence of soil aeration on the growth and absorption of nutrients by corn plants. Soil Sci. Soc. Amer. Proc. (1945) 10:263-8.

Lawton, K., and M. B. Tesar. 1958. Yield, potassium content and root distribution of alfalfa and bromegrass grown under three levels of applied potash in the greenhouse. Agron. J. 50:148-151.

Lawton, Kirk, and R. L. Cook. 1954. Potassium in plant nutrition.' Advance. Agron. 6:253-303.

Liebhardt, W. C., and John T. Murdock. 1965. Effects of potassium on morphology and lodging of corn. Agron. J. 57:325-328.

McLean, E. O., D. Adams, and R. E. Franklin, Jr. 1956. Cation exchange capacities of plant roots as related to their nitrogen content. Soil Sci. Soc. Amer. Proc. 20:345-347.

McLean, E. O. 1957. Plant growth and uptake of nutrients as influenced by levels of nitrogen. Soil Sci. Soc. Amer. Proc. 21:219-222.

Martin, George C., and Gerald E. Wilcox. 1963. Critical soil temperatures for tomato plant growth. Soil Sci. Soc. Amer. Proc. 27:565-7.

Mederski, H. J., and J. B. Jones, Jr. 1963. Effect of soil temperature on corn plant development and yield: I. Studies with a corn hybrid. Soil Sci. Soc. Amer. Proc. 27:186-189.

Miller, R. J., J. T. Pesek, and J. J. Hanway. 1961. Soybean yield responses to fertilizers. Soybean Digest 22(3):6-8.

Miller, Raymond J., and C. B. Davey. 1967. The apparent effect of water structure on K uptake by plants. Soil Sci. Soc. Amer. Proc. 31:286-287.

Moss, Barbara. 1957. Growth and chemical composition of mycorrhizal and non-mycorrhizal apples. Nature Lond. 174:922-924.

Rains, Donald W. 1967. Light-enhanced potassium absorption by corn leaf tissue. Science 156:1382-1383.

Robertson, W. K., L. C. Hammond, and L. G. Thompson, Jr. 1965. Yield and nutrient uptake by corn for silage on two soil types as influenced by fertilizer, plant population and hybrids. Soil Sci. Soc. Amer. Proc. 29:551-554.

Rovira, A. D. 1962. Plant root exudates in relation to the rhizosphere microflora. Soils Fert. 25:167-172.

Tanner, J. W. 1967. Roots - underground agents. Plant Food Review 13(1):11-13.

Tsai-Fau Chiu, and Lan-TiLi. 1965. Varietal difference in fertilizer response and absorption of rice in Taiwan. Potash Review, Subject 6, 28th Suite:1-10.

Weber, J. B., and A. C. Caldwell. 1964. Soil and plant potassium as affected by soil temperature under controlled conditions. Soil Sci. Soc. Amer. Proc. 28:661-667.

Wiersum, L. K. 1958. Density of root branching as affected by substrate and separate ions. Acta. Botan. Neer. 7:174-190.

Wilkinson, S. R., and A. J. Ohlrogge. 1962. Principles of nutrient uptake from fertilizer bands: V. Mechanisms responsible for intensive root development in fertilized zones. Agron. J. 54:288-291.

Potassium Nutrition of Tropical Crops

H. R. VON UEXKUELL
Kali Kenkyu Kai (Potash Research Association) Tokyo, Japan

I. INTRODUCTION

Less than 50 years ago the tropics were thought of as a region of nearly unlimited fertility and natural wealth. Tropical agriculture provided much needed industrial raw materials such as rubber, guttapercha, fibers, dyestuffs, and basic materials for pharmaceuticals in addition to exotic spices, stimulants, fruit, and sugar.

With the development of synthetic products the importance of tropical agriculture as a producer of industrial raw materials has diminished. At the same time food requirements have grown immensely because of the rapid population growth. Many countries, whose agriculture once was a major supplier of raw materials, depend on large scale food supplies from the developed countries. The USA is the main supplier of these food supplies.

For many reasons it is far more difficult to increase the production of food crops than it is to increase the production of cash or plantation crops. When discussing the potassium nutrition of tropical crops, more time will be devoted to the most important tropical food crop, rice, than to commercial crops whose significance, though important, is limited to a small section of the community.

Good crop nutrition is a prerequisite for high yields. The fact, however, that tropical countries account for more than 65% of the world's agricultural population and yet consume less than 4% of all potassium fertilizer indicates that there exist severe obstacles to the use of fertilizer. A discussion of the potassium nutrition of individual tropical crops may be purely academic unless consideration is given to the factors that have so far prevented a wider use of fertilizer in the tropics.

II. GENERAL PROBLEMS OF FERTILIZER USE IN THE TROPICS

Though the basic principles governing plant growth are the same all over the world, fertilizer problems, for many reasons, are not the same in the tropics as in temperate regions.

A. Socio-Economic Difficulties

Tropical countries are, without exception, industrially less developed. The majority of the population depends on agriculture as the main source of livelihood. Unlike the developed countries, farming is usually not a business but rather a way of life. Farmers are slow in responding to economic incentives and in adopting yield-raising methods of production that require capital input.

Prices for agricultural products are low and unstable because there is no strong industrial sector in the economy to support fair and stable prices. Fertilizer prices, on the other hand, are comparatively high, mainly because of inefficient communications and marketing systems and because of excessive interest rates. Yields are low and highly dependent on the uncontrolled environment.

It has been calculated that under the above complex of circumstances a farmer in Thailand, for example, would have to sell 18-times more of his per hectare production of rice than a farmer in industrialized Japan in order to buy the same amount of fertilizer (von Uexkuell, 1964). Problems of land tenure further complicate progress in agriculture in many countries.

This brief account may illustrate that changes in the socio-economic and institutional field are one of the prerequisites to making fertilizer more freely accessible to needy crops.

B. Tropical Soils

There is a widespread opinion that most tropical soils are rich in potassium and the vigorous growth of the natural vegetation has often been considered proof of the fertility of tropical soils. In fact, however, most tropical soils are very poor, and the equilibrium between plant-soil-climate that enables the amazing performance of the tropical rain forest is extremely fragile.

The climax vegetation in the wet rain forests flourish in the environment which it creates itself. The rain forest is surprisingly independent from the fertility level of the original soil. Under the forest, fertility will accumulate at the surface. Low exchange capacity of the mineral part of the soil is no problem because a small amount of well-balanced nutrients selected by the vegetation is rapidly and efficiently circulated through the turnover of the organic matter. Once this circle is broken by clearing the forest, all the disadvantages of a highly-weathered tropical soil may become evident. The organic matter is an important structural element, as a buffering agent and main source of the exchange capacity, and plant nutrients will rapidly disappear. This leaves behind a soil that in extreme cases consists of aluminum trihydrate, hydrated iron compounds, and a few other less soluble minerals.

Most tropical upland soils are very low in mineral reserves and in exchange capacity. Permanent cropping and heavy yields are not possible unless great efforts are made to control erosion and to balance the severe physical and mineral deficiencies. Management of upland soils requires more skill in the tropics than it does in temperate regions.

In contrast to the poor upland soils, soils in the river valleys and coastal plains are comparatively fertile. With a high ground water table and poor drainage, silica will remain in the profile, leading to the formation of secondary clay minerals and a higher exchange capacity. If utilized for rice culture these soils may maintain their fertility for long periods. Through irrigation water or flooding they benefit from the fertility losses of the neighboring upland fields. The water layer covering the soil during the rainy season acts as an efficient buffer against the adverse influence of the elements and conserves the organic matter. Irrigated rice, therefore, may be considered as the agricultural climax vegetation of the tropics (von Uexkuell, 1965).

C. Climatic Problems

Provided water is available, the tropical climate has the great advantage that crops can be grown 365 days of the year. If the main growing season in the tropics, the rainy season, is compared with the growing season in temperate regions, however, the tropical climate may show several defects. Short days combined with insufficient sunshine keep assimilation at a low level. Long, warm nights, on the other hand, result in high respiration rates. A small excess in foliage may result in severe mutual shading and a yield reduction. The range of the optimum leaf area index is, therefore, much narrower than under the long-day climate with cool nights found in temperate regions. Problems of plant type, spacing, and fertilizer application in relation to the available light are more critical in the tropics. Climate, soil, and plant types also make it more difficult to maintain a good balance between root and top growth and between the vegetative and reproductive phases. To be fully effective, fertilizer application must be more precise with regard to quantity, nutrient ratio, and timing than in cooler climates.

D. Variety Problems

Nature selected most crops in the wet tropics for adaptability to low levels of fertility and low light intensity, favoring leafy plant types that show vigorous vegetative growth but are poor in reproductive growth. Plants selected under such conditions will not be particularly responsive fertilizer to. For most crops systematic selection and breeding for fertilizer responsiveness and high yields started only recently. High yields require high population densities of plants that will make the best use of one factor man can least influence in a positive direction—light. Breeding of plants whose leaf type and leaf position will secure maximum utilization of this factor will, therefore, be of outmost importance. Nutritional problems change drastically with the introduction of new, high-yielding varieties. Many chapters written on the nutrition of tropical crops may require rewriting in the future.

III. TROPICAL CROPS

A. Rice

Rice is staple food for more than half of the world's population. More significant, it is staple food in areas of fast population growth and food shortage. A good nutrition of rice, therefore, has great importance for a large part of mankind. Yields average about 5,000 kg/ha in Japan against 1,000-2,000 kg/ha in the tropics. The difference is mainly due to varieties, fertilizer, and water control.

1. Nutrient Uptake and Nutrient Removal

Detailed investigations in Japan showed that on the average about 16.8 kg N, 3.8 kg P, and 21.9 kg K are absorbed per 1,000 kg of rough rice (Takashashi, 1960). Other studies indicate that the above quantities and ratios remain fairly constant over a wide yield range (Yamazaki, 1967).

Most potassium is contained in the straw. Tropical rice has usually a much wider grain/straw ratio than rice grown in cooler regions. It is, therefore, surprising that some data from the tropics indicate a similar or even lower rate of potassium absorption per 1,000 kg of grain than is found in Japan. There are indications that in the tropics potassium absorption during early stages of growth is sufficient or even excessive, but it may be deficient during later stages.

Under normal conditions the percentage of potassium is high at transplanting and decreases gradually according to growth. It increases again after flowering until complete ripening (Ishizuka, 1965).

Though highly mobile in the plant, the net translocation of potassium from old leaves to other parts is low (Murayama, 1965). According to Tanaka (1956), 48% of all K found in the ear had been absorbed from the medium after heading against 19% for N and 15% for P. For high yields, therefore a healthy root system and a continuous absorption of potassium until maturity is of great importance. At the present low yield levels in the tropics, the total amount of K absorbed by rice may range between 30-60 kg K/ha. At high yield levels, absorption easily exceeds 250 kg K/ha. More important, this amount is taken up during a short period of about 3 months.

In wide areas of the tropics removal of potassium with the crop is low because only the panicle is harvested. The straw remains on the field.

2. Effects of Potassium

Potassium has the following noteworthy effects on rice:

A. YIELD COMPONENTS. The yield components affected by potassium are mainly the number of grains per head and the 1,000-grain weight. The number of effective tillers is not greatly affected by potassium. Kiuchi and Ishizaka (1961) showed that the optimum K content of the plant for an increase in the number of grains is around 2% at the booting stage. The most favorable condition for a good seed-setting rate and a high 1,000-grain weight is a continuous absorption of K from the soil and maintenance of a K content in the plant of between 1 and 2% at heading.

B. RESISTANCE TO LODGING. Potassium is known to increase lodging resistance in many crops, and rice is no exception (Hashimoto, 1959). Potassium accelerates the lignification of sclerenchymatic cells and is important for an increase in the thickness of the culm walls, especially at the lower parts of the stem (Table 1).

C. RESISTANCE TO DISEASES. Rice plants well supplied with potassium are more resistant than deficient plants to brown spot (*Ophiobolus miya-*

Table 1—Breaking strength of the culm in relation to the N and K application rate (Kono and Takahishi, 1961)

N level	K level	K level at culm base	Breaking load	K level in straw	N level in straw
		%	g	%	%
Low	Low	1.7	430	1.7	0.7
	Medium	2.2	420	2.0	0.9
	High	2.8	460	1.9	0.8
Medium	Low	0.5	320	1.6	1.1
	Medium	1.9	400	2.3	1.1
	High	2.8	420	2.3	1.0
High	Low	0.3	280	1.2	1.5
	Medium	0.7	370	2.1	1.6
	High	2.7	410	2.7	1.4

beanus), stem rot and eye spot (*Helminthosporium sigmoideum*), bacterial leaf blight (*Xanthomonas oryzae*), and sclerotium disease (*Sclerotium oryzae*) (Noguchi and Sugawara, 1966; Matsuo, 1948; Corbetta, 1954; Ono, 1957; Okamoto, 1958; von Uexkuell, 1966). In the case of blast (*Piricularia oryzae*), the most devastating disease, positive as well as negative effects of potassium have been observed (Kosaka, 1963; Okamoto, 1958). This problem should perhaps be further studied with special consideration of a possible interaction of potassium with silicon and magnesium.

D. PHYSIOLOGICAL DISEASES. Physiological diseases are usually the result of disturbances in the root zone. Potassium is the element whose uptake is most frequently affected by such disturbances, resulting in nutritional disorders. Potassium has proved to be effective in the case of a number of physiological diseases such as "akagare," "aogare," "akiochi," "suffocation disease," some types of "bronzing," and "straighthead" (Baba and Harada, 1954; Baba et al., 1965; Chang, 1961; Chin, 1961; Ponnamperuma, 1958; Tisdale and Jenkins, 1912).

E. POTASSIUM DEFICIENCY. Potassium-deficient plants are stunted with small, dark green leaves and a short, thin stem. The older leaves become chlorotic and dry off, starting from the tip. Dark brown spots appear, first at the leaf tip, spreading later over the whole leaf. The spots develop on the row of stomata cells first, then near vascular bundles but rarely close to motor cells. Since the appearance of brown spots, considered to be a typical symptom of potassium deficiency, is also related to the presence of ample nitrogen and ferric iron, Tanaka et al. (1966) suggest that the spots may be caused by excessive iron uptake under conditions of potassium deficiency. Photosynthesis is reduced and respiration is increased, particularly in lower leaves. More water is transpired per unit dry weight, and the moisture content of the plant is lower, particularly during earlier stages of growth. Ammonia, amides, and amino acids (valine, aspartic acid, glutamic acid, arginine, and putrescine) accumulate in the leaves. Total sugars increase whereas total carbohydrates decrease. Pollen from deficient plants may contain hardly any starch and do not germinate, causing a high percentage of empty grains. The activity of phosphorylase is low but the activity of B-amylase, catalase, peroxydase, and cytochrome oxydase is high in potassium deficient plants.

A rice plant is considered deficient in K if the K content of the straw at harvest is below 0.8% (Ishizuka and Tanaka, 1951). Noguchi and Sugawara (1966) give further details.

3. *Natural Supply of Potassium*

A. PADDY SOILS. Most paddy soils in the tropics have a far better fertility status than the upland soils. There are, however, conditions under which the soil supply of potassium may be critically short. This will apply to:

1) Soils that went through an intensive weathering and leaching (laterization) process before they were turned into rice paddies
2) Hydromorphic organic or peaty gleyed soils as found in Madagascar, parts of Malaysia, and Indonesia
3) Soils developed from highly calcareous materials
4) **Light sandy soils**
5) Soils where the absorption of potassium is retarded due to the presence of toxic substances, leading to physiological diseases.

B. IRRIGATION WATER. Irrigation water may supply between 3-45 kg K/ha. At low levels of production, irrigation water may therefore be a significant source of potassium for rice. On the other hand, irrigation water will increase leaching losses, especially on soils rich in potassium. Irrigation has thus a leveling effect on the potassium status of paddy soils(Takahashi, 1965).

C. EFFECT OF FLOODING ON THE AVAILABILITY AND FIXATION OF POTASSIUM. The availability of soil potassium increases with increasing soil moisture. With upland crops, however, potassium uptake will be retarded if root oxygen supply is short. Rice roots can function normally under extremely low oxygen levels and are very efficient in extracting nutrients from the soil solution (Okajima, 1965). There is hardly any fixation of potassium under flooded conditions (Chang and Chen, 1960; Chan and Feng, 1964). Generally potassium is more easily available to paddy rice than to other crops (Table 2).

"Rice", states Pendleton (1943), "is doubtless the most adaptable food crop man grows, and if enough water remains on the soil, it can produce at least a little grain on soils that are unbelievably poor in plant nutrients."

4. Potassium Fertilizer in Rice Production

A. HIGH POTASSIUM USE IN JAPAN. On an acerage of 3,200,000 ha, Japan is using substantially more potassium fertilizer in rice production than the rest of the world is using on a combined acreage of about 120,000,000 ha. This tremendous difference in the pattern of potassium usage is not due to

Table 2—Natural supply of nutrients to rice, wheat, and barley in Japan (Shiroshita, 1963a)

Treatment	Paddy rice	Upland rice	Wheat & Barley
	(Relative yield)		
NPK	100	100	100
PK	83	51	50
NK	95	84	69
NP	96	75	78
None	78	38	39
No. of sample places	1,161-1,187	117-176	822-841

a low natural supply of potassium in Japanese soils, but due to the interactions between a highly-developed industrial economy, intensive research, and progressively minded farmers. All these factors are lacking in the rice-growing tropics. Farmers in Japan have been using fertilizer responsive varieties and advanced techniques for many years. They have constantly increased yields by eliminating limiting factors, and under such conditions application of potassium has become a necessary component for both maintaining and increasing yields on nearly all soils. High yield content winners have applied up to 434 kg K/ha (Shiroshita, 1963b).

B. Low Potash Usage in the Tropics. Apart from the socio-economic reasons mentioned in the first part of this paper, agronomic factors have also been limiting the use of potassium fertilizer.

The combination of relatively unresponsive varieties, a climate that favors excessive vegetative growth, soils that are very often deficient in many other elements, and poor management practices has severely limited the effectiveness of potassium fertilizer in tropical rice culture. With unimproved varieties, potassium, especially if applied as basal dressing on poor soils, may contribute towards a more vigorous early vegetative growth, prolong the "lag period", and add towards mutual shading with the result that sometimes more straw and less grain is produced. Based on unsatisfactory responses to potassium, many noted researchers such as Angladette (1966) have expressed the opinion that potassium has only a minor role in rice production. Girst (1959) writes, "Some evidence exists that where there is no lack of potash, further additions may have a small negative effect on yields, and therefore the inclusion of potash in fertilizer mixtures is not warranted merely on an 'insurance' or 'balancing' principle unless it is demonstratably in deficit."

This holds true under the present set of conditions in the tropics. The effect of applied potassium is restricted to locations where deficiency of potassium is limiting yields even at low levels of production.

It is therefore surprising that even under such conditions remarkable responses to potassium have been obtained in many tropical countries. In Ceylon, potassium was found to be as effective as nitrogen and phosphorus (Weerawickrema and Constable, 1967). Reviewing the results of 2,800 experiments carried out by the Government of India, Panse and Khanna (1964) state that the average response to 18.6 kg K/ha (22.4 kg K_2O/ha) in the presence of equal quantities of N and P_2O_5 (9.8 Kg P) amounted to a 255 kg yield of paddy rice. This is in conformity with the results reported by Mukerjee (1955) for the state of Bihar and by Gouin and Chadha (1958) and Chadha (1966) for other states of India. Striking responses to potassium have also been reported from Pakistan (Vermaat, 1964), N.E. Thailand (Rice Department 1961), Madagascar (Dufournet et al., 1966), and other regions (Mukerjee, 1965). With improved varieties and better management techniques responses to potassium can be expected to

Table 3—Response of different rice varieties to K when grown on Bohol coralline calcareous soil (Navasero and Estrada, 1967)

Variety	Fertilizer treatment kg/ha of N - P - K	Height cm	Straw wt. kg/ha	Grain kg/ha	Yield index	% P	% K	% Ca
Taichung Native 1*	50-22-0	44	3,000	1,400	100	0.2	0.9	1.2
	50-22-250	56	5,500	4,800	343	0.2	1.4	0.8
Tainan N 3†	50-22-0	61	3,000	1,800	100	0.2	1.0	1.0
	50-22-250	82	7,900	4,400	244	0.2	2.5	0.8
Peta‡	50-22-0	55	6,700	1,100	100	0.2	1.0	1.2
	50-22-250	85	7,800	2,400	218	0.2	2.7	0.9
Caintí‡	50-22-0	69	5,900	1,700	100	0.2	0.8	1.3
	50-22-250	99	m. a.	1,900	112	0.2	2.4	1.0

* A fertilizer responsive indica type variety from Taiwan.
† A fertilizer responsive japonica type variety from Taiwan.
‡ Fertilizer unresponsive indica type varieties, commonly used in the tropics.

be of far greater magnitude, and responses will also occur on more soils. This view is also expressed by Doyle (1966).

C. VARIETAL DIFFERENCES IN THE RESPONSE TO POTASH. It is a well-established fact, that varieties differ greatly in their responsiveness to nitrogen. Little is known about potassium in this regard. Recent experiments in Taiwan (Chin et al., 1965) and Japan (Ogiwara, 1960; Motomatsu, 1964) have shown that the various fertilizer-responsive varieties used in these countries differ in their responsiveness to potassium, though the differences are not large. However, if improved, fertilizer-responsive varieties are compared with the varieties that are in common use in the tropics, the differences in response are striking indeed (Table 3).

Explaining the need for better varieties in the tropics, Beachell and Jennings (1965) said, "... the tall, vigorous growing, late maturing photo-sensitive indica varietal types that have persisted for centuries in S.E. Asia because of their ability to survive relatively deep water and to compete with weeds at low fertility levels offer little hope for substantial yield increases through the use of (nitrogen) fertilizer." Introduction of high-yielding, fertilizer-responsive varieties is not only a prerequisite for a more effective utilization of nitrogen fertilizer, but for potassium as well. Such varieties may eventually have a similar impact on rice production in the tropics as the introduction of hybrid corn had in the USA.

In this light the breeding work done at the International Rice Research Institute in the Philippines (1963, 1964, 1965) deserves particular attention. Some of the existing varieties may also become very fertilizer responsive if the planting season is changed.

D. TIMING OF POTASH APPLICATION. Many studies have been made on the optimum timing of nitrogen fertilizer, but little attention has been paid to potassium in this respect. Influenced by the results with upland grains, most researchers have considered a basic application of potassium together with phosphate as satisfactory. There is growing evidence that in the case of paddy rice topdressing of potassium in 2 or 3 split applications may prove superior to a basic dressing in the following instances:

1) In warmer areas
2) On lighter soils with a low cation-exchange capacity
3) On poorly-drained soils with a tendency to accumulate toxic substances that affect root activity
4) With high applications of nitrogen.

The most effective time for topdressing will differ by circumstances, but it is believed that in most instances the first application should be made at the maximum tillering stage (about 2 weeks after transplanting) and the second about 50 days before heading.

In experiments reported by Yamazaki (1967) potassium applied as basal dressing only, increased the straw yield by 29% but had hardly any effect on grain yield. Applied in three split doses, the same amount of K increased the straw yield by 23% and the grain yield by 19%. This tendency was even more pronounced in the presence of silica. Many similar results have been obtained in Taiwan (Chin et al., 1965).

E. Sources of Potassium for Rice. So far no significant differences have been observed as to the effectiveness of different potassium sources. Tests have been made with potassium chloride, potassium sulfate, potassium bicarbonate, potassium metaphosphate, and potassium silicate. There are cases where the sulfate anion is undesirable because of the danger of H_2S toxicity. On the other hand, many tropical soils are short in sulfur, and with the advance of concentrated fertilizers low in sulfur, potassium sulfate may have advantages. Under normal conditions potassium chloride is the most economical source of potassium for rice. The standard quantities used in Japan range between 50-100 kg K/ha. In the tropics the quantities rarely exceed 45 kg K/ha. With improved varieties, better management, and fertilizer application techniques (split application), quantities of 50-75 kg may soon become economical.

F. Potash and High Yields. High yields have been reported from many countries (Chandler, 1963). But unfortunately it is only in Japan where the practices of the contest winners have been studied in detail (Shiroshita, 1963b; Honya, 1967; Matsuo, 1966; Matsushima, 1966).

According to Shiroshita (1963b) who investigated the techniques used by the 20 first and special prize-winners in Japan from 1944 to 1962, these farmers applied an average of 237 kg of N, 79 kg of P, and 235 kg of K for an average yield of 11,680 kg of rough rice per ha. This corresponds to an NPK ratio of 100:33:100. Without exception the prize-winning farmers have used very large amounts of potassium fertilizer.

For comparison, yields in Japan 1960-1963 averaged 4,940 kg rough rice/ha with an average fertilizer application of 89 kg N, 27 kg P and 61 kg K/ha. This is an NPK ratio of 100:30:68.

High-yield contest winners in Korea (average yield over 8,000 kg/ha) applied an average of 179 kg N and 144 kg K/ha. This corresponds to an N/K ratio of 1:0:80.

Matsuo (1963) found that high-yielding rice absorbed N and K at a fairly constant rate of 1:0:87.

There is no doubt that nitrogen is the most important single element for high yields. But to be effective, the plants must also be able to absorb large quantities of other elements such as silica, magnesium, and trace elements in addition to phosphorus and potassium. Yamazaki (1967b) proved experimentally that if plants absorbed large quantities of silica, potassium, magnesium, phosphorus, and trace elements (through liberal application of above elements), a plant developed that was more fit for an efficient utilization of fertilizer nitrogen. Application of potassium by itself will not lead to high yields. However, in combination with other factors, it is an essential component in high-yield rice culture.

B. Sugar Cane

1. Potassium Requirements

Sugar cane uses large amounts of potassium. Over 750 kg K/ha may be found in the above-ground parts of a good crop. Though the larger part of this will be contained in the tops and the trash and will be returned to the soil, removal with the cane is considerable. Depending mainly on the age of the cane at harvest, the amount of K removed per ton of millable cane may vary from 0.4 to 2.5 kg.

Potassium uptake is slow in the early stages, but as soon as roots and shoots have developed to a certain extent, the rate of absorption increases markedly, reaching a maximum between 3 to 7 months after planting. Thereafter absorption slows down but still remains considerable till harvest.

Figure 1 demonstrates the tremendous demand made by this crop on the soil for potassium. Being a member of the grass family, sugar cane is rather efficient in utilizing soil potassium provided that moisture and oxygen conditions (aeration) are correct.

The total amount of potassium contained in the leaves increases during the first 6 months until a maximum canopy is developed after which the quantity in the leaves remains practically constant until the number of leaves per stalk decreases in the period of ripening.

The potassium content of leaves on the same stalk decreases according to the age with the basal part of each leaf being the highest in potassium. The same trend is found in the cane with a K content of over 5.75% in the top of the stalk and below 0.75% in the basal internodes (Humbert, 1963).

There is some evidence that varieties may differ considerably in their po-

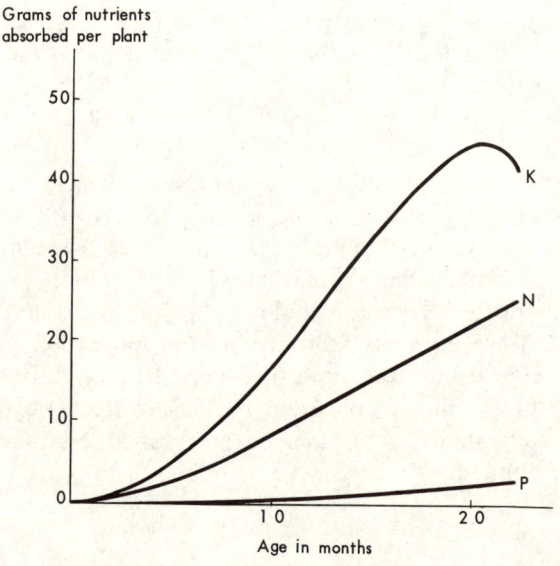

Fig. 1—Nutrient uptake by a complete cane stool. (Adapted from van Dilewijn, 1952)

tassium requirement. Water culture experiments in Hawaii (Martin, 1940) provided evidence that among four different varieties tested, H 109 suffered most from potassium deficiency whereas the variety 32-8560 showed rather good growth under similar conditions.

2. *Effects of Potassium*

Potassium stimulates the activity of invertase, peptase and catalase, while that of amylase is reduced (Hartt, 1934). It promotes the formation and translocation of sugars, improves the purity of the juice, and betters the rendement. Increases in tonnage of cane, due to potassium are usually accompanied by an increase in the sucrose content (Mauritius Sugar Cane Res. Sta., 1954). Humbert (1953) has shown that plants having too low a potassium content in the leaves, cannot take up the large quantities of water required for vigorous growth.

Cane suffering from potassium deficiency has a greatly reduced vigor when used as planting material (du Toit, 1955), and increased stalk mortality has been attributed to potassium deficiency in the Philippines (Victoria Milling Co., 1955).

According to van Dillewijn (1952) experiments in Java have shown that potassium could counterbalance adverse effects of both supraoptimal and suboptimal applications of nitrogen and phosphorus. Sugar losses resulting from high nitrogen at average potassium levels are not apparent when high nitrogen is accompanied by extra potassium (Borden, 1940).

Lee and Martin (1928) reported from Hawaii that potassium increased resistance to eye spot (Helminthosporium saccari). It has also been suggested that the large cavities in the cortex of the roots and the poor development of root hairs in potassium-deficient plants may be the casual factor for the root failure complex or the Lahaina disease (Moir, 1930). Potassium has also proved effective in correcting a physiological disorder, known in Java as Kalimati disease. It is believed that the cause of the disease is iron toxicity (Koningsberger, 1931). In this connection it is interesting to note that in rice plant potassium deficiency and iron toxicity also seem to be closely related (Tanaka, 1967; *personal communication*).

3. Potassium Deficiency

Potassium-deficient plants show depressed growth. Young leaves are dark green in comparison with the old leaves that develop an orange-yellow color with numerous necrotic spots developing from the tips and margins. As the spots coalesce, general browning of the leaves results. The epidermal cells on the upper surface of the midribs develop a reddish discoloration. Both young and old leaves appear to have developed from a common point—a characteristic of a cane plant that is not growing (Humbert and Martin, 1955). Potassium deficiency has been studied in detail by Hartt (1929).

4. Response to Potassium Fertilizer

With a potassium uptake that can reach over 1,000 kg K/ha, it is not surprising that sugar cane usually responds well to potassium fertilizer. But because of its efficiency in utilizing soil potassium, plant cane sometimes does not respond as well as one would expect. This is particularly true for well-textured, light soils, low in total potassium. If the plant cane is not fertilized with potassium, a competely exhausted soil may be left behind that is not able to support a good ratoon crop. With regard to conditions in Hawaii, Clements (1953) stated, "In Hawaii potash fertilization is indeed an important aspect of our program. Once adequate fertilization of this important material was worked out through sheath analysis, yields of ratoons have been higher than those of the plant cane yields. It is an unfortunate fact that plant crops yield well, even though their K-index may be lower than specified on the log, but it is the following ratoon that suffers as a result ."

Leaf analysis has proved to be a reliable method for determining potassium needs. For details reference is made to the paper by Samuels (1960) which discusses in detail the merits of various methods of foliar diagnosis used in different countries. Average rates of potassium application range from 83 to 166 kg K/ha. But in Hawaii, rates of 300 kg K/ha are common for high-yielding stands.

C. Tea

1. *Role of Potassium in the Physiology of the Tea Bush*

The harvested part of the tea plant is a young, vegetative organ. The "development of the garnered crop is not subject to the vagaries of floral differentiation, or complicated by the translocation of food reserves from one specialized organ to another of a different category with a different function" (Eden, 1958).

The young flush, representing the potential tea crop, enjoys preference in the distribution or redistribution of absorbed and stored nitrogen, potassium, and carbohydrates. To maintain high yields, it is essential that the plants accumulate sufficient carbohydrates and nutrients in order to support a vigorous flush after each pruning, tipping, and plucking and at the same time keep the roots healthy and well supplied with carbohydrates.

If potassium is deficient, the young flush will drain the potassium reserves from the older leaves. The leaves' net assimilation will be adversely affected with the result that available assimilates will be short in supply for both the new shoots and the roots. The roots will not receive the amount of carbohydrates needed for balanced growth and active water and nutrient uptake. This aggravates the damage in the foliage which finally leads to a complete exhaustion of the carbohydrate reserves, a breakdown of the root system, and the death of the plant.

Differences in yield between high-potassium and low-potassium plots are often not so much due to differences in the production of individual bushes, but rather due to an increased mortality rate of the plants in the low-potassium plots. Experiments in India (Jayaraman and de Jong, 1955) have shown that the tea bush population in three pruning cycles was reduced by 46.1% in the no-potassium plots. Depending on soil, climate, spacing, nitrogen rates, and the pruning and plucking intensities adopted, the time lapse between planting and the first noticeable potassium effect can vary greatly. In the famous experiments in Ceylon (Portsmouth, 1950) and in India (Jayaraman and de Jong, 1955), it took 12 and 8 years respectively, before the first effect of potassium became manifest.

Once potassium deficiency shows up, the effect on yield is very marked and irreparable damage may have been done to the bushes, with the root system severely weakened and unable to absorb applied potassium and moisture in sufficient quantities to restore normal growth. Venkataramani (1957)

Table 4—Effects of fertilizer treatment on the structure of tea leaves attacked by *gloeosporium* (after Nagata, 1954)

Treatment	Thickness of leaves	Hardness of leaves*	No. of leaves attacked
	mm		
N	0.22	12.0	16.5
NP	0.22	15.1	21.5
NK	0.25	19.6	3.0
NPK	0.25	18.9	3.7

* Pressure required in grams to puncture a leaf with a needle.

reports that a 1% foliar spray of KCl applied at a rate of 49 kg K/ha (about 60 kg K_2O) restored growth and significantly increased yield and girthing of branches.

2. Potassium Deficiency

Potassium deficiency (de Haan and Schoorel, 1940; Portsmouth, 1953) is indicated by a dark-green foliage, showing necrotic brown to purplish-brown margins. Many leaves will be shed prematurely, and the twigs are long and thin with extremely long internodes. The frame of the bushes is poor and recovery after pruning is very slow, partly because of poor callus formation in potassium-deficient bushes.

3. Effects on Disease Resistance

Potassium deficiency is reported to reduce resistance to brown blight. (*Colletotrichum camelliae*), gray blight (*Pestalezzia theae*) (de Gues, 1950), blister blight (*Exobasidium bezans*) [H. C. Paterson, 1950. *Exobasidium vexams* in tea: A series of notes and letters and recommendations for combating the disease. Ceylon], red rust (*cephaleuros virescens* Kunze), and anthracnose (*gloeosporium theae*). According to Nagata (1954) the lower disease resistance of potassium-deficient tea might be associated with the decreased hardness observed in potassium-deficient leaves (Table 4).

4. Effects on Quality

Considering the high nitrogen content in the harvested shoot (3.5-4.5% N) it is not surprising that adequate potassium (1.4-1.9% K) is needed for good quality. Eden (1958) reports that at the London market, tea treated with potassium was preferred because of its better quality. Tests in Japan (Kawai, 1959) revealed that top quality (aggregate of flavor, color of extracted solution, taste, and color of leaves) was obtained only from those treatments that included potassium. Mukasa and Kawai (1956) found no clear relationship between potassium application and tanin or caffeine con-

Table 5—Nutrient removal by a crop of 1,000 kg of manufactured tea (Eden, 1958)

Part of the plant	N, kg	P, kg	K, kg
Young shoots (harvested)	40.20	3.71	13.28
Wood (removed)	23.60	3.06	15.52
Foliage (dropped)	27.20	2.01	10.79
Total	91.00	8.78	39.60
Permanent loss (1 + 2)	63.8	6.77	28.81

tent, but potassium seemed to reduce the amount of water-soluble nitrogen and the content of reducing sugars while increasing nonreducing sugars.

5. *Potassium Levels in the Leaves*

Leaf K in healthy mature leaves ranges between 0.7 and 1.8% K. According to Zhurbitzky and Strausberg (1964) the optimum K content in the young flush was around 1.65%. A similar figure is given by Lin (1962, *Personal communication*) from Taiwan. In a pot experiment reported by Kawai (1959) an increase of the K content in the leaves from 1.89 to 2.95%, with the levels of N, P, Ca, Mg remaining almost unchanged, resulted in an increased leaf weight per pot from 49.4 to 83.2 g.

6. *Nutrient Removal and Fertilizer Application*

Compared with other crops potassium removal by tea is moderate (Table 5).

The amount of potassium used in tea gardens in the tropics rarely exceeds 66 kg K/ha (80 kg K_2O) with the average in the range of 33-50 kg K/ha (40-60 kg K_2O). Progressive planters practicing dense planting (45,000 plants/ha) and intensive utilization have been using quantities of 125-165 kg K/ha (150-200 kg K_2O). With the exception of sulfur-deficient soils, as frequently found in East Africa, potassium chloride is the standard form of potassium used in tea culture.

D. Coffee

1. *Potassium Uptake and Removal*

It has been estimated that a coffee plantation yielding around 1,000 kg of made coffee needs in total about 100 kg N, 10.5 kg P, and 125 kg K (Anstead and Pittock, 1913). The amount of nutrients removed per 1,000 kg of clean coffee (equivalent to about 1,700 kg dry berries or 5,000 kg fresh cherries) is in the range of 30-35 kg N, 22-31 kg P, and 33-42 kg K. In absolute terms, this is not exceedingly high, but as the bulk of nutrients will be required over a rather short period prior to maturing, the nutrient stress on parts of the tree easily becomes excessive.

2. *Potassium in the Physiology of the Tree*

Heavy bearing branches carry a bundle of fruit for every pair of leaves. As the nutrients usually cannot be absorbed at a pace demanded by the developing fruit, they are drawn from the adjacent leaves. If the potassium level falls below a certain limit (usually around 0.3% K) a vicious cycle may be initiated. The drain of potassium into the fruit results in a narrowing of the K/Ca ratio in the leaves, causing them to age prematurely, to lose moisture, to develop marginal scorch, and to drop. At the same time the rate of photosynthesis per unit leaf area decreases while respiration rate increases, thus sharply reducing net assimilation.

As the developing fruit has an absolute priority in the distribution of assimilates the roots will suffer and decrease in activity, thus aggravating the damage in the leaves and twigs. Finally, roots as well as branches may die. A heavy-bearing coffee tree may die or be severely damaged due to a temporary potassium shortage, even on a fairly rich soil.

It seems that one of the most serious effects of potassium deficiency in heavy-bearing trees is an insufficient energy transfer to the roots. Because of its effect on the root system, severe potassium deficiency is rather difficult to correct. Pruning, combined with sprays containing biuret-free urea or potassium chloride may be the safest way to save trees, exhausted from overbearing. The most economic measure is to prevent occurance of potassium deficiency by potassium applications and mulching. Where fertilizer is not available, consideration should be given to growing coffee under shade where there is less danger of physiological disorders.

3. *Effects of Potassium on Quality*

Fruits from potassium deficient trees are low in quality. In India a high percentage of floats was found to be associated with potassium deficiency. Malavolta et al., (1962) report that fruits from potassium-deficient trees soon become dark because of fungus attack.

4. *Leaf Potassium and Yield*

A very close relation has been found between leaf potassium and yield in Brazil by Medcalf et al. (1955) (Fig. 2). Haag and Malavolta (1960) consider a leaf K content (third pair of leaves) between 1.72 and 1.90% as adequate. Similar data are given by Loué (1958) for robusta coffee on the Ivory Coast. According to Loué's findings the optimum cation balance for

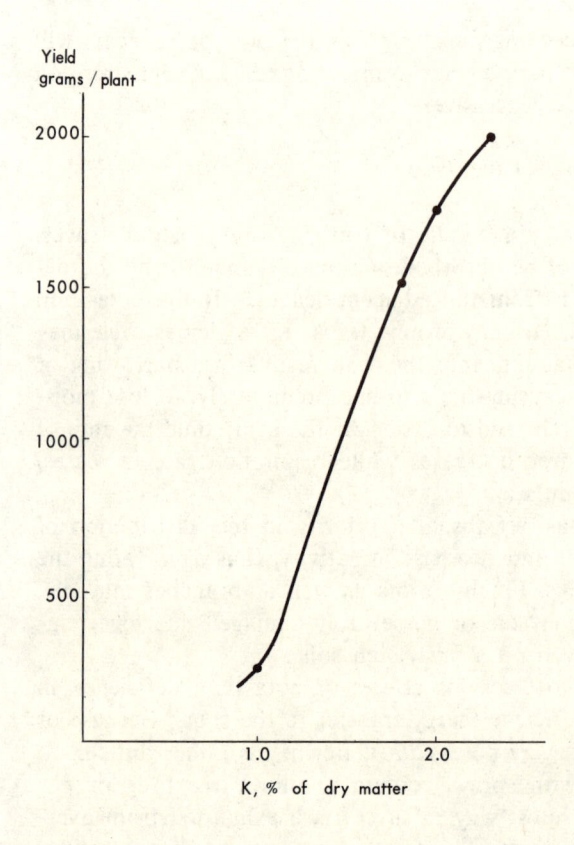

Fig. 2—Relationship between leaf K and yield of *C. arabica*. (Adapted from Medcalf et al., 1955)

robusta coffee is between 46-42-12 and 50-38-12 (K + Ca + Mg = 100). Considering that the sum of K + Ca + Mg is fairly constant around 3.8% dry matter, the optimum concentration for K would be 1.75-1.90%, Ca 1.45-1.61% and Mg 0.46%.

Abruna et al. (1965) found that leaf K and N values between 2.5 and 3.0% were associated with high yields.

5. *Fertilizer for Coffee*

The importance of potassium in coffee culture has been recognized in most coffee growing areas. Fertilizer analyses like 10-5-20 (10%N-2.2% P-16.6% K), 10-8-18 (10% N-3.8% P-15% K), and 13-13-20 (13% N-5.7% P-16.6% K) are most widely used. In Hawaii, for high yielding, sun-grown coffee, quantities of over 250 kg N, 33 kg P, and 250 kg K are recommended (Goto and Fukunaga, 1956).

For details, reference is made to the monographs by Coste (1955) and Wellman (1961).

E. Cocoa

1. General

Cocoa is one of the most unstable crops physiologically. Its root system is shallow and rather inefficient in utilizing soil moisture and nutrients. Cocoa leaves, on the other hand, have rather high transpiration quotients. Furthermore, there are indications that leaching losses of potassium and magnesium through leaf wash from rain are higher than in other crops.

Lemee (1955) showed that photosynthesis, growth, and transpiration are already markedly reduced when soil moisture drops below 60-70% of the available range. When cocoa leaves lose about 1/6 of their water content, necrotic areas, resembling potassium deficiency develop (Alvim, 1965). According to Murray and Maliphant (1965) the tolerable range in nutrient levels is more restricted in cocoa than in most other crops. Small imbalances may lead to premature leaf fall.

In most cases cocoa is grown under reduced light intensities, sacrificing yield in order to maintain a healthy balance between water uptake and transpiration, nutrient uptake and dry matter production, and between vegetative and reproductive growth.

If shade is removed, cocoa trees easily overbear due to the excessive impulse for generative reproduction caused by the combination of moisture stress and the increased rate of photosynthesis. Unless this increased rate of transpiration and photosynthesis is met by a correspondingly larger water and nutrient intake, shade removal will result in a short, sharp increase in production, followed by severe leaf fall, dieback, and a deterioration of the stand. Where water and nutrient requirements can be met, removal of shade can be the most effective measure in increasing yields (Fig. 3).

In order to meet the increased water and nutrient requirements, unshaded cocoa requires a much larger active root surface in relation to the leaf surface than shaded cocoa. Shade removal must therefore be done carefully and on poorer soils should be combined with some pruning and mulching in order to adjust the root to leaf area. This is also true for coffee.

2. Potassium Requirements

According to Humphries (1939) 600 kg of dry beans (corresponding to about 1,000 kg dry pod weight or 5,000 kg of fresh fruit) remove around 16 kg N, 2.6 kg P, 23.5 kg K, 3.6 kg Ca, and 4.8 kg Mg. Substantially larger quantities of nutrients are absorbed and fixed in the wood or circulated through the leaves. But considering that average yields in the main growing areas seldom exceed 600 kg of dry beans/ha (van Dierendonck, 1960), nutrient removal by cocoa is moderate indeed. Inspite of this low removal,

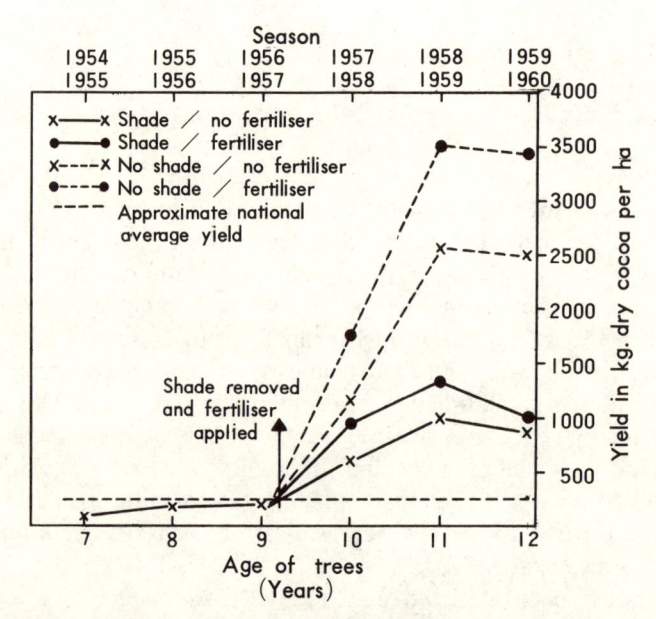

Fig. 3—Effect of shade removal and fertilizer application on the yield of Amelonado cacao in Ghana. (Adapted from Cunningham and Arnold, 1962)

most cocoa gardens respond quite well to potassium applications of up to 65 kg K/ha. In Trinidad (Pound and de Verteuil, 1936) shaded cocoa showed a progressive response up to 46 kg K/ha. Within this range the average response to 1 kg K was 6 kg of dry cocoa. In contrast to nitrogen, the potassium effect on cocoa is independent of light intensity within a certain range, though naturally more potassium will be required where under full light intensity high yields are obtained. Under shadeless cultivation potassium uptake is not always satisfactory which has in some cases led to the impression that shadeless cocoa is less responsive to potassium.

3. *Potassium Deficiency*

Potassium deficiency manifests itself in a marginal necrosis of the leaves, and sometimes there is a yellow zone near the midrib. Usually the leaf outside the necrotic zone is quite green. Older leaves may show no signs of necrosis, but they turn orange-yellow and drop prematurely. Length of shoots of potassium-deficient plants is less affected than girth. Loue (1962) studied the second and third branches from the first fully green sprouts on five branches. The studies indicate that when a leaf contains 1.2% K it is K deficient yet shows no visible symptoms. Visible disorders will appear in the neighborhood of 0.8% K. Acquaye, Smith, and Lockard (1965) report from Ghana that visible symptoms appeared when leaf K was 0.5-0.6% and when exchangeable K was below 0.20 meq/100 g soil. Hardy (1936) drew attention to the possibility of potassium deficiency appearing on soils of normal potassium status but too high in calcium and magnesium. Potassium deficiency may be one of the causal factors for cherelle wilt (Humphries, 1959). Alvim (1954) postulated that a carbohydrate strain brought about either by a competition between young flushes and fruit or by depression of carbohydrate translocation could be the main cause of cherelle wilt.

F. Rubber

1. Changes in Potassium Requirements

The rubber tree is adapted to life in a humid, tropical environment, where soils are heavily leached and low in pH. Rubber, therefore, proved to be the ideal crop for utilization of poorest soils, not suited for more demanding crops like banana, coffee, or cocoa. Most prewar data indicate no response to potassium and little importance was attached to potassium fertilizer in rubber culture.

Recent reports, however, show that potassium deficiency is commonly found in mature rubber in Ceylon (Constable, 1953), Vietnam (Beaufils, 1955; Compagnon, 1959), the Ivory Coast (IRCA, 1961), Nigeria (Ward, 1964), and Malaya (Rubber Research Institute of Malaya, 1964). Application of potassium has become a most important factor in rubber growing for the following reasons:

1) Most prewar rubber was planted on virgin forest soils, containing some accumulated fertility that had disappeared when the second generation of rubber was planted.

2) Early plantings were made with unselected seedlings, rarely yielding over 500 kg/ha of dry rubber against yields of up to 3,500 kg/ha obtained from high yielding secondary clones. Shorrocks (1965c) estimates that the nutrient requirement of more vigrous clones will be 20-25% higher than for average clones and much higher than for the old type seedlings.

3) More intensive use of nitrogen, phosphorus, and magnesium and the more intensive tapping has substantially increased the need for potassium.

4) Though there is no proof to it, it seems most likely that the intensive tapping and yield stimulation practiced in modern rubber growing, will in the long run reduce root development and root growth and thus increase the demand for easily available potassium.

2. Nutrient Uptake and Nutrient Cycle

A stand of 267 trees at an age of 33 years contains about 1,779 kg N, 277 kg P, 1,233 kg K, and 417 kg Mg (Shorrocks, 1965 a,b). Less than 10% of above nutrients are contained in the green branches and leaves. The average annual uptake per tree has been estimated at 187 g N, 23 g P,

142 g K, 37 g Mg, 206 g Ca, and 22 g S. The figures do not include the amount of K lost through rainwash of the leaves which may reach the surprisingly high figure of 20 kg K/ha (Shorrocks, 1965 b) at 1,800 mm rainfall a year. Also not included is the removal of K with the latex which amounts to about 10 kg K per 1,700 kg dry rubber. Considering the fact that many soils where rubber is grown contain less than 300 kg/ha of acid-soluble K in the top 46 cm (Shorrocks, 1965b), it is not surprising that K deficiency has become a widespread disorder in rubber. In Vietnam, on soils known to be deficient in copper, a broadcast application of 12 kg /ha of copper sulfate increased the yield and the utilization of applied potassium (Compagnon, 1959). A positive correlation between leaf potassium and boron has been reported by Cocci (1960) and Rosenquist (1963, *personal communication*) observed that boron and potassium were suspiciously low near the tapping cut, whereas the levels of nitrogen and magnesium were high.

Few crops seem to be as sensitive to nutrient imbalances as the rubber tree. Potassium deficiency as well as a small relative excess may have immediate effects on growth and yield. Many earlier reports where potassium showed negative effects (Owen et al., 1957) can be explained by the depressing effect of potassium on leaf magnesium.

3. Effects of Potassium on Wind Damage and Seed Production

Rosenquist (1960) suggested that wind damage may be associated with potassium deficiency. Middleton et al. (1965), on the other hand, found that potassium reduced wood strength. It seems likely that at some points of non-equilibrium additional potassium will increase resistance to wind damage; whereas, at other configurations it may reduce such resistance. High potassium levels in the leaves seem to be negatively correlated with seed production. In experiments where potassium significantly increased yield, it reduced seed production (Watson and Navayanan, 1965). Heavy fruiting seems to be favored by a wide N/K ratio in the leaves. Potassium deficiency affects the time and length of the wintering period, and this in turn may have an effect on seed setting.

4. Determination of Potassium Needs by Leaf and Soil Analysis

Leaf analysis is widely used in rubber culture for determining fertilizer requirements. The Rubber Research Institute of Malaya has worked out the following values for shade leaves with midribs (in percent dry matter):

	N	P	K	Mg
Response likely	3.30	0.21	1.30	-
Response unlikely	3.70	0.27	1.50	0.28%

Fig. 4—Theoretical relationship between soil potassium (acid extractable) and the annual rate of KCl application required for good growth of rubber. (Shorrocks, 1965c)

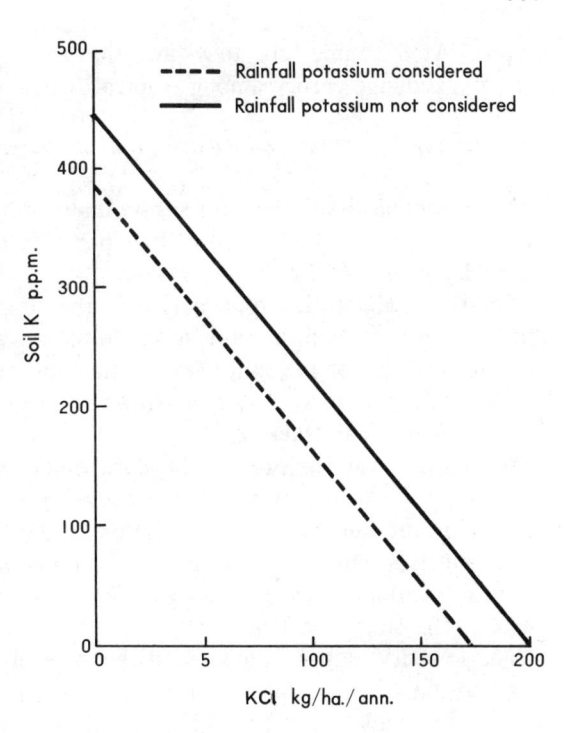

These figures will require refinement, as clonal variations may be considerable. There are, for example, indications that for some clones optimum magnesium levels for maximum growth are higher than those for maximum yield.

According to Guha and Pushparajah (1966) satisfactory leaf levels of K can be expected, when total soil K is over 500 ppm. Young rubber responded to applied K when soil K was below 120 ppm. Shorrocks (1965b) offered a tentative proposal shown in Fig. 4 on how to assess the potassium requirement on the basis of acid-extractable soil potassium.

Depending on the age of the tree Shorrocks (1964) recommends applications of about 55 to 450 g K/tree for correcting potassium deficiency. For maintenance application rate of 35-40 kg K/ha should be sufficient in most cases. With new clones and better tapping techniques the above rates may soon become too low.

G. Bananas

1. General

Bananas are the most widely grown tropical fruit. Because of their high food value and good taste they are an important addition to the diet in the tropics. At the same time they are the most important export item and foreign exchange earner among tropical fruit.

2. Potash Absorption and Distribution in the Plant

Most banana tissues contain substantially more potassium than the tissues of other plants. A good stand of bananas may contain about 200-280 kg N, 22-33 kg P and 600-1,000 kg K/ha.

Potassium absorption follows closely the course of dry matter production, reaching its maximum around flower initiation (around the 15th leaf stage). After emergence of the inflorescence the rate of potassium uptake decreases and a redistribution of potassium from the leaf sheaths, petioles, and leaves into the fruit stem takes place.

Among the various organs, the corm maintains the most constant potassium content. According to Martin-Prevel and Montagut (1966) the corm has important functions as a regulator in the selective nutrient uptake and as a "nutrient pump." Leaf sheaths and petioles are more passive organs for translocation. They also act as reservoirs of potassium for the development of the stem and fruit.

The low dry matter content and the very high relative potassium content of a "normal" banana plant indicate a very high rate of mobility of potassium within the plant (Table 6). This great mobility combined with the large storage volume explains why the banana has a very wide range of "hidden hunger" and is so responsive to potassium.

3. Potassium Deficiency

Growth rate, net assimilation, and translocation are greatly affected long before any visual symptoms appear. Once the potassium contained in the storage organs is exhausted, however, deficiency symptoms appear suddenly, starting with a yellowing of the tips and distal margins of older leaves, closely followed by necrosis and desiccation. The yellowing and necrosis spreads rapidly in a proximal direction until the whole leaf has withered standing in a normal position. Splits develop parallel to the secondary veins and the lamina fold downwards while the midrib bends and fractures, leaving the distal half of the leaf hanging. Purplish-brown patches appear at the base of the petioles and in severe cases the center of the corm may show areas of brown, watersoaked, disintergrated cell structures. The above process may proceed very fast, changing an apparently healthy leaf into a completely withered condition within a period of 7 days. The rate of new leaf formation slows down, leaving ultimately the bunch at a pseudo-stem where only brown, desiccated leaves are hanging down. Fruits are badly shaped, poorly filled and unsuitable for marketing (Tai, 1958; Murray, 1959; Freiberg and Steward, 1960; Osborn and Hewitt, 1963; Martin-Prevel and Charpentier, 1964).

Table 6—Potassium content of various parts of the banana plant at various stages of growth (K in percent of dry matter) (After Martin-Prevel and Charpentier, 1964)

Part of the plant	5th leaf stage Dry Matter	5th leaf stage K	8th leaf stage Dry Matter	8th leaf stage K	15th leaf stage Dry Matter	15th leaf stage K	Flowering stage Dry Matter	Flowering stage K	Cutting stage Dry Matter	Cutting stage K
Rhizome	12.20	4.45	9.41	4.40	11.78	4.23	10.87	4.34	12.04	4.70
Rhizome of offset	---	---	10.50	4.07	13.70	4.00	12.60	4.93	12.30	3.73
Roots	5.60	8.40	---	---	---	---	---	---	---	---
Sheaths	3.00	10.90	3.38	9.70	4.42	9.00	4.82	7.20	5.88	6.80
Petioles	5.40	9.30	5.96	8.70	6.54	7.80	7.28	6.60	6.99	6.70
Midribs	7.66	5.90	8.37	5.60	9.30	5.20	10.10	4.70	9.51	5.20
Laminae	17.50	3.84	18.60	3.67	19.10	3.80	19.60	3.43	19.70	3.03
Immature leaves and inflorescence	---	---	---	---	---	---	4.92	8.53	---	---
Stem	---	---	---	---	---	---	---	---	3.61	14.00
Fruit bunch exported	---	---	---	---	---	---	---	---	17.50	4.12
Pads	---	---	---	---	---	---	---	---	8.64	9.50
Peduncules	---	---	---	---	---	---	---	---	8.04	9.50
Skin	---	---	---	---	---	---	---	---	9.24	6.90
Flesh	---	---	---	---	---	---	---	---	28.17	1.30

Table 7—Range of nutrient content of the banana leaf (*Lacatan bananas,* 3rd fully opened leaf at time of shooting.

Nutrient status	% of dry matter				
	N	P	K	Ca	Mg
Adequacy	2.60	0.22	2.75	1.00	0.36
Severe deficiency below	1.5	0.09	2.08	0.54	0.12

4. Potassium Levels in the Leaves Associated with Adequacy and Deficiency

Table 7 gives data worked out by Hewitt (1955) in Jamaica. More recent work by Hewitt and Osborne (1962) indicates that there is good reason to examine again the adequacy level for potassium, which is too low. For plantains, Caro-Costas et al. (1964) give adequacy and deficiency levels at 3.96 and 2.76% K, respectively. Our own work in Taiwan showed that under some conditions the plants still responded to applied K at leaf levels over 5% K. (Taiwan Potash Research Foundation, *unpublished data*). It also seems that the optimum potassium level is higher under conditions of moisture stress than under conditions of abundant water supply. For further details reference is made to the work by Dumas (1958, 1960) and Martin-Prevel (1963).

5. Effects of Potassium

A. EFFECTS OF FLOWERING AND MATURITY. Potassium stimulates early shooting and significantly shortens the time required for fruit maturity. This is of particular importance for countries affected by tropical storms. Earlier fruit is often higher priced.

B. EFFECTS ON YIELD. Provided the plants have adequate levels of the other elements and the moisture requirements can be met, bananas respond to extremely large doses of potassium fertilizer (Fig. 5 and 6).

C. EFFECTS ON QUALITY. Yield increases by potassium are often accompanied by improvements of quality. With an exceptionally potassium-responsive crop like banana it is therefore not surprising that effects on quality are large (Chu, 1960).

Potassium increased the contents of soluble solids, affected the acid/sugar ratio, improved the flavor and significantly improved keeping qualities which may be associated with the thicker rind and the higher acid content of fruits from plants high in potassium.

D. EFFECTS ON DISEASE RESISTANCE. It has been suggested that the incidence of banana wilt (Panama disease, *Fosarium oxysporum f.* Cubense) may be associated in some cases with a decreased potassium uptake under the influence of excessive moisture and insufficient aeration (Risbeth, 1960).

Hasselo[1] (1961) reports a highly significant effect of potassium on the

---

[1] H. N. Hasselo. 1961. The soils of the lower eastern slopes of the Cameroon Mountains and their suitability for various crops. Thesis. Wageningen, Netherlands.

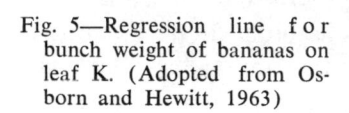

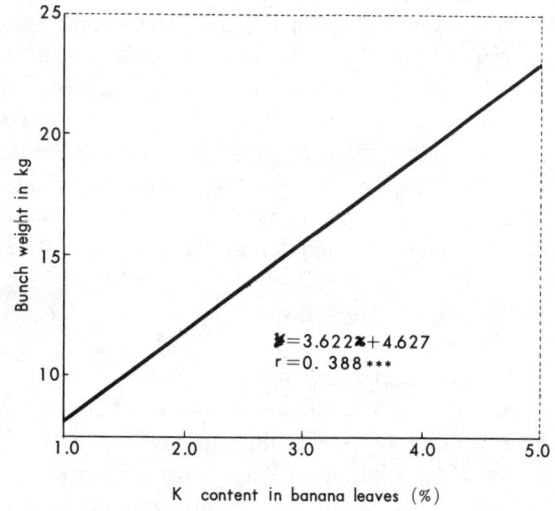

Fig. 5—Regression line for bunch weight of bananas on leaf K. (Adopted from Osborn and Hewitt, 1963)

$y = 3.622x + 4.627$
$r = 0.388$ ***

Fig. 6—The relation between the levels of applied K and fruit yield.

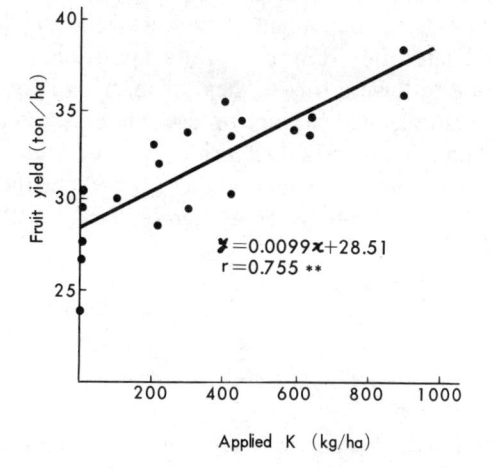

$y = 0.0099x + 28.51$
$r = 0.755$ **

water economy and resisance to siga toka (*Cerospora*). Application of potassium reduced drought losses from 48% to 12%, and from 16% to 5% for Siga toka losses. Total losses were reduced from 56% to 17%.

Premature yellowing of 8-to-10-month old plants, heavily infected by eelworms was corrected by potassium application but accentuated by the application of nitrogen (Hasselo, 1961).

6. Potash Fertilizer Requirements

Nutrient removal per ton of fruit bunches has been estimated at 2 kg N, 0.22 kg P, and 5 kg K (Martin-Prevel, 1964). For new plantings, under most conditions, fertilizer application according to expected removal is inadequate because very few soils will be able to supply the total amount of nutrients taken up by the plants. The plants may remove 4 to 5 times more nutrients than the bunches. Usually it is the ratoon crop that suffers most from inadequate potassium supply. Reasonable rates for establishment are in the range of 250 to 500 g K/plant for giant types (about 1,100 plants/ha) and 100-200 g K/plant for dwarf types (2,500 plants/ha). For maintenance of the following sucker crops, lower rates according to the production level can be adopted, as the nutrients contained in the mother plant will be gradually returned to the soil.

7. Timing of Application

Applied potassium is usually most effective when given in 2 to 3 doses during the earlier stages of growth from the fifth leaf stage to flowering. The number of potentially effective hands per bunch is already determined at the flower initiation stage so a good nutritional status at this stage is of outmost importance. Good nutrition after flower emergence is, of course, important to prevent premature finger drop and to fill the fruits properly. A well-nourished plant, however, will usually not respond to fertilizer applied late after flowering. Late applications, however, are important to get strong followers for the next crop. This is of particular importance in growing districts with a cool or dry winter season.

Though potassium chloride in most cases will be the economical source of potassium for bananas, there are slight indications that at very high rates potassium sulfate gives a slightly darker yellow color of the fruit flesh.

H. Oil Palm

1. Potassium Uptake and Requirement

All palms seem to have high potassium requirements though this is not reflected in the potassium levels of the leaves. There is general agreement that for the oil palm, the most productive tree in the palm family, potassium is the most important plant nutrient.

Potassium requirement is low during the first 2 to 3 years after planting. It increases rapidly after the 3rd year to the 7th year when a full crown has been developed and bunch production is reaching its maximum. Fully-grown stands may contain about 840 kg K/ha, the larger part of which is contained in the bunches (Tinker and Smilde, 1963). Removal of potassium with the harvested bunches is about 4.5 kg K/1,000 kg. With yields varying from 2,000 kg/K in the natural dura groves in Africa to well over 30,000 kg/ha for the best progenies of new hybrid material, the loss of K in harvested fruit varies from 10-140 kg/ha.

Potassium fertilizer requirements, however, usually far exceed the amount removed by the fruit. With high-yielding hybrid material, it was found necessary in many areas of Malaya to apply as much as 5.4-6.8 kg of KCl palm/year to maintain satisfactory leaf K levels. At 145 palms/ha this amounts to about 780-890 kg of material or 390-495 kg K/ha. On peat soils as much as 9 kg of KCl/palm or 650 kg of K/ha may be necessary to obtain satisfactory growth and yield (R. A. Bull, 1967; *Personal communication*).

2. Potassium and Yield

In contrast to many other crops the oil palm seems to be responsive to potassium at a very wide range of yield. On poor lateritic soils at the Ivory Coast, application of 1 kg KCl/palm/year raised the yield from 2 to 12 tons of bunches/ha per year. Within a few months after potassium application, chlorotic leaves became green and the assimilation surface doubled through an increase in the length and breadth of leaflets and an increase in the rate of emergence and length of lifetime of new fronds. The improvement in the foliage was followed by a sharp increase in the number of bunches per tree and the average bunch weight increased from 13 to 22 kg. Over a period of 16 years, a continuous application made-up solely of potassium chloride had not induced any nutritional imbalances (Boye, 1963). At low production levels potassium had a remarkable residual effect.

A single dose of 1.5 kg KCl/palm applied in 1951 gave a yield of 67 kg bunches/palm against 17 kg for the control. With high-yielding hybrids, which have a much faster turnover of absorbed potassium, such residual effects can hardly be expected. It seems that a potassium-deficient palm builds up the potassium reserves and leaf surface before increasing the yield. Conversely, when the supply of potassium is stopped, the tree uses up all potassium reserves before production falls. In most growing districts fertilizer application is based on leaf analysis. Critical values most widely used are (in percent dry matter) 2.50 N, 0.15 P, 1.00 K, 0.60 Ca, and 0.24 Mg (IRHO, 1961). For high-yielding hybrids in Malaya the following values are used: 2.7-2.8 N, 0.18-0.19 P, 1.2 K, 0.30 Mg, and 0.5 Ca (R. A. Bull, 1967; *personal communication*).

3. Role of Potassium in the Physiology of the Oil Palm

It is generally accepted that potassium plays an important role in photosynthesis, respiration, translocation, and protein formation but little is known about its influence on oil formation. However, the presence of large amounts of potassium in oil palm fruit suggests that potassium is an important factor, perhaps as an accelerator of enzyme activity.

Nitrogen and potassium are important factors in determining the sex ratio of the inflorescences. Excessive nonprotein nitrogen in relation to available carbohydrates seems to favor development of male and retard development of female inflorescenses. Through its effects on the enzyme system, photosynthesis, protein synthesis, and water economy, potassium has importance in favorably regulating the sex ratio (J. D. Ferwerda, 1955. Questions relavant to replanting in oil palm cultivation. Thesis Wagenigen, Netherlands); Sparnaaij, 1960; Hasselo, 1961[2]. The positive effect of potassium with regard to the moisture status of the palm, particularly in low rainfall areas, has been shown by Ochs (1963), and Prevot and Ollagnier (1958).

4. Potassium and Disease Resistance

Potassium deficiency has been associated with increased incidence of the following diseases: Vascular wilt (*Fusarium oxysporum*) (Haines and Beuzian, 1956; Prendergast, 1957), Frecle disease (*Cerospora elaeides*) "Little leaf" and genoderma (Forde et al., 1966). The department of agriculture in Malaya has suggested that potassium might give some resistance to upper stem rot, caused by *Fomes noxius* (R. A. Bull, 1967, *Personal communication*).

IV. CONCLUSIONS

Tropical crops differ widely in their potassium uptake, potassium removal, their efficiency in utilizing soil potassium and their response to applied potassium fertilizer (Table 8).

Actual potassium fertilizer usage, however, depends, at present, more on the rate of commericalization of the crop than on its responsiveness to potassium. Plantation crops like oil palm, pineapple, sugar cane, rubber, etc., are therefore the main consumers of (potassium) fertilizer, whereas subsistence crops, even if highly responsive to potassium are hardly fertilized.

But even with plantation crops, fertilizer use at present is aimed generally at maintenance, rather than maximum production. A new approach will be needed for a full exploitation of the potential offered by the tropical environment in combination with the rapidly improving genetic capabilities of newly developed varieties, hybrids, and clones.

With high yields particular attention must be paid to maintain a healthy root system. Because of the faster metabolic turnover in the tropics, the strain on the root system under high yield conditions is larger than under cool climate conditions. For the same reasons nutritional imbalances are more detrimental to yields and vigor of the crops than in cool regions. This is particularly true for all tree crops such as tea, coffee, cocoa, rubber, and oil palm. Potassium requirements of those crops are increasing progressively

---

[2] Ibid.

Table 8—Potassium levels in leaves of tropical crops associated with deficiency symptoms, response to K and adequacy

Crop	Visual deficiency symptoms	Response to K likely	Response unlikely	Remarks
Rice	< 0.4	< 0.8	> 1.2	Straw at harvest
Sweet potato	< 0.8	< 3.5	> 4.0	1st mature leaf, incl. petiole
Sugar cane	< 0.6	< 1.5	> 1.8	Center part of 4, 5 and 6 leaf without midrib at 6 month
Rubber	< 0.6	< 1.3	> 1.5	Basal leaf from a terminal whorl Shade leaves incl. midrib
Oil palm	< 0.6	< 1.2	> 1.5	Leaflets from frond No. 17
Coconut Palm	< 0.35	< 0.7	> 1.1	Leaflets from frond No. 9
Coffee (Arabica)	< 0.7	< 1.8	> 2.2	3rd pair of leaves on a branch at flowering
Tea	< 0.5	< 1.1	> 1.4	5th leaf on a newly shooting branch
Cocoa	< 0.6	< 1.4	> 2.0	2nd and 3rd leaves on fully green sprouts
Banana	< 1.2	< 4.0	> 4.8	Center part of 3rd fully opened leaf at shooting
Pineapple	< 0.5	< 3.5	> 4.2	White basal tissue of semi-matured leaves during main growing season

with more intensive exploitation and high yields. For annual crops such as rice and sweet potatoes potassium requirements follow a more linear course with increasing yields. Semi-permanent crops (plant crop and ratoons) like sugar cane, pineapple, and bananas take an intermediate position in this respect.

REFERENCES

Abruna, F., J. Vincente-Chandler, L. A. Becerra, et al. 1965. Effects of liming and fertilization on yields and foliar composition of high-yielding, sun-grown coffee in Puerto Rico. J. Agr. Univ. Puerto Rico 49:413-428.

Acquaye, D. K., R. W. Smith, and R. G. Lockard, 1965. Potassium deficiency in unshaded Amazon cocoa (*Theobroma cacao L.*) in Ghana. J. Hort. Sci 40:100-108.

Alvim, P. de T. 1954. Studies on the cause of cherelle wilt of cocoa. Turrialba 4:72-78.

Alvim, P. de T. 1965. Eco-physiology of the cacao tree. Congr. Int. sur les Rech. Agron. Cacaoyeres, Abdijan, p. 23-35.

Angladette, A. 1966. Le Riz. Maisonneuve & Larose, Paris, 930 p.

Anstead, R. D., and C. K. Pittock, 1913. The varying composition of the coffee berry at different stages of its growth and its relation to the manuring of coffee estates. Plant Chron. 8:455-460.

Baba, I., and T. Harada, 1954. Physiological disease of rice plants in Japan. Rep. 5th meeting Int. Rice Inst. Working party in Rice Breeding, p. 101-150.

Baba, I., K. Inada, and K. Tajima, 1965. Mineral nutrition and the occurrence of physiological disease. In The mineral nutrition of the rice plant. Int. Rice Res. Inst. Symp. John Hopkins Press, Baltimore, p. 173-195.

Beachell, H. M., and P. R. Jennings, 1965. Need for modification of plant type. *In* The mineral nutrition of the rice plant. Int. Rice Res. Inst. Symp. The John Hopkins Press, Baltimore, p. 29-36.

Beaufils, E. R. 1955. Mineral diagnosis of some *Hevea brasiliensis*. Arch. Rubber Cult. 32:1.

Bordon, R. J. 1940. Nitrogen-potash-sunlight relationships. Hawaiian Planters' Record. 45:131-146.

Boye, P. 1963. Inorganic nutrition and potassium deficiency of the oil palm. Potash Review, Subject 27, 39th suite, p. 8.

Caro-Costas, R., D. Abruna, and J. Vincente-Chandler, 1964. Response to fertilization of strip-cultivated plantains growing on a steep latosol in the humid mountain region of Puerto Rice. J. Agr. Univ. Puerto Rico 48:312-317.

Chadha, T. R. 1966. Fertilization extension. IPSA's contribution to agricultural extension with special reference to potash responses to paddy. Indian Potash Supply Agency, Madras.

Chandler, R. F. Jr. 1963. An analysis of factors affecting rice yield. Int. Rice Comm. News letter. Vol. XII. No. 4.

Chang, S. C. 1961. Control of suffocating disease of rice plants in Taiwan. Soils and Fertilizers in Taiwan. p. 1-4.

Chang, S. C., and W. K. Chen. 1960. Effect of fertilizer treatments on the exchangeable bases of lowland paddy soils. (In Chinese, Engl. summary) Mem. Col. Agr. NT Univ. Taipei. Vol. No. 5:4.

Chang, S. C., and M. P. Feng, 1964. The fixation accumulation and depletions of potassium in lowland rice soils. (In Chinese, Engl. summary) Mem. Col. Agr. NT Univ. Taipei, Vol. 8, No. 4.

Chu, C. C., 1960. The effect of potash fertilizer on the production of banana. Soils and Fertilizers. Taiwan. p. 39-41.

Chin, T. F. 1961. Plant nutritional studies on physiological disease of rice in Taiwan. Soils and Fertilizers in Tiawan, p. 5-9.

Chin, T. F., W. L. Huang, and L. T. Li. 1965. Experiment on fertilizer response of rice varieties of Taiwan (part 1). (In Chinese, Engl. summary) J. Agr. Assn. China, 51: 37-46.

Clements, H. F. 1953. Crop logging of sugar cane, principles and practice. Proc. Int. Soc. Sugar Cane Tech. Conf. Barbados. 79p.

Cocci, J. 1960. Sur une Corrélation entre bore et potassium observée dans les feuilles d'hévéa. Proc. Nat. Rubber Res. Conf. Kuala Lumpur. p. 102-116.

Compagnon, P. 1959. Fumure potassique de l'hévéa. Potassium Symposium, 1958. p. 345-349.

Constable, D. H. 1953. Manuring replanted rubber. Rubber Res. Inst. Ceylon Quart. Circ. 29. p. 1-2.

Corbetta, G. 1954. Concimacione potassica e azotata e malattie parasitarie del riso. Il Riso 3 (2): 20-24.

Coste, R. 1955. Les caféiéres et les cafès dans le monde. Editions Larose, Paris. 381 p.

Cunningham, R. K., and P. W. Arnold. 1962. The shade and fertilizer requirements of cacao (*Theobroma cacao*) in Ghana. J. Sci. Food & Agr. 13: 213-221v.

de Geus, J. G. 1950. Rubber en Thee. Bodemgesteldheid en bestrijding van Ziekten en plagen. Plant en Bodem. 5-6: 21-59.

de Haan, K., and A. F. Schoorel. 1940. Kaligebrek in de Theecultuur. Arch, Theecult. 14: 43-81.

Doyle, J. J. 1966. The response of rice to fertilizer. FAO Rome. 69p.

Dufournet, R., P. Roche, Rabetrano, Rekotondrainibe, J. Velly, and C. B. Ngo. 1966. Problems of fertilizer usage on the high plateau of Madagascar. Fertilite No. 26: 7-36.

Dumas, J. 1958. Détermination d'une feuille-origine pour l'étude des bananiers cultivés. Fruits 13:211-224.

Dumas, J. 1960. Contrôle de nutrition de quelques bananeraies dans trois territoirs africaines. Fruits 15: (6) 375-386.

du Toit, J. L. 1955. Cited by A. Jakob and H. von Uexkuell: Fertilizer Use, 3rd ed. Verlagsgesellschaft f. Ackerbau, Hannover 1963. p. 171.

Eden, T. 1958. Tea . . Longmans. 1960. 2nd ed. London. p. 74.

Forde, M. C., M. J. P. Leyritz, and J. M. G. Sly. 1966. The importance of potassium in the nutrition of oil palm in Nigeria. Potash Review, Subject 27, 46th suite. 8p.

Freiberg, S. R., and F. C. Steward. 1960. Physiological investigations on the banana plant. III. Factors which affect the nitrogen compounds of the leaves. Ann. Bot. 24 (93): 147-157.

Girst, D. H. 1959. Rice. Longmans, London. 464p.

Goto, Y. B., and E. T. Fukunaga. 1956. Coffee: How to grow seedlings. Hawaii Col. Agr. Extension Svc. Circ. 345. 10p.

Gouin, L., and T. R. Chadha. 1958. NPK fertilizer experiments on paddy in Madras, Andhra and Bombay States. Potascheme. Int. Potash Inst. Bern.

Guha, M. M., and E. Pushparajah. 1966. Responses to fertilizer in relation to soil type. Rubber Res. Inst. Malaya. Planters Bull. 87:178-183.

Haag, H. P., and E. Malavolta. 1960. (cited by Malavolta, E. 1963, in Cultura e adubacao do cafeeiro, Inst. Brasileiro de Potassa, Sao Paulo, p. 155.

Haines, W. B., and B. Benzian. 1956. Some Manuring experiments on oil palms in Africa. Emp. J. Exp. Agr. 24 (94): 137-160.

Hardy, F. 1936. Manurial experiments on cocoa in Trinidad. 6th Ann. Rep. Cocoa Res.

Hartt, C. E. 1929. Potassium deficiency in sugar cane. Bot. Gaz. 88:229-261.

Hartt, C. E. 1934. Some effects of potassium upon the amounts of protein and amino forms of nitrogen, sugars and enzyme activity of sugar cane. Plant Physiol. 9: 453-490.

Hashimoto, T. 1959. Interrelationship between potassium application and Young's moduli of crop plants. Investigations about lodging. Japanese potassium symposium. Int. Potash Inst. Bern. p.6-25.

Hasselo, H. N. 1961. Premature yellowing of Lacatan bananas. Trop. Agr. Trinidad. 38:29-34.

Hewitt, C. W. 1955. Leaf analysis as a guide to the nutrition of bananas. Emp. J. Exp. Agr. 23:11.

Hewitt, C. W., and R. E. Osborne. 1962. Further field studies on leaf analysis of Lacatan bananas as a guide to the nutrition of the plant. Emp. J. Exp. Agr. 30 (119): 249-256.

Honya, K. 1967. High yielding rice cultivation in Tohoku district. Japanese Potassium Symposium. Int. Potash Inst., Bern. (English edition in print).

Humbert, R. P. 1953. Basic problems in sugar cane nutrition: II. Applying basic facts in sugar cane nutrition. Agr. Dept. HSPA Exp. Sta. Hawaii 44.

Humbert, R. P. 1963. The growing of sugar cane. Elsevier Amsterdam. 710p.

Humbert, R. P., and J. P. Martin. 1955. Nutritional deficiency symptoms in sugar cane. Hawaiian Planters, Record 55:95-102.

Humphries, E. C. 1939. Growth rate and mineral intake by the pod. Leaf flush and mineral intake by the root. 9th Ann. Rep. Cocoa Res. Trinidad.

Humphries, E. C. 1959. Wilt of cocoa fruits (*Theobroma cacao L*). Seasonal variations in potassium, nitrogen, phosphorus and calcium of the bark and wood of the cocoa tree. Ann. Bot. 14:149-164.

Int. Rice Res. Inst., Los Banos. Ann. Rep., 1963, 1964, and 1965.

IRCA, 1961. Institut de Recherches sur le Caoutchouc in Afrique., Rapp. IRCA 1961:35.

IRHO, 1961. Institut de Recherches pour les Huiles et Oléagineux. Rapports annuales 1950-1960.

Ishizuka, Y. 1965. Nutrient uptake at different stages of growth. *In* Int. Rice Res. Inst. (ed.) The mineral nutrition of the rice plant. John Hopkins Press, Baltimore, p. 199-217.

Ishizuka, Y., and A. Tanaka. 1951. Studies on the nitrogen, phosphorus and potassium metabolism of the rice plant. The influence of the potassium concentration in the culture solution on the growth of the rice plant, especially on the amount of potassium in the plant. (In Japanese) J. Sci. Soil and Manure. Japan. 22(2):102-106.

Jayaraman, V., and P. de Jong. 1955. Some aspects of nutrition of tea in Southern India. Trop. Agr. 32 (1): 58-65.

Kawai, S., 1959. Effects of potassium on the tea plant. Japanese potassium symposium. Int. Potash Inst., Bern. p. 77-92.

Kiuchi, T., and H. Ishizuka, 1961. Effects of nutrients on the yield constituting factors of rice (potassium). (In Japanese) J. Sci. Soil and Manure. Japan. 32(5): 198-202.

Koningsberger, V. J. 1931. Nogmals over de oorzaak de Z-gn. "Kalimati-Ziekte." Arch. Suikerind. Ned.-Indie 39: 161-166.

Kono, M., and J. Takahashi, 1961. Study on the relationship between breaking strength and the chemical components of the paddy stem. (In Japanese) J. Sci. Soil and Manure. Japan 32(4):149-152.

Kosaka, T. 1963. Control of blast by cultivation practices in Japan. Symp. on the rice blast disease. Int. Rice Res. Inst. John Hopkins Press, Baltimore. p. 421-440.

Lee, H. A., and J. P. Martin. 1928. Effect of fertilizer constituents on the eye-spot disease of sugar cane. Ind. Eng. Chem. 20:220-224.

Lemée, G. 1955. Influence de l'alimentation en eau et de l'ombrage sur l'économie hydrique et la photosynthèse du cacaoyer. Agronomie Tropicale 10:592-603.

Loué, A. 1958. Mineral nutrition of robusta coffee plant and its manuring in Ivory Coast. Fertilité 5: 27-68.

Loué, A. 1962. Nutrient deficiencies in cacao trees in the plantations of the Ivory Coast. Fertilité 14:42-52.

Malavolta, E., H. P. Haag, F.A.F. Mello, and M.O.C. Brasil Sobro. 1962. On the mineral nutrition of some tropical crops. Coffee. Int. Potash Inst. Bern. p.43-67.

Martin, J. P. 1940. Varietal differences of sugar cane in growth, yields and tolerance to nutrient deficiencies. Hawaiian Planters' Record. 45:79-91.

Martin-Prevel, P. 1963. Le bilan minéral base d'interprétation du diagnostic foliaire. Summary in Fruits. 18 (10): 455-467.

Martin-Prevel, P. 1964. Nutrient elements in the banana plant and fruits. Fertilité. 22:3-14.

Martin-Prevel, P., and G. Montagut. 1966. Essais sol-plante sur bananiers: Les interactions dans la nutrition minérale du bananier. Fruits 21 (6): 283-294.

Martin-Prevel, P., and J. M. Charpentier. 1964. Symptoms of depreviation of six nutrient elements in the banana. Fertilité. 22: 15-50.

Matsuo, T. 1948. On the influence of potash dificiency in soils upon outbreak of the sesame leaf spot of the rice plant. (In Japanese) Ann. Phytopath. Soc. Japan. 13: 10-13.

Matsuo, H. 1963. Hight yield of paddy rice and potash. Potash Review, Subject 9.

Matsuo, H. 1966. Fertilizer application techniques and water management for rice. In Assn. of Agr. Relations in Asia (ed.) Development of paddy rice culture techniques in Japan. Tokyo.

Matsushima, S. 1966. Crop Science in Rice. Fuji publishing Co., Ltd. Tokyo. 365 p.

Mauritius Sugar Cane Research Station. 1954. Sugar Cane research in Mauritius. Rev. Agr. Maurice 33: 28-43.

Medcalf, J. C., W. L. Lott, P. B. Teeter, and I. R. Quinn. 1955. IBEC Res. Inst. Bull. 6. Experimental Programme in Brazil.

Middleton, R. K., T. T. Chin, and G. C. Iyer, 1965. A comparison of rock phosphate with super phosphate and of ammonium sulphate with sodium nitrate as sources of phosphorus and nitrogen for rubber seedlings II. Association with abnormal growth and effect on wood strength. J. Rubber Res. Inst. Malaya 19 (2):108-119.

Moir, W. W. G. 1930. The plant food problem. Rept. Assn. Hawaiian Sugar Tech. 9: 175-203.

Motomatsu, T. 1964. Varietal differences in the response of paddy rice to phosphorus and potassium. (In Japanese) Kali Kenkyu, Tokyo, No. 9.

Mukasa, Y., and S. Kawai. 1956. Effects of phosphate and potash fertilizer on chemical components of tea leaves. Study of tea. 15: 81-84. Cited by Kawai, 1959.

Mukerjee, H. N. 1955. Potash response in supposed unresponsive soils, determined by a new technique of experiments on cultivators' fields. Potassium symposium. Int. Potash Inst. Bern. p. 259-291.

Mukerjee, H. N. 1965. Fertilizer tests in cultivators' fields. In Int. Rice Res. Inst. (ed.) The mineral nutrition of the rice plant. Symposium. John Hopkins Press, Baltimore; p. 329-354.

Murayama, N. 1965. Mineral nutrition and characteristics of plant organs. In Int. Rice Res. Inst. (ed.) The mineral nutrition of the rice plant. John Hopkins Press, Baltimore. p. 147-172.

Murray, D. B., 1959. Deficiency symptoms of the major elements in the banana. Trop. Agr. Trinidad 36 (2):100-107.

Murray, D. B., and G. K. Maliphant. 1965. Problems in the use of leaf and tissue analysis in cacao. Conf. Inter. sur les Rech. Agron. Cacaoyères, Abdijan p. 36-38.

Nagata, T., 1954. Studies on anthracnose of the tea plant. (In Japanese) Bull. Tea Div. Tokai-Kinki Agr. Exp. Sta. 2: 97-131.

Navasero, S. A., and Estrada, 1967. Amelioration of coralline calcareous soil in Bilar, Bohol for lowland rice cultivation. Int. Rice Res. Inst.

Noguchi, Y., and T. Sugawara, 1966. Potassium and Japonica rice. Int. Potash Inst. Bern. p. 102.

Ochs, R. 1963. Recherches de pédologie et de physiologie pour l'étude du problème de l'eau dans la culture du palmier à huile. Oléagineux 18: 231-238.

Ogiwara, T., 1960. Findings on K-deficiency of paddy rice. (In Japanese) Special Bull. Fukuoka Agr. Exp. Sta.

Okajima, H., 1965. Environmental factors and nutrient uptake. In Int. Rice Res. Inst. (ed.) The mineral nutrition of the rice plant. Symposium John Hopkins Press, Baltimore. p. 63-74.

Okamoto, H. 1958. Relation between rice blast and sesame leaf spot and potassium. Japanese Potassium Symposium. Int. Potash Inst., Bern. p. 76-89.

Ono, K. 1957. The relation between rice disease and potassium. Japanese Potassium Symposium. Int. Potash Inst. Bern. p. 71-87.

Osborn, R. E., and C. W. Hewitt. 1963. The effect of frequency of application of nitrogen, phosphate and potash fertilizer on lacatan bananas in Jamaica. Trop. Agr. Trinidad 40 (1): 1-8

Owen, G., D. R. Westgarth, and G. C. Iyer. 1957. Manuring Hevea: Effects of fertilizers on growth and yield of mature rubber trees. J. Rubber Res. Inst. Malaya 15: 29.

Panse, V. G., and R. C. Khanna. 1964. Response of some important Indian crops to fertilizers and factors influencing this response. The Indian J. Agr. Sci. Vol. 34, no. 3.

Pendleton, R. L. 1943. Land use in Northeastern Thailand. Geog. Rev. 33 (1): 15-31.

Ponnamperuma, R. W. 1958. Lime as a remedy for a physiological disease of rice associated with excess iron. Int. Rice Comm. News Letter. 7: 10-13.

Portsmouth, G. B. 1950. Potash requirements of tea. Tea quarterly 21 (1): 18-22.

Portsmouth, G. B. 1953. Potash deficiency in tea. Tea quarterly 24 (4): 70-81.

Pound, F. J., and J. de Verteuil. 1936. First year observations in an experiment designed to test the gross effects of applications of nitrogen, potash and phosphorus on the cocoa tree. Trop. Agr. Trinidad 13: 9.

Prendergast, A. G., 1957. Observations on the epidermiology of vascular wilt disease of the oil palm (*Elaies quinensis J.*) J. W. African Inst. Oil Palm Res. 2 (6) 148-175.

Prévot, P., and M. Ollagnier, 1958. La fumure potassique dans les régiones tropicales et subtropicales. 4th Potassium Symposium, Int. Potash Inst. Bern. p. 277-313.

Rice Department, Bankok. 1961. Simple fertilizer experiments on cultivators' fields, 1958-59 and 1959-60, Bankok Thailand.

Risbeth, J. 1960. Factors affecting the incidence of banana wilt. (Panama disease) Emp. J. Exp. Agr. 28 (110): 109-113.

Rosenquist, E. A. 1960. Manuring of rubber in relation to wind damage. Proc. Nat. Rubber Res. Conf. Kuala Lumpur: 81-88.

Rubber Research Institute of Malaya. 1964. The effect of fertilizer on growth and yield of mature rubber. Ann. Rep. 1963: 16.

Samuels, G. 1960. The relative merits of various methods of foliar diagnosis for sugar cane. Proc. 10. Congr. ISSCT: 529-537.

Shiroshita, T. 1963a. Theory and practice of growing rice Chapt. 4. Theory and practice of fertilizer application. English edition. Fuji Publishing Co., Ltd. Tokyo. p. 192.

Shiroshita, T. 1963b. Fertilization of the fields of high rice yield contest winners. Japanese Potassium Symposium. Int. Potash Inst. Bern. (English edition in print).

Shorrocks, V. M. 1964. Mineral deficiencies in *hevea* and associated cover plants. Rubber Res. Inst. Malaya, Kuala Lumpur. 76p.

Shorrocks, V. M. 1965a. Mineral nutrition, growth and nutrient cycle of *Hevea brasiliensis*. I. Growth and nutrient content. J. Rubber Res. Inst. Malaya 19 (1): 32-47.

Shorrocks, V. M. 1965b. Mineral nutrition, growth and nutrient cycle of *Hevea brasiliensis* II. Nutrient cycle and fertilizer requirements, J. Rubber Res. Inst. Malaya 19 (1): 48-61.

Shorrocks, V. M. 1965c. Mineral nutrition, growth and nutrient cycle of *Hevea brasiliensis* IV. Clonal variation in girth with reference to shoot dry weight and nutrient requirement. J. Rubber Res. Inst. Malaya 19 (2): 93-97.

Sparnaaij, L. D. 1960. The analysis of bunch production in the oil palm. J. W. African Inst. Oil Palm Res. 3 (10): 109-180.

Tai, E. A. 1958. Ann. Rept. Jamaica Banana Board Res. Dept. 1956. 7. 3-8.

Takahashi, J. 1960. Potash and the cultivation of rice. Fertilité. no. 11. p. 13-22.

Takahashi, J. 1965. Fertilizer tests in cultivators' fields. *In* Int. Rice Res. Inst. (ed.) The mineral nutrition of the rice plant. Symposium. John Hopkins Press, Baltimore p. 320-334.

Tanaka, A. 1956. Studies on the characteristics of the physiological function of leaves at definite position on stems of rice plants. 4. Relation between phosphorus and potassium content and physiological function of leaves at definite positions. (In Japanese) J. Sci. Soil and Manure Japan 27 (6): 223-28.

Tanaka, A., R. Leo, and S. A. Navasero. 1966. Some machanisms involved in the development of iron toxicity symptoms in the rice plant. Soil Sci. and Plant Nutr. 92 (4): 32-38.

Tinker, P. B. H., and K. W. Smilde. 1963. Dry matter production and nutrient content of plantation oil palms. Plant and Soil 19: 350-363.

Tisdale, W. H., and J. H. Jenkins, 1912. Straighthead of rice and its control. USDA Farmers Bull. 1212: 1-16.

van Dierendonck, F J. E., 1960. The manuring of coffee, cocoa, tea, and tobacco. Centre d'Etude de l'Azote, 3, Geneva: 57-78.

van Dillewijn, C., 1952. Botany of sugar cane. Chronica Botanica Co., Waltham, Mass. pp. 371.

Venkataramani, K. S. 1957. Foliar application of fertilizers to tea. Fertilité 1: 17-22.

Vermaat, J. G. 1964. Expanded program of technical assistance, FAO no. 1887. Report to the government of Pakistan on soil fertility investigations.

Victoria Milling Co. 1955. When to fertilize your soil with potash and how much potash; fertilization tests with sugar cane in the Victoria Milling District. Release no. 11. VMC Exp. Sta. Bull. 2; 10.

von Uexkuell, H. R. 1964. Obstacles to using fertilizers for rice in S. E. Asia. World crops, March, 1964.

von Uexkuell, H. R. 1965. Problems in development of modern tropical agriculture. Japanese Potassium Symposium. Yokendo Press, Tokyo, p. 26-38.

von Uexkuell, H. R. 1966. Rice disease and potassium deficiency. Better Crops I (3): 28-35.

WAIFOR, 1959. West African Institute for Oil Palm Research, cited by Werkhoven.

Ward, J. B., 1964. cited by Shorrocks, 1964.

Watson, G. A., and R. Navayanan. 1965. Effect of fertilizers on seed production by *Hevea brasiliensis*. J. Rubber Res. Inst. Malaya. 19 (1): 22-31.

Weerawickrema, S. K. A., and D. H. Constable, 1966, 1967. The Australian-Ceylon FFHC soil fertility-fertilizer project. 4 and 5. Field report to the government of Ceylon.

Wellman, F. L. 1961. Coffee. (Botany, cultivation and utilization) Leonhard Hill Ltd. London, 488p.

Werkhoven, J. 1966. The manuring of the oil palm. 2nd edition. Verlagsgesellschaft fuer Ackerbau, Hanover. Green Bull. no. 18, 52p.

Yamazaki, T. 1963. Analysis of rice growing techniques of high yield contest winners with three noteworthy problems involved. Japanese Potassium Symposium, Int. Potash Inst., Bern. (English edition in print).

Yamazaki, T. 1967. Possibilities of obtaining high rice yield through increased fertilizer application. Symposium on present situation and the future of high yield techniques of paddy rice. (In Japanese) Tottori University. p. 109-121.

Zhurbitsky, Z. I., and D. V. Strausberg. 1964. Foliar diagnosis of tea shrubs. Plant analysis and fertilizer problems. Proc. 4th Colloq. p. 392-397.

ature-livestock research programs are integrated and directed toward high
Potassium Nutrition of Forage Crops with Perennials

ROY E. BLASER and E. LAMAR KIMBROUGH

Virginia Polytechnic Institute
Blacksburg, Virginia

I. INTRODUCTION

Since forages are produced primarily for ruminants, this paper intentionally relates to some aspects of animal nutrition. There is a critical need for interrelating and improving the production of forages and their utilization by animals. Feed stuffs fed through animals for later human consumption are much less efficient than direct utilization of plants by humans. Also, certain foreign countries, New Zealand and Australia, are producing livestock products at low costs, accounting for large exports to the USA. Pasture-livestock research programs are integrated and directed toward high production per hectare and animal. Research in this country has been directed primarily toward high outputs per animal. Outputs per hectare and per animal must be wisely compromised in today's competitive agriculture.

Perennial grasses and grass-legume mixtures make up a tremendous acerage in the USA. The large acreages of sod crops are attributed to factors such as: steep topography, shallowness, stoniness, poor structure and poor drainage of soils, wind erosion under cropping, poor rainfall distribution, and rotations for reasonable control of soil and water. Sod crops on such soils furnish forages (pasture, silage, and hay) that are excellent sources of proteins, minerals, and vitamins for ruminants. However, sod crops in today's aggressively competitive agriculture cannot compete with intensive row cropping on soils suitable for intensive farming. This contention is supported by two primary principles: (i) perennial grasses and legumes produce less dry matter per unit land area than annuals such as corn and sorghums and (ii) the utilizable energy from perennial forages for ruminants, as shown by meat and milk production, is suboptimum as compared to the energy from grain feeds or with the forage corn silage.

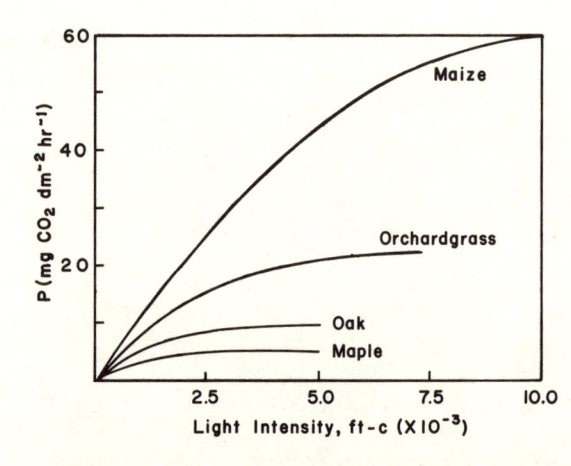

Fig. 1—Relationship of light intensity to photosynthesis of four species. Data from Hesketh and Baker (1967).

The low potential production of perennial cool season grasses as compared to corn is attributed to lower net assimilation rates (NAR). Individual leaves of cool season perennials become light saturated at 3,000 to 6,000 ft-c, but NAR of corn continues to increase up to the maximum light energy from the sun of more than 10,000 ft-c according to Hesketh and Baker (1967) (Fig. 1) and Brown et al. (1966) (Fig. 2). The warm season bermudagrass leaves had higher NAR values than alfalfa and orchardgrass. Morphologically, the display of leaf-stem canopies of perennials utilize light less efficiently than for corn. Legumes such as alfalfa and the clovers with near horizontal leaf displays intercept nearly all of the incident light with a leaf area index (LAI) of about 3 (3 square units of leaf area, one side, per unit of soil area) as determined by Donald (1963). Perennial grasses with semi-erect leaves require an LAI of 4-7 to intercept most of the light. Corn, with leaves displayed on tall stems and the upper leaves semi-erect requires a much higher LAI than perennials to intercept nearly all of the light. Thus, corn with a large functioning leaf area per unit soil area has higher dry matter production potentials. In addition, the NAR of shaded and/or aging leaves in perennial grass-legume canopies apparently drops rapidly as compared to corn. High summer temperatures apparently retard photosynthesis of cool season sod crops as compared to summer annuals. Thus, the yield potentials of crops differ, and practical yields that are obtainable by Virginia farmers are about: corn silage 18,000; alfalfa alone or in mixtures 9,000; red clover 7,800; nitrogen fertilized perennial grasses 9,000; and pastures 4,500 kg dry matter/ha.

Starting in 1949 at the Virginia Forage Research Station, the programs with production and utilization of perennial grasses and legumes stressed two primary objectives: (i) quality of herbage with management systems in terms of high per animal outputs (meat or milk per animal) and (ii) high yields of animal products per hectare (the combined effect of output per animal and animal units per hectare). We found that output per animal of

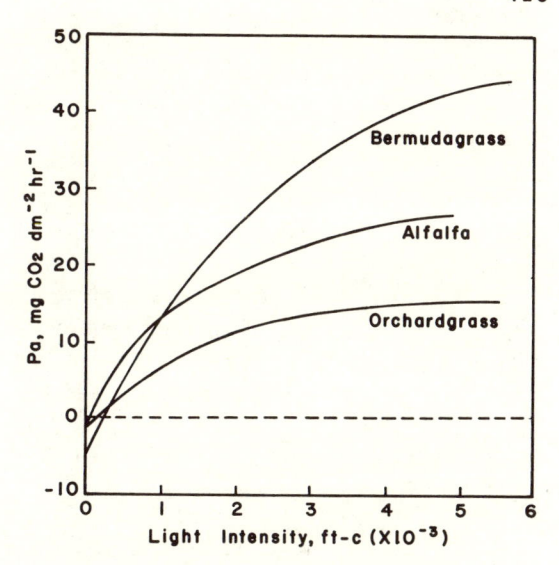

Fig. 2—Apparent photosynthesis (Pa) for leaves of three forage species at different light intensities. Data from Brown et al. (1966).

any perennial forage system used could be improved by energy supplements (ground shelled corn). When animals were restricted to perennial forages, the suboptimum outputs per animal were attributed to insufficient utilizable energy for two reasons: (i) low intake and (ii) low digestibility of feeds from perennials as compared to the utilizable energy from grain or corn silage. For example, steers fed relatively high dry matter corn silage without supplements gained from 50 to 75% faster than those fed our best quality alfalfa hay or wilted silage. Animals require 10 to 15 times more digestible energy than digestible protein for high outputs per beast. When considering the daily utilizable energy per animal and the potential animal outputs, meat production possibilities in Virginia are approximately: corn silage 1,680-2,240, alfalfa mixtures as silage and hay 560-780, and pastures 300-500 kg/ha of liveweight gains.

High yields, 18,000 to 22,500 kg/ha of alfalfa, have been reported in some states, but such yields are not easily nor readily attainable as compared with usual high total forage yields from corn crops. The economic suitability of a crop depends on yields and on production and utilization costs. Perennial grasses and legumes have low utilization cost advantages when grazed, but the variable seasonal production with wasted pasture and lack of quality control causes reduced animal products per hectare. Mechanical utilization costs with perennial grasses and legumes are high because of harvesting 3 to 8 low-yielding growths as compared with one large yield with corn silage. With mechanical harvesting, dry matter losses for corn silage are low as compared with perennial grasses and legumes and quality control for producing a high quality feed from corn silage is a certainty as compared with ensiling or haying perennial sod crops.

Research in the future must be directed to overcome the adverse characteristics of cool season perennial grasses and legumes. Sod crops do furnish feed suitable for beef herds and will continue to be of major importance because there are no economic alternatives for the large land areas where forages are now used.

II. POTASSIUM COMPOSITION AND YIELD

It is difficult to make precise positive linear associations of dry matter yields with potassium absorbed because: (i) stages of plant growth often affect potassium content more than fertilization rates, (ii) competition for potassium among plants in mixtures alters the rate of potassium application and potassium content as compared to pure stands, (iii) luxury potassium absorption, (iv) relative magnitude of cations in the soils and other factors. The latter two items will not be discussed.

A. Growth Stage Effects

The dry matter yield and the mineral composition of perennial herbaceous grasses and legumes change sharply with maturity or date of cutting because of shifts in morphological and physiological characteristics. Rates of dry matter production increase as cutting frequencies are lengthened, then decline sharply and finally there is a net dry matter loss because of foliage

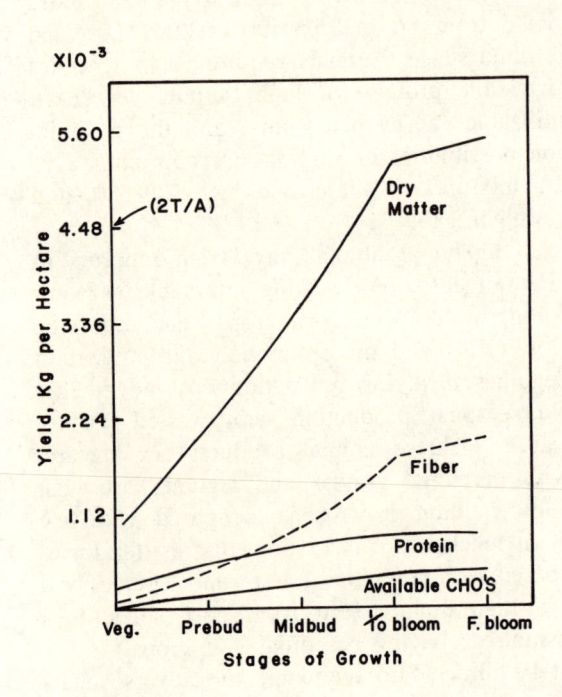

Fig. 3—Yield of alfalfa and its composition at various stages of growth. The available carbohydrates include starch and sugars. Data from Smith (1959).

losses (Smith, 1959) (Fig. 3). With advanced maturities there are rapid declines in leafiness and in succulence, and sharp inclines in cell wall materials (cellulose, starch, fructosan, lignin, hemi-cellulose, etc.). Total protein and mineral constituents generally increase in herbage until somewhere in flowering stages, pending rates of leaf loss. However, the percentage composition of ash and protein drops drastically during shifts from leafy to stemmy morphologies of plants. For two grass-legume mixtures the K content of the four species ranged from 3.40 to 3.69% dry matter for leafy growth 15-25 cm high as compared to only 0.85 to 1.58% at seed maturity (Fig. 4). Thus, stage of plant growth often influences potassium content in plant tissue more than its availability in soils. The K content of third cutting alfalfa for various K treatments in New Jersey was 23% lower in 1960 than in 1961 (Table 1). The favorable moisture in 1960 increased the dry matter yields and apparently advanced the growth stage, which caused a dilution effect (Markus and Battle, 1965). It is clear that for potassium rate experiments, the potassium values will be much higher for frequent than infrequent cuttings. Alfalfa cut in a prebloom stage was 62% higher in K than at full bloom (Fig. 4).

Data on potassium composition should consider practical stage of cutting as applied to animal nutrition. Quality of herbage (digestibility and con-

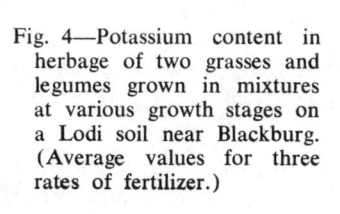

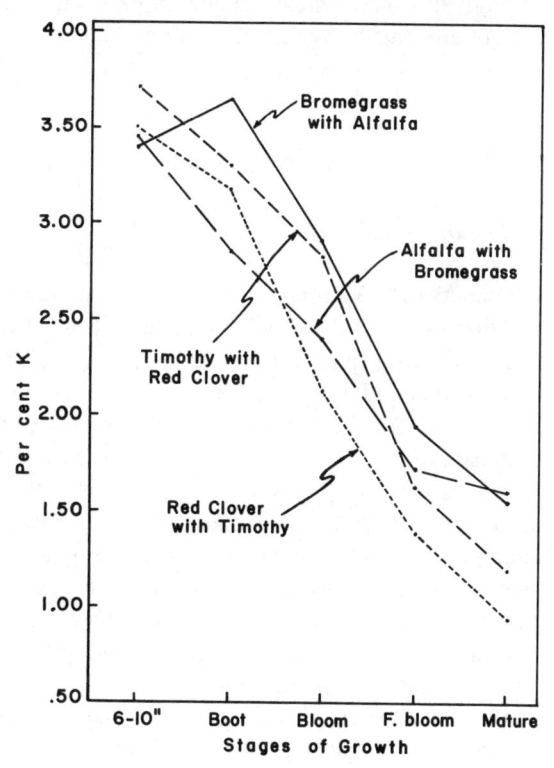

Fig. 4—Potassium content in herbage of two grasses and legumes grown in mixtures at various growth stages on a Lodi soil near Blackburg. (Average values for three rates of fertilizer.)

Table 1—The persistence, yield, and K content of alfalfa with rates of K (Markus and Battle, 1965)

K applied	Alfalfa stand 3rd yr.	Alfalfa stand 9th yr.	Yield during 8 years	K content of 3rd cutting 1960	K content of 3rd cutting 1961
kg/ha	—— % ——		tons/acre/yr	—— % ——	
0	17	0	1.83 d	----	----
46.5	26	0	2.67 c	----	----
93	34	18	3.45 b	0.99	0.90
186	55	49	3.98 ab	1.05	1.34
280	59	43	4.22 a	1.40	1.80
372	66	70	4.36 a	1.59	2.15

sumption) declines rapidly as herbage matures or ages. For example, for uninterrupted growth during spring, for various perennial forages, there was an average decline in digestibility of 0.48 unit for each day as cutting was delayed (Reid, 1959). From the viewpoint of feeding value, it is not possible to obtain maximum yield and quality. Frequent cutting increases quality but reduces yields of morphological erect, easily defoliated plants, such as alfalfa and red clover. Thus, yield and quality must be wisely compromised, cutting as frequently as possible to obtain high yields of utilizable energy. The trend of cutting to obtain young and leafy growth for better feed quality means higher potassium values of herbage. For example, K contents will be higher for 3 to 5 than for 2 to 3 yearly cuttings of forage plants.

B. Competition in Plant Associations

1. Legume-grass Mixtures

Donald (1963) reports, "It seems that when two fodder species are grown together they give no advantage in terms of yield of dry matter over the higher yielding pure culture. This represents a simple picture of competitive relationships. One species is the aggressor, able to exploit more than its "share" of the factors of the environment, while the other is suppressed because it is able to secure only a lesser part of the light, water, or nutrients." This contention applies best over short periods while excellent stands of an aggressive species are maintained. In environments where alfalfa is well adapted, it is as productive in pure stands as in any grass association. However, grasses are usually used with most pasture and forage legumes, because such perennial mixtures generally yield more dry matter over a period of years since legumes are more vulnerable to injury or stand losses from adverse biotic, climatic, and soil effects. Including grasses with legumes minimizes adverse effects such as bloat and makes mixtures more suitable for flexible utilization, such as silage, hay, or grazing.

Table 2—Potassium content of alfalfa and orchardgrass when the two were grown together in a mixture at Blacksburg, 1951 *

K, kg/ha	K, percent of dry weight – Orchardgrass	K, percent of dry weight – Alfalfa	K content in alfalfa relative to grass
372	3.85	3.53	92
93	4.01	1.78	42
46.5	3.46	1.21	35
00	2.71	0.70	26

* Lime, phosphorus, and boron were applied uniformly on this Lodi loamy soil.

Because nitrogen fertilizer is a method for intensifying livestock farming, nitrogen fertilized perennial grasses are used extensively. However, grass-legume mixtures will continue to be used because the low yields from many nitrogen fertilized grasses add to production costs per unit of dry matter. The better herbage quality from legumes or grass-legume mixtures as compared to grasses fertilized with nitrogen is an important consideration. Data quite consistently show higher output per animal from legumes or legume-

Fig. 5—The dry matter yields of an alfalfa-orchardgrass mixture and the yields and K content of the botanical components for rates of K fertilization. Yields for a 7-year period on a Cecil soil are low because of two dry years. Data from Blaser (1961-62).

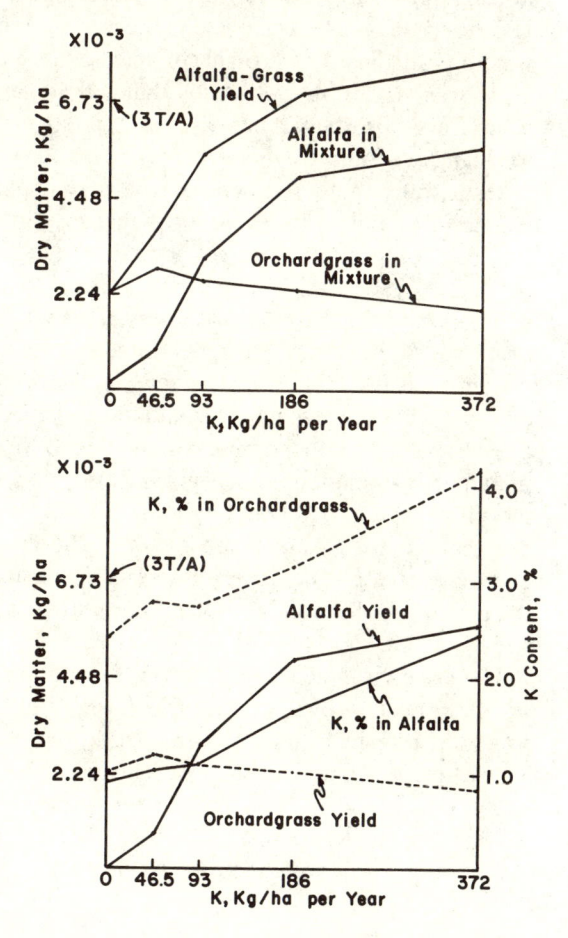

grass mixtures than for nitrogen fertilized grasses (Blaser et al., 1956; Heinemann and VanKeuren, 1958; Murdock et al., 1958; VanKeuren and Heinemann, 1958; Browning, 1963; and Hogg and Collins, 1965). Other research shows that calves per cow and weaning weights of calves are improved by legumes in grass pastures as compared to nitrogen fertilized grass pastures (Williams, 1967). Thus, it appears that grass-legume associations will continue to be important forages, even though the competitive effects of plant associations complicates fertilization and management.

The aggressive potassium competition among an alfalfa-orchardgrass mixture on a Lodi loam that was medium-low in potassium during the first productive year is shown in Table 2. With liberal K (372 kg K/ha) alfalfa had only 8% less K than grass. As fertilizer potassium was reduced, the alfalfa absorbed much less K than the grass. Without applying K, the alfalfa had 74% less K in herbage tissue than the grass associate. The yield and composition data for the 7th year of an experiment on a Cecil soil with potassium fertilization of an alfalfa-orchardgrass mixture are given in Fig. 5 (Blaser, 1961-62). The yield increases of the mixture with rate of potassium are attributed to the sharp increases in the alfalfa component. Orchardgrass was higher in potassium than alfalfa at all potassium rates, but orchardgrass absorbed relatively more potassium than alfalfa with low than for high potassium treatments.

Data from another experiment on a similar Lodi soil show the potassium content of alfalfa and bromegrass grown in association with a high and low rate of phosphorus and potassium (Fig. 6). Growth during this first productive year was excellent. With 167.4 kg K, alfalfa during early spring was higher in the K than bromegrass; however, with increased growth and less available K the grass associate finally absorbed more K per unit of dry matter than the alfalfa. Conversely, with low potassium, bromegrass absorbed much more potassium than alfalfa, except during the early spring vegetative growth. The higher potassium absorption of grasses than for legumes grown in association under potassium stress is not well understood. Root morphology and relative physiological activity under various environments are thought to be important factors (Blaser and Brady, 1950). The base exchange of the root tissue of species associations has also been considered important in differential potassium uptake among species (Drake et al., 1951).

Under environments where cool season grasses produce high yields and good seasonal growth, they have been observed as aggressive competitors for potassium in legume associations if soils are low in available potassium. White and Ladino clovers, red clover, and alfalfa stands and yields have declined sharply as compared to the grasses in cool season perennial plant associations under potassium stress. These legumes were invariably much lower in potassium content than the grass associates. On the other hand,

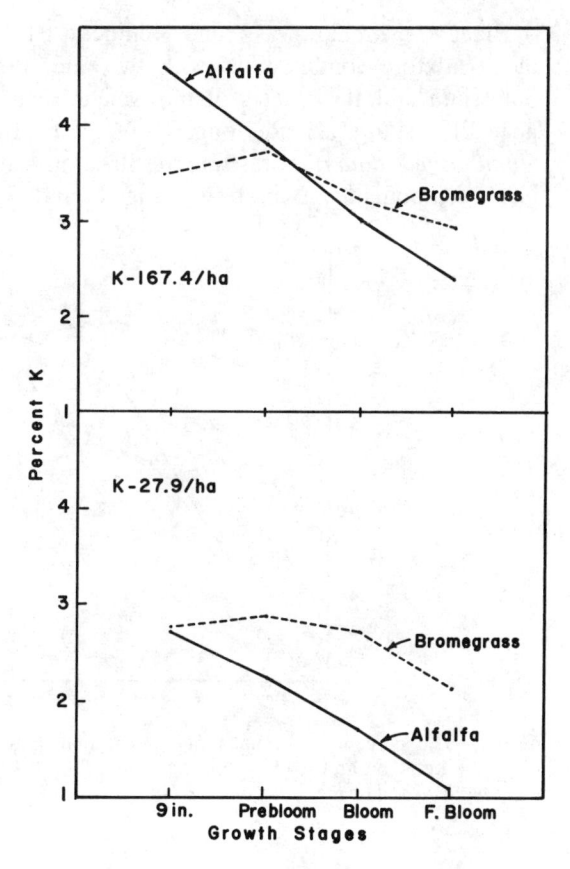

Fig. 6—Potassium content in herbage of plants in an alfalfa-bromegrass mixture as influenced by low and high P-K fertilization (0-14.8-27.9 and 0-88.8-167.4 kg/ha) and growth stages.

Birdsfoot trefoil yields and stands have been as persistent and productive as the Kentucky 31 fescue associate under very low soil potassium on a Lodi soil near Blacksburg.

The generally aggressive competition for potassium by grasses grown with most of the cool season legumes, indicates that it is necessary to maintain higher levels of available soil potassium for mixtures than for pure stands of legumes. Likewise, it appears necessary to maintain more potassium in tissue to assure fast growth recovery after defoliation for maintenance of legumes grown with aggressive grasses as compared to legume monocultures.

2. Grass Associations

There is considerable evidence of competition and differential potassium requirements and absorption among the grasses in grass associations. Marriot (1961) concluded that orchardgrass needed higher maintenance levels of potassium than timothy or Kentucky bluegrass, since he found higher tissue potassium content and relatively larger yield responses with potassium and a stand reduction of orchardgrass without potassium fertilizer.

Bluegrass, orchardgrass, and Kentucky 31 fescue were grown alone and in all mixture combinations with two and three species with four rates of potassium and three rates of nitrogen in a field experiment on a Lodi soil near Blacksburg (Henderlong, Paul, R. E. Blaser, and R. Worley. 1963. *Unpublished data*). Potassium fertilization caused large differences in the yield components of mixtures (Fig. 7 and 8). For a fescue-orchardgrass

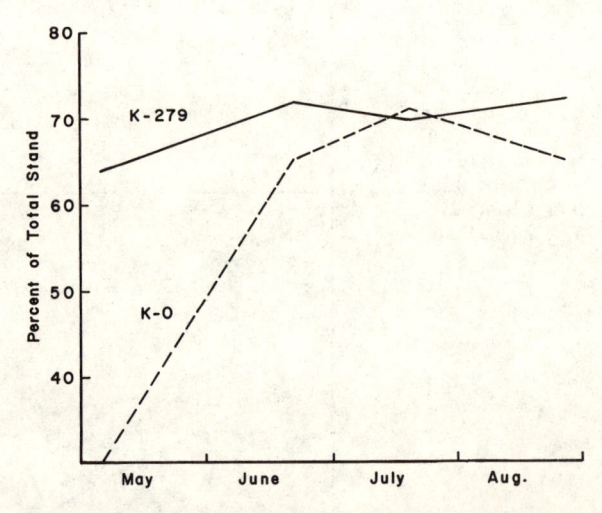

Fig. 7—The percent of orchardgrass herbage in a Kentucky 31 fescue mixture grown at 279 kg K/ha and no K. (Averages for 3 N rates) (Henderlong, Blaser and Worley, *unpublished data*)

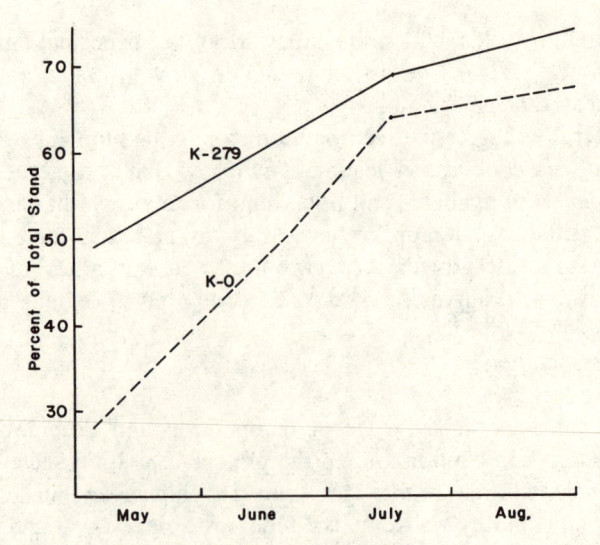

Fig. 8—The percent of Kentucky 31 fescue herbage in a Kentucky bluegrass mixture grown at 279 kg K/ha and no K. (Averages for 3 N rates) (Henderlong, Blaser, and Worley, *unpublished data.*)

mixture, orchardgrass was much more aggressive with liberal than with low potassium in this association. After several years, fescue became more aggressive and orchardgrass almost disappeared with low potassium fertilization. For a Kentucky 31 fescue-bluegrass association, fescue was aggressive toward bluegrass with high as compared with low K fertilization (Fig. 8). Of the three grasses, Kentucky bluegrass was the most and orchardgrass the least aggressive and persistent under low soil potassium. Other observations show that bluegrass commonly invades plots low in potassium in long-time experiments with sod crops.

3. Legume Associations

The competition for K among legume associations is difficult to study. The differences in morphology, rate of regrowth, and cutting frequency for high yields make it difficult to maintain legume species in mixtures because of light competition. Inherent differences in longevity of stands of differential seasonal growth also complicate competition studies. Preliminary data show significant interactions among a red clover-alfalfa association at several potassium rates. During the second year, red clover dominated under potassium stress, but alfalfa made up most of the botanical composition with liberal potassium.

4. Weeds in Associations

It is well known that weeds are aggressive competitors for light, water, and nutrients in crop and sod communities. In a degenerated meadow of timothy, quackgrass, and weeds, Ladino clover invasion was encouraged by fertilizing. Fertilization along with adequate potassium increased the Ladino clover regrowth as compared to broadleaf weeds and grasses. When averaging all potassium and nitrogen treatments for the season for one of the experiments,

Fig. 9—Seasonal K content of Ladino clover, grasses, and weeds grown in association with 112 kg/ha N and also for Ladino clover without N. (Averages for all K treatments) Data from Blaser and Brady (1950).

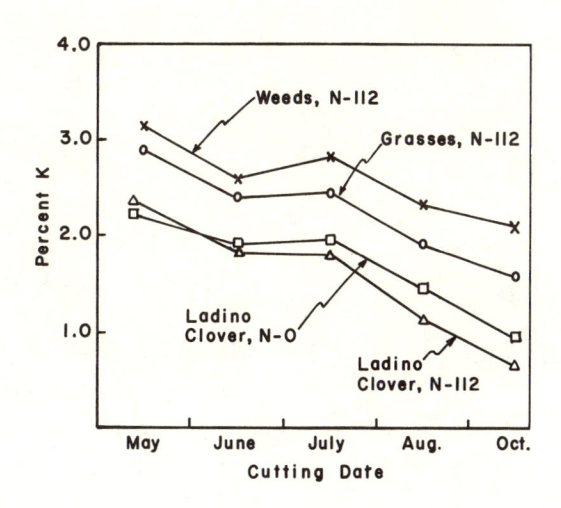

the K content of weedy and cultivated grasses was 2.28% as compared with 2.63% for broadleaf weeds and 1.57% for Ladino clover (Blaser and Brady, 1950). In this association for all potassium rates with grasses and clover, the weeds were aggressive competitors for potassium during the entire season (Fig. 9). Note that competition was most severe toward the end of the growing season when the fertilizer and exchangeable soil potassium were lowest. During late season, Ladino clover for this series of comparisons had only 0.65% K and was severly retarded because of K deficiency. Without potassium fertilizer, the Ladino clover was even lower in potassium and more severly retarded because of competition for potassium. The second application of potassium for the season made in June increased the potassium content of all three botanical components for the July harvests. The presence of weeds in sods apparently makes it necessary to use higher applications of potassium. Potassium nutrition is also an important factor in controlling plant succession in sods.

C. Potassium Effects on Growth

1. Morphological and Physiological

The increases in dry matter production as potassium stress is alleviated are not well understood, but are apparently attributed to morphological and physiological interrelationships. A factor causing the higher yields as the potassium content of alfalfa leaf tissue increased (Fig. 10) was the faster rate of new leaf development (Cooper et al., 1967). Other morphological

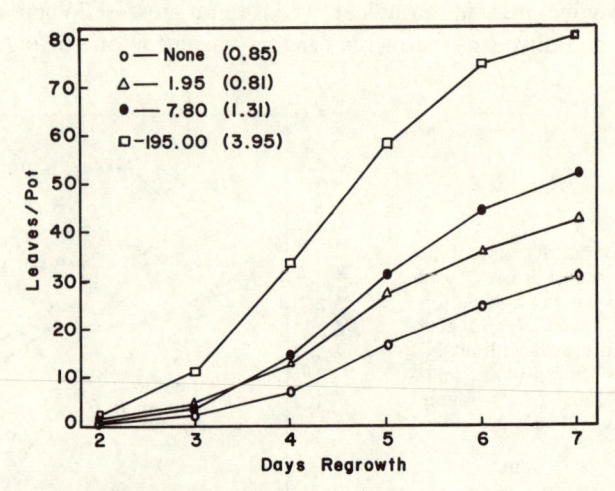

Fig. 10—Leaf accumulation rates 2 days after complete removal of leaves at various potassium levels. Numbers in parenthesis are percent potassium in leaves; other values are ppm of K. Data from Cooper et al. (1967).

Fig. 11—Effect of potassium content in leaves on photosynthesis (Pn), the correlation coefficient was 0.61. Data from Cooper et al. (1967).

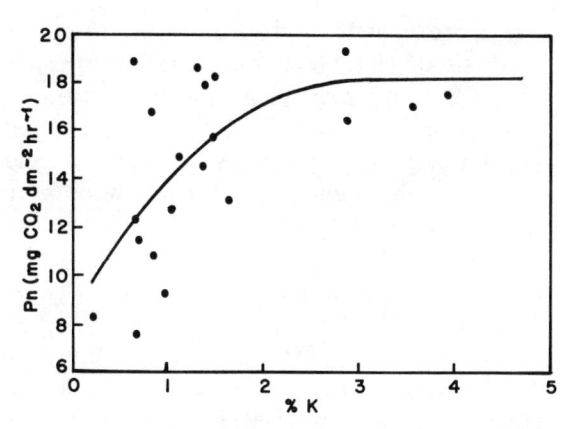

characteristics associated with the increased K content in tissue were larger leaves, more and larger epidermal cells per leaf, more weight per unit area of leaf, and more stomates per unit leaf area with larger apertures. High potassium nutrition caused increases in net photosynthesis (Fig. 11). Also, leaves high in potassium had lower CO_2 compensation points, indicating higher efficiency of CO_2 assimilation under limiting CO_2 environments.

Grotelueschen reported increased new tillering, longer tillers, more roots, and more herbage of timothy during the autumn season with increases in potassium fertilization and concurrent increases of K/g in fresh tissue (K. D. Grotelueschen. 1966. The development and loss of cold resistance in timothy (*Phleum pratense* L.) and associated changes in carbohydrates and nitrogen as influenced by nitrogen and potassium fertilization. Ph.D. Thesis. Univ. Wisconsin, Madison). Fertilization with potassium also increased the weight of individual tillers and carbohydrates per gram of fresh tissue. These potassium relationships generally occurred only when nitrogen was applied.

The efficiency as related to potassium content of maize leaves is illustrated in Table 3 (Moss and Peaslee, 1965). The leaves were numbered in sequence from the top of the plant. The seventh leaf of both plants was designated as the ear leaf. In the well-fertilized plant the rate of photosynthesis differed little between the second and the eleventh leaf although there

Table 3—Potassium contents and photosynthesis of maize leaves (9,000 ft-c; 30C) (Moss and Peaslee, 1965)

Leaf no. (from top)		K content µg/g fresh wt.	Photosynthesis mg $CO_2 dm^{-2} hr^{-1}$
Well-fertilized	2	6,100	40
	4	5,500	38
	7	5,000	36
	11	4,350	36
K-stressed	2	2,150	33
	6	800	15
	7	600	14
	11	250	1

was approximately 45 days difference in age between the two leaves. Photosynthesis of leaf 11 on the potassium-stressed plant was only 3% of that of leaf 2 on the same plant, which in turn, was only 80% of the rate of leaf 2 of the normal plant. Leaf 11 on the stressed plant showed prominent potassium deficiency symptoms but leaf 2 appeared normal and vigorous. All leaves on the well-fertilized plant were normal and vigorous (Also see section on longevity of plants).

2. Yields and Potassium Content

Stage of plant growth, plant association, and season of harvest are some of the factors discussed in other sections that confound associations of potassium content with yields. Most of the available data on potassium content are a mass bulking of all plant tissue with extreme variance in morphology, anatomy, and physiological functions. Also, most of the analysis on potassium content in plant tissue has been taken when crops were harvested and cannot be associated with the rates of dry matter production during regrowth. Forage plants should be sampled for growth rate and potassium content at intervals during regrowth to make valid associations of potassium content with rate of dry matter accumulation. When the physiological functions of potassium in plant nutrition are more clearly established, growth rate potentials from potassium that are associated with specific plant tissues will make deductions more precise. The associations of dry matter production with potassium content in mass tissue may serve only as rough approximations.

As potassium or any nutrient is increased from nil to high amounts, the increases in dry matter production may be associated with a series of overlapping zones: (i) severe deficiency, (ii) hidden hunger, and (iii) maximum yield (Fig. 12) according to Blaser et al. (1958). The yields of orchard-

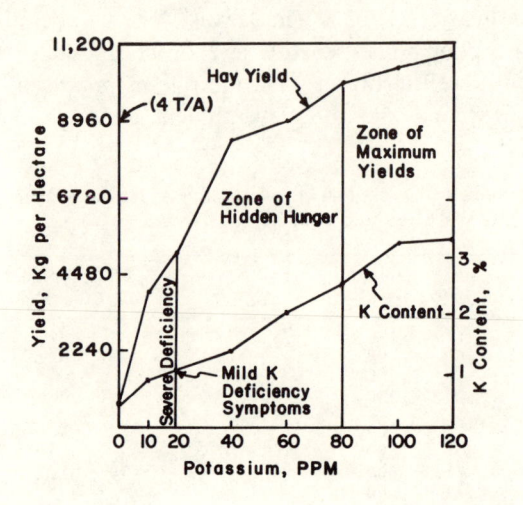

Fig. 12—The yield (16% moisture) and K content of uninterrupted orchardgrass growth as influenced by K fertilization. Data from Blaser et al. (1958).

grass increased to 10,900 kg/ha (4.86 tons/acre) as K in mass herbage tissue increased to 3.34%. However, with high amounts of soil available potassium, luxury intake may result with the smaller increases in dry matter production.

McNaught (1958) found the K level for optimum growth to be approximately 1.6% for ryegrass and orchardgrass. Kresge and Younts (1963) found that orchardgrass with 1.0% K showed severe potassium deficiency symptoms. The potassium level in forage, at which a significant decrease from the maximum occurred, increased with higher rates of nitrogen. The percent K was 2.15 for 56 and 112 kg N/ha as compared to 2.68% for higher N rates.

Chandler et al. (1945) reported that yield responses usually resulted from potassium applications, if the potassium content of alfalfa at the early bloom stage was less than 1.25%. They found that significant yield responses could be realized if more than 15% of the alfalfa plants showed deficiency symptoms. Gerwig and Ahlgren (1958) found that the "critical percentage" of K in alfalfa varied between 1.42 and 1.84% depending on season, and Margus and Battle (1965) reported similar values for the same experiment. Seay et al. (1949) found superior winter survival and yields of alfalfa, if concentrations of K in the plant were 1.25% or more. Stivers and Ohlrogge (1952) found that 0.9 to 1.1% of K was necessary for good alfalfa growth. These rather low K values may be associated with rather advanced maturity as only three cuttings were made yearly. McNaught (1958) found 0.89 and 2.06% K in K-deficient and healthy alfalfa plants, respectively. Blaser (1961-62) attained maximum alfalfa yields with K contents between 2.0 and 2.5%.

McNaught (1958) found the critical level, or minimum level for near-maximum growth, was approximately 1.8% K for whole leaves (immature plus mature) of white and red clover at grazing height. Clover leaves with deficiency symptoms, showed potassium levels below 0.7%. Brown (1957) during a 6-year experiment with Ladino clover, reported ranges in K content of 1.4 to 3.3% without K as compared to 2.9 to 3.0% when 186 kg/ha of K was applied yearly in split applications. In a Virginia experiment, the Ladino clover growing with grass was K deficient when the K content of herbage was 1.02% or less. Best clover growth occurred when the K content was 2.6% or higher (Blaser et al., 1958). When 140 kg/ha was applied in New York, the Ladino clover for all cuttings averaged 2.16% K (Blaser and Brady, 1950). Clover growth was poor when K values were around 1%. It appears that K content of white clover for maximum growth should be higher than for red clover or alfalfa, as the ratio of leaflets to herbage growth is much higher for Ladino clover than for legumes with stemmy herbage. (*See also* sections on longevity of plants and potassium fertilization for maximum yields.)

3. Longevity of Plants

Stand reductions or losses of hardy persistent perennials, such as alfalfa, orchardgrass, and Coastal bermudagrass under K stress are common. On the other hand, short lived legumes in perennial sods because of their vulnerability to injury from drought, high or low temperatures, and diseases have not generally responded to favorable fertilization or management. Many short-time experiments show that potassium fertilization, under low soil potassium, increases white and Ladino clovers in grass sods or its survival (Blaser and Brady, 1950; Parsons et al., 1953; Doll et al., 1959; and Hunt and Wagner, 1963). However, Ladino and white clovers are readily injured or exterminated by many growth factors and are short lived even under favorable fertility. In five experiments in different regions and soils in Virginia, Ladino clover stands were lost in two or three years, irrespective of the soil fertility factors. In three of the areas, alfalfa-grass and Ladino clover-grass mixtures were split-plot treatments with phorphorus and potassium subplot treatments. The alfalfa remained productive for the duration of the experiments (7 to 9 years). Ladino clover stands were lost in 2 or 3 three years, irrespective of the soil fertility status.

During a 7-year period, when an alfalfa-orchardgrass mixture was grown

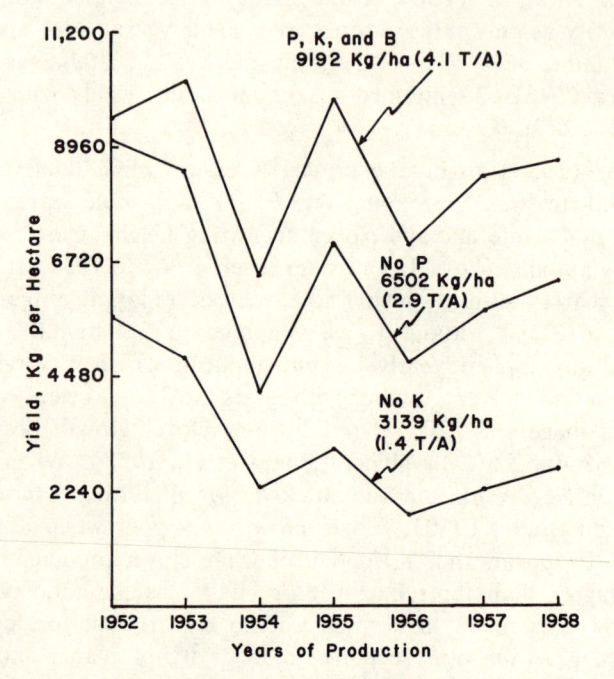

Fig. 13—Yield of alfalfa-orchardgrass hay as influenced by P and K, all plots were treated with lime and boron. Data from Blaser (1960).

on a Cecil soil in Virginia, the rapid decline in yield during the first 3 years without potassium fertilization was attributed to loss of the alfalfa component (Fig. 13 and 14) (Blaser, 1960). For the 7-year period, there was an average yield of 9,192 kg/ha (4.1 tons/acre) with liming and fertilization (P, K, and B) as compared to 3,139 kg/ha (1.4 tons/acre) when K was omitted from the fertilizer. The residual alfalfa plants per square meter and average yearly hay yields during 7 years for the different potassium treatments were highly associated. The stand of alfalfa and the yield of the alfalfa-grass hay increased as spring applications of potassium increased. When 186 kg of K was applied after the first cut or split into two 93-kg applications made in spring and after the second cut to reduce grass competition for K, the yields improved. Unfortunately stand data were not taken for the latter two treatments, but observations indicated very satisfactory alfalfa stands. When the experiment was terminated, the exchangeable soil K was very low and less than at the start of the experiment when less than 186 kg/ha of K was applied yearly.

Alfalfa stand losses due to stress factors also occur in monocultures. It required annual applications of 186 kg K/ha to maintain alfalfa during a 9-year period (Table 1) (Markus and Batile, 1965). The losses were attributed to depleted exchangeable potassium in the Nixon loam. Soil K accumulated in the surface soil when K was applied at rates higher than

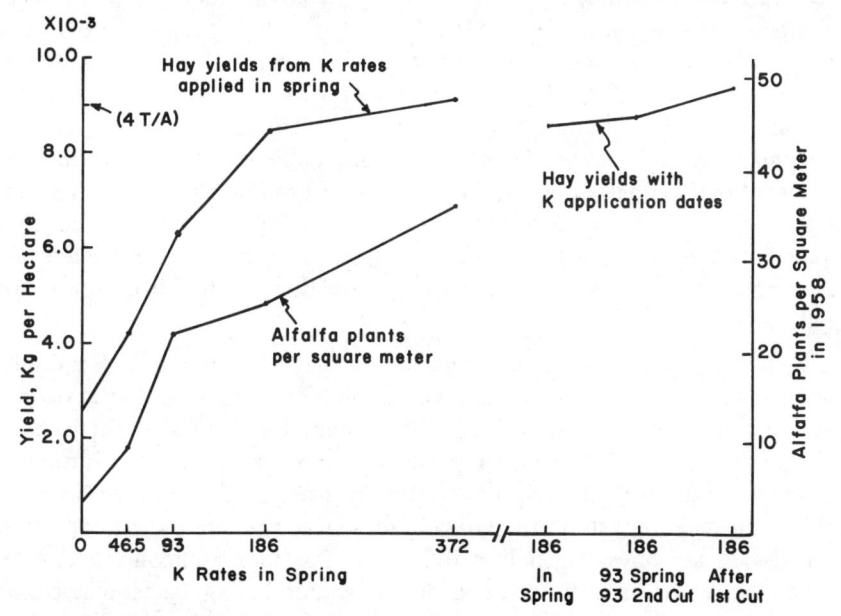

Fig. 14—The relationships between hay yields and alfalfa stands with K fertilization during the 7th year of harvesting, 1958. Lime, P, and B were applied uniformly and K was applied each year for the alfalfa-orchardgrass mixture on a Cecil soil in Virginia.

280 kg K/ha. Parks and Safley (1961), with alfalfa on a Dickson soil with annual rates of K during 4 years, found increased yields with K applications through 279 kg K/ha. The soil tests showed an initial K content of 178 kg K/ha. Without or with low K fertilization, the soil was depleted in K to about 45% of the initial. Even with 186 kg K/ha yearly, there was a decline in exchangeable K in the soil. Soil K was not depleted at rates of 279 and 372 kg K/ha yearly. Stivers and Ohlrogge (1952) found that alfalfa stands declined as much as 36% in 3 years under low potassium fertilization and that stands were directly correlated with yields. Better stands of alfalfa with potassium and other nutrient treatments have been attributed to better winter survival (Wang et al., 1953; and Jung and Smith, 1959). The improved survival may be associated with increased accumulations of starch, non-reducing sugars, and protein substances. However, under severe competition for potassium, as in nitrogen fertilized grass associations, alfalfa stands may be lost at any season.

Kentucky 31 fescue and orchardgrass grown under stress of K in various experiments on a Lodi soil near Blacksburg showed declines in yields and stands with time. Initially under K stress these grasses became dark green, grew slowly, and the leaf laminae wilted or curled during low soil moisture. Later the laminae turned brown along edges and the apex. Under such conditions, it appeared that stands were severely depleted during winter.

An increase in winter injury or mortality of plants under K stress occurs for grasses and legumes because small weak plants with poor roots heave and are then lost because of desiccation. With perennial grasses, it appears that vigorous new tillers that develop under good mineral nutrition during the autumn-winter months would be less vulnerable morphologically and physiologically to adverse microclimatic and biological factors during winter. This contention is supported by data with timothy that show more and bigger new tillers, more carbohydrate accumulation in tillers and more winter hardiness (measured by the percentage of electrolytes extracted after freezing) with high than low potassium nutrition when nitrogen was also applied.

Photosynthesis of cool season grasses is high during the cool autumn-winter season, even though cell division and expansion in top growth is slow (Powell et al., 1967). The rapid accumulation of carbohydrates in grasses when herbage dry matter production is static during late autumn is also evidence of high photosynthesis (Brown et al., 1963). Because available soil K is usually low late in the growing season and photosynthetic and basal plant tissues are physiologically active, low potassium commonly encountered in late season may be responsible for winter killing because necessary physiological organic processes are retarded. Legumes in semi-rosette stages during cool autumn periods are apparently active as carbohydrates in basal tissues increase, even though top growth is nil or slow. Since potassium influences the rate of photosynthesis, K applications during autumn, under low soil K, may improve rates of dry matter production the next spring because of better stands and physiological status of plants. (Also see section on morphology and physiology.)

4. Potassium Fertilization for Maximum Yields

Hay yields of 21,100 kg/ha (9.4 tons/acre), 88% dry matter, have been obtained during two years by using (i) excellent varieties, (ii) early and frequent cutting, (iii) late fall harvesting, (iv) adequate soil fertility, (v) good stands, and (vi) by minimizing field losses with natural rainfall (Hittle and Jacobs, 1966). The K content ranged from 1.4% for the third cut to 2.6% for the first cut. Alfalfa-grass hay yields of 18,600 kg/ha (8.3 tons/acre) on a large scale farm operation in Indiana, were obtained with normal rainfall, a good variety, cutting four times per year, and liberal fertilization (George et al., 1967). The yield increases of the mixtures in 3 successive years were attributed to more liberal fertilization each year. The yields of 18,600 kg/ha were obtained by applying 45 N, 31 P, and 83 kg K/ha in early spring and after the first and second cuttings. Of 36 cuttings made during 3 years, the K content of the four lowest harvest ranged from 1.50 to 1.99%; 23 of the 36 cuttings had more than 2.5% and 14 cuttings contained over 3% K.

For exceptionally high yields all potential factors that may restrict growth must be kept above optimum requirements to avoid stress. Thus, it appears that K content in tissue would need to be higher for 10-ton yields than it would for 5-ton yields. Also, since high yield potentials are associated with cutting more frequently and because potassium makes up more of the dry matter of young than older tissue, this would elevate potassium requirements. For maximum yields, regrowth after each hay cutting must be very rapid; hence, adequate availability of each soil nutrient can be assured only by applying somewhat more than optimum amounts needed.

III. POTASSIUM APPLICATIONS FOR PERENNIAL FORAGES

A. Establishment of Stands

The potassium requirements for establishing perennial forages depend on such factors as soils, companion crops, species, and mixtures. Soil tests and experience are useful for deciding what amounts to apply. Applications of potassium for establishment should generally be low to avoid leaching losses and interference with germination from high soluble salts. Fertilizer P:K ratios of 1:1 (2:2 ratio for $P_2O_5:K_2O$) have given good results.

B. Maintaining Stands and High Yields

The high potassium requirements of perennial forages are attributed to the high potassium content of young nutritious herbage, high yields, potassium competition among species, and herbage removal as for hay and silage utilization. The rapid expansion to all-year dry lot livestock feeding operations has caused large increases in nutrient removal from soils. Much of the potassium in animal excreta is lost and animal manures are not redistributed uniformly to all land areas. The rapid decline of exchangeable soil potassium under intensive crop removal makes it necessary to apply potassium annually at liberal amounts, except for soils unusually high in exchangeable potassium. It is possible that severe potassium depletion in the surface and subsoils may cause yield declines when tilled crops follow sod crops. For soils low to medium in exchangeable potassium, the annual maintenance fertilizers should be high in K, such as P/K ratios of 1:4 to 1:6 ($P_2O_5:K_2O$ ratios of 1:2 to 1:3).

As yields increase, it may become practical to apply potassium in several split applications during the year as pointed out in Fig. 9 and 14. Such split applications would assure a more even potassium nutrition, improve soil K in late season, and possibly produce more dry matter per unit K on soils that are medium to low in exchangeable potassium. Robinson et al. (1962) found that fall or spring applications of potassium fertilizer to orchardgrass resulted in herbage that was high in potassium in the spring but low in late summer and fall. An annual summer application of potassium greatly reduced the trend toward luxury consumption in the spring and potassium starvation in the fall and was relatively effective in maintaining a moderate level of available potassium during the season. Yield increases, from yearly split applications of potassium, would have to be high enough to offset the added costs of application.

The potassium fertilizer requirements for maintenance also depend on the plant associations. Legume monocultures require less potassium than a given legume in grass-legume mixtures. Nitrogen fertilization of the latter will add to the potassium requirements for two reasons: (i) yield increases and (ii) nitrogen makes grasses more aggressive toward legumes (Fig. 15) (Blaser et al., 1958). Nitrogen fertilizer decreased the amount of clover in the sod and the nitrogen-stimulated grass caused the K content of clover at each of the two potassium treatments to be much lower than for the clover-grass mixture without nitrogen. Note the rapid decline in potassium content of Ladino clover during late season with a very low potassium content in late season. Also, nitrogen fertilizer lowered the potassium content of Ladino clover because of grass competition (Fig. 9). Grass fertilized liberally with nitrogen may add to the potassium requirements (Kresge and Younts, 1963).

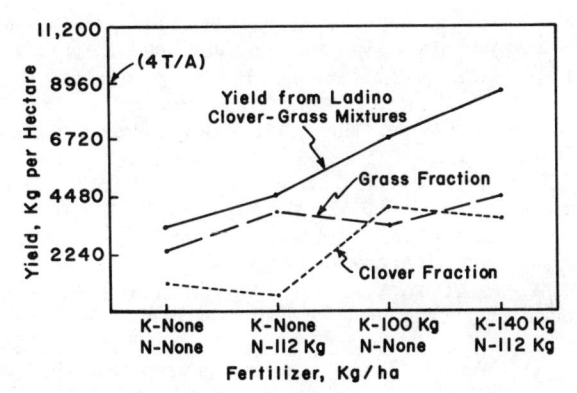

Fig. 15—The total mixed herbage yield (16% moisture) and botanical components as influenced by N and K fertilization on a limestone soil near Blacksburg. Data from Blaser et al. (1958).

When sod crops are grazed, the K requirements are considerably lower than for hay or silage utilization because of lower yields and less removal. Unfortunately, nutrient removal under most pasture conditions is high because of poor distribution of animal excreta on land areas.

REFERENCES

Blaser, R. E. 1960. Fertilizing silage and hay crops for profits. Better Crops With Plant Food. 44 (6):50-66.

Blaser, R. E. 1961-62. Yield and uptake (P_2O_5 and K_2O) in alfalfa fertilization. Better Crops with Plant Food. 46: 14-22.

Blaser, R. E., and N. C. Brady. 1950. Nutrient competition in plant associations. Agron. J. 42: 128-135.

Blaser, R. E., R. C. Hammes, Jr., H. T. Bryant, C. M. Kincaid, W. H. Skrdla, T. H. Taylor, and W. L. Griffeth. 1966. The value of forage species and mixtures for fattening steers: Agron. J. 48:508-513.

Blaser, R. E., C.Y . Ward, and W. W. Moschler. 1958. Hidden hunger in grasses and legumes. Better Crops with Plant Food. 42: 42-50.

Brown, B. A. 1957. Potassium fertilization of Ladino clover. Agron. J. 49: 477-480.

Brown, R. H., R. E. Blaser, and H. L. Dunton. 1966. Leaf-area index and apparent photosynthesis under various microclimates for different pasture species. Proc. 10th Int. Grassland Congress.

Brown, R. H., R. E. Blaser, and J. P. Fontenot. 1963. Digestibility of fall grown Kentucky 31 fescue. Agron. J. 55:321-324.

Browning, C. M. 1963. A comparison of irrigated Ladino clover, Coastal bermuda, and a Coastal bermuda-Ladino clover mixture for summer grazing for dairy cows. Inf. Sheet 834. Mississippi State Univ.

Chandler, R. F. Jr., M. Peech, and R. Bradfield. 1945. A study of techniques for predicting the K and B requirements of alfalfa: I. The influence of muriate of potash and borax on yield, deficiency symptoms, and K content of plant and soil. Soil Sci. Soc. Amer. Proc. 10: 141-146.

Cooper, R. B., R. E. Blaser, and R. H. Brown. 1967. Potassium nutrient effects on net photosynthesis and morphology of alfalfa. Soil Sci. Soc. Amer. Proc. 31:231-235.

Doll, E. C., A. L. Hatfield, and J. R. Todd. 1959. Effect of rate and frequency of potash additions on pasture yield and potassium uptake. Agron. J. 51: 27-29.

Donald, C. M. 1963. Competition among crop and pasture plants. Advance Agron. 15: 1-114.

Drake, M., J. Vengris, and W. G. Colby. 1951. Cation exchange capacity of plant roots. Soil Sci. 72: 139-147.

George, J. R., C. L. Rhykerd, and C. H. Noller. 1967. Luxury K use—fact or myth. Better Crops with Plant Food. 51 (2): 2-5.

Gerwig, J. L., and G. H. Ahlgren. 1958. The effect of different fertility levels on yield, persistence, and chmical composition of alfalfa. Agron. J. 50:291-294.

Heinemann, W. W., and R. W. VanKeuren. 1958. A comparison of grass legume mixtures, legumes, and grasses under irrigation as pasture for sheep. Agron. J. 50: 189-192.

Hesketh, J., and D. Baker. 1967. Light and carbon assimilation by plant communities. Crop Sci. 7: 285-293.

Hittle, O. N., and J. A. Jackobs. 1966. Ten tons of alfalfa. Better Crops with Plant Food. Fall issue. pp. 8-12. Amer. Potash Inst. Washington, D.C .

Hogg, P. G., and J. C. Collins. 1965. Beef production of Coastal bermuda with legumes and nitrogen on heavy clay soils in the Mississippi Delta. Info. Sheet 889. Mississippi State Univ.

Hunt, O. J., and R. E. Wagner. 1963. Effects of phosphorus and potassium on legume composition of several grass-legume mixtures. Agron. J. 55:16-19.

Jung, G. A., and D. Smith. 1959. Influence of soil potassium and phosphorus content on cold resistence of alfalfa. Agron. J. 51:585-587.

Kresge, C. B., and S. E. Younts. 1963. Response of orchardgrass to potassium and nitrogen fertilization on a Wickham silt loam. Agron. J. 55:161-164.

Markus, D. K., and W. R. Battle. 1965. Soil and plant responses to long-term alfalfa (*Medicago sativa* L.) Agron. J. 57: 613-616.

Marriott, L. F. 1961. Nitrogen fertilization of perennial grasses. Pennsylvania Agr. Exp. Sta. Bull. 688.

McNaught, K. L. 1958. Potassium deficiency in pastures. I. Potassium content of legumes and grasses. New Zeland Jour. Agr. Res. 1:148-181.

Moss, Dale, and D. E. Peaslee. 1965. Photosynthesis of maize leaves as affected by age and nutrient status. Crop Sci. 5: 280-281.

Murdock, F. R., A. S. Hodson, and H. M. Austenson. 1958. A comparison of orchardgrass-Ladino clover and orchardgrass as pasture for dairy cows. J. Dairy Sci. 42:1675-1685.

Parks, W. L. and L. M. Safley. 1961. Alfalfa fertilization on a Dickson soil. Bull. 330. Univ. of Tennessee.

Parsons, J. L., M. Drake, and W. G. Colby. 1953. Yield and vegetative and chemical composition of forage crops as affected by soil treatment. Soil Sci. Soc. Amer. Proc. 17:42-46.

Powell, A. J., R. E. Blaser, and R. E. Schmidt. 1967. Physiological and color aspects of turfgrasses with fall and winter nitrogen. Agron. J. 59:303-307.

Reid, J. T. 1959. Evaluation of energy in forage. Grasslands. p. 213-24. Publication no. 53. Amer. Assoc. for the Advance of Sci. Washington, D. C.

Robinson, R. R., C. L. Rhykerd, and C. F. Gross. 1962. Potassium uptake by orchardgrass as affected by time, frequency, and rate of potassium fertilization. Agron. J. 54: 351-353.

Seay, W. A., O. J. Attoe, and E. Truog. 1949. Correlation of the potassium content of alfalfa with that available in soils. Soil Sci. Soc. Amer. Proc. 14: 245-249.

Smith, Dale, 1959. Yield and composition of alfalfa for stages of growth. Wisconsin Agr. Exp. Sta. Res. Rept. 4.

Stivers, R. K., and A. J. Ohlrogge. 1952. Influence of phosphorus and potassium fertilization of two soil types on alfalfa yield, stand, and content of these elements. Agron. J. 44: 618-621.

VanKeuren, R. W., and W. W. Heinemann. 1958. A comparison of grass-legume mixtures and grass under irrigation as pastures for yearling stress. Agon. J. 50: 85-88.

Wang, L. C., O. J. Ottoe, and E. Truog. 1953. Effect of lime and fertility levels on the chemical composition and winter survival of alfalfa. Agron. J. 45: 381-384.

Williams, Mary. 1967. Clover boosts calf production.. Res. Rept. 12 (4): 4-5. Univ. of Florida.

Potassium Nutrition of Soybeans and Corn

JOHN PESEK

Iowa State University
Ames, Iowa

I. INTRODUCTION

Plants react favorably to improved nutrition expressed as the concentration of an element or its total accumulation in the plant community provided the element in question is at a low level of concentration and is limiting plant growth or crop yield as much as or more than the other nutrient elements and environmental factors. It is evident from this statement that, for the purposes of this discussion, nutrition will be defined to mean the level of supply of one of the essential mineral elements normally derived from the soil by absorption as either a cation or an anion. In this chapter the discussion will be directed almost entirely to the levels of potassium in the plant or its parts and these levels will be interpreted in terms of deficiency or sufficiency and the effect which the supply of potassium and other factors has upon it.

Plant communities, which we use to produce grain, vary in the numbers and age of individuals present and sometimes vary greatly in the genetic composition of the individuals. (In the crops concerned in this chapter, however, genetic diversity within a single community is relatively small and usually incidental.) Because of this variation in the members of the communities and within each community, it is reasonable to expect that the levels representing adequate nutrition may vary among communities and within the communities as each "ages" or the individuals change from vegetative to reproductive to senescent.

Plants also are unique in that once they accumulate mineral elements and translocate them to the aerial portions, the quantities of these elements are captives of the system. Except under certain extreme conditions, plants cannot excrete or dissipate unnecessary or excessive amounts of a particular element after it is absorbed, in fact, even the mechanism for rejecting these elements at time of absorption is highly imperfect. (Concentrations, however, sometimes are diluted with carbonaceous matter.) This has two significant results. First, there is an unavoidable serial correlation between the concentration or quantity of an element in a plant community at a given time and previous conditions of concentrations and quantities. Second, certain factors which cause or permit high or low concentrations of an element in plants tend to persist and continue to have the same relative effect. While other factors are subject to change, the result of a major change may be drastic or at least undesirable, e.g., imposing a drouth stress will reduce the phosphorus absorbed, thus reducing the phosphorus content but this is usually accompanied by reduced yields.

The second point just stated, results, because our crops are produced on soil areas. Soil has natural physical, chemical, and biological buffers to moderate the influence of external factors on nutrient supply. Another factor is that the production sequence (crop management) has certain physical imperfections which restrict the latitude within which the physical, chemical, or biological environment can be changed after the crop is growing. The situation is further complicated by the inadequate control over moisture supply and temperature.

Once plant nutrition with respect to mineral elements is understood, and in spite of the present imperfections in our knowledge of it, it can be predicted either by observing the soil and its management or by observing the nutritive status of previous or the current plant communities. After the expected nutritive status is predicted, it is possible to decide first upon the remedial actions needed, and then upon their intensity, quality, and timing. In potassium nutrition of annual grain crops we usually are limited to basing our actions upon a measurement of the potassium supply in the soil profile exploited by roots under given conditions, or perhaps upon a measurement of the nutritive status of a previously grown crop. The most common and most dependable remedial action available is the application of soluble potassium compounds or materials containing soluble potassium (mixed fertilizers, manures, etc.) to the soil at or prior to the time of planting the crop.

Better systems may be hypothesized or even utilized (hydroponics) but most of the grain production for some time to come will be under conditions and in systems similar to the ones used now and in the recent past. This chapter, then, will explore some of the ideas about nutrition of crops and report some observed relationships reported for potassium and indicate how these may be used favorably to influence the potassium nutrition of grain crops.

II. SOME GENERAL CONCEPTS

Two general ideas or concepts need to be set down in order to present, effectively, the subsequent parts of this chapter. The first of these is the concept of a relationship between the concentration of an element in a crop and crop yield. The second is the relationships among the initial condition in the soil, application of a soil amendment (fertilizer, lime, etc.) and the behavior of the plants in terms of nutrient uptake and yield.

A. Macy's Hypotheses

Macy (1936) brought together and integrated the concepts of the relationship between mineral concentration of an element in a plant and its yield which had been advanced by such investigators as Weinhold, Atterburg, Pfeiffer, and Remy, all of whom worked after von Liebig. Goodall and Gregory (1947) also credit these men with developing the early ideas and testing them.

In general, Macy (1936) suggests that the yield, y, is a function of the nutrient concentration, C, or

$$y = f(C) \qquad [1]$$

when this relationship is graphed with y on the ordinate and C on the abscissa. The result is a curve concave to the abscissa starting near the origin of the two axes and rising steeply and approximately parallel to the y-axis, then sloping off to the right and increasing at a decreasing rate, thus asymptotically approaching some limiting value of y, which would be the maximum yield.

The initial almost vertical part of the curve, Macy called the "minimum percentage." It is characterized by an increasing yield of the plant with little or no increase in the concentration of the element in the plant. The moderately rising part of the curve has been called the range of "poverty adjustment" and is characterized by increases in yield closely proportional, over most of the range, to the increase in concentration of the element in the plant. A term, "luxury consumption," was used to describe the upper relatively flat part of the curve asymptotic to the limiting value of y. It is described as a range in which the concentration of the element increases but this increase is not associated with any further significant increase in yield.

Particular attention should be called to the "point" where the curve changes from "poverty adjustment" to "luxury consumption." This point has been referred to as the "critical percentage" and would be defined as that percentage concentration of an element above which a further increase in concentration would not result in an increase in yield, all other factors remaining constant.

Conceptually, the representation of Macy is a very useful tool in understanding the relationship between the percentage of an element in a plant and its yield even through the concave curve asymptotic to the limiting value of y causes some uncertainty and awkwardness in estimating a particular critical percentage for a crop. The crux of the matter is, "Where, on a continous asymptotic curve, does yield increase cease?" Mathematically, it never does within the finite range, and statistically and experimentally convincing evidence that there is no further increase depends upon the precision (variance) with which yields are measured and the odds accepted to protect the investigator from drawing a "wrong" conclusion. This difficulty has its roots in the Mitscherlich-Spillman type of yield equation and perplexed Tyner (1947), Bennett (1952)[1], and Bennett et al. (1953) all of whom established arbitrary points on the "nutrient sufficiency curve" and assumed these were close to the critical percentage, e.g., Bennett et al. (1953) assumed that the critical percentage would be found at a yield equivalent to 95% of the limiting yield in terms of the Mitscherlich-Spillman equation.

This problem of identifying the critical percentage correctly mathematically in agreement with the conceptual definition can be avoided if it is assumed that the relationship between yield and nutrient concentration in equation 1 passes through a maximum, i.e.,

$$dy/dC = 0, \qquad [2]$$

and

$$d^2y/dC^2 < 0. \qquad [3]$$

Polynomials of the second order or higher can have these properties and the second approximation of Mitscherlich and other equations provide for these, too. A two-variable polynomial of the second order was first used by Dumenil (1958)[2] to estimate the critical percentages of N and P in corn and Miller (1960)[3] first used it with soybeans.

While Macy (1936) does not hold that the critical percentage is "absolute-

---

[1] Bennett, W. F. 1952. Nitrogen, phosphorus and potassium content of the corn leaf and grain as related to nitrogen fertilization and yield. Unpublished M.S. thesis, Iowa State Univ., Ames.

[2] Dumenil, Lloyd C. 1958. Relationship between the chemical composition of corn leaves and yield responses from nitrogen and phosphorus fertilizer. Unpublished Ph.D. disertation. Iowa State Univ., Ames.

[3] Miller, R. J. 1960. Soybean yield responses and plant composition as affected by phosphorus and potassium fertilizers. Unpublished Ph.D. dissertation. Iowa State University Library, Ames, Iowa.

ly invariable" he does say that growth factors which cause a variation in the percentage of the nutrient do not change the plant requirements. This position is generally supported by Lundegradh (1951) and Ulrich (1943), who present data showing only small variations. Others have taken exception with the "invariable" concept applied to the critical percentage and results of experiments will be cited which indicate the various factors affecting K composition and yield of crops. Nevertheless, the concept of a critical percentage is an important one in the future of diagnosing nutrient needs of crops.

B. Response to Added Fertilizer

The best way to improve the supply of a nutrient element at suboptimum levels as shown by plant analysis or plant yield is to apply compounds (fertilizers) of the element in short supply to the soil. For perennial crops, this nutrient may be added upon the basis of chemical analysis of part of the plant community present, but the opportunity to do this is severly limited for annual crops. The main reason is that by the time the plants are large enough to permit diagnosis, they may be too old to be significantly affected by an additional supply of nutrients. A second related reason is that root distribution, available application methods, weather variation, and fertilizer-soil-plant relationships do not permit effective application of some fertilizers after the crop has emerged.

The alternative for annual crops is to diagnose the nutrient needs by observing nutrient concentrations in previous crops or by observing the availability of the nutrient element in the root zone of the crop to be planted and applying the additional nutrient before or at the time of planting. Knowing that yield response, Δy, is a function of the fertilizer applied, i.e.,

$$\Delta y = f(F), \qquad [4]$$

at a given site, and that the failure of Δy to be the same at all sites, at all given times, and for all varieties is due to an interaction of other factors, S, with F, a general relation can be written expressing yield increase as a combined fertilizer effect and an interaction, or

$$\Delta y = f(F) + f(SF). \qquad [5]$$

Given equation [5] and knowing the values for factors S at a specific site and the prices of yields and fertilizers, it is possible to calculate the economically optimum level of F to apply by equating the first partial derivative of Δy with respect to F to the inverse price ratio.

Because a base yield, y_0, without treatment at a given site is related only to the factors S, the total yield, y, is given by adding equations [5] and

$$y_0 = f(S), \qquad [6]$$

[6] to give

$$y = y_0 + \Delta y = f(S) + f(F) + f(SF). \qquad [7]$$

If y is also related to C as given in equation [1], then,

$$y = f(C) = f'(S) + f'(F) + f'(SF). \qquad [8]$$

It should be possible therefore to predict the additional fertilizer required by using either generalized equation [7] or [8]. The former has been used most widely for fertilizer recommendations for annual crops while the latter has been used mostly for establishing certain nutrient level criteria or deficiencies but not much for recommending quantity of fertilizer to use.

III. POTASSIUM NUTRITION OF CORN

In corn, as in most other crops, potassium concentration in plant tissues and potassium accumulated in the aerial portions normally are exceeded only by nitrogen in the concentration and accumulation of essential elements derived from the soil. Actually, the amount of potassium exceeds the total of all the other essential elements in the ash. Representative data were presented by Nelson (1956) who made an extensive review of the total mineral nutrition of corn. Dumenil (1958)[4] made an exhaustive review of literature pertaining to the chemical composition of corn prior to publishing his findings on the critical levels of nitrogen and phosphorus in corn leaves (Dumenil, 1961).

A. Factors Affecting Potassium Concentration in Corn

Potassium fertilizer interacts with other fertilizers in affecting corn yields (Dumenil and Nelson, 1948) and interacts with other elements resulting in variations in the degree to which each is absorbed. For example, Tyner and Webb (1946) reported that potassium content was reduced in corn by the use of ammonium sulfate as a fertilizer. Calcium and magnesium uptake were found to be reduced by additions of potassium fertilizer (Stanford et al., 1941; Thomas and Mack, 1939). Others have reported evidence that potassium may interfere with the absorption of nitrogen causing more severe deficiency symptoms, reduced nitrogen concentration, or reduced responses (negative interactions of nitrogen and potassium) to applied fertilizers (Tyner and Webb, 1946; Dumenil and Nelson, 1948). Lawton (1945) observed that

---

[4] Ibid.

potassium absorption was reduced by low soil pore space (reduced aeration) more than absorption of nitrogen, phosphorus, calcium, and magnesium.

Voss (1962)[5] studied the potassium levels in the leaves of corn opposite and below the main ear shoot at time of silking in eighteen composite nitrogen, phosphorus, and potassium experiments in Iowa. He found that potassium content in the control plots was positively correlated with exchangeable potassium in layers of the profile down to 3 feet. Nitrifiable nitrogen had a negative relationship and not significant while the pH effect was negative and significant. Higher levels of population, later planting date, and potentially higher yielding varieties all had significant positive effects on potassium concentration while the number of stress days during all previous periods of growth had negative effects on potassium level.

When considering the leaf potassium concentration of all plots, Voss (1962)[6] again found negative correlations with drouth stress and pH and also found a negative correlation with applied N (NH_4NO_3) over part of the range, but this effect depended upon several other factors and the effect was positive in some cases at high levels of nitrogen. Applied potassium had a positive effect on potassium concentration but was affected by the soil supply of potassium at the respective sites. The K levels in this study ranged from a low of 0.36% to a high of 2.97%.

Hanway et al. (1962) reported on a potassium fertilization study of 51 experiments in the north central USA and found that exchangeable potassium in the soil had an influence on potassium absorption by corn and that uptake of fertilizer potassium was inversely related to exchangeable soil potassium. Plant population seemed to have no effect on potassium content of the corn crop. The yields of corn grain were correlated with potassium content of the corn leaves opposite and below the main ear shoot at time of silking and yields of corn ranged from 2,070 to 8,000 kg/ha (33 to 127 bu/acre). Total K absorbed ranged from 25 to 240 kg/ha (22 to 213 lb/acre) and 90% of this K was in the plants by time of silking. Potassium uptake seemed to be complete within 15 days after silking.

B. Total Potassium Accumulation by Corn

Sayre (1948) reported that potassium accumulated in corn until 3 weeks after silking and also showed how it was distributed among the different plant parts. Accumulation of K accelerated until a week or 10 days before silking and reached a rate of 3.7 kg/ha (3.3 lb/acre) per day. Under con-

---

[5] Voss, R. D. 1962. Yield and foliar composition of corn as affected by fertilizer rates and environmental factors. Unpublished Ph. D. dissertation. Iowa State Univ., Ames.

[6] Ibid.

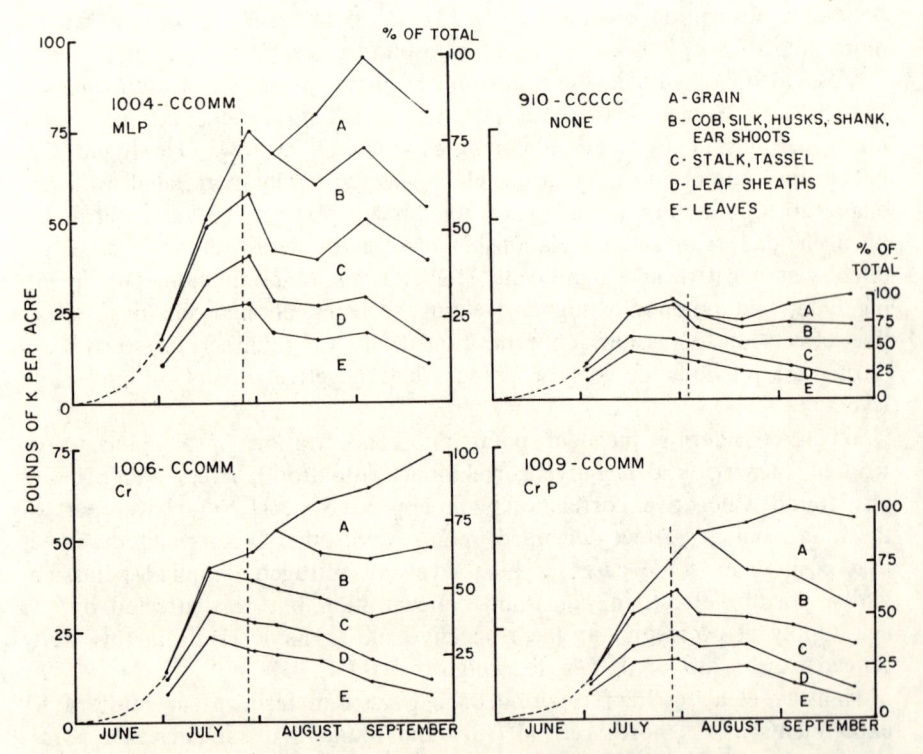

Fig. 1—Total K contents, in pounds/acre and in percent of total, of different parts of the corn plants taken from four plots on seven dates with the date of silking shown by the vertical dashed line (Reproduced from Hanway, 1962a)

ditions in Ohio, the crop contained a maximum of about 128 kg/ha (114 lb/acre) of K in the aerial portion, but this amount was reduced to 110 kg/ha (98 lb/acre) at maturity and yielded 6,410 kg/ha (5,723 lbs/acre) of grain. Approximately 40% of the K was found in the stem and the next highest quantity was in the leaves which contained about 25% of the total. This left 35% (38 kg/ha or 34 lb/acre) in the ear of which over one-half was in the grain.

Hanway (1962a) reported K accumulation in corn from plots which were treated so that 33, 65, 84, and 106 kg/ha (29, 58, 75, and 95 lb/acre) of K were absorbed. Accumulation of potassium continued at least for 40 days after silking in all cases (Fig. 1) and a significant loss of potassium from maximum accumulation to maturity was found in one case. This was from the plot on which 106 kg (95 lb) were accumulated and here the loss was close to 22 kg/ha (20 lb/acre) of K. Uptake was lowest on a nitrogen-deficient plot and highest on a plot regularly manured. A plot which had received phosphorus fertilization regularly accumulated appreciably less potassium than a plot which had not received supplemental phosphorus.

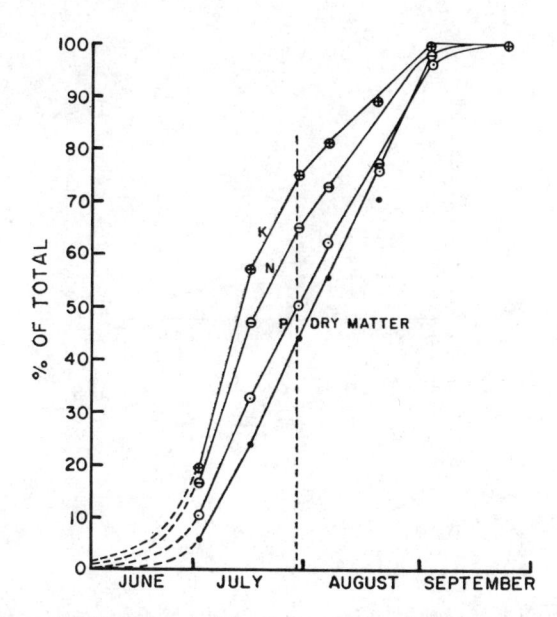

Fig. 2—Relative rate of accumulation of dry matter, N, P and K in corn during the growing season with vertical dashed line showing date of silking (Reproduced from Hanway, 1962c)

Figure 2, also from Hanway (1962a), shows that the relative rates of accumulation are in the increasing order of dry matter, phosphorus, nitrogen, and potassium. By the time of silking, 75% of the total K in corn on the three highest yielding plots had been accumulated. At this stage only half of the phosphorus had been adsorbed and a little over 40% of the total dry matter produced. This rapid early accumulation of potassium suggests that potassium must be in adequate supply early in the life of the corn crop and that it should be available in a relatively smaller volume of soil than nitrogen and phosphorus because it is absorbed mostly by a plant with a smaller root system.

C. Potassium Concentration Indicating Nutritive Status

Reporting on the same study, Hanway (1962b) concluded that nutrient deficiencies resulted in greater differences in concentrations of nutrients in the leaves than in the other plant parts. The basis for this conclusion is presented in part in Fig. 3. This figure shows a particularly wide spread in the potassium content of the leaves at silking. It also shows that there is a gradual decline in potassium concentration in leaves from the first sampling to the last. Contrast this with the more rapid decline of potassium in the leaf sheaths and stalks followed by a leveling off. The highest yielding plot had about 1.5% K in the leaves at time of silking.

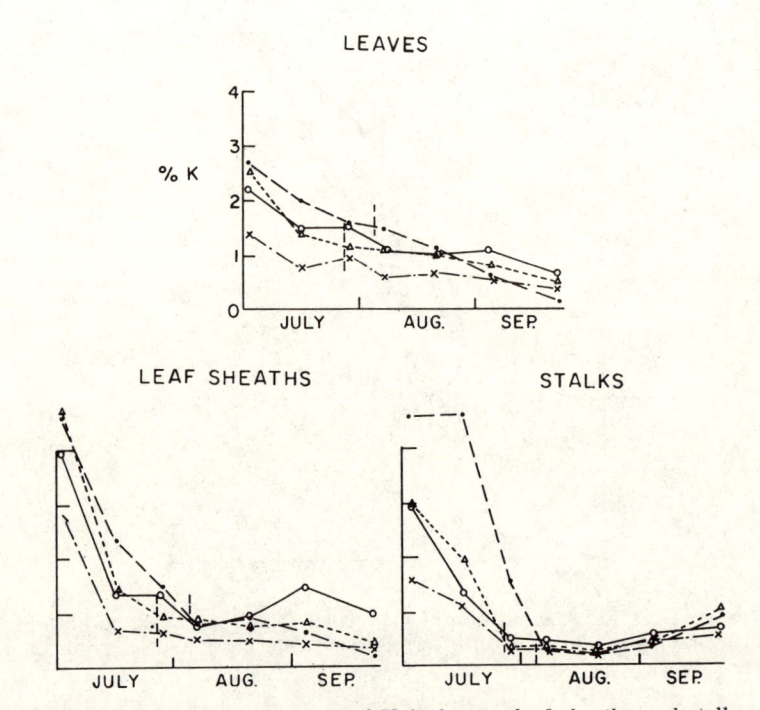

Fig.3—From left to right, percentages of K in leaves, leaf sheaths and stalks of corn from four plots at seven dates during the season with the silking date shown by the vertical dashed lines (Reproduced from Hanway, 1962a)

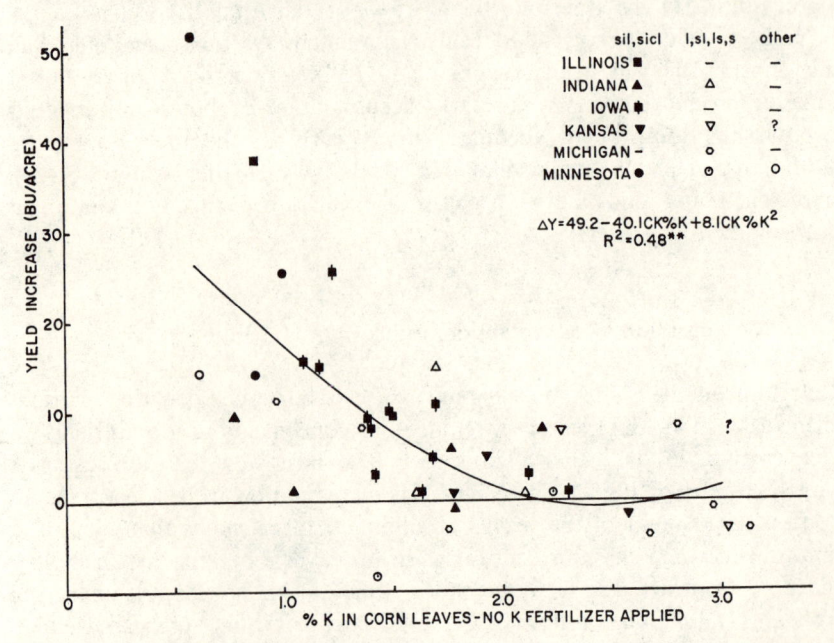

$$\Delta Y = 49.2 - 40.1 CK\%K + 8.1 CK\%K^2$$
$$R^2 = 0.48^{**}$$

Fig. 4—Relationship between the percent K in corn leaves at silking on control plots and the increase due to applied K fertilizer (Reproduced from Hanway et al. 1962)

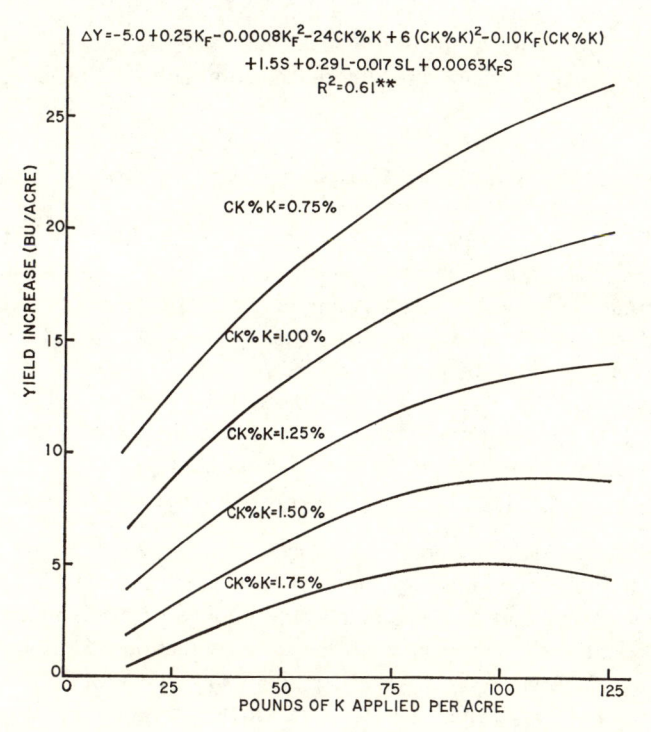

Fig. 5—Response of corn grain yields to amount of K fertilizer applied (K_F) as affected by CK%K, the K concentration of corn leaves at silking time in untreated control plots (reproduced from Hanway et al., 1962).

Tyner (1947) recognized the difficulty of using a whole plant sample to determine the critical percentage of nutrients in corn and proposed the use of a plant part. He presented a detailed argument why a leaf near the middle of the corn plant taken at the time of silking and pollen shedding should be ideal for reflecting the nutrition of the crop. Based on his experiments he concluded that the critical percentage of K was 1.4% for the sixth corn leaf present and counted from the base of the plant. Subsequent investigators have generally settled on the leaf opposite and below the primary ear shoot taken when 75% of the plants show silks to be the standard plant part to sample for determining the nutritive status with respect to potassium (Hanway et al., 1962; Voss, 1962[7]; Dumenil, 1961).

Data collected and summarized by Hanway et al. (1962) in Fig. 4 indicate that a response to K fertilization of corn would not be expected, on the average, when the unfertilized plants had 2% or higher K in the leaves just described. Figure 5 presents a different view of the same situation. The lower the percent potassium in the leaves of the control plot the greater is the response to potassium fertilizer and higher rates are needed to give maximum response. The graph suggests that economic responses to potas-

---

[7] Ibid.

sium are not profitable if the untreated plots have as much as 1.75% K even though maximum yields are not reached until the leaf-K concentration is 2%.

D. Supplying Potassium Requirements

Bray (1944) devised a procedure for estimating the amount of potassium fertilizer needed to satisfy the requirements of corn for potassium in Illinois. He found that there was a relationship between the level of exchangeable potassium in the plow layer and the relative yield of corn. (The relative yield was expressed as a percentage of the yield of a plot with adequate potassium at the same locations.) He then plotted the relative yields, Y, on the ordinate and the exchangeable potassium, k, on the abscissa and fitted a Mitscherlich curve to the results to give the equation.

$$\text{Log } (100-Y) = \log 100 - 0.0055\, k. \qquad [9]$$

Because relative yields do not have prices, Bray estimated, from his observations, that the economic optimum yield would occur at about 95% of the maximum yield and he determined the amount of potassium needed to secure this yield level. For example, when k in equation [9] was 55 pp2m, the relative yield was 53% and 67 kg/ha (60 lb/acre) of K is needed to secure the 95% relative yield. On the other hand, 150 pp2m k secures 90% relative yield and only 31 kg/ha (37 lb/acre) more K is needed. It is estimated that most tables for writing recommendations for potassium fertilization of corn now are based on the method of Bray or minor modifications of it.

A procedure for estimating the optimum amount of potassium fertilizer to apply for corn in absolute terms is based on the relationship between yield or yield response and amount of potassium applied as affected by the soil potassium present and other environmental factors affecting corn response to potassium. One such equation is given in Fig. 6 (Hanway et al., 1962) and a similar equation previously was used by Dumenil et al. (1959). Given this equation and the soil and environmental properties one estimates the most profitable level of potassium fertilizer to use by differentiating y with respect to potassium fertilizer, equating to the inverse price ratio (i.e., price of potassium/price of corn) and solving for the amount of potassium to apply. This procedure has the advantages of (i) giving a direct solution reflecting prices, (ii) permitting environmental and treatment factors other than soil potassium to affect the yield response and (iii) avoiding some possible errors in the basic assumptions associated with relative yields.

According to the results of Hanway et al. (1962), exchangeable K in excess of 140 pp2m will lead to at least 1.5% K in corn leaves at silking on silt loam soils of the Corn Belt. It takes 240 pp2m K in the soil to produce corn plants with the sampled leaves at 2% K. Soils with 100, 150, and 200 pp2m exchangeable K have to be treated with over 140, 112, and 39

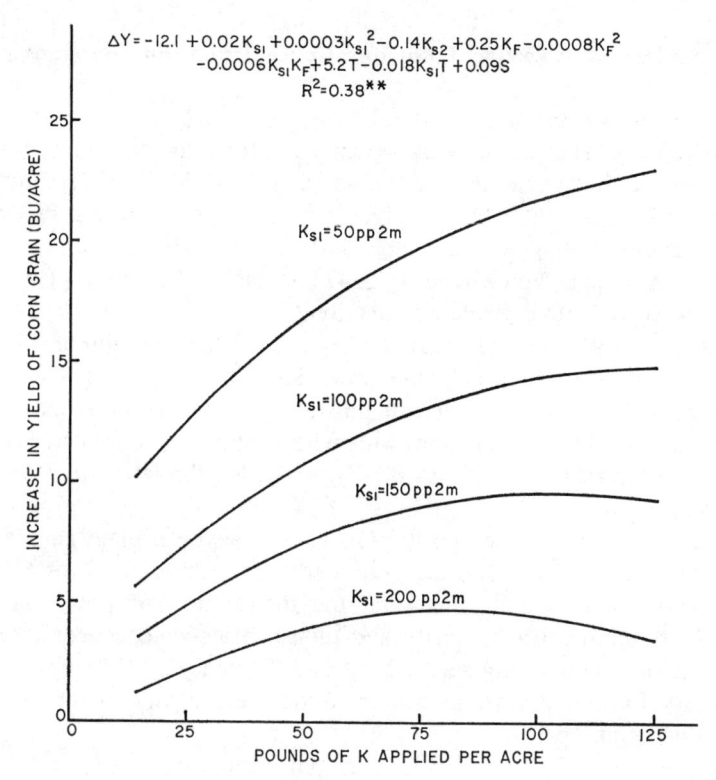

Fig. 6—The response of corn to K applied as affected by exchangeable K, K_{S1}, in the surface soil, where K_{S2} is the exchangeable K in the 6-12 inch layer, S is plant population in thousands of plants per acre (13.3), and T is soil texture (silt loam = 5) (Reproduced from Hanway et al. 1962)

$$\Delta Y = -12.1 + 0.02 K_{S1} + 0.0003 K_{S1}^2 - 0.14 K_{S2} + 0.25 K_F - 0.0008 K_F^2 - 0.0006 K_{S1} K_F + 5.2 T - 0.018 K_{S1} T + 0.09 S$$
$$R^2 = 0.38**$$

kg/acre (125, 100, and 35 lb/acre) of K, respectively, to produce 2% K in the leaves. A level of 1.75% K can be achieved with about 101, 39, and 0 kg/ha (90, 35, and 0 lb/acre) of applied K given the same exchangeable soil K values. In other words, about 200 to 212 kg/ha (180 to 190 lb/acre) of readily available K are needed in the plow layer to assure 1.75% K in the corn leaves at time of silking and close to 280 kg/ha (250 lb/acre) are needed to produce corn at 2% K.

IV. POTASSIUM NUTRITION OF SOYBEANS

Responses to potassium fertilization of soybeans can reflect in yield of grain, yield of dry matter, the content of K, and the content of other mineral elements. Ohlrogge (1960) has made an exhaustive study of the literature pertaining to the mineral nutrition of soybeans. This section will deal only with the part played by potassium.

A. Factors Affecting Potassium Concentration in Soybeans

The soybean was brought into the USA in the last part of the 19th century as a forage crop. Hence, many of the analyses for mineral composition were oriented toward its value as feed (Borst and Thatcher, 1931; Shuster and Graham, 1927). No thought was given to the mineral nutrition of this crop in these studies although von Liebig had already formulated his "law of restitution" (Goodall and Gregory, 1947). Usually the whole plant and/or the seed were the subjects of investigation.

Ohlrogge (1960) reported that the lower and upper limits of K concentration in prebloom soybean plants were 0.30% and 5.7%. These values were found in a potassium-deficient nutrient culture and in a dialized clay-sand mixture in Missouri. Field-grown soybean plants were reported, generally, to fall within this range (from 0.50% to 4.2%, the latter value being for the leaves).

Studies on the mineral composition of various soybean plant parts sampled at different stages of development were reported by Piper and Morse (1923). The concentration of potassium and the distribution of potassium varied as the plants approached maturity and then continued to concentrate in the young pods and developing seed. Borst and Thatcher (1931) observed that total potassium and potassium concentration were higher in the seeds than in any other plant part.

Nelson et al. (1945) observed that potassium increased yields of soybeans fourfold on soils low in potassium and magnesium while it decreased the concentrations of magnesium, calcium, and phosphorus in leaves and petioles, but total magnesium absorbed increased. On the other hand, Smith and Hester (1948) observed the lowest calcium contents in potassium-de-

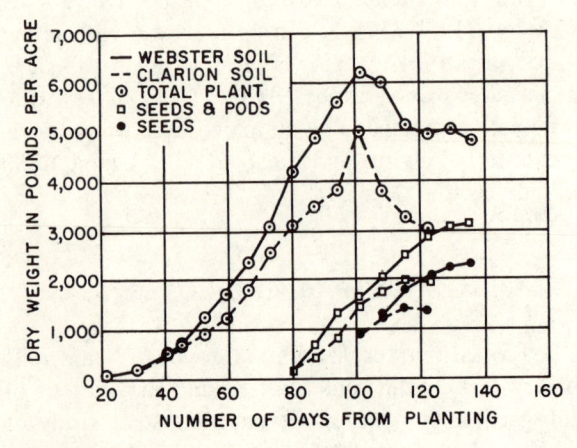

Fig. 7—Dry matter accumulation of soybeans grown on Webster and Clarion soils in Iowa (After Hammond et al. 1951)

ficient soybean plants, suggesting that the deficiency of one element might reduce the absorption rate of another.

Hammond et al. (1951) grew soybeans in experiments on Webster and Clarion soils in Iowa and followed the accumulation of dry matter and potassium and of the concentration of potassium in the plant parts. Fig. 7 and 8 are presented to illustrate the effect of the two soils on the characteristics measured. It will be noted that while dry matter accumulated more slowly from the beginning on the Clarion soil, K accumulated at different rates on the two soils but reached the same total quantity by about 75 days after planting. Total K accumulated throughout the first 100 days after planting and then leveled off at a stage at which about half of the seed had been produced.

The rate of accumulation per unit time is given by slopes of the line segments connecting the sampling dates (the first derivative of the accumulation curves). This relationship is shown better in Fig. 9 for the Webster soil where the units are reduced to pounds per acre per day. There was an acceleration in uptake of potassium for the first 90 days while peak rates of uptake of phosphorus, calcium and magnesium were reached about 12 days earlier.

Acceleration of dry matter accumulation (Fig. 7) observed by Hammond et al. (1951) lasted only about 72 and 80 days on Clarion and Webster soils, or approximately until significant pod formation started. One half of the maximum dry matter accumulation took about 72 days in both experiments. On the other hand, half of the maximum potassium accumulation

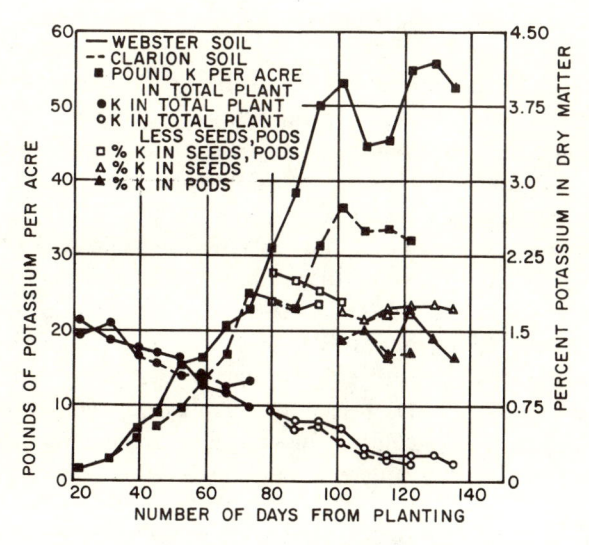

Fig. 8—Percentage of K in plant parts and accumulation of K in soybeans grown on Webster and Clarion soils in Iowa (After Hammond et al. 1951)

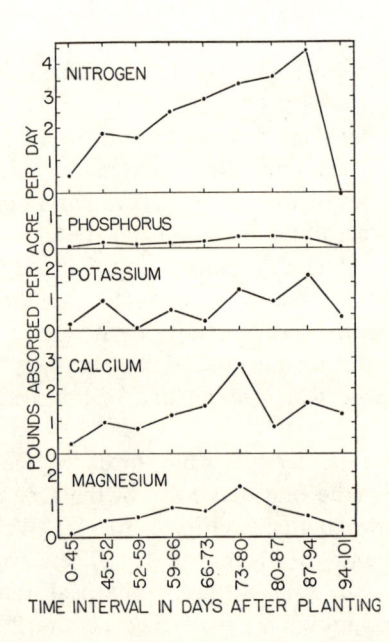

Fig. 9—Rate of accumulation of mineral elements during the life of a soybean community on Webster silt loam in Iowa (Reproduced from Hammond et al. 1951)

occurred at about 68 days on the Clarion soil and at about 75 days on the Webster. This suggests that the relative rates of accumulation of K and dry matter may not bear a constant relationship to each other. It may be significant, however, that the concentration of potassium in the plants and plant parts for the two experiments were quite similar.

B. Effect of Potassium on Seed and Seed Yield

Potassium supply also has an effect on the quality of soybean seed. For example, Shuster and Graham (1927) report that while potassium alone did not influence the soil content of soybean seed, combinations of phosphorus and potassium, and nitrogen and potassium increased the oil content on unlimed soil. Adams et al. (1937) also found the highest oil content with a combination of nitrogen and potassium fertilizer. Cartter (1940), working with soils of medium fertility did not observe these effects of applied potassium. The study of Nelson et al. (1945) showed that the potassium treatment of 112 kg/ha (100 lb/acre) of K doubled the number of seed pods and reduced the percentage of shriveled seed from 35 down to 3.

Miller et al. (1961, 1964) observed responses of seed yield of soybeans up to 1,810 kg/ha (27 bu/acre) due to 178 kg/ha (160 lb/acre) of K added in the presence of adequate P. deMooy (1965)[8] also observed maxi-

---

[8] de Mooy, C. J. 1965. Differential responses of soybean varieties to application of phosphorus, potassium and calcium carbonate materials with respect to leaf composition and yield. Unpublished Ph.D. dissertation. Iowa State Univ., Ames.

Fig. 10—Influence of P concentration in leaves of soybeans at growth stage 7 on the relationship of yield to K concentration (Reproduced from Miller et al. 1961)

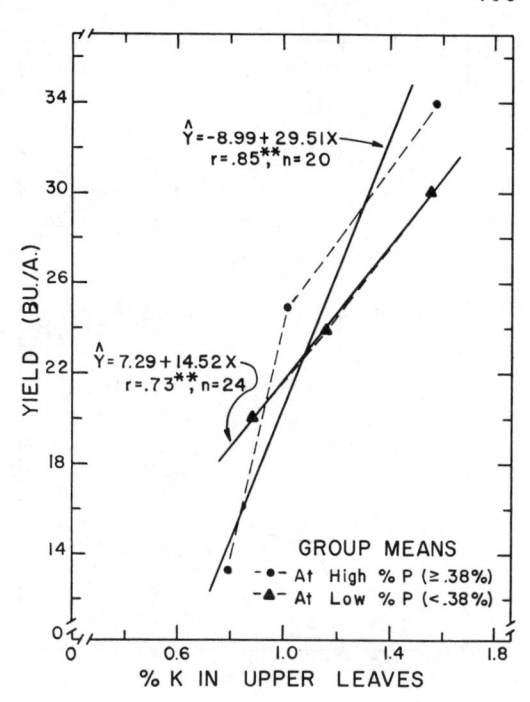

mum yield responses to occur at approximately 562 kg/ha (500 lb/acre) of K in the presence of 224 to 336 kg/ha (200 to 300 lb/acre) of P applied in a test involving four varieties. There was a fertilizer by variety interaction and the maximum predicted yields within limits of the treatments varied from 2,550 to 3,080 kg/ha (38 to 46 bu/acre) for different varieties.

In a study involving four experiments at nine rates of phosphorus and potassium, Miller (1960)[9] reported an interaction between the content of phosphorus and potassium in the upper mature leaves of soybeans at growth stage 7 (Kalton et al., 1949) on the yield of grain. In general, the increase in yield was greater as a result of the increased potassium content on plots with a higher leaf concentration of phosphorus. A detailed statistical analysis involving the relationship of yield to P and K concentrations in the petioles, leaves and stems in the upper and lower halves of the plants at stages 3, 5, 7, and 9 led Miller to conclude that the most meaningful relationships would be obtained from an analysis of the leaves in the upper half of the plant at growth stage 7. The possibility of utilizing one plant part to measure the deficiency of one element and another plant part to measure the deficiency of another was not explored in this study; however, there is a suggestion that perhaps the petioles in the upper part of the plant might provide a better potassium index.

---

[9] Miller, op. cit.

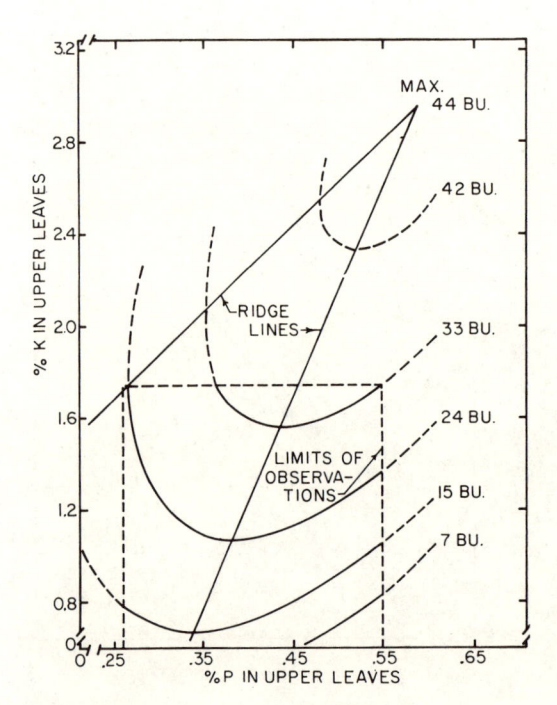

Fig. 11—Yield isoquants showing the simultaneous dependence of yield on the K and P concentrations in soybean leaves at growth stage 7 (Reproduced from Miller et al. 1961)

Figure 10 from Miller et al. (1961) shows the rather sharp increase in yield of soybeans due to increases in potassium concentration in the upper mature leaves at growth stage 7. It also shows the significant interaction of the concentration of phosphorus and potassium on seed yield. This interaction relationship is shown even better in Fig. 11. It is evident that the highest rates of fertilizer used, i.e., 127 kg P and 240 kg K/ha (113 lb P and 213 lb K/acre) were not high enough to define maximum yields.

deMooy (1965)[10] extended this study in part and increased the rate of fertilization up to 900 kg K and 450 kg P/ha (800 lb K and 400 lb P/acre). Maximum levels of K in the soybean leaves were 1.85% in the Harosoy variety and about 2.15% in the Hawkeye variety. These levels were observed at almost 900 and 675 kg/ha, respectively (800 and 600 lb/acre), or applied K. The P applied had relatively little effect on K concentration under 1.7% K; above that, there was a slight interference of P with K absorption.

C. Other Effects and Critical Percentages

deMooy and Pesek (1966) studied the effect of potassium, phosphorus, and calcium additions to soil cultures on the nodulation of different soybean varieties. These were grown in direct sunlight during the normal growing

---

[10] de Mooy, op. cit.

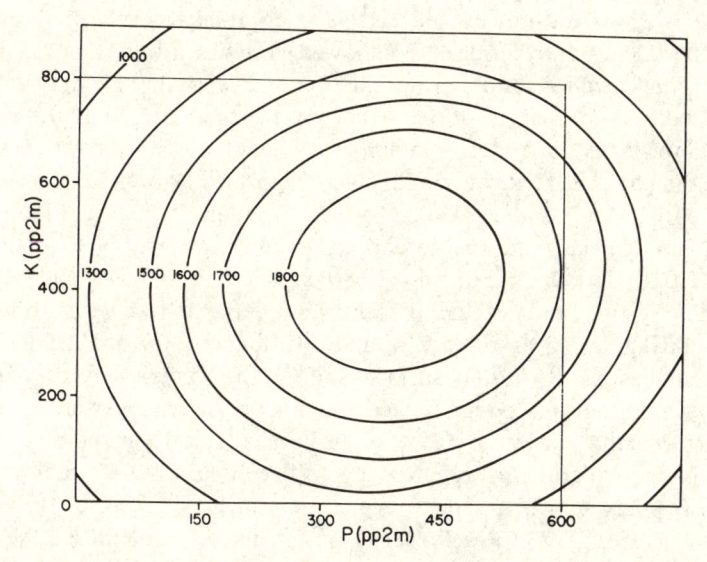

Fig. 12—Isoquants for nodule numbers per culture at end of flowering of selections from Plant Introduction Line 88805-2 as affected by P and K added (Reproduced from deMooy and Pesek 1966)

season. The fresh weight, number, and leghemoglobin activity of the nodules were affected by potassium level and by the variety although the phosphorus effect dominated and the phosphorus-potassium interaction was important in some varieties at certain stages of growth. Figure 12 illustrates the kind of results found showing an effect of both phosphorus and potassium on nodule numbers of one of the varieties grown. These data and other data from this test indicate that maximum numbers of nodules tend to occur at somewhat lower concentrations of potassium in the soil (about 400 pp2m) than are needed to secure maximum potassium concentration in the leaves. They tend to coincide more with the potassium and phosphorus levels in the field which result in maximum yields of seed.

Maximum concentrations of K in leaves and petioles reported by Miller (1960)[11] exceeded 3% and 5%, respectively, for selected experiments, stages of growth, and treatments. However, the observed upper limit of the data for upper leaves in growth stage 7 was only 1.77% and maximum yields were not reached within the treatment range of these experiments. The combined analysis (Miller et al., 1964), therefore, indicated that the critical levels of K and P at this stage were greater than 1.77 and 0.55%, respectively. The extrapolated data suggest that the critical level might be as high as 2.7% K and 0.74% P.

The results of deMooy (1965),[12] however, show that K concentrations

---

[11] Miller, op. cit.
[12] de Mooy, op. cit.

in leaves at stage 7 did not exceed 2.15% even at high levels of K applied in field experiments. Furthermore, he showed varieties to differ in maximum potassium concentration and in the relation between change in yield and change in potassium concentration. It is evident, therefore, that the critical percentage for K is limited by an upper value of 2.15%, at least in the varieties studied. The critical percentages reported by deMooy (1965)[12] were 1.80, 1.91, 1.87, and 1.99% K for the varieties Chippewa, Blackhawk, Harosoy, and Hawkeye in that order and were associated with estimated yields of 2,610; 2,680; 3,150; and 2,680 kg/ha (39, 40, 47 and 39 bu/acre). These yields and critical percentages were secured with 560, 650, 583, and 605 kg/ha (500, 580, 520, and 540 lb/acre) of applied K.

The exchangeable K in the surface soil of the experiment of deMooy (1965)[12] was about 65 pp2m; hence the maximum yields were achieved with approximately 675 kg (600 lb) of easily available K in the plow layer. The soils in the experiments reported by Miller et al. (1964) ranged from less than 60 pp2m to 125 pp2m K in the plow layer. Therefore, these sites all had over 338 kg/ha (300 lb/acre) of readily available K in the plow layer and the data show that this quantity was not adequate to give highest K concentration in the leaves of soybeans at stage 7, nor maximum yields. Until better evidence is presented, it is estimated that highest yields of soybeans cannot be achieved in the north central USA with less than 500 pp2m K in the plow layer.

While the maximum yields of soybeans are achieved with high rates of potassium, the most profitable levels to apply are considerably less in many cases. Bray (1945) estimated the additional K required for soybean yields equivalent to 95% of the maximum yields given the available K present in the plow layer. With available K at 60 pp2m he estimated a need for 77 kg/ha (69 lb/acre) of K. At 100 pp2m and 150 pp2m, the amounts needed were 57 and 34 kg/ha (51 and 30 lb/acre), respectively, of K while a soil with 200 pp2m exchangeable K in the plow layer was estimated to yield 97% of the maximum. The mean responses to K of the four varieties reported by deMooy may be given approximately by

$$\Delta y = 4.2K - 0.5K^2 \qquad [10]$$

where Δy is in bushels and K in units of 100 lb/acre. Assuming K at $0.06/lb and soybeans at $2.50/bu, the optimum rate of K is about 190 kg/ha (170 lb/acre). This would provide for about 258 kg/ha (230 lb/acre) readily available K needed considering the initial soil level of just over 60 pp2m in the plow layer.

REFERENCES

Adams, J. D., H. M. Boggs, and E. M. Roller. 1937. Effect of fertilizers on composition of soybean hay and seed and of crop management on carbon, nitrogen and reaction of Norfolk sand. U.S. Dept. Agr. Tech. Bull. 586.

Bennett, W. F., G. Stanford, and L. Dumenil. 1953. Nitrogen, phosphorus and potassium content of the corn leaf and grain as related to nitrogen fertilization and yield. Soil Sci. Soc. Amer. Proc. 17:252-258.

Borst, H. L., and L. E. Thatcher. 1931. Life history and composition of the soybean plant. Ohio Agr. Exp. Sta. Bull. 494.

Bray, R. H. 1944. Soil-plant relations: I. The quantitative relation of exchangeable K to crop yields and to crop response to potash additions. Soil Sci. 58:305-324.

Bray, R. H. 1945. Soil-plant relations: II. Balanced fertilizer use through soil tests for potassium and phosphorus. Soil Sci. 60:463-473.

Cartter, J. L. 1940. Effect of environment on composition of soybean seed. Soil Sci. Soc. Amer. Proc. 5:125-130.

deMooy, C. J., and John Pesek. 1966. Nodulation responses of soybeans to added phosphorus, potassium and calcium salts. Agron. J. 58:275-280.

Dumenil, Lloyd. 1961. Nitrogen and phosphorus composition of corn leaves and corn yields in relation to critical levels and nutrient balance. Soil Sci. Soc. Amer. Proc. 25:295-298.

Dumenil, Lloyd, J. Hanway, H. R. Meldrum, and J. Pesek. 1959. Your corn may need potassium. Iowa Farm Sci. 13:(11) 3-4.

Dumenil, Lloyd, and L. B. Nelson. 1948. Nutrient balance and interactions in fertilizer experiments. Soil Sci. Soc. Amer. Proc. 13:335-341.

Goodall, D. W., and F. G. Gregory. 1947. Chemical composition of plants as an index of their nutritional status. Imp. Bureau of Hort. and Plantation Crops Tech. Communication 17.

Hammond, L. C., C. A. Black, and A. G. Norman. 1951. Nutrient uptake by soybeans on two Iowa soils. Iowa Agr. Exp. Sta. Res. Bull. 384.

Hanway, J. J. 1962a. Corn growth and composition in relation to soil fertility. II. Uptake of nitrogen, phosphorus and potassium and their distribution in different plant parts during the growing season. Agron. J. 54:145-148.

Hanway, J. J. 1962b. Corn growth and composition in relation to soil fertility. III. Percentages of N, P, and K in different plant parts in relation to stage of growth. Agron. J. 54:222-229.

Hanway, J. J., S. A. Barber, R. H. Bray, A. C. Caldwell, M. Fried L. J. Kurtz, K. Lawton, J. T. Pesek, M. Reed, and F. W. Smith. 1962. North Central regional potassium studies. III. Field studies with corn. Iowa Agr. and Home Econ. Exp. Sta. Res. Bull. 503.

Kalton, R. R., C. R. Weber, and J. C. Eldredge. 1949. The effect of injury simulating hail damage to soybeans. Iowa Agr. Exp. Sta. Res. Bull. 359.

Lawton, K. 1945. The influence of soil aeration on the growth and absorption of nutrients by corn plants. Soil Sci. Soc. Amer. Proc. 10:263-268.

Lundegardh, H. 1951. Leaf Analysis. London, England. Hilger and Watts, Ltd. (Translation DIE BLATTANALYSE, by R. L. Mitchell.)

Macy, P. 1936. The quantitative mineral nutrient requirement of plants. Plant Physiol. 11:749-764.

Miller, R. J., J. T. Pesek, and J. J. Hanway. 1961. Relationships between soybean yield and concentrations of phosphorus and potassium in plant parts. Agron. J. 53:393-396.

Miller, R. J., J. T. Pesek, J. J. Hanway, and L. C. Dumenil. 1964. Soybean yields and plant composition as affected by phosphorus and potassium fertilizers. Iowa Agr. and Home Econ. Exp. Sta. Res. Bull. 524.

Nelson, L. B. 1956. The mineral nutrition of corn as related to its growth and culture. Advance. Agron. 8:321-375.

Nelson, W. L., L. Buckhart, and W. E. Colwell. 1945. Fruit development, seed quality, chemical compositions and yield of soybeans as affected by potassium and magnesium. Soil Sci. Soc. Amer. Proc. 10:224-229.

Ohlrogge, A. J. 1960. Mineral nutrition of soybeans. Advance Agron. 12:229-263.

Piper, C. V., and W. J. Morse. 1923. The soybean. McGraw-Hill Book Co:, Inc., New York, N. Y.

Sayre, J. D. 1948. Mineral accumulation in corn. Plant Physiol. 23:267-281.

Shuster, G. L., and J. M. Graham. 1927. Effect of various fertilizers and lime on composition of soybeans. J. Amer. Soc. Agron. 19:574-576.

Smith, G. E., and J. B. Hester. 1948. Calcium content of soils and fertilizers in relation to composition and nutritive value of plants. Soil Sci. 65:117-128.

Stanford, G., J. B., Kenny, and W. H. Pierre. 1941. Cation balance in corn grown on high-lime soils in relation to potassium deficiency. Soil Sci. Soc. Amer. Proc. 6:335-341.

Thomas, W., and W. B. Mack. 1939. The foliar diagnosis of *Zea mays* subjected to differential fertilizer treatments. J. Agr. Res. 58:477-491.

Tyner, E. H. 1947. The relation of corn yields to leaf nitrogen, phosphorus, and potassium content. Soil Sci. Soc. Amer. Proc. 11:317-323.

Tyner, E. H., and J. R. Webb. 1946. The relation of corn yields to nutrient balance as revealed by leaf analysis. J. Amer. Soc. Agron. 38:173-185.

Ulrich, A. 1943. Plant analysis as a diagnostic procedure. Soil Sci. 55:101-112.

Potassium Nutrition of Tree Crops

ROBERT C.J. KOO

University of Florida Citrus Experiment Station
Lake Alfred, Florida

I. INTRODUCTION

The mineral nutrition of orchard crops is different from that of annual crops in certain problems and relationships because of the large size of fruit trees, their long life in the same locations, and the more complex interactions with climate and soil factors. Productive orchard trees require large, healthy root systems which probe into the subsoil more deeply and for longer periods of time than do the roots of most annual crops. They grow in a substrate which encompasses a broad gradient in nutrient availability, temperature, and physical soil structure. In addition, tree crops present certain unique problems because both rootstock and scion variety may influence the mineral nutrition of fruit trees.

Potassium and nitrogen play dominant roles in the mineral nutrition of tree crops. Unlike nitrogen, the potassium requirement of fruit trees is often not as easily determined as that of the nitrogen, because many tree crops can tolerate a wide range of potassium content without showing apparent influence on fruit production or vegetative growth. Potassium deficiency will affect tree growth, tissue composition, fruit production, and fruit quality. Prevalence of potassium deficiency may vary between tree crops and locations, but symptoms are similar for most crops in all fruit growing regions of the world. "Marginal scorch" of older leaves, reduced shoot growth, and leaf size are typical symptoms of potassium deficiency of most tree crops. On cherry, plum, peach, and pear, an upward lateral curling of leaves is evident before the scorch appears. On apple, the upward curling is not so prominent, but a darkening or browning appears on the upper side of the leaf margin. On citrus, chlorosis, curled and puckered leaves, will not appear unless the trees are extremely deficient in potassium.

Potassium has more influence on fruit quality than any other element. The fruit for most tree crops may be affected throughout the range from potassium deficiency to excess. Fruit production, on the other hand, is not usually affected by potassium except when it is at a very low level.

Leaf analysis to evaluate the potassium and other nutrients in orchard trees has been widely used in the past 2 decades. It has helped researchers to obtain an enormous amount of information on slow-growing tree crops in a relatively short time.

Potassium sulfate and potassium chloride are the principal sources of potassium used in tree crop fertilization. The preference of one source over the other is usually determined by crop preference, soil salinity problems, and economic advantage. Potassium nitrate is sometimes applied as a foliar spray to trees planted on fine-textured alkaline calcareous soil where soil application of potassium is sometimes ineffective. There is no report at present on the use of slowly soluble potassium sources such as potassium metaphosphate on tree crops under field conditions.

The present review reports some of the more recent studies of potassium fertilization in different tree crops and points out certain similarities, unique problems, and relationships revealed by these studies. Literature on potassium research varies from abundant in apple, peach, and citrus to almost nonexistent in mango and some of the edible nuts. It is hoped that this review will be of general interest to workers investigating potassium nutrition of tree crops under widely different environmental conditions.

II. POTASSIUM AND TREE CROPS

A. Apple

The apple tree, is a long-lived deciduous perennial grown in the temperate zone throughout the world. There probably has been more research done on the apple than on any other tree crop. A thorough review of the mineral nutrition of apple was made by Boynton and Oberly (1966).

Symptoms of potassium deficiency were first noticed by Wallace (1928-29) for apple trees grown in pots. He subsequently diagnosed a leaf scorch that was widespread in English orchards as due to potassium deficiency. Following his discoveries, workers in other parts of the world have associated this system with potassium deficiency. There seemed to be general agreement about the appearance of the scorch symptoms. These symptoms develop on older basal leaves first and gradually progress toward the younger leaves as the season advances. In severe cases, trees may be stunted and produce small leaves, fruit may fail to size normally, and the color may be poor. Such trees are subject to winter injury. Symptoms of potassium deficiency have been described and illustrated in several publications (Davis and Hill, 1941; Blake et al., 1937; Wallace, 1951; and Sprague, 1964).

The effects of potassium deficiency on growth of apple trees have been investigated by Batjer and Degman (1940). They measured the growth response of young York Imperial apple trees under greenhouse conditions to varying concentrations of potassium in the nutrient solution. There were significant increases in the weight of trees as the K content of the nutrient solution was increased from 0 to 60 ppm in four increments, and there was a suggested further increase in weight between 60 and 117 ppm. This was reflected in the mean height of trees, total length of lateral branches, total linear growth, trunk diameter, and leaf size. Total potassium in the leaves was closely correlated with potassium supplied in the nutrient solution. At the end of the experiment, leaves from the 0-K trees showed a leaf K content of .33% and those from the 117 ppm-K treatment contained 2.31% K.

Fruit production of trees showing scorch leaf symptoms will be increased by adequate potassium fertilization, but yield response to K fertilization in trees not exhibiting leaf scorch is inconsistent. Increased fruit production and fruit size resulting from the use of KCl or K_2SO_4 have been reported by Forshey (1963) and Shadmi et al. (1966) on trees that did not show scorching symptoms but were making poor growth. Barden and Thompson (1962) made heavy annual applications of potassium for 7 years and found no significant difference in fruit production or fruit size. It is fair to assume that fruit production response can be expected from potassium fertilization in trees low in potassium but not showing deficiency symptoms.

There is general agreement that the acid content of apples increases with the potassium levels in the tree (Fisher and Kwong, 1961; Mori and Yamazuki, 1960; Barden and Thompson, 1962). Mori and Yamazuki (1960) also noted that while the acid content increased as potassium increased, there was no effect on the soluble solids of the fruit, thus resulting in a decrease in the soluble solid/acid ratio of the fruit. Potassium fertilization improves the development of red color of fruit in orchards that show severe potassium deficiency. Fisher and Kwong (1961) showed K fertilizer improved the color of McIntosh apples only when the leaf K content was low (0.5%). Weeks et al. (1958) in a nitrogen and potassium rate study on McIntosh apples, reported in some instances potassium fertilization may offset the depressing effect of moderate nitrogen levels on fruit color. Potassium seems to have little or no effect on the storage life of apples (Barden and Thompson, 1962). Application of potassium spray also has shown little effect on storage quality of apples (Rasmussen, 1966).

Potassium absorption by apple trees may be influenced by the particular combination of scion variety and rootstock. Hoblyn (1940) reported observations of long-term trials of several scion varieties and rootstocks on a potassium-deficient soil and concluded that the reduction in growth due to potassium deficiency was related to the vigor of the rootstocks. The greatest reduction in growth occurred in the most vigorous rootstocks. Batjer and Magness (1939) in a survey of apple orchards from four important regions in the USA reported the leaf potassium content of Delicious averaged considerably higher than that of York, Jonathan, and Rome varieties.

B. Avocado and Mango

The avocado is a native of Central America and adjacent areas of North and South America. It is cultivated throughout the subtropical and tropical regions of the world. A number of investigations of the mineral nutrition of avocado have been reported by workers in California and Florida but most of them were concerned primarily with other nutrient elements.

Potassium deficiency in avocado was reported by Furr et al. (1946) in sand culture. Symptoms included narrow subnormal size leaves with numerous light-brown specks scattered over the surface. As the leaves aged, the brown specks coalesced into large irregular, reddish-brown necrotic areas between large veins, along the margin or the tips. The trees were stunted and some dieback occurred. Similar symptoms were observed by Haas (1939) using avocado seedlings in solution culture. No potassium deficiency has been reported from field grown avocado trees. Lynch and Goldweber (1954, 1956) found no relation between K in mixed fertilizer and tree growth and yield over a range of leaf K content between 0.58-2.51%. Embleton and Jones (1964) found no beneficial effects on yield by increasing the leaf K content from 0.9-1.3% over a 12-year period. The potassium content of avocado leaves reported in Florida was somewhat higher than that found in California and was found to be inversely related to the size of the fruit crop (Popenoe et al., 1961).

There is considerable evidence to indicate the influence of different varieties and rootstocks on the mineral composition of leaves. The different degrees of tolerance to chloride in the soil by different races of avocado are of direct interest to potassium fertilization. In general, the West Indian rootstock and scion varieties are more tolerant to high chloride conditions in the soil than the Guatemalan varieties, and the Guatemalan varieties, in turn, are more tolerant than the Mexican varieties, as indicated by a lower accumulation of chloride in the leaves and less tip burn (Kadman, 1963; Embleton et al., 1962). The different degree of chloride tolerance is an important consideration in choosing the source of potassium fertilizer.

Mangos originated in the Indo-Burma region and are now grown throughout the tropical and subtropical regions of the world. Nowhere have they attained the importance they have in India where almost 2 million acres of mangos are grown (Mukherjee, 1958). Investigations on the mineral nutrition of mangos are few. Smith and Scudder (1951) described symptoms of potassium deficiency in trees grown in sand culture as having irregularly distributed yellow spots in older leaves. These leaves are smaller and tapered out to a very fine point. In advance stages, the spots enlarge and produce necrotic areas along the margin of leaves. Affected leaves do not absciss until completely dead.

In a 4-year fertilizer trial with Kent Mango in Florida, Young et al. (1965) found heavy nitrogen fertilization substantially increased fruit production. A smaller increase in yield resulted from heavy potassium fertilization with KCl. Incidence of soft-nose, a physiological disorder in the fruit, increased with increased nitrogen, but tended to decrease with increased potassium fertilization. Heavy application of calcium also reduced incidence of soft-nose (Young et al. 1962). A combination of heavy nitrogen, potassium, and calcium fertilization resulted in a 35% increase of marketable fruit per year. The leaf K content between $+K$ and $-K$ trees in that study was 0.59% and 0.38%, respectively. Roy et al. (1951), in a fertilizer trial in India, found potassium or phosphorus when used singly or in combination had little effect on growth, but when used together with nitrogen gave better results in growth and fruiting.

C. Cherry

Cultivated cherries originated from two wild species, *Prunus avium* L. (sweet cherries) and *P. cerasus* L. (sour cherries). Both species are natives of south-central Europe and Asia Minor. Although cherry culture is widely scattered throughout the world, the USA and Europe are the principal cherry producing regions.

The cherry, in common with other deciduous fruit trees, is subject to a number of mineral deficiencies. In potassium-deficient soils, the cherry leaves are bluish-green, with the margins tending to roll forward parallel with the midrib. The leaves are chlorotic, followed by scorching or necrosis (Wallace, 1951). Correction of leaf curl or leaf scorch with potassium fertilization has been reported by Boynton (1944) and Lanford (1939).

The response of cherry trees to application of potassium varies widely, depending upon the soils in which the trees are grown. Westwood (1966) summarized several potassium investigations in Europe that showed improved tree growth and yield due to potassium fertilization of soils that were apparently low in potassium. Several workers have reported the effects of potassium on fruit quality. Kwong (1965) found K_2SO_4 fertilization increased the potassium content of leaves and increased fruit size but had no effects on the soluble solids or acid content of fruit. Curwen et al. (1966), investigating the effects of three rates of NPK on the firmness and pectic composition of cherry, found increasing potassium resulted in softer fruit having a higher juice loss upon pitting and a reduced water insoluble pectic content. The effect of nitrogen was just the reverse while phosphorus had no effects. Cline et al. (1965, 1966) compared two rates of KCl and K_2SO_4 with and without phosphorus. Leaf potassium content was increased with potassium treatments, but no significant difference in growth, yield, or fruit quality was found.

Leaf analysis is probably the most satisfactory method of determining whether a cherry tree needs additional potassium. Christensen and Walker (1964) sampled leaves from 79 Bing cherry orchards in Utah and found a K range of 1.0-1.8%. It is generally agreed that a leaf K content below 1% would indicate a possible K deficiency.

D. Citrus

All important citrus varieties belong to one genus, *Citrus*. The principal species are sweet orange, sour orange, grapefruit, mandarin, lemon, and lime. Citrus is grown throughout subtropical and tropical regions of the world. It thrives on a wide variety of soil types and in regions varying from abundant rainfall to irrigated desserts. Many of the citrus-producing areas apparently have soils with considerable natural potassium reserves. Some potassium is applied to certain localities in Africa, Australia, and the Caribbean Islands. In California, potassium is only used in limited areas, although reports by Chapman (1964) and Embleton et al. (1964) indicate the need for potassium may be more widespread than is generally believed. Potassium is used extensively in Florida because most of the citrus is planted on sandy soils that are inherently low in potassium.

The potassium requirement of young growing citrus was carefully studied by Chapman et al. (1947) in California using large solution chambers, and in Florida by Smith et al. (1953) using outdoor sand culture. In both studies, trees with K contents in leaves between 0.5 and 2.40% showed no difference in growth. Those with a leaf K content around 0.4% showed slight deficiency. The deficient trees had smaller leaves, finer branches, and a more compact appearance. Trees with a lavish supply of K grew poorly, and they showed a leaf K content in excess of 3.5%.

Foliage symptoms of potassium deficiency are seldom found in field-grown citrus. This is because the effects of low potassium on fruit production, fruit size, and premature fruit drop precede any leaf symptoms. Foliage symptoms of potassium deficiency were described by Haas (1948) in solution culture for orange. Deficiency symptoms of field-grown orange (Koo and Spencer, 1959) and grapefruit (Smith and Rasmussen, 1961) were observed in experimental orchards in Florida and for lemon in California (Embleton et al. 1964). The symptoms first appear on old leaves with undefined areas of yellowing of the leaves. Most of the chlorotic leaves show burnt edges and may curl toward the lower surface. In lemon, the chlorotic pattern is usually more clearly defined than in orange and grapefruit. Small leaves and dieback of twigs usually accompany the chlorosis in leaves. The K content of chlorotic leaves is usually not more than 0.25%.

Most of the studies indicate that potassium levels above deficiency range do not influence yield. Investigation in Florida indicated a level of leaf K content between 0.9 and 1.1% for orange (Koo, 1961; Reitz and Koo, 1960; and Deszyck et al. 1958) and grapefruit (Smith and Rasmussen, 1960; Deszyck and Koo, 1957), above which little or no yield response could be expected from K treatments. Acid fruits such as lemon and lime appeared to respond to potassium treatments at a higher level. Increased yield responses to K treatments have been reported at leaf K contents between 1.40 and 1.50% for lime (Young and Koo, 1967). In California, the range of leaf K in which Valencia orange will respond to K treatment seems to be between 0.5 and 0.7% (Embleton et al., 1956, 1966), although increased yield response to K treatments at leaf K content of 1% was obtained in 1 out of 3 years by the same authors.

Citrus fruit quality may be affected throughout the range from K deficiency to excess. Many workers have shown a direct relationship between potassium supply of the tree and fruit size and rind thickness (Chapman et al., 1947; Deszyck et al., 1958; Embleton et al., 1956; Reitz and Koo, 1960; Smith and Rasmussen 1960). The fruit becomes larger and coarser as the potassium level increases. Low level of potassium reduces fruit size of all varieties of citrus, but high potassium may not always increase fruit size indefinitely. Apparently other factors, such as size of fruit crop, moisture supply, rootstocks, other nutrients, and diseases may become limiting. Potassium nutrition seems to have little effect on the total soluble solids of the juice, but the acidity of the juice is consistently increased with potassium fertilization. Vitamin C (ascorbic acid) content also increases with potassium fertilization (Sites and Deszyck, 1952). Juice, expressed as percent by weight, usually decreases with increasing potassium level in the tree. Some of the effect of potassium on internal quality is undoubtedly secondary to its effect on fruit size. Large oranges and grapefruit have lower percent juice, soluble solids, and acid contents than small fruit. However, the acid content of the juice is often increased by K at levels above which fruit size may be limited by other factors (Smith and Rasmussen, 1960). The best quality orange and grapefruit then would be one with a moderate to low potassium content. For lemon and lime where high acidity is preferred, more potassium should be used. Embleton et al. (1964) and Koo (1963) have shown that high potassium fertilization can increase the acid content of lemon about 15% when compared to low potassium fertilization.

Low potassium in the tree has been associated with several rind disorders of orange and mandarins since potassium is directly related to rind thickness and textures. It is not known whether low potassium is a direct cause or may be one of the contributing factors to these disorders. Koo (1961) reported a highly significant inverse correlation between fruit splitting and the potassium content of leaves for Hamlin orange. Jones et al. (1967) significantly reduced the incidence of creasing by increasing the potassium levels in the trees.

Among the potassium sources used for citrus, KCl is the most common source. Potassium sulfate is preferred in areas where soil salinity is a problem. Potassium nitrate has been used on a limited scale as a foliar application in places where soil application of potassium is ineffective.

It has long been thought that potassium will induce greater cold hardiness in citrus trees. Observations made in Florida after the 1957 and 1962 freezes did not substantiate this theory. Smith and Rasmussen (1958) found no difference in freeze tolerance at different potassium levels. In fact, trees high in potassium appeared to be more susceptible to freeze injury than those with moderate potassium. Unpublished work of the author showed no difference in tree injury following the severe 1962 freeze as long as the leaf K content was above 0.6%. Trees with K content less than 0.6% suffered much more severe injury. Haas and Halma (1931) in California found increased levels of foliar potassium in citrus had no effect on freezing point depression. Semakin et al. (1937), as cited by Vasiliyev (1961), found potassium alone had little effect on cold hardiness of citrus. They found it was the overall nutritional regime and the resultant growth factors that determined the hardiness of plants.

Both rootstock and scion varieties may influence the absorption of potassium (Smith and Reuther, 1949; Wallace et al., 1952). The commonly used rootstocks of rough lemon, Cleopatra mandarin, and sour orange are efficient users of potassium as indicated in the leaf potassium content of the scion. Trifoliata and most of the citranges appeared to be less efficient in potassium absorption.

E. Edible Nuts

The temperate zone edible nuts include almond, walnut, pecan, and chestnut. Most of the edible nuts in the USA are grown in the western part of the country, but pecan production is centered in the Gulf states and adjacent areas. Literature dealing with potassium nutrition of nut crops is scarce. Potassium deficiency symptoms are usually exhibited by a marginal scorch of leaves and a drooping and rolling of leaf blades which give affected trees a wilted appearance. In pecan, poor filling of nuts usually results from potassium deficiency (Sprague, 1964).

Gossard and Hammer (1962) found increased fruit production from potassium fertilization over a 6-year period on a potassium deficient soil planted to pecan. Leaf potassium was increased with potassium fertilization but was inversely influenced by crop size. Sharpe et al. (1954) observed in Florida greater cold-hardiness in pecan where high potassium rates were used. Leaf analysis has been used to evaluate the potassium status of commercial walnut orchards in California. (Proebsting, 1946). A range of leaf K content from 1.3-2.4% was found. For most western orchards planted to nut crops, potassium deficiency is not a common problem. In a few almond orchards showing potassium deficiency symptoms, light applications of K_2SO_4 have corrected the deficiency (Lilleland, O. 1951. *Unpublished,* cited from Fruit Nutrition. Chap. XI. p. 265. Somerset Press, Inc., Somerville, N. J.).

F. Olive

The olive is grown in the arid regions of temperate and subtropical zones of the world. It requires a winter chilling to produce inflorescences and flowers, but the tree itself can be killed by temperatures below 15F. Olive is adapted to a wide range of well-drained soils and is one of the few tree fruit crops that is tolerant to saline soils. Olive trees have a relatively high potassium requirement, and potassium deficiency has been reported in California (Lilleland, O., and J. G. Brown. 1947. *Unpublished.* Cited from Fruit Nutrition. Chap. X. p. 254. Somerset Press, Inc. Somerville, N. J.), France (Prevot and Buchmann, 1960), and Greece (Demetriades and Gavalas, 1963). Leaves with potassium deficiency usually show dead area at the tip and along the lateral margins. They also show a tawny, light green color somewhat resembling nitrogen deficient leaves. Under severe potassium deficiency, premature leaf drop and actual dying of branches may occur. In sand culture, Hartmann and Brown (1953) reported a marked reduction in total tree growth when potassium was omitted.

The relationship between fruit production and potassium has been reported by several workers. Caumel (1958) showed a positive correlation between yield and leaf potassium. Buchmann (1962), in a fertility study of olives, found potassium in the tree was closely related to high yield. This was noticeable in the second year of the 2-year production cycle, when the olives were being formed. Fox et al. (1964) compared the leaf and soil analyses of 11 olive orchards in Turkey selected on the basis of differences in yield. They found a close relationship between tree productivity and the potassium content of leaves. The highest yielding orchards had a leaf K content of about 1.50%, while the leaf K content in the lowest yielding orchards ranged from 0.70-0.95%. Neither nitrogen or phosphorus levels in the leaves were related to yield. They also found leaf composition to be better related to yield than the nutrient element levels in the soil.

The only work on the effects of potassium fertilization on fruit quality was reported by Hartmann (1953) in California when he compared potassium deficient trees with trees which received potassium fertilization. Potassium fertilized trees showed increased shoot growth, fruit size, yield, and had four times the number of canning sized fruit when compared to trees that did not receive potassium fertilization.

G. Peach

Peaches are one of the most widely grown tree fruits in the world. They are native to China but have been distributed and grown under widely different conditions in Europe, Asia, South Africa, South America, Australia, and almost every state in the USA. Among the major fertilizer elements, nitrogen has been the one most frequently needed in peach orchards. Use of potassium fertilizers has often produced response in peach trees, especially on light sandy soils of low organic matter in the humid regions. Potassium deficiency symptoms are most noticeable in late summer just prior to fruit harvest. The leaves are usually lighter green in color and soon changed to pale yellow. Affected leaves become necrotic at tips and margins, and finally curl upward and inward. Trees affected with potassium-induced leaf scorch usually have a greatly reduced set of fruit buds.

Potassium fertilizers usually have no effects on fruit production except where trees are severly deficient in potassium, but in fruit quality, the effects are much more pronounced. Brinkley et al. (1958) compared four rates of K_2SO_4 and found no difference in yield at the end of 4 years. The K content of leaves had a range of 2.90-4.41%. Cummings (1965) comparing KCl and magnesium fertilizers found no difference in the yield of Elberta peaches during the first 5 years but showed a significant increase thereafter. The increase was both in the number of fruit and in size and weight of fruit for potassium fertilizer but only in size and weight of fruit for magnesium. Red color of fruit was improved by potassium fertilization while magnesium diminished the redness. Lilleland et al. (1962) in California using different rates of K_2SO_4 obtained a positive relationship between fruit size and K content of leaves up to 1.0%. No relationship was observed at higher leaf potassium content. Kwong and Fisher (1962), treating potassium deficient trees with K_2SO_4 in New York, found improved leaf color, shoot growth, and fruit size. Potassium fertilization also resulted in higher leaf potassium content and the titratable acid content. No response in the soluble solids content of the fruit was found. Stembridge et al. (1962) found desirable skin and flesh color of peaches was associated with low leaf nitrogen and high leaf potassium contents.

Several workers have reported the increase of cold resistance with potassium fertilization in peaches. McDaniel (1951) commenting on the winter injury in a long-term Elberta peach experiment in Illinois (Kelley, 1951), showed trees which received both potassium and nitrogen suffered less than trees that did not receive these elements. Also, trees that were fertilized with phosphorus but not with potassium had much greater wood damage than where both elements were applied. Trees receiving nitrogen and phosphorus but no potassium gradually declined over the course of the experiment and eventually were frozen to the ground. Scott and Cullinan (1946) in Maryland found the fruit buds of peach trees low in potassium were more susceptible to cold damage when artifically frozen than those on trees with adequate potassium supply. A similar finding was also reported by Archilbald (1965) in Canada under field conditions when more fruit buds from the high potassium trees survived the severe winter of 1962-63.

H. Pear

Most of the pears grown in the USA are horticultural varieties of the European species *Pyrus communis*. Like the apple, the pear is a temperate zone tree, but it is not as widely adaptable as the apple. This is because the important commercial varieties are less hardy than most apple varieties. Also, the susceptibility of pears to fire blight limits the industry to areas where the disease is not present or may be controlled by alert management.

Literature on potassium research in pear nutrition is very limited. Potassium deficiency symptoms in pears are essentially the same as in other deciduous fruit trees, involving a marginal leaf scorch appearing first on older leaves (Wallace, 1951). The levels of potassium present in leaves were comparable to those of the apple in commercial orchards (Kenworthy, 1950). Reuther (1941) found less than 0.5% K in leaves of Bartlett pear trees showing scorch, whereas healthy trees receiving manure contained more than 1% K. Lilleland and Urie (1966) using leaf analysis found several areas of low potassium in the pear producing districts of California. Potassium fertilization has resulted in an increase in leaf K content from 0.3-1.0%. It has increased fruit size, marketable fruit from 20-75%, and yield from 60-190 lb/tree. Fruit on potassium deficient trees matured earlier and showed lower pressure tests and higher total solids when compared to fruit from trees with adequate potassium. Increase in fruit size resulting from K fertilization was also reported by Fisher et al. (1959). The response corresponded to an increase in leaf K from 0.7-1.0%. No effect on fruit size was noted when the level of leaf K was close to 1% prior to fertilizer application.

I. Plum, Prune, and Apricot

The principal commercial varieties of plums are the European and the Japanese. Plums which have fruit of sufficiently high sugar content for drying are called prunes. The apricot is a native of Asia. It is grown in the USA but only in Washington, California, and Utah. Plums, prunes, and apricots will be considered in one section because there are many similarities among them and also the literature on potassium nutrition on these tree fruits is not plentiful.

The fruit of plum, prune, and apricot trees has a high requirement for potassium, and potassium deficiency has been related to excessive bearing in California (Lilleland and McCollam, 1961). Annual light applications of potassium fertilizer partially corrected leaf scorch and dieback but was not sufficient to correct the malnutrition in years of heavy fruit crop. A heavy application (25 lb/tree) of K_2SO_4, together with elimination of excessive bearing would control K deficiency for 5 years or more.

Several workers (Cain and Boynton, 1952; Sanvard, 1962; Johansson, 1962) in the USA and Europe reported plums and prunes have responded to moderate potassium fertilizer applications. The responses usually are represented by improved tree vigor, better bloom and yield, and increased fruit size. Wade (1956) in Tasmania found a relation between the potassium status of apricot tree and its susceptability to brown rot. An annual application of 2 lb KCl/tree was effective in controlling the disorder.

Leaf analysis is used extensively to assess the potassium status of the tree. Ystaas (1966) found K deficiency symptoms appear at 0.6-0.8% leaf K and suggested 2.0% K as being optimum for plum trees. Lilleland (1961) maintained apricot leaves should have at least 2.0% K for sufficiency. For prunes and plums, the leaf K sufficiency level can be lower.

III. LEAF ANALYSIS AND POTASSIUM NUTRITION

The use of leaf analysis has been widely reported in literature dealing with nutritional and fertility research in tree crops in recent years. It has provided a vast amount of knowledge on the behavior of slow-growing tree crops in a relatively short time. Leaf analysis has provided a common basis for evaluating the potassium nutritional level and correlating this with tree growth, fruit production, and fruit quality in different soils and climate throughout the world.

Some of the reasons why leaf analysis is preferred over soil analysis by researchers in tree fruits are:

First, the widely spreading and deep tree root systems may extend into several different soil horizons, including those of the subsoil. The sum total of nutrients extracted from the entire soil profile penetrated by the roots is reflected in the above ground parts of the tree. Soil analysis of samples collected at certain, usually shallow, depths does not necessarily reflect the mineral content of the tree. Soil analyses have seldom been closely correlated to response of trees to fertilizers.

Second, most fruit trees are propagated vegetatively and are a combination of different scion and rootstock. The scion-rootstock combination often affects the absorption of nutrients from the soil.

Third, it is difficult to find an analytical method that is satisfactory to all soils when dealing with different soil types which contain different clays with different exchange capacities and varying degrees of base saturation.

Workers in tree crops have placed much emphasis on leaf analysis as a guide to nutrient status and fertilizer requirements. It should be emphasized, however, that leaf analysis is only one of the tools used in attempting to understand tree behavior. It is not the only tool. It should be used together with observation of leaf and fruit symptoms and with some soil analysis, together with all other historical information about prior fertilizer programs (Koo, 1962). Leaf analysis is futile if soil conditions, parasites, and diseases go uncontrolled or if soil, physical conditions, and environmental conditions do not permit good growth.

Extensive reviews of the literature on leaf analysis have been made by Chapman (1966); Kenworthy (1964); Kenworthy and Martin (1966); Bould (1966); and Smith (1966). Most of the reviews include discussions of methods of sampling, sample preparation, standards for mineral composition, and interpretation of results. Some of the points pertaining to potassium nutrition will be briefly discussed.

Leaf analysis can now be used to detect and diagnose potassium deficiency in nearly all fruit trees. This is especially important in crops such as the citrus where visible leaf symptoms of potassium deficiency do not appear until the trees are in extreme deficiency. Leaf analysis also gives important information even if the potassium level is above the deficient range, because the fruit quality of several tree crops is affected by potassium level over the entire range from deficiency to excess. One of the greatest values of leaf analysis is the possibility of early detection of developing deficiencies so that preventive fertilizer programs can be utilizeable rather than corrective. Thus, trends in the leaf potassium content, over a series of years, may determine if the supply of potassium is inadequate, satisfactory, or unnecessarily high. The study of Reitz and Koo (1960) showed how a downward nutritional trend can be detected before severe deficiency develops, enabling timely corrective measures to be taken.

In order to obtain valid leaf analyses, certain sampling and analytical procedures should be followed consistently. The potassium content in leaves of most tree crops decreases with age, so for comparative purposes, all samples should be of leaves approximately the same age. For most tree crops, sampling of 4-7 month old leaves is preferred. Sample preparation and analytical methods for leaf analysis are fairly well standardized.

Information for leaf potassium standards is plentiful in certain crops but lacking in others. Table 1 contains tentative potassium standards for several tree crops, compiled from literature reviewed in this chapter. To obtain data for these potassium standards, results from field experiments have been supplemented with results from solution and sand cultures. Surveys of com-

Table 1—Standards for K content of leaves of tree crops

Crop	Deficient less than	Low	Medium	High	Excess more than	Reference
			K, % dry weight			
Almond	0.50	--	0.70 - 1.80	--	--	Proebsting (1966)
Apple	0.70	0.70 - 1.10	1.10 - 1.70	--	--	Batjer & Degman (1940)
Apricot	1.00	1.50	2.00	--	--	Lilleland (1961)
Avocado	0.35	--	0.75 - 2.00	--	3.00	Goodall et al. (1965)
Cherry	0.75	--	1.00 - 2.20	--	--	Boynton (1944) Zubriski & Swingle (1950)
Citrus	0.70	0.70 - 1.10	1.20 - 1.70	1.80 - 2.30	2.40	Smith (1966) Young, et al. (1965)
Mango	--	0.30 - 0.45	0.45 - 0.65	--	--	Lilleland (1958)
Olive	0.30	0.70 - 0.95	1.10 - 1.50	--	--	Fox, et al. (1964)
Pear	0.50	--	0.70 - 2.10	--	--	Reuther (1941) Kenworthy (1950)
Peach	0.35	--	0.85 - 1.45	1.70 - 2.20	--	Lilleland, et al. (1962) Kwong & Fisher (1962)
Pecan	0.35	--	0.45 - 1.00	--	--	Hammar & Hunter (1949) Gossard & Hammar (1962)
Plum & Prune	0.75	--	1.50 - 2.00	--	--	McCollam (1953) Ystaas (1966)
Walnut	0.50	--	1.00	--	--	McCollam (1953)

mercial orchards provide some information on range of potassium content in leaves, but survey studies are often inconclusive and may even be misleading because most variables are not controlled. All final leaf analysis standards should be calibrated in long term, well designed, field fertility experiments. The lack of information from controlled fertility field experiments in certain tree crops made the information in Table 1 vague for some crops.

There are a number of factors that will affect the potassium content of leaves. In addition to the age of leaves, position on shoot, rootstock-scion combination, size of the fruit crop, and year to year variations may also effect the K content of leaves. Probably none of these factors is as important as the antagonistic effects among potassium, calcium, and magnesium on each other. Potassium deficiency in tree crops is sometimes induced by an inbalance of cation nutrients. High calcium and magnesium may cause potassium deficiency just as high potassium may induce magnesium deficiency. All of these factors should be considered in interpreting the results of leaf analysis for potassium.

In conclusion, it should be emphasized again that leaf analysis is a most useful tool, but its use should be supplemented by other tools to most accurately evaluate the potassium status of the tree.

REFERENCES

Archibald, J. A. 1965. N-K treatments for peaches. Better Crops with Plant Food. March-April. 49(2):30-32.

Barden, J. A., and A. H. Thompson. 1962. Effects of heavy annual application of potassium on Red Delicious apple trees. Amer. Soc. Hort. Sci. Proc. 81:18-25.

Batjer, L. P., and E. S. Degman. 1940. Effects of various amounts of nitrogen, potassium and phosphorus on growth and assimilation in young apple trees. J. Agr. Res. 60:101-116.

Batjer, L. P., and G. R. Magness. 1939. Potassium content of leaves from commercial apple orchards. Amer. Soc. Hort. Sci. Proc. 36:197-201.

Blake, M. A., G. T. Nightingale, and O. W. Dairdson. 1937. Nutrition of apple trees. New Jersey Agr. Exp. Sta. Bull. 626.

Bould, C. 1966. Leaf analysis of deciduous fruits. In Fruit nutrition. p. 651-684. Somerest Press, Somerville, N. J.

Boynton, D. 1944. Responses of young Elberta peach and Montmorency cherry trees to potassium fertilization in New York. Amer. Soc. Hort. Sci., Proc. 44:31-33.

Boynton, D., and G. H. Oberly. 1966. Apply nutrition. In Fruit nutrition. p. 1-50. Somerest Press, Somerville, N. J.

Brinkley, A. M., E. Rogers, and F. M. Green. 1958. Foliar analysis of Elberta peach in relation to K fertilization. Colorado Agr. Exp. Sta. Tech. Bull. 64.

Buchmann, E. 1962. La fumure de l'olivier. (The manuring of olive trees.) Potassium symposium. 1962. p. 497-589. (English summary). International Potash Inst., Berne, Switzerland.

Cain, J. C., and D. Boynton. 1952. Fertilization, pruning and mottled leaf condition in relation to the behavior of Italian Prunes in western New York. Amer. Soc. Hort. Sci., Proc. 59:53-60.

Caumel, E. 1958. L'elude de la fumure des oliveraies irriguees. (A study on the manuring of irrigated olives.) Fruits et Prim. 28:113-114. (Hort. Abstr. 29: 346).

Chapman, H. D. 1964. Potash in relation to California citrus. California Citrog. 49: 454-455, 464, 466, 468.

Chapman, H. D. 1966. Diagnostic criteria for plants and soils. p. 1-793. University of California.

Chapman, H. D., S. M. Brown, and D. S. Rayner. 1947. Effects of K deficiency and excess on orange trees. Hilgardia. 17:619-650.

Christensen, M. D., and D. R. Walker. 1964. A nutritional survey of sweet cherry orchards in Utah. Amer. Soc. Hot. Sci., Proc. 85:112-117.

Cline, R. A., and J. A. Archibald. 1965, 1966. The effect of potassium and phosphorus fertilizer on yield and quality of Montmorency cherry. Rept. Ont. Hort. Exp. Sta. Prod. Lab. p. 28-35.

Cummings, G. A. 1965. Effect of potassium and magnesium fertilization on the yield, size, maturity and color of Elberta peaches. Amer. Soc. Hort. Sci., Proc. 86:133-140.

Curwen, D. F., J. McArdle, and C. M. Ritter. 1966. Fruit firmness and pectic composition of Montmorency cherries as influenced by different NPK treatments. Amer. Soc. Hort. Sci., Proc. 89:72-79.

Davis, M. B., and H. Hill. 1941. Apple nutrition. Dom. Canada Dept. Agr. Pub. 714, Tech. Bull. 32.

Démétriadès, S. D., and N. A. Gavalas. 1963. La carence potassique de l'liver en Grèce (Potassium deficiency of olive trees in Greece.) Potassium symposium. 1962. p. 395-400. International Potash Inst., Berne, Switzerland.

Deszyck, E. J., and R. C. J. Koo. 1957. K fertilization in a mature Duncan grapefruit orchard. Proc. Florida Soil and Crop Sci. Soc. 17:302-310.

Deszyck, E. J., and R. C. J. Koo, and S. V. Ting. 1958. Effect of K on yield and quality of Hamlin and Valencia oranges. Proc. Florida Soil and Crop. Sci. Soc. 18: 129-135.

Embleton, T. W., and W. W. Jones. 1964. Avocado nutrition in California. Proc. Florida State Hort. Soc. 77: 401-405.

Embleton, T. W., W. W. Jones, and A. L. Page. 1964. Potash hunger. Better Crops with Plant Food. 48(4):2-5.

Embleton, T. W., W. W. Jones, and A. L. Page. 1966. Potassium deficiency in Valencia orange trees in California. Hort. Sci. 1:92.

Embleton, T. W., J. D. Kirkpatrick, W. W. Jones, and C. B. Cree. 1956: Influence of application of dolomite, potash and phosphate on yield and size of fruit on composition of leaves of Valencia orange trees. Amer. Soc. Hort. Sci., Proc. 67:183-190.

Embleton, T. W., M. Matsumura, W. B. Storey, and M. J. Garbers. 1962. Chlorine and other elements in avocado leaves as influenced by rootstock. Amer. Soc. Hort. Sci., Proc. 80:230-236.

Fisher, E. G. and S. S. Kwong. 1961. Effects of potassium fertilization on fruit quality of the McIntosh apple. Amer. Soc. Hort. Sci. Proc. 78:16-23.

Fisher, E. G., K. G. Parker, N. S. Luepschen, and S. S. Kwong. 1959. The influence of phosphorus, potassium, mulch and soil drainage on fruit size, yield and firmness of the 'Bartlett' pear and on the development of fire blight disease. Amer. Soc. Hort. Sci., Proc. 73:73-90.

Forshey, C. G. 1963. Potassium-magnesium deficiencies of McIntosh apple trees. Amer. Soc. Hort. Sci., Proc. 83:12-20.

Fox, R. L., A. Aydeniz, and B. Kacar. 1964. Soil and tissue tests for predicting olive yields in Turkey. Empire J. Exp. Hort. 32(125):84-91.

Furr, J. R., P. C. Reece, and F. E. Gardner. 1946. Symptoms exhibited by avocado trees grown in outdoor sand culture deprived of various mineral nutrients. Proc. Florida State Hort. Soc. 59:138-145.

Goodall, G. E., T. W. Embleton, and R. G. Platt. 1965. Avocado fertilization. California Agr. Ext. Ser. Leaf. 24.

Gossard, A. C., and H. E. Hammar. 1962. Some effects of potassium fertilization and sod culture on peach trees performance and nutrition. Amer. Soc. Hort. Sci., Proc. 81:184-193.

Haas, A. R. C. 1939. Avocado leaf symptoms, characteristic of potassium, phosphate, manganese and boron deficiency in solution cultures. California Avocado Year book 24:103-109.

Hass, A. R. C. 1948. Symptoms of low K in leaves of citrus. California Citrog. 33(12): 520, 530, 532, 534-535.

Hass, A. R. C., and F. F. Halma. 1931. Sap concentrations and inorganic constituents of mature citrus leaves. Hilgardia 5(13):407-424.

Hammer, H. E., and J. E. Hunter. 1949. Influence of fertilizer treatment on the chemical composition of Moore pecan leaves during nut development. Plant Physiol. 24:16-30.

Hartmann, H. T. 1953. Olive production in California. California Agr. Exp. Sta. Manual 7.

Hartmann, H. T., and J. G. Brown. 1953. The effect of certain mineral deficiencies on the growth, leaf appearance and mineral content of young olive trees. Hilgardia 22(3):119-130.

Hoblyn, T. N. 1940. Manurial-trials with apple trees at East Malling 1920-1939. J. Pom. Hort. Sci. 18:325-343.

Hammer, H. E., and J. E. Hunter. 1949. Influence of fertilizer treatment on the chemical composition of Moore pecan leaves during nut development. Plant Physiol. 24:16-30.

Jones, W. W., T. W. Embleton, M. J. Garber, and C. B. Cree. 1967. Creasing of orange fruit. Hilgardia 38 (6):231-244.

Johansson, E. 1962. Manurial trials with plums at Alnarp and Ugerup. Medd. Trädgardsfôrs. Alnarp 140: 18. (Hort. Abstr. 33:45).

Kadman, A. 1963. The uptake and accumulation of chloride in avocado leaves and the tolerance of avocado seedlings under saline conditions. Amer. Soc. Hort. Sci., Proc. 83:280-286.

Kelley, V. W. 1951. Peach fertilization tests. 1950. Trans. Illinois State Hort. Soc. 84:140-148.

Kenworthy, A. L. 1950. Nutrient-element composition of leaves from fruit trees. Amer. Soc. Hort. Sci., Proc. 55:41-46.

Kenworthy, A. L. 1964. Fruit, nut, and plantation crops, deciduous and evergreen. A guide for collecting foliar samples for nutrient-element analysis. Memo. Rept. Hort. Dept. Michigan State Univ., East Lansing, Mich.

Kenworthy, A. L., and L. Martin. 1966. Mineral contents of fruit plants. In Fruit nutrition. Chap. XXIV. pp. 813-870. Somerest Press, Inc. Somerville, N.J.

Koo, R. C. J. 1961. Potassium nutrition and fruit splitting of Hamlin orange. Florida Agr. Exp. Sta. Ann. Rept. 223-224.

Koo, R. C. J. 1962. The use of leaf, fruit, and soil analysis in estimating potassium status of orange trees. Proc. Florida State Hort. Soc. 75:67-72.

Koo, R. C. J. 1963. Nitrogen and potassium rates study on Lemon. Florida Agr. Exp. Sta. Ann. Rept. 233-234.

Koo, R. C. J., and W. F. Spencer. 1959. Potassium studies. Florida Agr. Exp. Sta. Ann. Rept. 1959. 208-209.

Kwong, S. S. 1965. Potassium fertilization in relation to titratable acids of sweet cherries. Amer. Soc. Hort. Sci., Proc. 86:115-119.

Kwong, S. S., and E. G. Fisher. 1962. Potassium effects on titratable acidity and the soluble nitrogenoeus compounds of Jerseyland peach. Amer. Soc. Hort. Sci., Proc. 81:168-171.

Lanford, L. R. 1939. Effect of potash on leaf-curl of sour cherry. Amer. Soc. Hort. Sci., Proc. 36:261-262.

Lilleland, O. 1961. Potassium deficiency of fruit trees in California. Proc. 15th Int. Hort. Congr. Nice, 1958. p. 168-172.

Lilleland, O., and M. E. McCollam. 1961. Fertilizing western orchards. Better Crops with Plant Food. 45(4):1-5, 46-48.

Lilleland, O., and K. Urie. 1966. Potassium responses in pear orchards in California. Proc. XVII. Int. Hort. Conq. 1:33.

Lilleland, O., K. Urie, T. Muraoka, and J. Pearson. 1962. Relationship of potassium in the peach leaf to fruit growth and size at harvest. Amer. Soc. Hort. Sci., Proc. 81:162-167.

Lynch, S. J. and S. Goldweber. 1954. Some effects of nitrogen, phosphorus and potassium fertilization on the yield, tree growth and leaf analysis of avocados. Proc. Florida State Hort. Soc. 67:220-223.

Lynch, S. J., and S. Goldweber. 1956. Some effects of nitrogen, phosphorus and potassium fertilization on the yield and tree growth of avocados. Proc. Florida State Hort. Soc. 69:289-292.

McCollam, M. E. 1953. Potassium status of fruit trees. Western fruit growers. 7(10): 31-32.

McDaniel, J. C. 1951. Peach tree post-mortem. Orchard topics. June. p. 1-2.

Mori, H., and Ta. Yamazuki. 1960. Studies on the potassium nutrition of apple trees. 1. Response of apple trees to different potassium supply in nutrient solution of water culture (English summary). Tohoku Nat. Agr. Exp. Sta. Bull. 18:44-56.

Mukherjee, S. K. 1958. The origin of mango. Indian J. Hort. 15:129-134.

Popenoe, J., P. G. Orth, and R. W. Harkness. 1961. Leaf analysis survey of avocado groves in Florida. Proc. Florida State Hort. Soc. 74:365-367.

Prevot, P., and B. Buchmann. 1960. Foliar diagnosis of irrigated olive trees. Evolution of concentrations in the course of the year. Fertilite 10:3-11.

Proebsting, E. L. 1946. Edible nuts. In Fruit nutrition. p. 262-275. Somerest Press, Somerville, N. J.

Rasmussen, P. M. 1966. Forskellige naeringsstoffers indflydelse pä frugtens holdbarhed pä 1 ager. (The effects of various nutrients on the storage quality of fruits.) Tiddsskr Planteavl. 1966. 69:532-537. (English summary) (Hort. Abstr. 37:48).

Reitz, H. J., and R. C. J. Koo. 1960. Effect of N and K fertilization on yield, fruit quality and leaf analysis of Valencia oranges. Amer. Soc. Hort. Sci., Proc. 75: 244-252.

Reuther, W. 1941. Studies concerning the supply of available potassium in certain New York orchard soils. Cornell Agr. Exp. Sta. Memo. 241.

Roy, R. S., R. C. Mallik, and B. N. De. 1951. Manuring of the mango. (Mangifera Indica, Linn.) Amer. Soc. Hort. Sci., Proc. 57:9-16.

Sanvard, K. 1962. Manurial trials with cherries and plums (English summary). Tidskr. Planteavl. 66:609-642. (Hort. Abstr. 33:241).

Scott, D. H., and F. P. Cullinan. 1946. Some factors affecting the survival of artifically frozen fruit bud of peach. J. Agr. Res. 73:207-236.

Semakin, K. S., Ye. S. Moroz, and V. K. Abashkin. 1937. Effects of potassium fertilizer on citrus hardiness. Sov. Subtropiki, no. 12(40), 71-74. Cited by I. M. Vasilyev. 1961. Wintering of plants. Roger and Roger, Inc. Washington, D.C.

Shadmi, A., M. Hofman, and J. Hagin. 1966. The effect of potassium fertilization in an apple orchard. Potash Rev. Subj. 8. Dec.

Sharpe, R. H., G. H. Blackman, and N. Gammon, Jr. 1954. Relation of potash and phosphate to cold injury of Moor pecan. Better Crops with Plant Food. 38(1): 17-18, 48-49.

Sites, J. W., and E. J. Deszyck. 1952. The effect of varying amounts of K on the yield and quality of Valencia and Hamlin oranges. Proc. Florida State Hort. Soc. 68:65-72.

Smith, P. F. 1966. Leaf analysis of citrus. In Fruit nutrition. p. 208-228. Somerest Press, Somerville, N. J.

Smith, P. F., and G. K. Rasmussen. 1958. Relation of fertilization to winter cold injury of citrus. Proc. Florida State Hort. Soc. 71:170-175.

Smith, P. F., and G. R. Rasmussen. 1960. Relationship of fruit size, yield and quality of Marsh grapefruit to K fertilization. Proc. Florida State Hort. Soc. 73:42-49.

Smith, P. F., and G. R. Rasmussen. 1961. K deficiency symptoms in grapefruit under field conditions in Florida. Amer. Soc. Hort. Sci., Proc. 78:169-173.

Smith, P. F., and W. Reuther. 1949. The influence of rootstock on the mineral content of Valencia orange leaves. Plant Physiol. 23:455-461.

Smith, P. F., W. Reuthen, and G. T. Scudder, Jr. 1953. Effects of differential supplies of N, K, and Mg on growth and fruiting of young Valencia orange trees in sand culture. Amer. Soc. Hort. Sci., Proc. 61:38-48.

Smith, P. F., and G. T. Scudder, Jr. 1951. Some studies on mineral deficiency symptoms in mango. Proc. Florida State Hort. Soc. 64: 243-248.

Sprague, H. B. 1964. Hunger signs in crops. David McKay Co., Inc., New York, N. Y.

Stembridge, G. E., C. E. Gambrell, H. J. Sefick, and L. O. van Blaricom. 1962. Effects of high rates of nitrogen and potassium on the performance of peaches. Amer. Soc. Hort. Sci., Proc. 153-161.

Vasilyev, I. M. 1961. Wintering of plants. Royer and Roger, Inc., Washington, D.C.

Wade, G. C. 1956. Investigations on brown rot of apricots caused by *Sclerotinia fruiticola* (Wint.) Rehm: II. The relationship of the potassium status of apricot trees to brown rot susceptibility. Australian J. Agr. Res. 7:516-526.

Wallace, A., C. J. Naude, R. T. Mueller, and Z. I. Zidan. 1952. The rootstock-scion influence on the inorganic composition of citrus. Amer. Soc. Hort. Sci., Proc. 59:133-142.

Wallace, T. 1928-29. Leaf scorch on fruit trees. J. Pom. Hort. Sci. 7:1-31.

Wallace, T. 1951. The diagnosis of mineral deficiencies in plants by visual symptoms. A color atlas and guide. H. M. Stationary Office, London. 1943. (Revised 1951).

Weeks, W. D., F. W. Southwick, M. Drake, and J. E. Steckel. 1958. The effects of varying rates of nitrogen and potassium on the mineral composition of McIntosh foliage and fruit color. Amer. Soc. Hort. Sci., Proc. 71:11-19.

Westwood, M. H. 1966. Cherry nutrition. In Fruit nutrition. p. 158-173. Somerest Press, Inc., Somerville, N. J.

Young, T. W. and R. C. J. Koo. 1967. Effects of nitrogen and potassium fertilization on Persian limes on Lakeland fine sand. Proc. Florida State Hort. Soc. 80:337-342.

Young, T. W., R. C. J. Koo, and J. T. Muier. 1962. Effects of nitrogen, potash and calcium fertilization on Kent mangoes on deep acid sandy soil. Proc. Florida State Hort. Soc. 75:364-371.

Young, T. W., R. C. J. Koo, and J. T. Muier. 1965. Fertilizer trials with Kent mango. Proc. Florida State Hort. Soc. 78:369-375.

Ystaas, J. 1966. (Fertilizer trials with potassium and nitrogen on plum trees.) Forsk, Fors. Landbr. 17:281-295. (Soil and Fert. 30(3):310).

Zubriski, J. C., and C. F. Swingle. 1950. Potassium content of Montmorency cherry leaves in relation to curl-leaf and to exchangeable soil potassium. Amer. Soc. Hort. Sci., Proc. 56:34-39.

22

Potassium Nutrition of Vegetable Crops

ROBERT E. LUCAS

Michigan State University
East Lansing, Michigan

I. INTRODUCTION

The use of potassium fertilizers for vegetables has long been a recognized necessity. Over 124,000 metric tons of K were applied on about 2 million ha in the USA in 1964. The average rate of application was about 100 kg/ha (90 lb/acre).

Vegetable growers use potassium fertilizers where needed to improve yields, quality of produce, color, strength of stalk, disease resistance and cold tolerance.

In this paper, the following topics will be discussed: (i) Potassium composition and uptake of some vegetable crops, (ii) response of vegetables to potassium when grown on several soil types, and (iii) effect of excess potassium on vegetables.

II. POTASSIUM COMPOSITION AND UPTAKE

Vegetables include crops that have a broad range of potassium uptake. We can site the extremes of potassium requirement of celery and radish. Information about potassium nutrition for certain vegetables is meager. On the other hand, numerous papers have been published on such crops as beans, tomatoes, and potatoes.

Vegetables can have as much as a two-fold difference in composition in which there is no apparent difference in quality or quantity of the product. Fortunately, we have such elasticity in composition so as to help compensate for varietal, soil, weather, and fertility differences. Data in Table 1 report

Table 1—The approximate K utilization by vegetables and some suggested percent K sufficiency ranges of dry petioles or midribs near harvest

Vegetable	Good yield--fresh weight	Percent K sufficiency range	Total K uptake
	tons/ha		kg/ha
Asparagus	4	1.5-2.5	70
Snap beans	7	2-3	60
Lima beans	3	2-3	80
Broccoli	13	3-5	125
Cabbage	45	2.5-4	100
Cauliflower	22	3-5	150
Carrots	45	3-5	250
Celery	70	5-8	500
Cucumbers	22	2.5-4	90
Leaf lettuce	33	2.5-4	100
Head lettuce	60	2-4	100
Melons	15	4-8	80
Onion bulbs	60	3-4	150
Parsnips	30	2-4	70
Peas	4	2-3	60
Peppers	4	2-3	60
Irish potatoes	40	6-9	400
Sweet potatoes	20	3-5	120
Sweet corn	10	1.5-2.5	120
Table beets	30	3-5	160
Turnips	20	2-3	100
Field tomatoes	70	3-5	300
Greenhouse tomatoes	140	4-7	500

some sufficiency values in tissue analyses and the approximate utilization of potassium by the plant. Much of the data was obtained from a compilation by Knott (1957).

Howard et al. (1962) have reported on the nutrient content for the edible portion of fresh grown vegetables for California. The potassium content usually ranged from 0.2 to 0.35% K. Head lettuce, peppers, onion bulbs, and cucumbers ranged from 0.1 to 0.2% K. Sims and Volk (1947) reported on the dry weight basis for Florida conditions. The K content for cabbage heads averaged 3.0%; green beans, 2.9%; celery, 7.0%, and tomatoes, 4.5%. James (1951) reported broccoli contained 3.6% K and collards 3.3% K. Fleming (1956) reported that the edible parts of cabbage tested 2.5%, snap beans, 2.3%, and sweet potatoes, 0.9% K (dry weight basis). Carpenter and Murphy (1965) report that dried potato tops near harvest tested 5% K and the tubers 2.0% K. The plant uptake of K by the tops and tubers was about 1 kg/100 kg of fresh tubers. The literature contains numerous other reports on nutrient composition with similar potassium values.

III. PLANT RESPONSE TO POTASSIUM FERTILIZER

Obviously the need for potassium fertilizer for crops is related to the supplying ability of the soil. Many soils, particularly in arid regions, are well

Table 2—Effect of K fertilizer on exchangeable K in the soil and the yield of pickles growing on a Hillsdale sandy loam (3-year average, 1958-60)

Average annual K	Average yield	K soil test at end of 3rd year
kg/ha	tons/ha	pp2m
0	18.6	78
25	20.6	96
66	23.8	110
132	24.3	158

supplied with potassium reserves. Numerous fertilizer trials are carried out over the nation. Reports on the results generally are not published in society proceedings but are reported at state fertilizer conferences and in experiment station reports and extension publications. This is the situation in Michigan. The tests are necessary to keep us updated on changing technology and for making more reliable fertilizer recommendations.

One good example of changing technology is pickle production. Formerly, this crop was planted in 1.5-2 meter row spacings and was hand harvested about 10 times during a season. Growers are now changing to close row spacing, high plant population, semidwarf hybrids, and a once-over machine harvest. With wide rows, fertilizer efficiency can be greatly increased by band placement. Today, with close row spacings, growers prefer to broadcast most of their fertilizer. Our recommendations now need to be increased about two-fold.

The data in Table 2 illustrate a typical response of pickles to potassium when grown with 1.2-meter row spacings. At the start of the trials in 1958, the soil tested 90 pp2m of exchangeable K. Except for the "O" treatment, all plots received 25 kg K/ha (22 lb/acre) in a band. The additional K needed for the 66 and 132 (60 and 120 lb) rates was applied broadcast. The yield response on this soil proved profitable up to about 100 kg K/ha (90 lb/acre).

Numerous fertilizer trials have been carried out on potatoes. When soil tests for sandy loams are less than 300 pp2m of exchangeable K, we can expect response to K fertilization. Typical results of trials in Michigan are reported in Table 3.

The data in Table 3 report on the effect of placement and rate on specific gravity. Normally in placement studies, we prefer up to 80 kg K/ha (72 lb/acre) in a band. If additional amounts are needed, then the potassium should be applied broadcast. Excessive rates of potassium cause yield depression and lower the specific gravity reading.

In the last decade tomato yields have nearly doubled. This increased production has greatly upped potassium needs. The response and need is similar to potatoes. A physiological fruit disorder called "grey wall" has become serious and is related to low potassium levels.

Table 3—Effect of rate and method of K application on yield and specific gravity of Sebago potatoes on Hodunk sandy loam in 1967 (Unpublished data by Valverde and Doll, Michigan State Univ.)*

Rate of K kg/ha	Method of application	Yield tons/ha	Specific gravity
0	--	28.0	1.084
50	Banded	39.2	1.083
100	Banded	36.6	1.081
200	Banded	42.4	1.079
400	Banded	39.7	1.072
100	Broadcast	37.0	1.080
200	Broadcast	47.4	1.079
400	Broadcast	45.0	1.074

* Soil test: pH (water), 5.9; P (Bray P_1), 90 pp2m; K (exchangeable), 150 pp2m; Ca (exchangeable), 646 pp2m; Mg (exchangeable), 131 pp2m.

An interesting experiment was carried out by the writer and John Downes of the Horticultural Department on tomatoes growing on a Genesee sandy clay loam at the Sodus Horticultural Farm in Berrien County, Michigan. The original purpose of the experiment was to correlate potassium levels by soil test and petiole analysis with quality and yield. The location was a river bottom alluvial soil, which is not a typical soil used for vegetable production. The site was chosen because potassium soil tests were low and water from the St. Joseph River was handy for irrigation.

In the experiment, plots received broadcast up to 500 pp2m of K in the fall of 1966. The field was plowed shallow so as not to bury the potassium. Both transplants and direct seeded Heinz 1350 tomatoes were grown. Plots were replicated four times and received ample nitrogen and phosphorus. Petiole samples were collected at 4-week intervals late in June, July, and August. The portion used was the top, fully-developed petiole from which the leaflets were stripped. The plant material was oven-dried, ground, and potassium determined with a flame photometer. The results of the experiment are illustrated in Fig. 1.

Normally an application of 500 pp2m of K is excessive for tomatoes. Yet on the Sodus Farm experiment, by late August all petioles contained less than 0.5% K. Wilcox (1964) reports that 2.3% K is minimal for top production. A study of the yield trends also indicates that the tomatoes could have used more potassium fertilizer.

Soil samples were collected from each plot and were tested for exchangeable potassium, calcium, and magnesium by the 1N ammonium acetate method. The soil samples were collected in May, 1967 before the tomatoes were planted and again in October after harvest. The data in Table 4 show the results. "Grey wall" disorder and cracked shoulders in the fruit account for most of the culls.

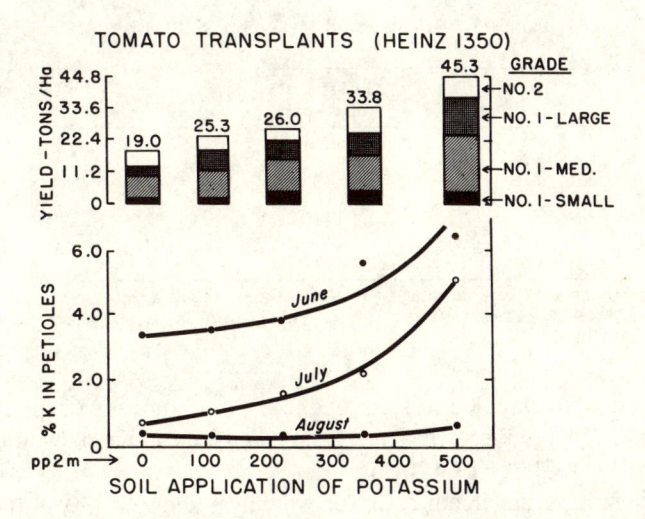

Fig. 1. Effect of broadcast K fertilizer on petiole composition, yield and grade for irrigated tomatoes growing on a Genesee sandy clay loam that tested 104 pp2m in exchangeable K.

Table 4—Effect of K rates on soil tests, yields, and cull tomato fruit when grown on a Genesee sandy clay loam*

K applied	K soil test--pp2m Before planting	After harvest	Total yield	Percent culls in fruit
kg/ha			tons/ha	
0	104	98	36.7	48
33	112	106	38.7	44
110	124	116	43.4	41
220	124	118	42.2	32
350	130	124	51.6	34
500	144	132	58.5	22

* The soil is alluvial and tested 5000 pp2m exchangeable calcium and 400 pp2m of magnesium and had a pH of 6.8-7.2. Mechanical analysis of the soil showed 31% clay, 18% silt, and 51% sand. About 60% of the clay fraction was vermiculite material.

The surprising result of the soil test was the low recovery of exchangeable potassium. For the 500 pp2m application the soil test changed from 104 to 144 pp2m. This amounts to only 8% recovery.

To substantiate the field tests, wetting and drying tests were carried out in the laboratory with 125, 250, and 500 ppm of K treatments. The recovery was 16% after the first incubation period for all three K levels. The recovery on the second wetting and drying was 10% and 8% for subsequent incubation periods. Thus laboratory studies confirmed the field tests.

Further laboratory evaluation showed that the clay fraction was mostly vermiculite which accounted for the high fixation of the potassium. For high

Table 5—Effect of K rates on soil tests*

K applied	Soil K test at harvest†	Percent recovery of K
pp2m	pp2m	
0	90	--
50	111	42
100	144	54
200	243	76
400	400	78

* Soil -- Granby sandy loam, a humic gley soil; Crop - head lettuce.
† Soil test - pH = 6.2, organic matter 10%, exchangeable Ca, 9,000 pp2m and Mg, 824 pp2m.

tomato production clay loam soils ought to test above 250 pp2m of exchangeable K. To obtain such a level on the Sodus Field would require about 3,800 kg KCl/ha (3,420 lb/acre).

Generally most sandy loam soils for Michigan show 30-50% recovery of applied K as determined by the ammonium acetate method. A soil that showed high recovery was a Granby sandy loam. Data in Table 5 report the results. In the test, the KCl was applied broadcast and disked in before planting.

Organic soils are popular for growing certain vegetables in the northern states and in Florida. Vegetables commonly grown are celery, onions, lettuce, carrots, radishes, and potatoes. Organic soils have low fixation properties. They also hold exchangeable potassium less strongly than clay minerals. However, in their natural state organic soils are notoriously deficient in potassium. The potassium requirements for vegetables have been fairly well determined. Table 6 reports yields for several crops growing on plots which annually received the same potassium rate. Approximately 200 kg K/ha (216 lb K_2O/acre) are needed annually for most vegetables. An exception would be celery, which requires about 400 kg K (432 lb K_2O) annually.

Table 6—Effect of K fertilizer on the yield of vegetables grown on Houghton muck, Michigan Muck Exp. Farm

K applied annually*	K soil test- ppm at end of 15th year†	Yield -- tons/ha				
		Potatoes	Carrots	Cabbage	Cucumbers	Table beets
kg/ha						
0	72	5.4	18.0	2.6	17.2	4.6
83	112	26.8	48.6	44.8	48.2	23.0
166	260	31.6	53.2	48.2	57.8	27.8
249	728	33.0	54.2	48.8	64.0	30.8
332	880	32.8	54.4	48.6	--	31.6
No. years in test		(8)	(5)	(4)	(1)	(3)

* All plots received 44 kg/ha of phosphorus annually.
† Soil on this farm has a bulk density of about 0.3g/cc.

Table 7—Effect of K rates on soil tests before and after onions were grown on organic soils in Ohio*

Plot no.	K soil test in spring	K applied	Total soil K plus K fertilizer	K soil test at harvest	K change	Yield
	kg/ha	kg/ha		kg/ha		kg/plot
			1964			
1	244	0	244	297	+ 53	252
2	323	168	491	500	+ 9	253
3	494	578	1,072	1,101	+ 23	258
4	633	986	1,619	1,335	−284	245
5	690	1,155	1,845	1,448	−397	220
			1965			
1	214	0	214	294	+ 70	267
2	392	168	560	506	− 54	296
3	771	578	1,349	1,088	−261	260
4	1,192	986	2,178	1,512	−666	244
5	1,385	1,155	2,540	1,500	−1,040	216

* Data from Hort. Mimeo Series 321, Ohio State Univ., 1966.

Table 6 also reports the soil test at the end of 15 years of cropping. Similar values were obtained at the end of the 5th and 10th year. Thus, leaching losses and crop removal had established a balance.

Work at the Ohio State Muck Experimental Farm at even higher potassium rates shows a remarkable leveling off of soil test values for each rate of application. Values in Table 7 illustrate the relationship. Note the similar soil test values at harvest for both 1964 and 1965. One would expect a marked increase when 1,155 kg K were applied to a soil that tested 1,385. Yet the soil tests for 1964 and 1965 are nearly identical. The loss of 1,040 kg K through leaching and crop removal certainly points out highly inefficient use. Yet there are soil testing organizations recommending such levels simply because organic soils have high exchange capacity values (150-180 meq/100 g). With the possible exception of celery, we have not observed K response to vegetables growing on organic soils when the soil test exceeds 400 kg/ha (800 ppm).

Greenhouse production of tomatoes, lettuce, and cucumbers is an intensive form of agriculture. Yields of 125-150 tons tomatoes/ha are not uncommon. Such production requires over 400 kg K/ha (432 lb K_2O/acre). Soil test levels to meet such demands usually need to exceed 500 pp2m of exchangeable K.

IV. EFFECT OF EXCESS POTASSIUM ON VEGETABLES

When suggesting rates of fertilizers, one needs to place himself in the position of a vegetable grower. Generally, when it comes to such items as potassium, the grower wants to use liberal rates. His other inputs are too great to chance a potassium deficiency reducing yields. Thus, excessive rates of potassium can be a factor. Some of the excess amounts come about because of long time applications of rates based on soils of low fertility. Periodic soil and plant tests can help one in correcting the rate.

Excess soil potassium sometimes indicates too much salt. A soluble salt determination should be made. Spurway (1943) reported that greenhouse soils testing over 300 ppm were generally considered excessive. In recent years where K carriers low in chloride and sulfates were used, soil tests over 500 ppm (Lucas et al., 1960) were acceptable providing Ca and Mg were in balance.

Such disorders as magnesium and calcium deficiency in vegetables are often related to the potassium nutrition. This is because of the reciprocal relationship of the cations in plant tissue. Some disorders in vegetables which are related to potassium nutrition are:

1) *Blossom End-Rot in Tomatoes*—This disorder was described by Geraldson (1957) to be related to cation balance and calcium deficiency. According to Ward (1963), when the fifth leaf was less than 1% Ca, the problem would be induced.

2) *Blossom End-Rot of Peppers*—Hamilton and Agle (1962) report that the fruit needs to contain over 0.24% Ca to prevent the disorder.

3) *Black Heart in Celery*—This disorder was first identified as calcium deficiency by Geraldson (1954). Bergman (1960) reports that when mineral soils contained over 425 pp2m K celery would show much higher incidence of blackheart. The critical point between no symptoms and the disorder was when the potassium level was above 18% of the sum total of the available calcium, magnesium, and potassium (weight basis).

4) *Inner Tipburn of Cabbage*—Walker et al. (1961) found that tipburn of the inner leaf marginal tissue of tipburn cabbage was consistently lower in calcium and higher in potassium. Excess soil moisture accentuated the disorder probably by rapid movement of potassium as compared to the relative slow movement of calcium ions.

5) *Carrot Cavity Spot*—This disorder was present in carrots when Ca in the petiole tissue was less than 1% and potassium over 6% (Maynard, et al., 1963).

6) *Brown Checking in Celery*—According to Yamagucki et al. (1958), low boron and high potassium levels in petioles produced a brown checking in celery stalks.

7) *Brownheart in Escarole*—This problem was associated with the crop when calcium was below 0.4% in the tissue (Maynard et al., 1962).

8) *Magnesium Yellowing in Celery*—Johnson et al. (1957) reported that potassium fertilizer individually and jointly with sodium fertilizer tended to increase the incidence of magnesium deficiency. Burdine (1959) also showed that high levels of potassium and calcium significantly increased magnesium chlorosis of celery.

Potassium has a marked interrelationship to sodium particularly for such vegetables as table beets, spinach, Swiss chard, and celery. This relationship was well illustrated by Harmer and Benne (1945).

Some of their values for celery are:

Fertilizer treatment	Cation content of plant				
	Ca	Mg	Na	K	Total
	meq/100g				
0-8-0 + NaCl(500)	110	36	172	16	334
0-8-24	166	28	27	63	284
0-8-24 + NaCl(500)	129	22	106	59	316
0-8-24 + NaCl(1000)	124	18	130	67	339
0-8-48	139	23	36	112	310

Semb (1965) reports Mg deficiency symptoms are common in Norway when soils are acid, have high K values, and have less than 30 ppm exchangeable Mg. If soils exceed 60 ppm Mg, symptoms occur only if K is excessive. These values are in close agreement with numerous unpublished field trials in Michigan. Excessive K levels occur when K exceeds Mg (equivalent basis). Bear et al. (1961) report the minimum value for exchangeable Mg to be about 6% of the exchange capacity. This is a much higher requirement.

Excessive rates of potassium can be a factor in crop production, but for most situations, the growers are not using enough for full production. Field trials and soil and plant tests need to be followed in order to improve the use of potassium materials for vegetable crops.

REFERENCES

Bear, F. E., A. L. Prince, S. J., Toth, and E. R. Purvis: Magnesium in plants and soils. New Jersey Agr. Exp. Sta. Bul. 760.

Bergman, E. L. 1960. Celery blackheart and its control in Pennsylvania. Pennsylvania Agr. Exp. Sta. Progress Rept. 215.

Burdine, H. W. 1959. A study of magnesium deficiency chlorosis in certain varieties of green celery. Amer. Soc. Hort. Sci., Proc. 74:514-525.

Carpenter, P. N., H. J. Murphy. 1965. Effects of accumulated fertilizer nutrients in a Maine soil upon the yield, quality and nutrient content of potato plants. Maine Agr. Exp. Sta. Bull. 634.

Fleming, J. W. 1956. Factors influencing the mineral content of snapbeans, cabbage and sweet potatoes. Arkansas Agr. Exp. Sta. Bull. 575.

Geraldson, C. M. 1954. The control of blackheart of celery. Amer. Soc. Hort. Sci. Proc. 63:353-358.

Geraldson, C. M. 1957. Factors affecting calcium nutrition of celery, tomato and pepper. Soil Sci. Soc. Amer. Proc. 21:621-625.

Hamilton, L. C., and W. L. Agle. 1962. The influence of nutrition on blossom end rot of pimento peppers. Amer. Soc. Hort. Sci. Proc. 80:457-461.

Harmer, P. M. and E. J. Benne. 1945. Sodium as a crop nutrient. Soil Sci. 60:137-148.

Howard, F. D., J. H. McGillivray, and M. Yamaguchi. 1962. Nutrient composition of fresh-grown California vegetables. California Agr. Exp. Sta. Bull. 788.

James, B. E. 1951. Composition of Florida grown vegetables. Florida Agr. Exp. Sta. Bull. 488.

Johnson, K. E. E., J. F. Davis, and E. J. Benne. 1957. Control of magnesium deficiency in Utah 10B celery grown on organic soil. Soil Sci. Soc. Amer. Proc. 21:528-532.

Knott, J. E. 1957. Handbook for Vegetable Growers. John Wiley and Sons, Inc.

Lucas, R. E., S. H. Wittwer, and F. G. Teubner. 1960. Maintaining high soil nutrient levels for greenhouse tomatoes without excess salt accumulation. Soil Sci. Soc. Amer. Proc. 24:214-218.

Maynard, D., N. B. Gersten, and H. F. Vernell. 1962. The cause and control of brownheart of escarole. Amer. Soc. Hort. Sci., Proc. 8:371-375.

Maynard, D., N. B. Gersten, and H. F. Vernell. 1962. The cause and control of of plant maturity on the calcium level and the occurrence of carrot cavity spot. Amer. Soc. Hort. Sci., Proc. 83:506-510.

Semb, G. 1965. Magnesium deficiency and magnesium content of Norwegian soils. Sci. Rept. Agr. College of Norway 44:NR19.

Sims, G. T., and G. M. Volk. 1947. Composition of Florida grown vegetables. Florida Agr. Exp. Sta. Bull. 438.

Spurway, C. H. 1943. Soil fertility control for greenhouses. Michigan State College Spec. Bull. 325.

Walker, J. C., C. L. V. Edington, and M. U. Nayuder. 1961. Tipburn of cabbage—nature and control. Wisconsin Agr. Exp. Sta. Res. Bul. 230. 1963.

Ward, G. M. 1963. Application of tissue analysis to greenhouse tomato nutrition. Amer. Soc. Hort. Sci. Proc. 83:695.

Wilcox, G. E. 1964. Effect of potassium on tomato growth and production. Amer. Soc. Hort. Sci. Proc. 85:484-489.

Yamagucki, M., F. D. Howard, and P. A. Minges. 1958. Brown checking of celery, a symptom of boron deficiency. Amer. Soc. Hort. Sci. Proc. 71:455-467.

GLOSSARY

Common and Scientific Names of Crops Referred to in this Book

alfalfa	*Medicago sativa* L.
almond	*Prunus amygdalus*
apple	*Malus sylvestris* Mill.
apricot	*Prunus armeniaca*
artichoke	*Cynara scolymus*
asparagus	*Asparagus officinalis*
avocado	*Persea americana* Miller
banana	*Musa paradisiaca sapientum*
barley	*Hordeum vulgare* L.
bean, bush	*Phaseolus vulgaris* L.
bean, lima	*Phaseolus lunatus*
bean, snap	*Phaseolus vulgaris* L.
bean, velvet	*Mucuna deeringiana* Small
beet, edible	*Beta vulgaris* L.
beet, sugar	*Beta vulgaris* L.
bentgrass, creeping (Seaside)	*Agrostis palustris* Huds. var. Seaside
bermudagrass	*Cynodon dactylon* (L.) Pers.
bluegrass	*Poa pratensis* L.
broccoli	*Brassica oleracea* var. *botrytis*
bromegrass, smooth	*Bromus inermis*
brussel sprouts	*Brassica oleracea* var. *gemmifera*
buckwheat	*Fagopyrum esculentum*
cabbage	*Brassica oleracea* var. *capitata*
carrots	*Daucus carota* L.
cauliflower	*Brassica oleracea* L.
cherry, sour	*Prunus cerasus* L.
cherry, sweet	*Prunus avium* L.
chestnut	*Castanea dentata*
chive	*Allium schoenoprasum*
clover, ladino	*Trifolium repens* L.
clover, red	*Trifolium pratense* L.
clover, white	*Trifolium repens*
celery	*Apium graveolens* L.
chard	*Beta vulgaris* var. *cicla*
cocoa	*Theobroma cacao* L.
coconuts	*Cocos nucifera*
coffee	*Coffea arabica* L.
collards	*Brassica oleracea* var. *acephala*
corn	*Zea mays* L.
cotton	*Gossypium hirsutum* L.
cucumber	*Cucumis sativa* L.
eggplant	*Solanum melangena* var. *esculentum*
endive	*Cichorium endivia*
escarole	*Cichorium endivis*
fescue, red (Pennlawn)	*Festuca rubra* (L.) var. Pennlawn
fescue, tall (Kentucky 31, Alta)	*Festuca arundinacea* Schreb.
flax	*Linum usitatissimum*
garlic	*Allium sativum*
gladiolus	*Gladiolus* spp.
grape	*Vitis vinifera*

grapefruit	*Citrus paradisi*
horseradish	*Amoracia lapathifolia*
kale	*Brassica oleracea* var. *acephala*
kohlrabi	*Brassica oleracea* var. *gongylodes*
larkspur	*Delphinum ajacis* L.
lemon	*Citrus limon*
lettuce	*Lactuca sativa* L.
lime	*Citrus aurantifolia*
lupine	*Lupinus* spp.
mandarin	*Citrus reticulata* Blanco
mango	*Mangifera indica* Linn.
melon	*Cucumis melo*
millet, foxtail	*Setaria italica*
mint	*Mentha* spp.
mustard spinach	*Brassica perviridis*
narcissus	*Narcissus* spp.
oats	*Avena sativa* L.
oil palm	*Elaeis guinenis* J.
okra	*Hibiscus esculentus*
olive	*Olea europaea*
onion	*Allium cepa* L.
orange, sour	*Citrus aurantium* Linn.
orange, sweet	*Citrus sinensis* Osbeck
orchardgrass	*Dactylis glomerata* L.
parsley	*Petroselinum crispum latifolium*
parsnip	*Pastinaca sativa*
pea	*Pisum sativum* L.
peaches	*Prunus persica* L.
pear	*Pyrus communis*
pecan	*Carya illinoensis* Wang.
pepper	*Capsicum* spp.
pineapple	*Ananas comosus*
plantain, banana	*Musa paradisiaca*
plum, European	*Prunus domestica*
plum, Japanese	*Prunus salicina*
potato	*Solanum tuberosum* L.
potato, sweet	*Ipomoea batatas* L.
pumpkin	*Cucurbita pepo*
quackgrass	*Agropyron repens* L.
radish	*Raphanus sativus* L.
rhubarb	*Rheum rhaponticum*
rice	*Oryza sativa* L.
rubbertree	*Hevea brasiliensis*
rutabaga	*Brassica napobrassica*
rye	*Secale cereale* L.
ryegrass	*Lolium multiflorum*
ryegrass, Italian	*Lolium multiflorum*, Lam. Tifton #1
safflower	*Carthamus tinctorius* L.
shallot	*Allium ascalonicum*
sorghum	*Sorghum bicolor* Pers.
soybean	*Glycine max* L.
spinach	*Spinacia oleracea*
squash, summer	*Cucurbita moschata*
strawberry	*Fragaria ananassa*
sudangrass, Piper	*Sorghum sudanensis,* Stapf.
sugarcane	*Saccharum officinarum* L.
tea	*Camillia sinensis*
timothy	*Phleum pratense* L.
tobacco	*Nicotiana tabacum* L.

tomato	*Lycopersicon esculentum* Mill.
trefoil, birdsfoot	*Lotus corniculatus* L.
turnip	*Brassica rapa* L.
walnut	*Juglans regia*
wheat	*Triticum (aestivum L.) spp.*
wheat, winter	*Triticum vulgare*

SUBJECT INDEX

Amorphous material, 82
Anoxia, 213
Apples
  K deficiency of, 470-471
  K effect on acidity, 471
  K uptake in, 471
Apricot
  K deficiency symptoms, 480
ATPase, 206-208
Avocado
  K deficiency symptoms, 472

Bananas, 407-412
  absorption of K, 408
  effects of K on, 410
  K deficiency in, 229-230, 408
    diseases caused by, 229-230
  use of K on, 411-412
Barley
  K deficiency in, 225
    diseases caused by, 225
Bermuda leaf spot disease, 235
Blossom end-rot, 496
Body fluids
  concentration of K in, 205-206
Brines, 17-18, 19, 29-32
Brown patch disease, 236

Cabbage
  K deficiency in, 232
    diseases caused by, 496
    nonparasitic diseases of, 232
  tipburn, 496
Carbon-14 labelling patterns, 127-128
Carrots
  diseases of, 496
  yield response to K, 261
Carbohydrate fraction, 148-149
  effect of K deficiency on, 148-149
Carbohydrate metabolism, 147-148
Cation-anion
  balance, 347-348
  ratios, 167-168
    relation to organic acids, 167-168
Cations
  and organic acids, 342-347
    effect of N on, 344-347
    effect of nutrient deficiency, 342-344
  influence of aeration on, 336-337
  influence of compaction on, 336-337

influence of soil moisture on, 336-337
micronutrients and, 340-341
  influence of K and B on, 340-341
  relation of K and Mo, 341
  relation of K and An, 341
osmotic pressure and, 339-341
physical-chemical characteristics, 322-323
relation among crops, 325-326
  K fertilization, 326
  Na fertilization, 326
relation among ions, 323
replacement of K by, 277-278
  dioctahedral mica-vermiculite, 277
  trioctahedral mica-vermiculite, 277-278
soil temperature influence, 340-341
Cation exchange
  and K absorption, 311-318
    experiment on, 312-318
  in the soil, 270-274
    effect of charge and density, 271-273
    exchangeable and soil K, 273-274
    kinds of, 270-274
Celery diseases, 496
Cells
  concentration of K in, 204-205
  accumulation of K in, 205-206
Cesium-137, 217-218
Charge density
  effect on K selectivity, 96-97
Cherry
  K deficiency symptoms of, 473
  response to K, 473-474
Citrus fruits, 250-251, 474-476
  effect of K on, 250-251
    juiciness, 251
    maturity date, 250
    on peel thickness, 250
    soluble solids, 251
  K deficiency in, 474-476
  K requirements of, 474
  sources of K for, 476
Cocoa, 400-404
  K deficiency in, 404
  K uptake and removal, 400-401
  use of K on, 403-404
Coconuts
  K deficiency in, 230-231
    diseases caused by, 230-231

Coffee, 400-402
  leaf K and yield, 401-402
  K uptake and removal, 400-401
  use of K on, 402
Corn
  K deficiency in, 223
    disease caused by, 223
  K nutrition of, 368-375, 452-453
    accumulation by, 453-454
    concentration in, 452-453
    effect on variety, 368-370
    nutrition value of, 455-458
    requirements of, 458-459
  K uptake of, 370-371
    influence on yield, 374-375
Cotton
  K deficiency in, 223-224
  influence of K on yield, 375

Dead Sea brines, 32
Dollar spot disease, 236
Donnan distribution, 313-316, 318
Drying, soil,
  release of nonexchangeable K by, 92-94

Enzyme systems
  activation by univalent cations, 191
  requiring K, 149-150
Escarole brownheart, 496
Exchangeable potassium
  effect of deficiency on vegetables, 256
  relation to soil solution K, 273-274
  effect of weathering on 85-87

Fertilizer
  effect on crop variety, 368-370
    corn, 368-370
    rice, 371
    soybeans, 370-371
  effect on species, 375-376
    legume vs nonlegume, 377-378
    root cation-exchange, 378-380
  effect of yield level on, 371-375
    corn, 374-375
    soybeans, 372-375
  from the TVA, 64-67
  in tropics, 386-388
    climatic problems, 387
    socio-economic difficulties, 386
    tropical soils, 386-387
    variety problems, 388
  on bananas, 407-412
  on cocoa, 403-404
  on coffee, 400-402
  on oil palm, 412-413
  on rice, 388-395
  on rubber, 405-407
  on sugarcane, 395-398
  on tea, 398-400
Fruit
  influence of K on quality, 244-253
    color, 244-245
    fruit size, 245-247
    fruit acidity, 247-248
    storage and shelf life, 249-250
    foliage and yield, 251-253
Fruit trees
  K deficiency in, 469-470
    effect on fruit production, 469-470
    influence on fruit quality, 469-470
  use of K on, 470-480
Fusarium patch disease, 238

Gas exchange components
  measurement of, 110-114

Halloysite, 82
Hargreaves process, 42-43

Illite, 81
Immunoelectrophoretic patterns, 195
Ion absorption, active and passive, 294
Ischemia, 213

Kaolinite, 82
Krebs cycle, 168
"Kurrol's salt", 64
Langbenite process, 41-42
Layer silicates, 83
Leaf analysis
  and K nutrition, 480-482
"Luxury consumption," 371

Macy's hypotheses, 449-451
Mango
  K and N effect on growth, 473
  K deficiency symptoms of, 473
Mannheim process, 43-44
Membrane ATPase, 206-209
Micas
  cations replacing K, 277-278
  effect of weathering on, 88-89, 98-100
Mineral structures
  descriptions of, 81-82
  influence on K reactions, 83-85
  significance, 80-81
Montmorillonite-saponites, 81
Muskmelon, yields response to K, 262

Negative charge
  effect on K selectivity, 96-97
  effect of weathering on, 90-91
Nonexchangeable potassium
  effect of weathering on, 275-276
  role of clay mineralogy, 275-276
Nonprotein nitrogen, 170-171
Nuts, effect of K on, 476-477

INDEX

Oil palm, 412-414
  K and disease resistance, 413-414
  K and yield, 413
  K uptake and requirements, 412-413
  role of K in physiology, 413-414
Olive
  K and fruit production, 477
  K deficiency symptoms, 477
Ophiobolus patch disease, 238-239
Organic acids, 165-179, 183-184
  cation-anion balance, 166-167
  cation relations and, 166-167, 344-347
    effect of N on, 344-347
    nutrient deficiency effects, 342-344
  fluctuation in pools, 168-170
  Krebs cycle, 168
  mechanisms for accumulation, 171-172
    K relationships to, 173-174
    studies of, 174-179
  nonprotein nitrogen, 170-171
Organic acid pools
  fluctuation in, 168-170

Pear, K deficiency symptoms, 479
Peas, yield response to K, 261
Perennial forages
  competition in plant associations, 428-433
  K content and growth stage, 426-428
  K deficiency in, 224
    disease caused by, 224
  use of K on, 441-443
    effects on growth, 434-441
    establishing stands 441-442
    maintaining stands, 442-443
    response to, 451-452
pH
  effect on K fixation, 278
  effect on K availability, 279
Photorespiration
  carbon-14 labelling patterns, 127-128
  effect of light intensity on, 121-125
  end product removal in, 129-130
  historical perspective, 114-117
  influence of other nutrients, 119-121
  role of K in, 117-130
    primary-secondary effects on, 118-119
  stomatal regulation, 128-129
  subcellular effects on, 125-129
Photosynthesis
  carbon-14 labelling patterns, 127-128
  effect of light intensity, 121-125
  end product removal in, 129-130
  historical perspective, 114-117
  influence of other nutrients, 119-121
  role of K in, 117-130
    primary and secondary effects, 118-119

stomatal regulation, 128-129
subcellular effects on, 125-127
Plant diseases
  Bermuda leaf spot, 235
  Brown patch, 236
  Dollar spot, 236
  effect of K on, 222-239
    field crops, 222-225
    horticultural crops, 225-229
    nonparasitic diseases, 231-239
    tropical crops, 229-231
  Fusarium patch, 238
  Ophiobolus patch, 238
  Red thread disease, 237-238
Plum, K deficiency symptoms, 480
"Potash," 1
Potassium absorption, see "potassium uptake"
Potassium adsorption
  effect of charge site and density on, 271-273
Potassium availability
  cation-exchangers in soil, 270-274
    effect of charge and density, 271-273
    exchangeable and soil K, 273-274
    kinds of, 270-271
  clay mineralogy, 275-276
    exchangeable-nonexchangeable K, 275-276
  in soil minerals, 85-87
  physical factors affecting, 280-283
    oxygen content of soil, 283-285
    temperature effects, 285-286
    water content, 280-283
  replacement of K by cations, 277-278
    dioctahedral mica-vermiculite, 277
    trioctahedral mica-vermiculite, 277-278
  soil reaction on K availability, 278-279
    effect of pH on availability, 279
    effect of pH on fixations, 278
    effect of weathering on, 87-88
Potassium calcium pyrophosphates, 64
  alterations in soil, 71
  crop response, 68-71
  germination effects, 67-68
  recovery of added potassium, 72-75
Potassium carbonate and bicarbonate, 48-49
Potassium chloride
  characteristics of fertilizer, 40-41
  flotation beneficiation, 33-36
  modification into fertilizer, 39-40
  nitrate fertilizers containing, 58-59
  solution-recrystallization process, 36-38
Potassium deficiency
  bananas, 408
  cocoa, 404

effect on tree growth, 469-470
  effect on fruit production, 469-470
  influence on fruit quality, 469-470
K loss from body, 214-216
  gastric juices, 214
  intestine and colon, 214
  kidney, 215
  measuring body K, 216
  saliva, 214
  therapeutic dosage of K, 216-217
K loss from tissue cells, 212-213
  anoxia and ischemia, 213
  breakdown in tissues, 213
  exercise or activity, 213
  K with liver or glycogen, 213
  low plasma K, 212
sugarcane, 397
tea, 399
vegetables, 255-256
Potassium feldspars, 82, 86
  effect of weathering on, 91-92
Potassium fertilizers
  analyzing amount of K in, 60
  effect on plant growth, 434-441
  for maximum yields, 441
  materials used in, 54
    in bulk blends, 56-57
    in fluid fertilizers, 58
    in granular fertilizers, 55
  on bananas, 411-412
  on cocoa, 403-404
  on coffee, 402
  on corn, 452-459
    accumulation by, 453-454
    concentration in, 452-453
    nutrition value, 455-458
    requirements of, 458-459
  on perennial forages, 441-443
    establishing stands, 441-442
    maintaining stands, 442-443
    response to, 451-452
  on rice, 391-395
  on rubber, 405
  on soybeans, 460-464
    concentration in, 460-462
    effect on seed and yield, 462-464
  on sugarcane, 395-398
  on tree crops, 470-480
  on vegetables, 489-497
    composition and uptake, 489-490
    effect of K on, 495-497
    response to, 490-495
  Preparation of, 55-58
    adding to bulk blends, 56-57
    adding to slurries, 55-56
    granular fertilizers, 55
    in fluid fertilizers, 58
  production of, 33-50
    percent consumed as fertilizer, 24

producing countries, 24
tonnage, 24
Potassium fixation, 94-96
  effect of water on, 281-282
Potassium fluxes, 209-212
  influences of Ca^{2+}, 311-318
  potassium outflux, 209-211
  sodium influx, 210-212
Potassium hydroxide, 47-48
Potassium nitrate
  characteristics of fertilizer, 46-47
  production of, 45-47
    Chilean production, 45
    from KCl and HNO_3, 46
    from KCl and $NaNO_3$, 46
Potassium phosphates, 49-50
Potassium polyphosphate, 63-64
  alterations in soil, 71
  crop response, 68-71
  germination effects, 67-68
  recovery of added potassium, 72-75
Potassium production
  percent consumed as fertilizer, 24
  producing countries, 24
  tonnage, 24
Potassium release, see "potassium availability"
Potassium reserves
  definition of, 3
  in Africa, 14-15
  in Asia, 15-16
  in Europe, 10-14
  in North America, 5-10
  in Oceania, 16-17
  in South America, 10
  projected, 17
Potassium selectivity
  effect of hydrogen ions on, 102-103
  effect of wedge zone on, 102-103
  of micas-vermiculites, 96-104
    hydroxy-Al interlayers, 100
    initial cation saturation, 101
    interlayer features, 97-98
    negative charge on, 96-104
    nonuniform weathering, 98-100
    surface features, 97
  specific sorption sites for K, 101-102
Potassium sulfate
  characteristics of fertilizer, 44
  production of, 41-44
Peaches
  effect of K on, 253
    relationship of foliage and yield, 251-253
  K and cold resistance, 478-479
  K deficiency symptoms, 478
Parenchyma
  breakdown of, 156-162
    carbohydrate fraction on, 157-159

INDEX

nitrogen fraction, 159-161
  senescence, 161-162
Potassium uptake
  effect of cation exchange on, 311-318
    experiment on, 312-318
  effect of clay suspensions on, 301-302
  effect of Donnan distribution, 313-316
  effect of growth stage on, 426-428
  effect of oxygen on, 283-285
  effect on plant growth, 434-441
    for maximum yields, 441
    longevity of plants, 438-441
    morphological-physiological, 434-435
    yields and, 435-437
  effect of plant spacing on, 362-363
  effect of plant variety on, 368
  effect of temperature on, 285-286
  exchange diffusion and, 302-303
  from soil, 303-309
  influence of calcium on, 298-299
  in plant association competition, 428-434
    grass associations, 431-434
    legume associations, 432-434
    legume-grass mixtures, 428-431
  of corn, 368-370, 452-453
    accumulation of K by, 453-454
    concentration of K in, 453-454
    effect of variety on, 368-370
    effect on yield, 374-375
  of rice, 371
  of soybeans, 370-371, 460-462
    concentration of K in, 460-462
    critical percentages of, 464-466
    effect of variety on, 370-371
    effect on yield, 372-374
  of tropical crops, 388-414
  of vegetables, 489-490
  quantity-intensity measurements, 300-301
  relation to concentration, 297-298
  relation to exchange capacity, 299-300
  role of root systems in, 355-357
    effect of rhizosphere on, 366-367
    extent of roots, 358-361
    maximal ion uptake, 357-358
    placement of K in, 361-363
    soil aeration, 363-364
    soil temperature, 364-366
Potatoes
  K deficiency in, 227, 256
    diseases caused by, 227
  K uptake and requirement, 256-258
  nonparasitic diseases of, 232
  plant response to K, 250-252, 258-259
    effect on chip color, 259
    effect on tuber characteristics, 259
    effect on yield, 258-259

Prune, K deficiency symptoms, 480

Recovery of K from
  as a byproduct, 19
  associated constituents of brines, 19
  brines, 17-18, 29-32
  raw materials, 24-25
    methods of shipment, 51
  shaft mining, 26-27
  solution mining, 27-28
  underground deposits, 18, 25
Red thread disease, 237
Respiration
  historical perspective, 130-131
  role of K in, 131-139
    CO_2 release and O_2 uptake, 133-135
    energy conversion, 137-139
    respiratory control, 132-133
    respiratory substrate use, 135-137
Rice, 232-233, 388-390
  effects of K on, 389-390
  fertilizer use on, 391-395
    effect on variety, 371
  K deficiency of, 232
    diseases caused by, 231
    nonparasitic diseases caused, 231-232
  K uptake of, 371
  natural supply of K, 390-391
  nutrient uptake and removal, 388-389
Root systems
  effect on K uptake
    cation-exchange capacity, 378-380
    effect of rhizosphere on, 366-367
    extent of roots, 358-361
    Maximal ion uptake, 357-358
    placement of K in, 361-363
Rubber, 404-407
  determining K needs, 406-407
  effects of K on wind damage, 406
  nutrient uptake and cycle, 405-406
  use of K on, 405

Salduro Marsh (Bonneville) brines, 31
Searles Lake brines, 29-31
Senescence, 161-162
Shaft mining, 26-27
Smectite, 81
Snap beans
  K deficiency in, 256
  yield response to K, 261
Soil aeration, effect on K uptake, 363-364
Solution mining, 27-28
Sources of K, 3-4, 17-18, 18, 19, 24-25, 25, 26-27, 29-32, 51
Soybeans
  K nutrition of, 368-375
    concentration in, 460-462

effect on seed and yield, 372-375, 462-464
  effect on variety, 370-371
  K uptake of, 370-371
  influence of yield level, 372-374
Stomatal regulation, 128-129
Sugarcane, 395-398
  K deficiency in, 397
  use of K on, 395-398
    effects of K, 396-397
    K requirements, 395
    response to K, 397-398
Supply and demand, K, 19-20
Sweet corn
  K deficiency in, 227-228
  diseases caused by, 227-228
Sweet potatoes, yield response to K, 262

Table beets
  K deficiency in, 226-227
  diseases caused by, 226-227
Tea
  K deficiency in, 399
  role of K in physiology, 398-399
  use of K on, 398-400
    effects on disease resistance, 399
    effects on quality, 399-400
Temperature
  effect on K equilibrium shifts, 285
  effect on K uptake, 285-286, 364-366
Tobacco, K deficiency in, 222
Tomatoes
  blossom end-rot of, 496
  K deficiency in, 225-226, 256
  diseases caused by, 225-226
  K uptake and requirement, 256-257
  plant response to K, 260-262
    effect on acidity, 261
    effect on juice, 261
    effect on ripening, 260-261
    effect on yield, 259-260
Translocation
  age of plant and, 332-334
  effect of K drymatter and, 154-156
  forces which start, 151-153
  general patterns of, 151
  of labelled compounds, 153-154
Trona process, 42
Tropical crops
  bananas, 407-412
    absorption of K, 408
    deficiency in, 408
    effects of K on, 410
    fertilizer requirements, 411-412
  cocoa, 403-404
    K deficiency on, 404
    K uptake and removal, 400-401

    leaf K and yield, 401-402
  coffee, 400-402
    fertilizer for, 402
    K uptake and removal, 400-401
    leaf K and yield, 401-402
  oil palm, 412-414
    K and disease resistance, 413-414
    K and yield, 413
    K uptake and requirement, 412-413
    role of K in physiology, 413-414
  rice, 388-390
    effects of K, 389-390
    K fertilization, 391-395
    natural supply of K, 390-391
    nutrient uptake and removal, 388-389
  rubber, 404-407
    changes in K requirements, 405
    determining K needs, 406-407
    effects of K on wind damage, 406
    nutrient uptake and cycle, 405-406
  sugarcane, 395-398
    effects of K on, 396-397
    K deficiency of, 397
    K requirements, 395
    response to K, 397-398
  tea, 398-400
    effects on disease resistance, 399
    effects on quality, 399-400
    K deficiency in, 399
    role of K in physiology, 398-399
Turfgrasses, effect of K on disease resistance, 233-234
TVA fertilizer products, 64-67
Two-component fertilizers, 59-60

Underground deposits of K, 18, 25
Univalent cations, 191-198
  protein conformation, 192
    enzyme stability, 194
    immunoelectrophoretic, 195
    kinetic observations, 192-193
    spectrophotometric studies, 195-198
    ultracentrifugation studies, 198
  protein subunit effect, 191-192
Vegetables
  K deficiency symptoms, 255-256
  K uptake and requirement, 256-258, 489-490
    potatoes, 257-258
    tomatoes, 256-257
  plant response to K, 258-262, 490-495
    potatoes, 258-259
    tomatoes, 260-261
  role of K in, 254-255, 489-497
    effect of K on, 495-497
Vermiculites, 81

Water
    effect on cation ratio in soil, 280
    effect on K diffusion, 280-281
    effect on K release and fixation, 281-282
    influence on K nutrition, 282-283
Weathering
    effect on mica, 88-89, 98-100
    effect on potassium feldspars, 91-100
    procedures on K bearing minerals, 87-88
    release of soil K by, 85-92, 275-276
        availability to plants, 85-87
Wheat
    K deficiency in, 224-225
        diseases caused by, 224-225

Zeolites, 82